Milk and Milk Products Technology
Second Edition

Milk and Milk Products Technology

Second Edition

Dr. S. Rajagopal
MVSc, Ph.D
Professor and Head (Retd.)
Dept. of Animal Nutrition and
Head, Dept. of Live Stock Product Technology
College of Veterinary Science and Animal Husbandry
Anjoro, Durg. C.G.

Dr. S.K. Roy
Dean (Retd.),
G.N. College of Dairy Science and Food Technology,
SDAU, Gujarat.

BS Publications
A unit of **BSP Books Pvt., Ltd.**

4-4-309/316, Giriraj Lane, Sultan Bazar,
Hyderabad - 500 095.

Milk and Milk Products Technology ***Second Edition***
by Dr. S. Rajagopal and Dr. S.K. Roy

Disclaimer: The authors and the publishe rs have taken due care to provide the authentic, reliable and up to date information related to the subject. However, neither the authors nor the publisher shall be responsible for any liability for any damage caused as a result of use of this book. The respective user must check the accuracy from other sources too.

BSP BS Publications
A unit of BSP Books Pvt., Ltd.

4-4-309/316, Giriraj Lane, Sultan Bazar,
Hyderabad - 500 095.
Phone : 040 - 23445600, 23445688
e-mail : info@bspbooks.net
website: www.bspbooks.net

ISBN : 978-93-89354-13-3

R.C. Maheswari
Vice Chancellor
S.D.A.U, Gujarat

Foreword

India is at a defining cross road with more than 50% of its 1.21 billion population under the age of 25. This places the country at an undeniable advantage against an aging yet resilient China and the recovering United States among other nations. India's youth - with their sheer number and undeniable potential - are entrusted with the responsibility of making her a Super Power. All these rests on a booming economy, improving living standards and accommodating the demands of the rising population. With a new career avenues opening up, the dairy industry is one glowing hot spot destined for a stellar growth.

India is the *numero Uno* milk producer of the world owing to "Operation Flood" and the surge in establishment of dairy co-operatives. Govt. of India's milk production target for the year 2010-2011 is 116.2 million tonnes (a rise of 3.6 million tonnes over FY 2009-10) to match the demand and supply of this vital food item. The economic survey estimates that by 2020-21 the country will require about 180 million tonnes of milk.

This would require setting up and mobilising of huge resources and manpower - Engineers/Scientists who will drive the growth. There is tremendous potential in the areas of production, R & D, supply chain management, cold storages and allied technology, increasing productivity and quality of milk and indegenous milk products as well as un tapped regions like Eastern States of India. Un-doubtedly bolstar India's dairying export which is very meagre percent despite producing world's largest milk volume.

But all this begins with sound basics being inculcated in students of this field when they are being groomed in college.

I, sometimes, find it difficult to relate to the college students of today. Most of them with earphones plugged in and yet a plethora of knowledge lying at their fingertips - through laptops and smart phones. Being a Vice Chancellor does give me the presidential authority over administration, course delivery and fostering of an intellectual atmosphere, but it is what happens inside the classrooms and, most importantly, inside the budding minds that really matters to me.

Good text books and great teachers - The combination of these two often defined the life of a college student a few decades ago. It still does but the impact has lessened. A textbook having all the right content - stimulating, relevant and to the point is always an important asset and guide in spite of digitization and Google results, a worthy text book is hard to come by Dr. S. Rajagopal and Dr. S.K. Roy, drawing on their rich dairying expertise and enviable academic experience spanning over 30 years have given us nothing short of an exemplary work.

This book will fill in the wide gap created owing to lack of quality texts in this prospering field. It will serve as a stimulating and enriching read - the right mix of industrial scenarios, in-depth textual content and, even, essential practicals for laboratory experiments. Albeit targeted at the students of Veterinary Science and Animal husbandry. This textbook can serve as a reference guide for the students of Dairy and Food Technology as well. The relevance of this book - in terms of modern day scenario and latest technologies - is huge.

The content is in line with the Veterinary Council of India's curricula and will serve as a torchbearer for students and faculty alike. I believe they will also be better equipped to understand and apply the concepts and inclusion of practicals makes this a wholesome repository.

The authors - both intellectuals and hugely popular among their students - bring their zest for teaching and fresh perspectives in writing this book. The text is comprehensive and every veterinary student will relish the large and the concepts which the authors have so beautifully portrayed in the coming pages.

I am reminded of an enlightening quote of Gautama Buddha :

"When the student is ready, the teacher will appear".

I believe that this bundle of knowledge, which you now hold in your hands, will guide you and, like a teacher, enrich you, and entice you to learn and contribute as well to the stirring world of Dairying.

R.C. Maheswari

Date and Place
5-5-2011
S.K. Nagar

Preface to Second Edition

The first edition of this book was published in the year 2014 after which many changes have taken place and more water has been passed through the river GANGA ultimately culminating in the ocean. In the same manner the Dairy and Food Industry has gone through several changes and taken a quantum leap. The FSSAI has taken a gigantic task for the revision of the standards for food products including milk and milk products in the country which is still under process.

Innovative ideas/automation/new product development has taken place substantially during this period. Looking into the above scenario the authors have thought that there is an urgent need to update the knowledge by incorporating the latest information.This might be an humble endeavour by the authors in order to fulfil the aspirations of the teachers and the students not only of B VSc & A H but also of Food Science and Dairy Science including the Agriculture as a whole. The authors also feel that this new edition may be helpful and the needs of the dairy industry especially under start up programme who intends to establish new endeavour in this area/field.

Once again the authors wishes to put it on record and express their indebtedness to the different scientific books/journals/periodicals as well as the publishers particularly to the Google Search in the internet media which provides vast information from where most of the useful and valuable information has been collected/taken.

-Authors

Preface to First Edition

Agriculture occupies a predominant position in the economy of our country. The contribution of the livestock sector has been substantial and is linked with the rural economy of our country in enhancing the income, national security, employment and reducing the incidence of poverty.

The dairy science has developed and the dairying has also progressed into a highly technical viable industry since/after 1920. This evolutionary process continues and it is desirable that the type of training should also keep pace with or even anticipate the changing conditions of the industry. Immense wealth of knowledge, that has been accumulated and its documentation has a direct link or impact on the progress and advancement of civilisation of the man and every step he made was a forward march or a leap for the humanity. The dairy science has developed in leaps and bounds and we can conclude today that milk is delicately balanced biochemical fluid.

This book of "Milk and Milk Products Technolgy" is an attempt and intended to provide the basic fundamentals relating to the milk and milk products to the students graduating from the different veterinary colleges of the country. It is a fact that books comprehensively covering all the aspects of milk and milk products technology either by the foreign/Indian authors are very limited or scant in nature.

In order to maintain the standards in veterinary education, the Veterinary Council of India, has formulated and brought out a uniform syllabus for the under graduate course leading to the degree of B V Sc., and A H. With the adoption and introduction of the new common syllabus there is an urgent need for a common coordinated teaching programme in the veterinary education. We have realized that it is high time and there exists an urgent need that a comprehensive text book on milk and milk products techonogy at under graduate level is an essentiality.

The purpose of this text book is a sincere venture and effort to provide a compact, simple, varied and vivid picture of milk and milk products technology in a concise manner covering the most important facts concerning all the aspects of the syllabus of the Veterinary Council of India to the B.V Sc and A.H. students in India maintaining the clarity on each topic. The subject matter has been presented lucidly through the tables and flow diagrams, thus enabling us to provide the vast accumulated data in the subject of dairy science with clarity and simplicity.

Apart from providing the basic information on milk and its composition, properties, proteins, carbohydrates, lipids, minerals, minor constituents, legal standards, nutritional importance; different dairy processes pertaining to the processing of milk manufacturing / technology of preparing different milk products, the topics on the microbiology of milk, quality control of milk and milk products themselves have a good buffering capacity thus maintaining a standard throughout the text book.

Necessary information has been provided in a comprehensive manner for the benefit of the students in the chapters on comparative aspects physico-chemical and biochemical approach to some dairy processes and analytical aspects.

We are indebted to the authors, editors, publishers of various scientific books, periodicals from which the information presented in the book has been gathered. In particular we have freely consulted the following.

1. M.P. Mathur, D. Datta Roy, P. Dinakar - Text book of Dairy Chemistry, ICAR Publication.
2. Edgar R. Ling - Text book of Dairy Chemistry
3. Prof. Lalat Chandar, Textbook of Dairy Plant - Layout and design, ICAR Publication.
4. M.K. Shrivastav, Hand book of Milk Microbiology, Daya Publishing Home Delhi.
5. M. Perumal, P.S. Mohan, M. Suresh, Dairy development and Income distribution in India, Ashipet Publication - Delhi.
6. CH. Eckles & EL, Anthony, Dairy Cattle and Milk Production, Biotech books, Delhi.
7. Pradeep Parihar & Leena Parihar, Dairy Microbilogy, A grobios India.
8. S.K. De, Outlines of Dairy Technology, Oxford, IBH Publishing Co. Delhi.
9. I.D.A Publications. R.K. Puram, IDA House, New Delhi.

Last but not the least the authors wishes to express their sincere gratitude to Smt. K. Aswani Kumari, for her patience and neat Desktop publishing work in typing the manuscript without which it would not possible for the authors to complete the work.

- Authors

Contents

CHAPTER 2

Milk Processing Plant Layout and Management

CHAPTER 3

Milk and its Constituents

Chapter 4

Factors Affecting the Composition of Milk

Chapter 5

Physico-chemical Properties

CHAPTER 8

Dried Milk or Milk Powder

CHAPTER 9

Dairy Products

Chapter 12

Quality Control of Milk and Milk Products

Chapter 13

Milk Residues

CHAPTER 14

Packaging

CHAPTER 15

Cleaning and Disinfection

Introduction

The Mammals primarily secretes the milk through the mammary gland to provide a nearly complete food to their off springs and such secretions normally and naturally occur after parturition under the influence of the inherent lactogenic harmones. The first secretion of the mammary gland after normal Parturition is considered as colostrum and later as milk. Colostrum is a repository of immunoglobulins and there by provides natural immunity to the newborn.

It is an Universally accepted fact that the mammalian off spring derives its nutrients whether they may be proteins, lipids, carbohydrates, enzymes, minerals, vitamins and other minor elements, essential, semiessential or non-essential amino acids through the milk provided by its mother.

The biological value which is a measure to assess the quality of proteins is very high for milk proteins and when supplemented with the cereals or with any other suitable substitutes the biological value gets enhanced and is known as the "Supplementary Value" of the milk proteins. The protein content of milk of any species is related to the days required to double the weight of any species. The stability of casein fraction has been attributed to their non-possession of tertiary structures and hence casein micelles are more stable than many other proteins.

Milk lipids exist in milk as fat in water emulsion in the shape of globule, which are stabilized by a membrane primarily a lipo-protein complex. The Triestors of glycerol and fatty acids are the main lipids of the milk which provide necessary energy/ calorie for the maintenance and growth of the off spring. The lower chain fatty acids and part of the medium chain fatty acids are bio-synthesized in the mammary glands, whereas the higher fatty acids rise from the depot fat of the body fat of the mammals.

Lactose is the principal sugar of milk and exists in true solution. Lactose apart from providing the necessary energy/calories for the maintenance and growth of the of spring, it also acts as the substrate for the microbes in providing many cultured

MILK AND ITS PRODUCTS

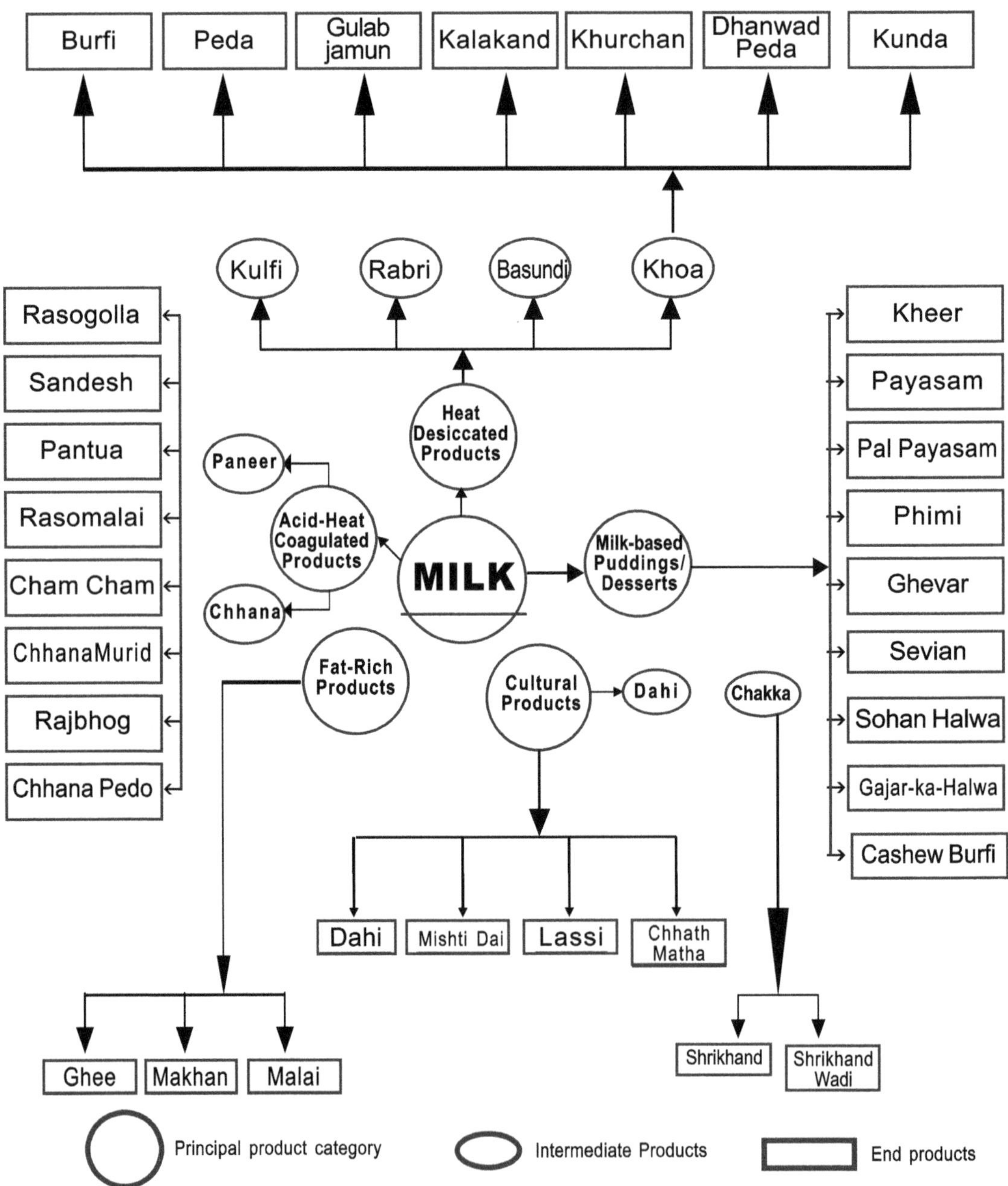

dairy products. Milk contains abundance of minerals which exists in true solution as well are in colloidal state. Trace minerals also have been present in milk. More than thirty enzymes have been found in milk, however, their exact role is yet to be well defined/established.

Fat soluble and water soluble vitamins including Vitamin C are also present in milk. Milk is highly perishable and susceptible to microbial attack. Quality control in the production of milk right from the farmer or producer level until it reaches the valuable consumer is top priority either in the case of liquid milk sales or the valued milk based products. From time immemorial milk has been converted to various products in which heat, acid and rennet have been used. Majority of the Indian dairy products are heat desiccated and in cases acid is also used. These heat desiccated and acid treated products are having a larger shelf life with good keeping quality.

Conversion of milk to different dairy products is mainly based on two aspects. (1) Reduction of the bulk and (2) Enhancement of the keeping quality. A practical and systematic quality control programme has to be carried out, during the flow of milk and milk products from the dairy plant premises, ensuring good quality before they reach the customer. In order to check any adulteration and contamination by the pathogenic organisms into the finished products, strict quality control measures, enforcement of standards are to be practised without any slackness. Compliance of legal standards bear a direct relationship with a practically oriented genuine quality control programme. Indiscriminate usage of pesticides, herbicides etc., by the farmers on the crops, leads to their accumulation in the crop residues which gain access into the milk, when such crops are fed to the milch animals thus posing a serious threat resulting in the health hazards of the human population.

Majority of the Indian dairy products are generally prepared by the "Halwali" or the "Sweet Meat" shop owners in the unorganised sector, who have an apathy and seant regard for the quality of the products that are being prepared or manufactured by them. The main objective of these men for the most part have an urge for more profit and least concern for aesthetic, safety and keeping quality of the finished product. Therefore particular emphasis has been given to the quality control, particularly to the analytical and legal standards in this text book. Further the Practical manual covering different analytical tests for milk and milk products have been given separately along with the text book for the convenience of the teachers and students.

Thus the preamble leads a glance to the fabulous landscape and the gorgeous depths in the literature/subject one can plunge and probe into.

Chapter 1

Dairy Scenario

1.1 History of Dairy Industry

The domestication of cattle and the use of their milk for human food began somewhere in Asia and North East Africa between 6000 B.C. and 8000 B.C. and before its domestication, probably it was hunted by the primitive man. Over the years the cow has been used as a source of food, an object of worship, a source of sacrificial offering a subject of mythology and as a beast of burden. Cows milk and its products have been used for food, sacrificial offerings, cosmetics and medicants.

The oldest written records of man are believed to go back to the Sumerians of Mesopotonia in approximately 6000 B.C. Dairying was highly developed at that time. A mosaic frieze of one of the oldest buildings unearthed in the Euphrates valley, dates back to at least 3100 B.C. shows the milking of cow from behind, the pouring of milk through a strainer into the vessel and the presence of two attendants wearing fleece petticoats that are believed to have been the dress of priests and priest kings. Egyptian records dates back nearly to 3000 B.C. indicate that milk, butter and cheese were used extensively.

Greek and Roman records go back to approximately 1550 BC and 750 B.C. respectively, and in both areas milk and cheese were important parts of the human diet. In ancient Greece most of the milk came from the goats, where as in Rome it came from sheep. Butter was used as food by some people in Europe, but it was not used extensively until the Eighth Century, when it became an important food in Norway.

References to cattle and milk have been made frequently in the Old Testament and the New Testament alike. The old testament has 40 references to cattle, milk and milk products are mentioned at least 50 times as desirable foods. Palastine is referred to approximately 20 times as a land flowing with milk and honey. Frequent references are found to cattle and their products, as more and more records became available to the historians. Gibbons, records "The Tartars (376 AD) took cattle with them and their compaings for milk and meat supplies. Each caravan of Tartars consisted about 20 to 30 oxen and planned to pasture them as they moved from one section of the country to the other."Gibbons" further states that the number of cattle owned by a ruler or person was indicative of his power and influence. Marcopolo during his travels over portions of Asia about the middle of the 13th century reported that he found cattle and their products being used by the people of these regions.

The people of India were raizers of cattle as early as 2000 BC. Butter was used as a holy offering to the gods. The butter was converted into ghee (butter oil or clarified butter) and the cow was considered as "holy" at that time. References have also been made in holy "Vedas" as well as in the great epic "Mahabhagavatham" pertaining to the usage of milk (ksheera) curd (dahi) and ghee (Aajya or Ghritam) in the performance of special rituals or "holy yagna or havan" by the saints. The Vedic hymns written about 3000 BC potrays the home life and the occupations of the early Hindu people. The sanskrit writings of India tell of milk and its value as food dating back to 6000 BC. Cattle constituted an integral part of India's Vedic culture (1800 - 600 BC), which is full of references to the milk of the cow.

The Indus valley civilization of the Harappa Mohanjodaro period and the recent archeological excavations at 'Mehrgarh' (Placed around 6500 BC) located about 150 miles to north west of Mohanjodaro at the foot of 'Bolan' pass in Beluchistan, have revealed signs of domestication of cattle earlier than other settlements of western Asia. Moreover, there is evidence to indicate that cattle were more significant to Mehrgarh's inhabitants than any other large domesticated animals a cultural trait that has been preserved over thousands of years down to the present day suggesting that the animal wealth as an indicator of the status and wealth of the people.

The traditional Indian dairy products reflect the intimate relationship that our ancesters established with the farm and animal. It has been stated that the five products from the cow either singly or in combination which are commonly called "Panchgavya" are necessary to maintain the ritual purity which is so important to the Hindus. These products are milk (Ksheera) curd (Dahi) ghee (Aajya or Ghritam) urine and dung. In the rituals carried out by the 'Purohits' or 'Archakas' in the temples, milk products were regularly used. The worship included 'Abhisheka' a ritual for the deity, on special occassions with milk, curds. ghee, banana, sugar, honey, tender coconut water and scented water. This kind of ceremony was called "Panchamritha Abhisheka" after which the deity was bathed with sanctified water. The mixture of the washed ingredients treated as 'teertha' or holyprasad will be consumed by the devotees.

The progress in the dairy industry was made possible by the use of science in dairy practices. This date marks the change from the art of dairying to the science of dairying. The great land marks in the history and the development of the dairy industry in U.S.A., U.K., Australia as well as in India has been ennumerated briefly hereunder.

1.2 Dairy Industry in U.S.A.

Cattle were first imported into the western hemisphere with Columbus's second voyage which were taken to either Cuba or South America. Prior to this there were no cattle in North America. The spanish expedition headed by Coronedo brought approximately 150 head of cattle to North America. The first importation of cattle by the settlers of the present U.S.A. occurred in the James town colony in 1611. Cattle were not brought over by the pilgrims on the May flower, but the cattle arrived in Plymouth colony in 1624. Later on many cattle importations have taken place.

The dairy industry from colonial times to the mid 1850's was based primarily on family cows, Most of the families having one or two cows, supplying milk for the family need. The major developments in the dairy industry from the begining of the christian era to the middle of 1850's occurred in Europe. Most of the currently important dairy cattle breeds in the U.S.A. and Europe originated there. Animals representing various improved breeds were well established during 1820 from the importations that were made from time to time either of an individual or a small group of animals. The first railshipment of milk by rail occurred in 1841 between Orange county New York and New York city allowing the farmers to ship fluid milk and cream as much as 80 miles to market. Initially the farmers shipped their own milk to the city where it was received and distributed by the dealers. Later on the milk receiving/collecting centres have been established in rural areas either by the dealers, independent shippers or co-operative groups of farmers.

Establishment of a dairy industry in any country depends mainly on three major factors.

1. A suitable market for milk and milk products.

2. Facilities and trained personnel to collect, process and distribute the milk and
3. A continuing supply of milk with satisfactory quality.

The factory system of dairy manufacture was introduced in 1851 which had the wide spread influence. The first cheese factory was established at Oneida, County New York, USA during this period by Williams.

1856 - Gail Borden received the first patent on condensed milk and the first successful condensary was built in 1857.

1860 - Formation of breeders associations for owners of purebred cattle were established.

1875 - Establishment of State Dairy Associations.

1880 - Introduction of artificial refrigeration and cold storage.

1890 - Co-operative dairy manufacturing and marketing of dairy products.

1895 - Expansion and strengthening of the State Dairy Associations.

1900 - Strengthening and expansion of the Co-op. movement for the manufacture and marketing of dairy products.

1905 - Organization of the first Cow test association in America.

The land grant Act 1862 established colleges of Agriculture in every state. However, the first college of agriculture had been established in Michigan in 1855. The colleges were strengthened by the Hatch Act 1887 empowering them to establish the Experimental Research Stations. The establishment of the extension services by the Smith-lever Act in 1914 resulted in creating more vistas/avenues for direct contact with the dairy farmers for passing the information collected by the scientists and the researchers accumulated in the scientific laboratories/research stations.

The centrifugal separator was invented almost simultaneously in 1871 by De lavel and by Winstrup and Nielson, providing a mechanical means for the separation of cream from milk. The first creamery was established in Lowa state by Stewart producing other dairy products in addition to cheese.

Development of a chemical test for the determination of fat in milk was perfected by S.M. Babcock and by Gerber simultaneously in 1890. The dairy industry has changed dramatically since 1900 and many developments/ inventions and discoveries have helped to promote the growth of industry more viable. These developments include, refrigeration, milking machines, and related equipment, homogenisation, pasteurization, churns, clarifiers, ice cream freezers, and bulk milk coolers etc.

1.3 Dairy Industry in U.K.

Man has consumed milk and milk products from the time immemorial, throughout the ages and the demand for these products, played a biggest role in the domestication of the animals without any doubt. Much importance has been attached to the cattle as the status and wealth of the man by the number of cattle he possess in biblical days. A careful survey of history shows quite clearly that not only milk has been consumed by man for the past 5-6 thousand years, but that cheese and butter have also been made and consumed for a similar period. The biggest development in the liquid milk trade came gradually from 1850 onwards, with the application of transport of milk through the railways. By 1945, it is correct to state that in Great Britan thousands of farmers merely put their milk in stands on the roads at their respective farm entrances, where it is collected by the lorries once a day in winter and twice daily in summer and taken to the creameries or the dairy plants.

1.4 Dairy Industry in Australia

In accordance with the available records, the first cargo fleet which reached Australia under Governor Philip consisted of four cows, which were intended to supply milk and to serve as the foundation stock for the future herds of cattle at Port-Jackson, the Rose hill. The foundations of the dairy industry were laid in the Illawara districts in New South Wales about 1820. The settlers of Richmond River district shifted to dairying in about 1890 and the industry was firmly established in that district within few years and extended to Northern River areas in Queens land to the Victorian border. In general, dairy farming has been confined to the coastal and near coastal regions, where the rainfall and topography are favourable. By 1936 dairying has been consolidated as one of the Australia's principal rural industries.

1.5 Dairy Industry in New-Zealand

Dairy farming in New-Zealand began from small beginings during the days of colonization by the Europeans. In 1814 the missionary Samuel Mars den introduced the dairy cows to the Bay of Islands from New South wales for the first time. From 1840's majority of the settlements had dairy farms with the short 'Horn' dairy cattle and these herds tended to be larger near the urban areas. In the year 1871 the first dairy cooperative was established at Otago peninsula. William Brown in 1881 has advocated and was instrumental in the establishment of the cheese and Butter factory at Flemington which was a great success. By 1920 there were 600 dairy processing factories 85% of which were generally owned by the Cooperatives and there are only 500 cooperatives by 1930. However, after world War II and with the improved processing technology, transportation and energy systems led to the trend of consolidation where the cooperatives merged together and became larger and fewer in number and by the year 1990 there were only four cooperatives.

Newzealand is the world's eighth largest milk producer with about 2.2% of world production. Traditional dairy production areas were confined to Waikato, Tazaanaki south and north islands and west land. The dairy production was entirely for local consumption before the advent of refrigerated shipping in 1880.

1.6 Dairy Farming in Canada

It is one of the largest agricultural sector and it has a significant presence in all the provinces and is one of the two agricultural commodities in seven out of 10 provinces. Dairy farming is still prominent in Canadian Society though the rate of dairy farmers have been dropping significantly since 1971 while making the average dairy farmer significantly weather in than the average Canadian family. On an average two-thirds of Canadian dairy produced is sold as raw milk whereas the remaining one-third (1/3) is refined into other dairy products such as milk, cheese and butter.

In Canada dairy farming is subject to the system of supply management. In this system the farmers manage their production such that it coincides with the forecasts of demands for their products over a predetermined period while taking into account certain imports that enter into Canada as well as some production which is shipped to export markets. The Canadian dairy farmers federation was founded in 1934 which later became Dairy Farmers of Canada in 1942. With its mandate to stabilize the dairy market and increase the revenue to the dairy farmer. Between 1950's and 1960's there was significant volatility in the dairy prices and the U.K was poised to enter the European common markets thus resulting in the loss of Canada's largest dairy partner. These challenges led to the creation of Canadian Dairy Commission whose mandate was to ensure the quality and supply of milk and that producers receive a fair return on investment.

The federal and provincial governments have a common interest in sharing the supply management jurisdiction. The basic idea is to manage the production so that the supply is imbalance copingup with the demand. The National milk marketing plan envisaged came into effect in 1983 to control the supply by setting up some guidelines for calculating the market sharing quota. Organic dairy farming is less prevalent eventhough both conventional and organic dairies exists across Canada because of the fact that due to the widely held misconceptions that organic farming is un-profitable and risky. On the contrary the costs are substantially lower with organic farming than the costs incurred by the conventional dairy farming. In contrast the lower economic supply is often off set by extra costs that are associated with the importing of fertilizers etc., making the conventional farming is no more economically profitable when compared with organic farming.

1.7 Global Dairy Scenario – Overview

Commercial dairy farms are typically one species enterprises eventhough any mammal can produce milk. All the dairy farms in almost all the developed and to some extent in the developing countries consists of high producing dairy cows followed by buffaloes, goats, sheep and camels. Milk will likely become one of the most volatile agricultural commodities in future. This is because of the fact that (a) small changes in the qualities available internationally have strong influence on world market prices (b) the length of time before there are increases in milk production as a result of price change and (c) delayed reaction of the demand to changing dairy commodity price.

The global dairy sector is currently going through a period of turbulence-slowing demand from China and Russia's trade embargo and the removal of E.U. milk quotas resulted in a period of excess supply and low prices. The key changes to making a reliable forecast of world market prices for milk are the nature of consumer reaction to rising prices of milk and the response of the dairy farmer with regard to the supply specially in low cost dairy regions. Another key determinant of milk price is feed which directly affects milk production through increased costs and indirectly higher land prices. Despite this the long term outlook for the sector remains positive. Rising populations and changing diets are increasing the demand for dairy products. As there is an increase in the income and nations become increasingly urbanised individuals tend to receive more of their calories from proteins including dairy as opposed to basic carbohydrates mostly the grains. Global demand for dairy is expected to increase by 2.5 percent per annum to 2020. Largely driven by increasing urbanisation and rising incomes in emerging markets.

The milk production levels in different parts of the world and the recent trends based on the surveys and on the secondary data from the F.A.O. The milk production volumes of all animals species have been standarised to "Energy Collected Milk" (E.C.M-4.0 percent and 3.3 percent protein). China, India and Pakistan alone are accounted for about 2/3 of all volume growth and most of the remaining growth was by eight countries approximately accounting for 85 percent of all milk volume growth. World milk production is set to reach 833.3 million tonnes in the year 2017 which is 1.4 percent more than in 2016. Much of the anticipated size is expected in Asia and America. While the sector might be stagnated in Europe and Africa and possibly face a decline in Oceana.

World trade in dairy products is forecast to increase by 1.3% in the year 2017 to 71.6 million tonnes making a modest growth for the second year in succession. On the demand side Asia is anticipated to drive the expansion with the imports by the region predicted at around 41 million tonnes by an increase of about 1.5 percent more than in the year 2016. Among the various products world trade in Cheese and S.M.P is expected to increase in 2017 while trade in butter and W.M.P may contract or show a decline. The major policy developments that took place in the year 2017 have been presented in the Table 1.1 and the milk and milk products statistics indicating the production, imports and exports globally have been presented in Table 1.2. The milk production trend across the countries globally has been presented in the Table 1.3 for information and to have a comparative study.

Table 1.1 Major policy developments

Country	Product	Data	Policy category/ Instrument	Description
Canada	Cheese	Aug-17	Tariff rate quota	Announced a new tariff rate quota (TRQ) administration policy applicable to the 16 000 metric tonnes (MT) of cheese that will be allowed to be imported under the Canada-EU Comprehensive Economic and Trade Agreement (CETA). Of the total TRQ, 60 percent will be allocated to small and medium-sized enterprises with the remaining 40 percent allocated to large companies. In both cases, 50 percent will be allocated to manufacturers and the balance to distributers and retailers. The 1700 MT of bulk EU industrial cheese TRQ will be allocated entirely to manufacturers of further processed food products.
	Cheese	Oct-17	Import ban	Banned, temporarily, imports of soft cheese products from the EU.
China (main land)	Dairy products	Oct-17	Food safety standards	Applied the Administrative Measures for the Registration of Recipes for Formula Powder Products for infants and Young Children, which mandate domestic and overseas producers to: (i) register their facilities with the China Food and Drug Administration (CFDA) and the Certificaton and Accreditation of the People's Republic of China (CNCA); and (ii) Limit the number of products and brands produced.
European Union	Dairy products	Sep-17	Free trade agreement	Approved the free trade agreement with Canada, which entered into force 21 September 2017, after its approval by EU Member States, expressed in the Council, and by the European Parliament.

Table 1.1 *Contd...*

Country	Product	Data	Policy category/ Instrument	Description
India	Dairy Products	Jun-17	Import ban extended	Extended the prohibition on import of milk and milk products (including chocolates and chocolate products, and candies/confectionary/food preparations with milk or milk solids as an ingredient) effective until 23 June 2018 or until further orders. The prohibition has been in place since 2008.
Indonesia	Dairy Products	Aug-17	Import policy	Introduced a requirement that US dairy-product establishments must pay a fee to undergo the review process to export dairy products to Indonesia. The review process consists of three phases: document review, onsite review and evaluation of onsite audits.
Mexico	Dairy products	Jul-17	Import ban	Suspended all dairy imports from Colombia after new outbreaks of foot-and-mouth disease were detected in the Andean country.
	Dairy products	Jun-17	Import ban extended	Extended until the end of 2018 the ban on the import of agricultural products including milk and milk products from the countries that applied economic sanctions against the Russian Federation.
Russian Federation	Dairy products	Aug-17	Tariff rate quota	Established tariff-rate quotas (TRQs) and respective volumes for 2018 imports of a number of products, such as whey, into the Eurasian Economic Union (EAEU). The volume of whey and modified whey products that will be allowed to enter into the Russain Federation is set at 15 metric 0.100 tonnes.

*A collection of major dairy policy developments starting in January 2012 is available at: http://www.fao.org/economic/est/est-commodities/commodity-policy-archive/en/?groupANDcommodity=Milk,%20Dairy%20products

Source: food.outlook. Milk and Milk Products Nov. 17 Internet

Table 1.2 Milk and Milk Products (thousand tonnes, milk equivalent)

	Production			Imports			Exports		
	2013-2015 average	2016 estim.	2017 f'cast	2013-2015 Average	2016 estim.	2017 f'cast	2013-2015	2016 estim.	2017 f'cast
Asia	**316 603**	**337 466**	**344 433**	**39 480**	**40294**	**40 908**	**6 709**	**6 658**	**6 554**
China	41 976	40 936	40 531	12063	11 998	12 500	75	52	49
India	146 501	162 964	169 320	89	124	136	706	276	213
Indonesia	1 265	1 230	1 250	2 573	2 839	2 863	87	55	50
Iran, Islamic Republic of	6 344	6 440	6 530	490	416	401	478	542	704
Japan	7 407	7 420	7 400	1 845	1 909	1 997	6	8	8
Korea, Republic of	2 159	2 126	2 083	922	1 022	1 115	21	23	25
Malaysia	84	86	87	2 061	2 169	2 137	612	693	689
Pakistan	50 233	53 000	53 700	484	638	717	66	35	33
Philippines	20	21	22	1 650	2 518	2 456	119	211	46
Saudi Arabia	2359	2 410	2 450	2 942	3 015	2 880	1 393	1 447	1 460
Singapore	-	-	-	1 787	1 576	1 585	629	573	531
Thailand	1 071	1 080	1 110	1 500	1 490	1 570	198	253	252
Turkey	18 719	19 900	19 980	214	160	115	541	930	873
Africa	**46 610**	**46737**	**46 833**	**10 234**	**10 098**	**10 451**	**1 151**	**993**	**979**
Algeria	4 206	4 612	4 700	2 776	2 895	3 265	3	-	-
Egypt	5 580	5 630	5 650	1 683	1 612	1 562	473	363	361
Kenya	4 882	4 830	4 820	61	69	103	14	8	10
South Africa	3 299	3 180	3 280	244	265	279	337	342	346
Sudan	7 616	7 540	7 440	231	258	247	-	-	-
Tunisia	1 222	1 235	1 265	95	80	85	47	31	29
Central America	**16 937**	**17 299**	**17 598**	**5 012**	**5 718**	**5 801**	**703**	**849**	**931**
Costa Rica	1 081	1 120	1 140	55	64	66	162	160	161
Mexico	11 321	11 780	12 027	3 048	3 693	3 789	186	326	387

Table 1.2 *Contd...*

	Production			Imports			Exports		
	2013-2015 average	2016 estim.	2017 f'cast	2013-2015 Average	2016 estim.	2017 f'cast	2013-2015 Average	2016 estim.	2017 f'cast
South America	**64 720**	**61 179**	**63 253**	**3 174**	**3 634**	**3 508**	**4 457**	**4 063**	**3 607**
Argentina	11 466	10 191	10 232	41	22	25	2 236	1 817	1 496
Brazil	34 810	32 725	34 427	854	1 659	1 462	293	180	194
Colombia	6 848	7 000	7 100	182	431	392	39	3	14
Uruguay	2 218	1 954	2 030	24	31	34	1 286	1 472	1 308
Venezuela	2 008	2 100	2 120	1 210	502	563	-	-	-
North America	**101 680**	**105 444**	**107 551**	**2 386**	**2 806**	**2 689**	**10 603**	**10 585**	**11 629**
Canada	8 551	9 100	9 450	661	666	640	530	613	895
United States of America	93 127	96 343	98 100	1 710	2 124	2 032	10 071	9 971	10 733
Europe	**217 471**	**222 070**	**223 103**	**7 366**	**6 669**	**6 727**	**23 262**	**24 788**	**25 369**
Belarus	6 824	7 169	7 290	170	204	128	3 634	3 930	3 722
European Union	158 867	163 800	164 500	1 427	1 310	1 247	17 347	18 480	19 348
Russian Federation	30 527	30 350	30 563	4 833	4 271	4 438	247	302	270
Ukraine	11 069	10 407	10 334	144	50	52	663	650	713
Oceania	**30 655**	**31 629**	**30 712**	**1 116**	**1 340**	**1 515**	**21 922**	**22 719**	**22 491**
Australia	96 68	9 991	9 301	720	889	1 098	3 274	3 341	3 280
New Zealand	20 897	21 568	21 341	201	281	245	18 645	19 374	19 208
World	**794 676**	**821 824**	**833 483**	**68 769**	**70 559**	**71 599**	**68 807**	**70 655**	**71 561**
Developing countries	411 533	428 668	438 163	55 160	56 890	57 671	12 598	12 140	11 639
Developed countries	383 143	393 156	395 795	13 609	13 670	13 928	56 208	58 516	59 921
LIFDC	247 389	267 126	274 090	5 958	6 219	6 398	1 381	875	827
LDC	33 448	33 784	33 671	3 990	4 016	4 055	174	138	136

[1]For production, the annual dairy cycle starting in April is applied

[2]For production, the annual dairy cycle starting in June is applied

[3]For production, the annual dairy cycle starting in May is applied

Note: Trade values that refer to milk equivalents were derived by applying the following weights: butter (6.60), cheese (4.40), skim/whole milk powder (7.60), skim condensed/evaporated milk (1.90), whole condensed/evaporated milk (2.10), yoghurt (1.0), cream (3.60), casein (7.40), skim milk (0.70), liquid milk (1.0), whey dry (7.6). The conversion factors cited refer to the solids content method. Refer to IDF Bulletin No. 390 (March 2004).

Source: food.outlook. Milk and Milk Products. Nov 17 Internet

Table 1.3 Global milk production across countries

Country	1970	1975	1980	1985	1990	1995	2000	2005	2010	2015
India	20.80	25.60	31.56	44.02	53.68	65.37	79.66	95.62	121.85	155.49
Afghanistan	0.75	0.85	0.84	0.72	0.83	1.37	1.66	1.73	1.72	1.62
Argentina	4.19	5.65	5.31	5.64	6.28	8.77	10.12	9.91	10.63	11.31
Australia	7.76	6.70	5.57	6.23	6.46	8.46	10.85	10.13	9.02	9.49
Bangladesh	1.07	1.18	1.16	1.31	1.59	1.99	2.14	2.62	2.02	2.09
Brazil	7.42	10.05	12.06	12.57	15.08	17.13	20.53	25.53	30.96	34.86
Canada	8.31	7.75	7.41	7.48	7.98	7.92	8.16	7.81	8.24	7.45
Chile	1.12	1.00	1.12	1.05	1.39	1.90	2.00	2.31	2.54	2.04
China	1.96	2.37	2.93	4.76	7.04	9.48	12.37	32.02	41.16	42.67
Denmark	4.48	4.92	5.12	5.10	4.74	4.68	4.72	4.58	4.91	5.36
Finland	3.31	3.16	3.28	3.08	2.82	2.47	2.45	2.43	2.34	2.44
France	22.85	24.72	27.89	28.40	26.81	26.09	25.74	25.71	24.21	25.93
Germany	28.18	28.75	32.10	33.63	31.34	28.63	28.35	28.49	29.65	32.71
Indonesia	0.17	0.19	0.25	0.40	0.60	0.73	0.79	0.85	1.48	1.46
Ireland	3.08	3.59	4.72	5.83	5.40	5.35	5.16	5.38	5.33	6.59
Mauritania	0.24	0.18	0.23	0.22	0.27	0.28	0.32	0.37	0.69	0.75
Mexico	4.11	6.24	7.23	7.47	6.27	7.54	9.44	10.03	10.89	11.61
Nepal	0.63	0.71	0.75	0.81	0.92	1.01	1.17	1.35	1.62	1.86
Netherlands	8.24	10.22	11.79	12.53	11.23	11.32	11.23	10.98	11.81	13.55
New Zealand	5.99	6.10	6.70	7.88	7.51	9.29	12.24	14.64	17.01	21.94
Norway	1.73	1.84	1.97	1.98	1.99	1.93	1.74	1.59	1.58	1.61
Pakistan	7.45	8.19	9.01	10.86	14.72	19.01	25.57	29.44	35.49	41.59
Poland	14.96	16.38	16.49	16.44	15.84	11.64	11.89	11.95	12.30	13.25
Romania	3.12	3.81	4.34	4.32	3.81	5.02	4.62	5.55	4.62	4.68
Russian Federation	0.00	0.00	0.00	0.00	0.00	39.31	32.28	31.15	31.84	30.79
South Africa	2.91	2.50	2.58	2.20	2.48	2.79	2.54	2.87	3.12	3.54
Sri Lanka	0.14	0.19	0.24	0.29	0.25	0.29	0.16	0.17	0.23	0.23
Sweden	2.93	3.17	3.47	3.67	3.51	3.30	3.35	3.21	2.90	2.93
Switzerland	3.20	3.40	3.68	3.87	3.88	3.93	3.91	3.96	4.11	4.07
Thailand	0.00	0.01	0.03	0.06	0.13	0.31	0.52	0.89	0.91	1.10
United Kingdom	12.97	13.93	15.97	16.02	15.25	14.84	14.49	14.47	14.07	15.45
United States of America	53.07	52.34	58.24	64.93	67.01	70.44	76.02	80.25	87.52	94.64
Viet Nam	0.02	0.03	0.04	0.05	0.06	0.07	0.10	0.23	0.34	0.75
World	**391.95**	**424.73**	**465.82**	**512.98**	**542.53**	**540.07**	**579.31**	**648.22**	**724.45**	**806.70**

Source: FAOSTAT

The dairy sector is currently dealing with a number of challenges favourable long-term consumer trends and developing technologies provide a number of opportunities. Looking at these challenges and opportunities at the producer, processor and the retailer levels suggesting some of the themes, how the dairy industry needs to adopt as a whole emerging technologies and datafication of agriculture and providing the farmers with the potential to obtain quantifiable information to continuously measure and monitor farm operations and react or take steps accordingly. The increasing demand for the clean label products and functional foods and the growth of protein consumption during the innovation in the sector. The ability to identify and exploit the consumer insights from local markets is also vital in generating increased value for consumers and higher margin products for producers.

The global dairy sector continues to face challenging trends. Milk prices have been on the decline most parts of the world and are currently rather low, while some degree of price recovery in the medium to long is expected. A considerable level of uncertainty remains due to the increased production recently witnessed in the E.U and U.S.A as well as the state of global economy and geographical landscape. Whereas the longterm global demand for dairy products continues to increase primarily as a result of population growth as well as the growth in the consumption of dairy products. The world wide increase in milk production is currently not matched by corresponding demand the overall demand for solids not fat is not increasing at the same rate as that of butter fat resulting in the growth of inventories of several dairy products especially the milk powder in particular. The global dairy trade followed suit growing significantly the equivalent of nearly 9% global milk output is currently traded internationally. However, demand from importing countries could not absorb the additional output on the market thus resulting in a sustained drop/decline in the prices of dairy products throughout the year mergers and acquisitions continue to intensity and are becoming increasingly global. International projections show a population of nearly 10 billion in 2050 which means increased demand for food in the coming decades. In 2015 the global per capita consumption of dairy products was estimated at 111.3 Kg. According to the OECD and FAO this number should increase to 12.5% by 2025. Consumption is developing countries should increase further. These trends are confirmed by the National experts who completed the "Global Marketing Trends" survey on predicted house hold dairy consumption in the coming decade.

Summing up the both traditional and evolving technical demands, the producers, processors and the retailers need to provide the consumers with nutritious, safe, convenient and value for money options which are required in the dairy context. Running dairy farms on professional basis has been a common place for some time in a number of countries. Regardless the scale of evolving technologies and improved farm management practices will play a greater role on dairy farms globally in the coming years. Successive farmers will need to focus on the key aspects to grow their enterprise in a sustainable and profitable manner.

Dairying plays an important role in agriculture in many countries globally across the world. The most important in some refers to milk production and processing industry. The future trends in the dairying can be split into three different areas and their interaction depending on each other influencing the future developments. The three different areas are namely the milk production, processing industry and the trade including the consumption of dairy products. Thus explaining their interdependency in the present situation and the likely tendencies in future such as the manufacture of equipment packaging, cleaning and senilisive agents of the industry which are different in different regions throughout the world. Milk is not only a valuable natural food but has become primarily an agropolitical and social problem, whether as surplus or scarcity.

Due to the surplus of milk and dairy products resulted in the Quota system in Western Europe whereas on the other hand the developing countries are lacking in milk and are trying with more or less success to alternate the situation by increasing their milk production. However, due to the increase in the population growth, the per capita milk production more or less stagnated, because, the increase in milk production is mainly due to the increase in the number of milk cattle and not due to the higher milk yield per cow since the feeding resources have not been improved correspondingly. The countries in transition will recover faster in milk production and will become an important factor in the world dairy market provided they will compete both with improved quality and price of the milk.

There is a great difference in milk production between the regions of the world to demonstrate how it is unjustified to generalize the situation in milk production and accordingly the future trends. They have to be estimated for different regions in the world separately for each country and even smaller areas. Milk yield will continue to increase upto the genetic potential of the breeds and the physiological capacity of the individual cows which is stimulated by favourable growth in prices and consequently higher concentrate input. In some countries showing lately an un-visual increase of the average milk yield the improvement is not due to the result of better feeding, genetic improvement, better management and card etc but simply caused by the replacement of the low yielding cows with high yielding variety of cows.

Depending on the conditions and the situation present/prevalent in each county area an alternative must be found for the farmers who for economical reasons have been forced out of dairy business. The worse situation would be to abandon agricultural production or activity on farms for which generations were working very hard. The policy and the solutions are in the hands of the government agencies and their services as well as in the hands of the farmers themselves. The sooner the better they find the other alternatives. Direct sale of milk from farms or investments in very small dairies that are occurring already in some areas which is a short term measure and a doubtful solution.

Payment of milk according to quality had a double effect and consequently the improvements obtained in a very short period of time are clearly demonstrating that only the strict and stringent policy of the dairies. Permanent advisory activities and control can lead to success. In the absence of any clear cut agricultural policy defining with regard to the alternatives to the farmers learning milk production will inflict difficulties to the governments respectively it would be pity to lose the high yielding herds of cows which are excellent nucleus for the general improvement of milk production solutions must be found in new ownership and management.

Trends in other countries in transition varies differently from country to country. Inspite of the latest improvements it will take time and great effort is needed to find a solution for the future. Obviously there are different conditions problems, solutions and trends. Milk is not only the most perfect food but it became primary on a geopolitical and social problem regardless if it appears as a surplus or scarcity commodity. The future international trade will be influenced by the GATT agreement which is implementing the reduction of export subsides and giving access to the imports. In the dairy industry the trend is and will be the concentration of few processing centres in each country which will cover not only the home market but aggressively invade the worldwide market. Careful studies will be necessary to foresee the trends in different markets keeping in mind that markets are not given but made and fought for concentration of experts from different fields of activities will result in greater efficiency provided they are integrated in the top management and their competance is equal to their responsibility. The trend goes toward high quality products with long shelf life. Packaging and wrapping material of high quality will be accepted only if recyclable low weight and efficiently cleanable with low concentration of cleaning and sanitizing agents.

There are tendencies to establish new small dairies by individual or group of farmers in which liquid milk and/or other dairy products are produced. The small entrepreneurs must be aware of the fact that small dairies have to meet the same regulatory and hygiene requirements as those prescribed for the large dairies. In addition these small units might have difficulties in marketing their products especially in case they are situated in the remote areas far from the markets. Same general trend which can be foreseen is in the consumption of the dairy products which is very variable. It is most cynical to launch a general campaign against milk consumption as even in the developed countries there are so many under nourished people who would be only too happy to get a glass of milk per day. Does the advocated danger from the dairy products apply to millions of starving people in different parts of the world among them are so many children. The dairy industry will certainly take part in the emerging market of neutraceutical foods in order to sell profitable intermediary products to dietetic and pharmaceutical industries. The consumption of dairy products in developing and under developed country will countries to differ which will be further influenced by poverty, possibly to produce milk in unfavorable conditions health reasons etc., where generalization is not possible.

The food and Agriculture organization of the United Nations in their report on the global dairy sector on sustainable production, processing and consumption of milk and milk products outlined some facts which may benefit people and the planet that can help to achieve the sustainable development growth (SDG). An attempt has been made to summerise and reproduce the facts for the benefit and better understanding of the entrepreneurs, dairy farmers as well as the students of the dairy science here under briefly.

1. Milk is one of the most produced and valuable agricultural commodities world wide milk ranked third by production tonnage and was the top agricultural commodity in value terms the world over. Milk contributes 27% to the global value added to livestock and 10% to that of agriculture.
2. Milk is a local as well as a global commodity. Basically milk is produced locally and consumed globally in almost all over the world and ranks among the top five agricultural commodities interms of both quantity and value. The milk and dairy products account for about 14% of the global annual trade with particular reference to the whole milk powder (WMP) and skim milk powder (SMP) as percentage production whereas the fresh dairy products with less than 1% are the less traded commodity on the basis of production.
3. The dairy sector is growing fast with the projected world milk production to increase by 177 million tonnes by 2025 at an average

growth rate of about 1.8% per annum. This is because of the sheer size of the dairy industry, the growth rates can produce big development pay offs for peoples livelihood, for the environment as well as to the public health also.

4. The dairy sector is heterogonous because of the fact that global milk production largely derives from cattle representing 82.7% followed by buffaloes 13.3%, goats 2.3%, sheep 1.3% and camels 0.4%. Milch animals are raised in a multitude production systems whose main objective is the production of milk.
5. Dairy animals are a popular asset in the rural areas. Globally there are more than 150 million farmers who are keeping atleast one milch animal including cows, buffaloes, goats and sheep farmers often keep mixed herds with more than one species of dairy animal and the most common dairy animal being the cow in developing countries.
6. Dairy animals support livelihood because of the fact that those animals are a regular source of food and cash for the farmers who either consume or sell milk and milk products every day. Dairy animals are a store of wealth and enhances the resilience by selling them for cash in needy times. The animals also contribute to crop productivity through animal traction/power and also provide social status and social capital thus facilitating a network system which is at the core of the effective market and supply chain of relations and alliances.
7. Dairy products are key to nutrition and health of the human beings. Milk and dairy products are dense nutrient foods supplying energy, protein and micro nutrients including calcium, magnesium, selenium riboflavin and vitamins B_5 and B_{12} which are essential to reduce the hunger and mal nutrition amongst the most vulnerable especially the pregnant women and children.
8. Milk production supports women empowerment where in majority of the developing countries the most popular asset among the rural women is the dairy animal which can be acquired very easily. Women however regardless their owning the milch animals play a major role in the dairy production system by their multiple activities in the care and management of these animals. Dairy often serves as the first stepping stone for the rural women to start a consolidating better place for themselves in the society.
9. The dairy industry creates jobs since the dairy producers are very often organised in coperatives or in liasion with other valued chain of actors to process and sell their products to the consumers. Employment is a major pathway out of poverty and job creation is a global challenger

with an estimated 150 million dairy farmers globally it is likey that the dairy sector supports the livelihoods of about one billion people world wide.

10. The dairy industry has a keynote in the public health. Consumption of dairy products even though is considered as a part of the healthy diet, the consumer must be aware of the zoonotic and food borne diseases which are originating from the dairy animals that are harmful to the public at large. The consumption of raw milk and other unpasteurized dairy products can result in the occurence of the food borne diseases. Excess intake of antibiotics by the milch animals can also contribute to anti microbial resistance in human beings.
11. The dairy industry relies on natural resources such as land, water, nutrients and energy. In order to feed the dairy animals which includes all species of a milch animals requires around 1 (one) billion ha of land or 7% of the total land on the planet earth. The global dairy herd consumes about 2.5 billion tonnes of dry matter feed annually. About 40% of the global livestock feed intake as around 77% constituting grass and straws suggesting that the global dairy herd is converting the materials which are not edible to human population into high quality protein and other essential micro nutrients.
12. The dairy herd contributes to green house gas emissions especially through rumination. Dairy animals produce around 3.1 giga tonnes of CO_2 equivalent per year or 40% of global livestock emissions with dairy cattle responsible and accounting for 75% of the same. Depending on the species and the production system that is being followed or adopted the enteric methane production represents 51 to 67% of the herd emissions. Compared with CO_2 which is a longlived climate pollutant methane even though is short lived but traps 84 times more heat than CO_2 over the first two decades after it is released into the air. Therefore the potential of reducing the negative impacts on climate through the increased productivity of ruminents is most important. Options aiming to reduce the emissions mainly dependent on target feed efficiency manure management and herd performances through improved animal health and husbandry practices.

The linkages between the dairy sector, perople and the planet are multiple and all emcompassing investments that promise and promote a sustainable development of the livestock industry can contribute to achieve these objectives which include:

Full productive employment and decent work for all reducing the inequality and to end poverty, hunger in order to achieve food security contributing to healthy lives. Achieve gender equality and partnerships for the goals b promoting sustainable consumption and production patterns to combat climate change. Protect and restore terrestrial eco-systems including bio-diversity and sustainable management of water and sanitation.

1.8 Dairy Industry in India

Dairying is practiced in the villages as a cottage industry meeting the house hold requirements which is scattered throughout the length and breadth of the country. There is no organised dairy industry. Semi commercial dairying has started with the establishment of the military dairy farms by the colonial British rule in order to meet the requirement of their perosnnel.

The first cooperative activity in dairy enterprise started in Uttar Pradesh with the organisation of "Katra" cooperative milk society (1913, 1917) at Allahabad, Lucknow (1938), Varanasi (1947), Kanpur (1948) and Meerut (1950). Presently about 55 districts of the state are covered under dairy development programmes in cooperative sector.

1.9 Mile Stones in Dairy Development

1862 - First veterinary school was opened at Poona in order to meet the requirement of the Remount and veterinary corps in the Army.

1886 - First military dairy farm was set up at Allahabad.

1889 - Establishment of Imperial Bacteriological Laboratory at Poona, which become the fore runner of Imperial Indian Veterinary Research Institute, Izatnagar, Bareilly, UP

1891 - Establishment of Indian Civil Veterinary Department later named differently as Animal Husbandry and Veterinary Services Department including the dairy development.

1907 - First importation of Ayrshire stock.

1914 - Studies were initiated on the milk production and composition of Indigenous cattle and attempts were made to establish pedigree herds of Indian Breeds.

1916 - Appointment of Imperial Dairy expert.

1920 - William Smith the imperial dairy expert recommended
(1) Scientific feeding, breeding and management

(2) Establishment of a training centre to meet the man power.

1923 - Establishment of Imperial Institute of Animal Husbandry and Dairying at Bangalore. The institute was renamed as Imperial Dairy Institute in 1941 which was shifted to Karnal in 1955 and renamed as NDRI. In 1923 two years diploma course (I DD) was started at Bangalore and Allahabad.

1928 - Appointment of Royal Commission Agriculture.

1929 - Setting up of Imperial Council of Agricultural Research which was renamed as Indian Council of Agricultural Research (ICAR) on 16th July 1929. It was reorganised twice in 1965 and 1973.

1930 - Shri Pestonji Edulji Polson established Polson Model Dairy at Anand for the Polson butter etc.

1936 - Dr. N.C. Wright, Director Hannah, Dairy Research Institute, U.K. suggested various methods for the improvement of dairying in India.

1937 - The Lucknow Milk Producers Co-operative Union Limited was established followed by the organisation of such unions in other districts and states.

1938 - Agmark act was enacted which laid down the standards for the grading of ghee and butter.

1944 - Establishment of Central Livestock and Research cum breeding station at Harringhata with an intention to supply processed milk to Calcutta.

1945 - Inauguration of Government Bombay Milk Scheme (GBMS) and setting up of the Aaray milk colony.

1948 - Establishment of Kaira District Co-op. milk producers union at Anand, which came to be known as AMUL - After independence both AMUL and GBMS set together a faster pace of dairy development with emphasis on developing techniques of processing and marketing under Indian conditions.

1950 - Pasteurised and bottled milk was sold for the first time to the common man in India in Bombay.

1955 - NDRI was shifted to karnal at the central cattle breeding farm.

1956-61- During the second five year plan, seven liquid milk plants were completed. Eight pilot milk schemes, three creameries and two milk product factories have been started.

1957 - Establishment of dairy science college at NDRI Karnal offering the B.Sc Dairying Degree.

1960 - Production of baby food and milk powder by Amul from buffalo milk for the first time in the world and of course in India with the help of CFTRI, Mysore.

1961-66 - During third five year plan 23 liquid milk plants, 27 Pilot milk schemes were in operation, 2 cheese factories and 4 cattle feed factories have been established and another 37 liquid milk plants have been started.

1965 - National Dairy Development Board was set up.

1970 - Establishment of Indian Dairy Corporation and launching of Operation Flood (OF1).

Establishment of National Co-operative Dairy Federation of India.

1975-76 - Evolution of National Milk Grid Marketing of Milk through rail tankers began, Anand - Delhi, Anand - Bombay, Jalgaon - Bombay, Dhule - Bombay, Erode - Madras etc.

Dairy educational and research activities in the country commenced with the establishment of the "Imperial Institute of Animal Husbandry and Dairying" on 1st July 1923 at Bangalore. William Smith, Asst. Director Military farms was appointed as the imperial dairy expert. The idea to establish the institute was conceived in 1916 when the then Govt. of India was implementing the recommendations of the Board of Agriculture. The military dairy farms at Bangalore, Karnal and Wellington (Nilgiries) were transfered to the Imperial Institute of Agriculture in 1923. During the year 1936 the military dairy farm at Karnal was handed over to the Imperial Institute of Agricultural Research at New Delhi. The Imperial dairy experts section along with the Bangalore Dairy Institute was seperated and placed under the erstwhile Dept. of Education, presently Ministry of Agriculture and at the same time re-naming the institute as Imperial Dairy Institute.

After Independence in 1954 the Govt. of India has decided to establish the Imperial Dairy Institute, together with the Dairy science college shifting the head quarters from Bangalore to Karnal in the premises of the cattle cum dairy farm and renaming the institute as National Dairy Research Institute, as per the recommendations of Prof. H.D.Kay, Director, National Institute for Research in Dairying Reading, U.K. The Institute was finally shifted to Karnal in the year 1955 and the under graduate B.Sc dairying course was started at NDRI in 1957. After the successful implementation of the operation flood programme in 1971, many Dairy Science colleges were started in the country.

1.10 Educational Institutions

The list of educational institutions which are imparting training in the field of dairy science have been listed below:

1. Dairy Technology College, Tirupathi, Andhra Pradesh
2. College of Dairy Technology - Kamareddy, Andhra Pradesh
3. Sanjay Gandhi Institute of Dairy Technology - Patna, Bihar
4. Seth M.C. College of Dairy Science - Anand, Gujarat

5. Dairy Science College, Hebbal, Bangalore, Karnataka
6. College of Dairy Science & Technology, Raipur, Chattisgarh
7. College of Dairy Science & Technology, Udgir, Maharashtra
8. G.N. College of Dairy Science & Food Technology, S.D.A.U, Gujarat
9. College of Dairy Science & Technology, Mannuthy, Thrissur.
10. Sam Higgin Bottom University of Agriculture and Technology & Science, Allahabad, (previously Allahabad Agricultural Institute Naini Allahabad), Uttar Pradesh
11. College of Dairy Technology, Warud, Pusad Dist., Maharashtra.
12. College of Dairy Science, Udaipur, Rajasthan.
13. Faculty of Dairy Technology to W.B.U.A & FS Mohanpur, Calcutta.
14. Dairy Science College, Chennai
15. Dairy Science College, Ludhiana
16. Dairy Science College, NDRI, Karnal
17. Dairy Science College, Amreli, Gujarat
18. Dairy Science College, Gulbarga
19. Mansinhbhai Institute of Dairy and Food Technology, Mehsana, Gujarat

The earliest attempts of the dairy development can be traced back to the colonial British rule where they have established the military dairy farms, with an intention to ensure the supply of milk and milk products especially the butter to their army personnel. The first of these farms was set up in Allahabad in 1913, followed by Bangalore, Ootacamund (Nilgiris) and Karnal etc., in a phased manner maintaining with the improved milch animals. This ap,proach did not have any effect/impact on the supply of milk to the urban civilian consumer as it was less important to the military authorities. With the increasing growth of the urban population, the consumer has to depend on the milk vendors who sell/supply the milk door to door either by collecting the milk from the nearby villages or by keeping the milch animals (Cattlesheds) in the close proximity of the cities/towns. To some extent the second world war gave an impetus to the private dairies with modestly modernised processing facilities like Polsons, Keventers and the Express way were some of the urban dairies supplying the processed milk table butter, and ice-cream though not on a large scale.

These dairies were not concerned with either the improvement of the breed or the milk production level of the breed/animals which are reared in the rural areas as they are interested and contended with the procurement of the milk from the producers because they are commercial in their nature.

These early modern systems did not bring about any significant changes/ shifts in the milk production and the dairying remained in an un-organised and under developed state.

1.11 Dairy Development

The milk production is predominantly the domain of small and marginal farmers and the land less, who keep 1-2 milch animals generally as a part of the mixed farming system. Small holders own about 70 percent of the milch animal population but these are widely dispersed in rural areas with poor infrastructure and limited access to services and markets. The cattle and buffaloes have a complimentary, supplementary and sustainable relationship with crops under mixed farming system prevalent in the country.

The domesticated dairy herd was of poor quality, which reflected the emphasis on draught but not on milk. Today the situation has changed dramatically with the milk-production increasing rapidly both absolutely and relative to the population. Dairy development fits most appropriately in the country's programme of increasing food production, rural development, and equitable distribution. Any effective programme for poverty alleviation would have to include dairying as its base to bring about maximum growth with minimum capital out lay. India's dairy industry is on the threshold of many good changes that would totally transform the dairy scenario and give the needed thrust for its rapid growth. Greater emphasis must be placed on the facilities in the areas of milk production, processing, marketing and research in order to meet the challenges ahead. Buffaloes constitute only one third of the total milk bovine population, but account for more than half of the total milk production. The low productivity of some 75 million non-descript cows and buffaloes is weighing down the milk enhancement programmes. The major challenge today is how to bring about even a small gain in their productivity.

The dairy development in India has been achieved through the **"Operation Flood"** programmes. Milk cooperatives in the villages are the collection points for locally produced milk. The main objective of dairy development programme is to increase the milk production level of the native non-descript cow/buffaloe which have been scattered in the rural areas for which the following steps/considerations are to be followed.

1. The availability of good quality and quantity of feed must be increased.

2. Improvement of management, housing, husbandry methods and veterinary care.
3. Genetic improvement, production and reproductive capacity of the cattle.

After Independence, the Govt. of India has launched its 1st Five year plan in the year 1951 giving the priority to the dairy industry with an intention to supply/provide hygenic milk to the urban population by organising/ strengthening the milk schemes in the large cities. In order to stimulate the milk production levels of the animals the Govt of India has launched and implemented the Integrated Cattle Development Programme (ICDP) and the Key Village Scheme (KVS). In the absence of stable and remunerative prices as well as good marketing facilities for the milk producers the milk production remained more or less stagnant.

Dairy development is conceived as a major development programme not only in the context of poverty mitigation plan but also as a measure to augment the milk production to meet the increased demand for milk from urban consumers. Distribution of milch animals and encouragement of milk production is one of the popular action programmes under the Integrated Rural Development Programme (IRDP) for alleviating the poverty in India.

In the early sixties (1960) the State Governments have adopted and tried different strategies for the development of the dairying as an industry by establishing the modernised milk/dairy plants and run by their own (trained staff) departments organising the various milk schemes. The Urban orientation of milk production lead to the establishment of cattle colonies at Mumbai - Aarey Milk Colony, Kolkata - Harringhatta and Chennai - Madhavaram Milk Colony. In the early stages, these government programmes had to face extreme difficulties in organising the procurement of milk from rural areas in order to run the milk schemes/plants economically because they failed to create an organised milk procurement system which the middle man and the contractors took advantage due to the perishable nature of the milk. The queue of the consumers were larger because the govt. dairies were not able to meet the urban demand while the rural milk producers were left at the mercy of the trader and the money lender.

Dairying shall continue to be a subsidiary occupation, and major source of income for the small, marginal farmers and landless labourers. In any scenario, the non-descript cows and buffaloes will remain as the back bone of tomorrow's dairying. The most significant contribution of the cooperative dairying is to bridge the rural producer and the urban consumer through a national milk grid. It has been able to bring to the door step of the dairy farmer, the lucrative urban market which is otherwise inaccessable to him. To ensure the orderly dairy development including the dairy equipment industry, which is relatively young in order to give a momentum of growth it is essential to identify the future needs, the technology gaps and priorities in the years ahead.

An effort has been made to present all the aspects pertaining to milk production which needs a careful consideration in a schematic manner for better understanding.

A concrete example of the services has been seen through the Integrated Dairy Development Programme (IDDP) or the Integrated Rural Development Programme (IRDP) launched by the National Dairy Development Board (NDDB) and the National Bank for Agriculture and Rural Development (NABARD) respectively. They provide various types of services to the cattle owners as well as to the consumers.

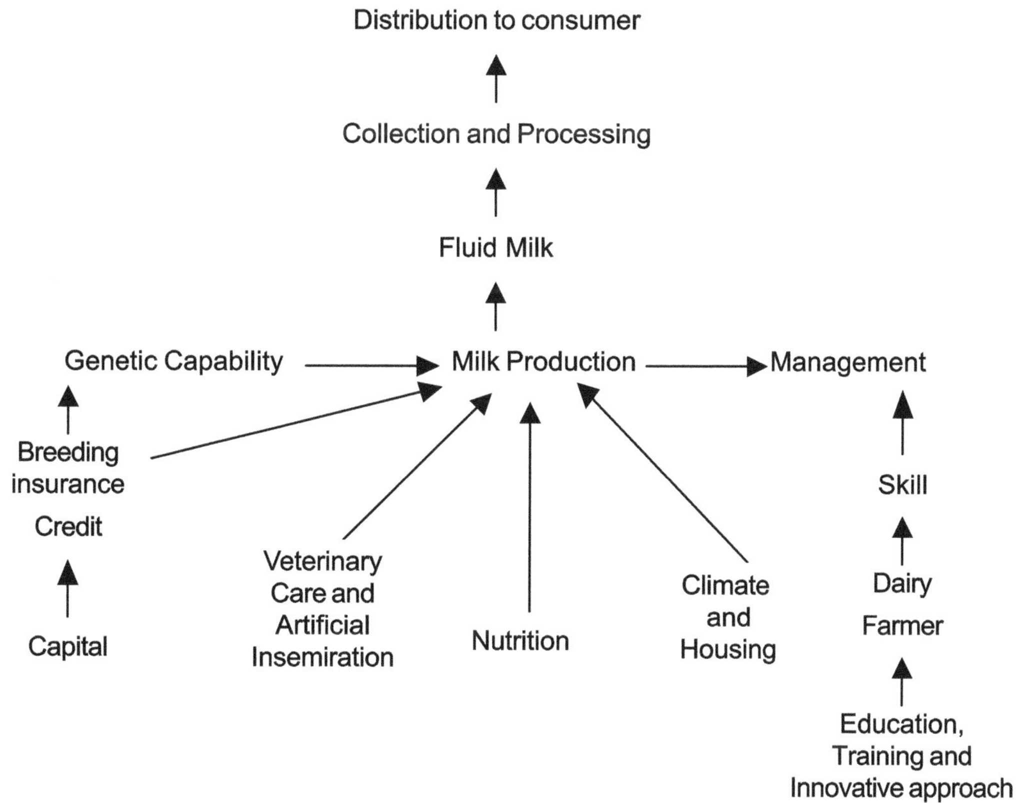

Source: M. Perumal, P.S. Mohan and M. Suresh

1.12 Evolution of Anand Model/Pattern

The major problem that is faced by the Indian Dairy Industry since Independence to till date is the procurement of milk and its marketing from the scattered rural villages due to lack of proper road and communication facility. Polson's - a private dairy which was established at Anand in the erstwhile Bombay state and present Gujarat State used to procure the milk from the milk producers of the nearby villages through the middle man/ contractors and the milk after processing was sent to Mumbai which is about 425 km away. In mid 40's the milk producers of Kaira Dist. who are supplying the milk to the Polson's Dairy plant demanded for a proportionate share of the trade margin and a modest increase in the procurement rate of the milk which was denied by the Polson's Management. Thereafter the milk producers refused to supply the milk and on the advise given by late Dy. Prime Minister Sardar Vallabhbhai Patel and formed milk producers society which is registered as the "Kaira Dist. Co-operative Milk Producers Union" which is now popularly known as "AMUL" (Anand Milk Union Ltd) in the year 1946 thus laying a foundation for the organised milk marketing in India.

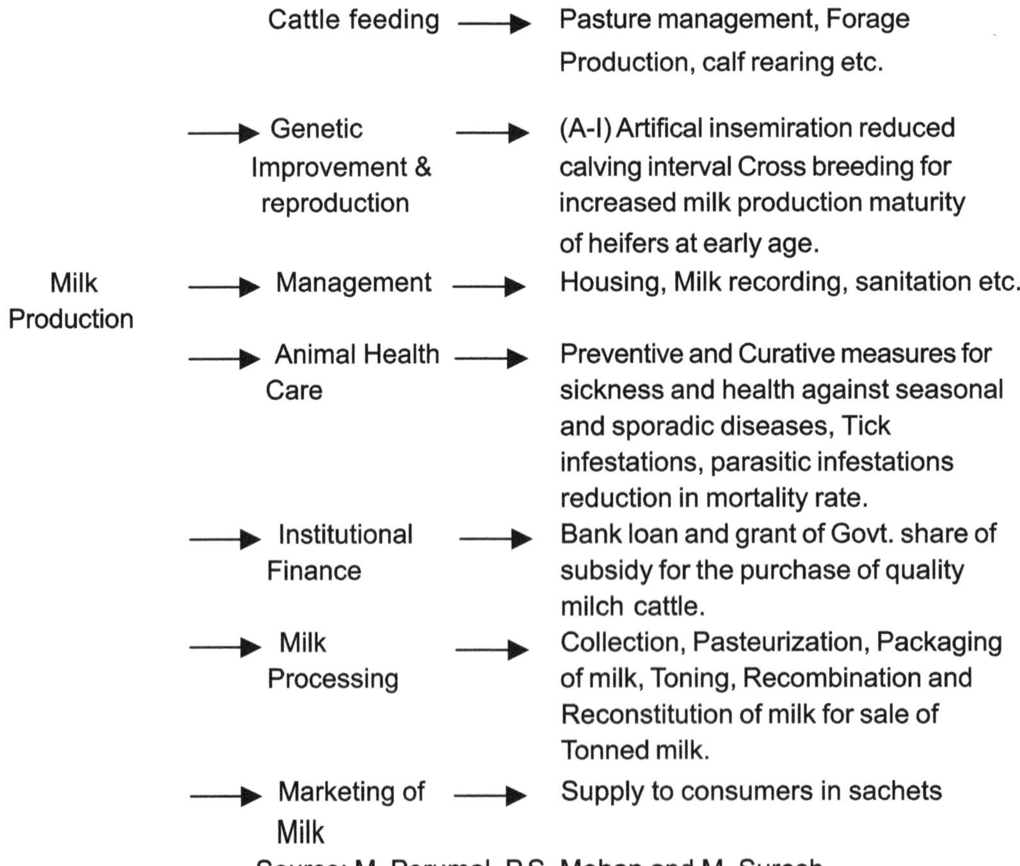

Source: M. Perumal, P.S. Mohan and M. Suresh

The three tier structure of Anand Pattern Cooperatives

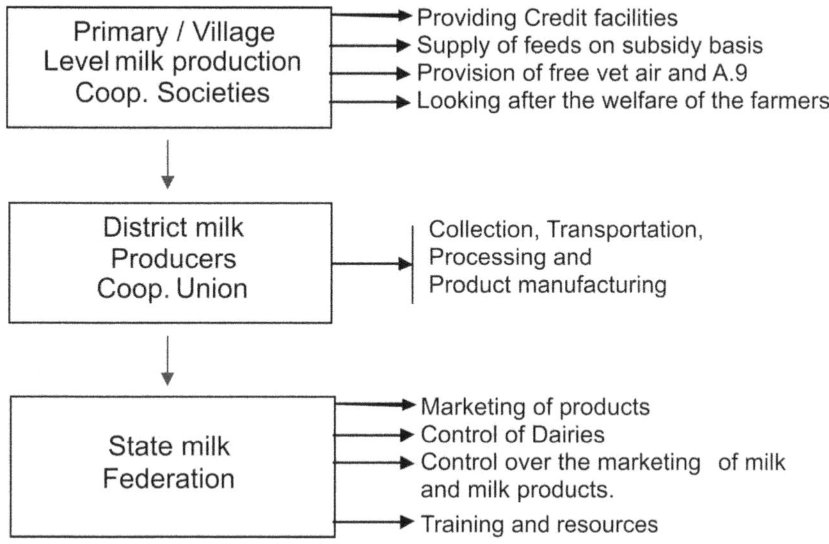

Source: M. Perumal, P.S. Mohan and M. Suresh

The Govt. of erstwhile Bombay state has cancelled the contract with the Polson's dairy and entered into a new agreement and handed over the entire business of supplying the milk to the AMUL in the year 1952. However, the AMUL has faced a major problem and difficulty in the disposal of the surplus milk in the flush/peak season. In order to overcome this problem the "Kaira Dist., Co-operative Milk Producers Union" has established the first and a new dairy plant at Anand for producing the milk products such as Butter, Ghee and Milk Powder in the year 1955. To cope up with the ever increasing demand of milk procurement the AMUL dairy has set up a second dairy plant and a product manufacturing unit in the year 1965 and 1971 respectively finally leading to the establishment of a fully automatic modern dairy in the year 1993 adjacent to the original dairy plant.

Thus the AMUL provided the basis for the Anand Model of Dairying the primary or the basic unit being the Milk Producers Co-operative Society at the village level. The village co-op. societies in the district combined together forming the district union which ideally has its own processing facilities. All the milk unions representing the different districts of the state are normally members of the State Federation whose prime responsibility is to provide the marketing facilities for the milk and milk products that are produced by the respective milk unions within the state as well as outside the state. There is also another organization at the national level - the National Co-op Dairy Federation (NCDF) which formulates the policies and programmes designed to safe guard the interest of all the milk producers of the country.

1.13 Role of NDDB in Dairy Development

The Prime Minister of India Late Lal Bahadur Sastry visited Anand to inaugurate the cattle feed plant in Oct. 1964 and spent a night as the guest of the Anand Milk Union Ltd. He was very much impressed by the socio-economic changes that were brought about by the milk co-operatives and expressed his desire to replicate the **"Anand Model"** through the national level organization throughout the country making available a multi-desciplinary professional expertise to the dairies in the public and co-operative sector. In the year 1965 the National Dairy Development Board (NDDB) was constituted which was registered under the Societies Registration Act, the Charitable Trust Act and Public Trust Act and located at Anand considering its model and mandate.

In the initial stages, the NDDB was supported and assisted financially by the Govt. of India, the Danish Govt. as well as by the Amul. NDDB also received some financial assistance and funds from UNICEF. NDDB was not equipped with any monitory powers to handle the government funds. Therefore, in the year 1970 the Govt. of India has established the Indian Dairy Corpn. (IDC) which was later merged with the newly constituted NDDB by an act of parliament passed in the October 1987.

NDDB began its operations with the mission of making dairying a vehicle to a better future for millions of grass root milk producers. Since its inception, the dairy board has planned and spear headed India's dairy programmes by placing dairy development in the hands of milk producers and the professionals they employ to manage their co-operatives. In addition NDDB also promotes their commodity based co-operatives, allied industries and veterinary biologicals in an intensive and nation wide basis.

1.14 Operation Flood (OF) 1970 - White Revolution OF - I

The strategy for the organised dairy development in India has actually been conceived in the early 60's which was strengthened with the establishment of the National Dairy Devt. Board (NDDB) in 1965. The operation flood (OF) is an unique approach conceived by Dr. V. Kurein, the founder Chairman of NDDB whose vision saw an opportunity to utilise the surplus milk products namely the milk powder and the butter oil that have been accumulated in the European Markets and have posed a major threat and a death knell for the developing dairy industry in India. The operation flood designed basically as a marketing project revolutionized the Indian Dairy Industry recognising the potential of the European surpluses as an investment in the modernisation of the Indian Dairy Industry with the assistance of the World Food Programme (WFP), Food Aid in the form of milk powder and butter oil obtained from the countries of European Economic Commission (EEC). The funds to the tune of 116.54 crores have been generated by the sale of 1,26,000 tons of skim milk powder (SMP) and 42,000 tons of butter oil for the implementation of the operation flood (OF). During this first phase of Operation Flood (OF) which was launched in 1970, and was popularly known as **"White Revolution".** The funds generated through the sale of WFP gifted commodities were utilised to develop 27 rural milk sheds covering 10 states and setup dairies called as **"Mother Dairies"** in the four metropolitan cities. Thus it laid the foundation to the modern dairy industry ultimately to meet the needs of the milk and milk products of the country. The commercial imports of milk powder which were around 60,000 mt prior to the launching of OF have shown a declining trend and were almost ceased completely by the year 1975-76 except for a couple of years, where the milk powder was imported commercially during the period of extreme drought conditions to meet the liquid milk demand.

OF - II (1981-85)

The second phase of OF was commenced in the year 1981 with an aim to build up a National Milk Grid (NMG) linking 136 rural milk sheds in 21 states and union teritories, with the urban demand thus creating the necessary infrastructure to support the viable dairy industry.

It was started in April 1981 with an intention of creating a viable dairy industry in order to meet the India's need for the milk and milk products. The world bank has provided about 150 million US $ in the form of commodity assistance from the EEC. The OF-2 integrated the Indian Dairy Association and all the dairy development projects that are being implemented in some of the Indian States and it was closed in March 1985.

OF - III (1985-96)

The third phase of OF which was commenced in 1985 aimed at the consolidation of the gains made under the earlier two phases thus achieving the self sustainability of the milk unions.

This project/programme aims at ensuing that the co-operative institutions become self-sustainable with an investment of US $ from world bank in the form of commodity and cash assistance from the EEC and NDDBs own internal resources.

India's white revolution has not only received support from the European community and the world bank but also from a number of Western Govt's, the United Nations (UN), Food and Agrl Organisation (FAO), the United

Nations Children's Educational Fund (UNICEF) and European Non-Government Organisations (NGOs) such as British Relief Agency etc.

Operation Flood (OF) IV - 1996-2006

This fourth phase of operation was aimed at the strengthening of the following objectives.

1. To create infrastructure and strengthening democratic values.
2. Strengthening the cooperatives by providing funds on 50:50 basis from central and state governments and
3. To increase the extension work in the fields of cooperative education, personnel training, marketing support, product development and improving the standards.

Efforts have been made to summarise the overall achievements of the OF Programmes of OF-1 to OF-4 and the achievements made by the Dairy Co-operatives state-wise upto 2009-10 during the OF Programmes in the Tables 1.4 and 1.5.

Table 1.4 Overall achievements under OF Programme

Item	Years				
	1971	1981	1985	1990	1996
1. No.of milk shed areas	5	39	136	170	170
2. No.of DCS	1600	13,300	34,500	60,800	70,000
3. Farmer membership	2811	7751	7361	701	93141
4. Average milk Procured LLPD	5.2	25.6	52.6	91.5	11.5
5. Total processing LLPD	168	64.9	122.8	178.2	22.6
6. Total marketing LLPD	NA	22.9	50.1	72.5	100
7. Investments	NA	116.54	277.17	411.59	1303.5

LLPD - Lakh Litres Per day

Source : India Dairy.com

1.15 Present Dairy Scenario

Dairying in India has been an eventful journey which has come a long way from being written off as a basket case a few decades back. Today India has emerged as the largest milk producer in the world with an annual production of more than 121.8 million tonnes in 2010-11. The success of India's dairy industry is not a story of the triumph of science and technology. Over the years milk has been transformed from insufficient production to self sufficient production, from ration to plentiful availability, from loose unhygeienic milk to milk that is pure from subjugation to a symbol of farmers economic independence. The requirements of our nation are on the move and must be met first with sustainable economic development. Based on the planning commissions estimates and subsequent corrections on account of consistantly higher growth in GDP, it is expected that the demand for milk is likely to be about 155 million tonnes by 2016-17 (end of 12th five year plan) and around 200 million tonnes by 2021-22. Indian dairying must address itself to issues of productivity, efficiency and response to consumer demands.

Table 1.5 Summary of (OF) Operation Flood achievements state wise by the dairy Co-operatives upto 2009 -10 *

Sl. No.	Name of the State / UT	No.of DCS Organized (Cumulative)	Farmer members ('000)	Women members ('000)	Milk Pro curement (Tkg pd)	Milk Marketing @TLPD
01.	AP	4911	841	177	1443	1409
02.	Assam	66	3	0	5	12
03.	Bihar	8299	441	64	736	437
04.	Chhattisgarh	751	31	7	24	35
05.	Delhi	---	--	--	--	3047
06.	Goa	179	19	3	36	75
07.	Gujarat	13890	2809	791	9053	3164
08.	Haryana	6881	313	73	522	384
09.	Himachal Pradesh	795	33	11	55	18
10.	Jammu & Kashmir	**	**	**	**	**
11.	Jharkhand	50	1	0	5	239
12.	Karnataka	11902	2052	660	3566	2468
13.	Kerala (#)	3637	775	162	769	1067
14.	Madhya Pradesh	5729	266	50	525	447
15.	Maharashtra	22217	1845	452	3151	2807
16.	Nagaland	47	2	0	9	3
17.	Orissa	3203	181	77	241	264
18.	Pondicherry	101	39	18	48	87
19.	Punjab	6904	381	54	952	723
20.	Rajasthan	15956	670	205	1654	1365
21.	Sikkim	287	10	1	11	14
22.	Tamil Nadu	10038	2122	874	2277	1990
23.	Tripura	84	6	1	2	13
24.	Uttar Pradesh	21343	971	265	518	402
25.	West Bengal	2962	210	70	262	655
	All India	140227	14021	4015	25864	21125

Source : India Dairy.com

Note:

1. * Refers to Provisional for not reported and NA for not available.
2. # Includes conventional societies and taluka unions formed earlier.
3. @ Co-operatives (State) and metro-dairies.

The Indian dairy industry is a top ranker in the global dairy band wagon. Currently around 46% of the milk is consumed in the form of liquid milk, 47% as traditional dairy products and 7% as western dairy products. The dairy in India is currently estimated to be about 130 million tonnes and is expected to grow at 4.5% per annum.

India is among the world's largest and fastest growing markets for milk and milk products. The average annual growth rate of milk, milk production has been 4 percent during the past decade. The country's per capita availability of milk is lower than the world's daily average of 285 grams though it has doubled from 124 gm in 1966 to 258 grams per day in 2009-10 and increased from 241 gm per day in 2005-06.

The world dairy is zooming in India for its rapidly growing markets that promise the 'moon'. World production of milk is projected to rise to 665 million tonnes by 2010 representing an average annual increase of 1.5 percent compared to an annual average growth rate of one percent during the 1990's. Of the estimated 16 million tonnes increase (2.6 percent) in the global milk production in 2005 over 2004. 75 Percent accounted by three countries - China (5.4 million tonnes), India (4.1 million tonnes) and USA (2.7 million tonnes).

The dairy sector need to be encouraged in order to fully exploit the complimentary, supplementory, synergistic and symbiotic relationship of animal productivity with crop production in the coming years where the food grain production has to increase and sustain itself to meet the demand of ever growing human population in our country. Enterprenurial development programmes hither to be carried out as innovative strategies for developing human resources, skills, and technologies with larger frequency and intensive extension programmes are to be revamped. India can take its pride in some of its major achievements in the field of dairying since 1971 or after launching the 'white revolution' or the Operation Flood Programme.

1. **Top milk producer yet low productivity**

   The total milk production of the country which was more or less stagnant around 23 million tonnes during the early sixties and seventies, has increased to about 107 million tonnes, during the year 2007-08 and is expected to reach to about 128 million tonnes during the year 2011-2012 maintaining the first rank in the world milk production. On the contrary it is very much disappointing to observe that the national average animal productivity is 1.5 ltrs per day as against 4.2 litres per day in buffaloes and 6.4 ltrs per day in cross breed cows in the operation flood areas.

Ideal productivity and milk production has been achieved and demonstrated at the institutional levels, however the average productivity levels exhibited in the actual field conditions are much too less/low showing variations around the mean in the entire population.

2. **Ranking first in world's milk production - yet low per capita availability and consumption**

   India's total milk production has increased from 23 million tonnes to 128 million tonnes during the year 2011-12 which is almost 4 to 4.75 times higher. Inspite of the fact that India has attained the first rank in the world's milk production, the per capita availability/consumption of milk even though has increased in a significant manner from 107 gm to 291 gm per day which is marginally higher than the world's average of 285 gm during the year 2011-12.

3. **High per capita availability - yet many under nourished**

   According to a survey conducted by the National Institute of Nutrition (NIN) the per capita availability of the milk in India is 258 gm per day which appears to be good. However, the actual intake of milk varies from a sparse or meagre 20 gm per day in many districts of Eastern Region to almost a litre per day in Punjab.

4. **Low purchasing power - less Consumption**

   The National sample survey organisation (NSSO) has conducted a survey on all the income groups both in urban and rural areas and recorded higher expenditure on milk and milk products. They concluded that low purchasing power denies consumption as per the dietary norms to majority of the Indians.

5. **Number one in Milk - but least support**

   Milk is India's number one farm commodity. Not withstanding its top place in the world's milk production, dairying has not received its due attention/recognition from the national planners, economists, social scientists as well as others.

6. Having about 12% of the cattle population of the world, India contributes about 11.5% only to the annual world milk production, despite the fact that more than half of the above milk production is

contributed by an additional strength of 39 million milch buffaloes in the country. It is also a fact that despite having only 2.4% of world land and also about 6 times the pressure of the human population on it when compared with the world average.

7. India has had the privilege to operate the largest world Food Programme aided dairy development project, which is popularly known as Operation Flood/White revolution, which has brought a remarkable, multi directional progress and achievements in the dairy field in several parts of the country during the three phases of its operation. The organised dairy sector is now commanding a major substantial share of the milk market in the four metropolitan cities, 770 other cities and towns of the country, immediately benefiting over 10.1 million producers in about 77000 villages covered under the field of multitier network of village milk producers cooperatives by assisting them round the year a remunerative milk market at their door step on one hand and good quality hygienic milk at a fair price to over 350 million urban consumers on the other in the country.

8. At present some 60% of the rural house holds, 80% of the rural farmers and 36% of the rural landless labourers have already adopted dairying as their important and sustainable profession. The operation flood has exhibited a remarkable multi directional progress in the form of concrete achievements in the field of dairy development in many milk shed areas of the country.

   The comprehensive role of the dairy cooperatives in this sector will be decernible if there exhists a strong network of cooperatives, institutions in other sectors of economy as well. Randomnly distributed patch work of these cooperatives in many states are too inadequate in the massive task of nutritional development.

9. The realistic progress has generally been confined at top percolating laterally in a slow manner in many of the dairy development programmes across the country, rather not percolating deeply, effectively and uniformly reaching right up to the bottom level of the population, thus leaving a tremendous scope to reach the optimal level of achievement.

10. Nearly 30% of the cultivated land of the country is under irrigation which probably the highest irrigated land average of the world. Most of the feed resources and roughages are available from crop residues,

byproducts of agricultural crops and agro industries including molasses, cakes and solvent extracts etc. However, in their actual utilization pattern the protein and energy deficiency does exist to the tune of 37% and 33% respectively on the animal nutrition front. Only some 10% of the required compound feed is being manufactured by the existing cattle feed factories in the organised sector. This clearly indicates the existing status of under nutrition/malnutrition prevalent in the bovine population in the country on overall basis.

11. On energy management front the dairy sector, continues to face the problem of energy shortage versus the growing demand for higher energy intensity competing with all other sectors of economy with growing industrialization and liberalisation of economic policies of the government.

12. Large dairy industry yet most milk left un procured and processed

Even though the dairy industry has made an enormous growth in our country, yet the dairy industry is not able to procure/collect all the milk that is being produced in the rural areas. The main difficulty in the procurement of this milk is due to the lack of proper transport and conveyance facilities as well as the scattered distribution of the villages. Further, dairy plants also have not evinced or shown any interest to reach to these areas by way of developing the infrastructure facilities such as milk cooling/chilling centres/plants etc for the collection of the milk from these distant/remote areas.

Bulk of the milk consumed in India is raw or unprocessed, which is an erroneous impression because of the fact that all the milk is boiled at house before its consumption. It is increasingly recognised that dairying could play a more constructive role promoting rural welfare and in the alleviation/mitigating the poverty.

Some of the areas which need to be focussed for more attention are :

1. The extent of collection and processing has to be enhanced. A target of enhancing milk processing from the current level of 16-17 percent to at least 25 percent.
2. Average collection of marketable surplus milk by organised dairies (both cooperative and private) is only to an extent of 30 percent except in Gujarat where it is about 85 percent. For enhancing production and

collection of liquid milk each state should create 2-3 new milk shed areas every year.

3. The national dairy plan envisages that for achieving the targeted growth in dairy sector the handling of marketable milk surplus by private and dairy co-operatives has to be increased essentially from the current 50 million kg per day to 160 million kg per day by 2020-21. To achieve this all the producers and processors should develop a good milk collection network.
4. At least 6 to 8 zones which have the potential for high milk collection and processing facilities should be identified and develop as export promotion zones.
5. Traditional Indian dairy products which have the huge potential for value addition and export, hitherto which have not been focused should be given more attention towards the creation of research and development facilities.

Rural dairying happens to be an integral and inter woven part of the traditionally diversified system of agricultural farming, which has now to stay in symbiotic, synergistic and sustainable relationship with it in the modern India. There is abundence of family labour and shortage of capital, while a declining contribution of agriculture to GDP values vis-a-vis live stock products shows a diversification trend of the agricultural base in the country. The annual value of India's anticipated milk production amounts more than Rs.1745 billion in 2009-10. Dairy Co-operatives generate employment opportunities for around 14 million farm families. Live stock contributes 22.2% to the GDP from the agriculture and allied activities. About 22.45 million people work in live stock sector which is around 5-8% of the total work force in the country. The present contribution of live stock sector to agriculture is around 25 to 30% and the dairy sector to live stock sector is around 66 to 70%. The dairy sector is the largest contributor to the agriculture GDP. In terms of output, milk is now the single largest agricultural commodity in India. The other attractions driving its growth are value added products like Ghee, Butter, Yoghurt, Paneer, Cheese, Flavoured milks, Ice creams and shredded and liquid cheese. Whereas the budget allocations to both dairy and live stock sectors by the Government in the Five year plans are proportionately less in relation to their respective contributions to the GDP.

The investment pattern on animal husbandry and dairying programme by the government shows the emphasis given to this sector for increasing the production and productivity. The total plan expenditure on agriculture and allied activities, animal husbandry and dairying and exclusively the dairy sector has been given in the Table 1.6.

Table 1.6 Investment/outlay on animal husbandry and dairying during various five year plan periods (Rs. millions)

Plans	Expenditure (Rs. in million)				Agri-culture as % of Total Plan	AH & dairy as % of Agrl. Sectors
	Total Plan	Agriculture and Allide	Animal Husbandry	Dairying		
First plan (1951-56)	19,600.0	2,900.0	160.0	77.8	14.8	5.5
Second plan (1956-61)	46,720.0	5,490.0	334.7	120.5	11.8	6.1
Third plan (1961-66)	85,770.0	10,890.0	770.0	336.0	12.7	7.1
Annual plans (1966-69)	66,254.0	11,071.0	597.0	257.0	16.7	5.4
Fourth plan (1969-74)	157,790.0	23,204.0	1,542.0	787.5	14.7	6.6
Fifth plan (1974-79)	394,262.0	48,665.0	3,189.8	944.5	12.3	6.6
Sixth plan (1980-85)	1,092,917.0	1,36.203.0	8,025.1	4,362.9	12.5	5.9
Seventh plan (1985-90)	2,202,163.0	2,79,611.0	12,033.0	5,887.9	12.7	4.3
Eighth plan (1992-97)	4,341,000.0	224,670.0	13,000.0	9,000.0	5.2	5.8
Ninth plan (1997-2002)	8,592,000.0	424,620.0	15,456.4	4,695.2	4.9	3.6
Tenth plan (2002-07)	1525639.0	58933.0	1707.7	285.8	3.86	2.90
Eleventh plan (2007-12)	3582768.0	162849.0	2907.1	576.3	4.55	1.79
Twelth plan (2012-17) (Outlay)	7669807.0	363273.0	11610.0	4976.0	4.72	3.20

Source: Dairy India year book seventh edition

The expenditure at current prices in the dairy sector has increased from Rs.77.8 million in the first five year plan to 9,000 million in the eighth plan. However, the plan outlay for the dairying sector in the ninth and tenth five year plans has been reduced drastically from Rs. 9,000 millions to Rs.4,695.2 and Rs. 2858 millions respectively despite the fact that the contribution of the dairying sector to the annual GDP is quite significant as compared to the agricultural sector.

An attempt has been made to furnish the upto date and to update the information pertaining to the budgetary allocation of funds to the animal husbandry department/ sector including the dairy and fisheries during the 11th five year plan by the Govt. of India has been presented in Table 1.7. The approach for the 11th five year plan for the livestock sector was aimed at achieving an overall growth rate of between 6 to 7 percent Res annum. An out lag of Rs. 8174 crore has been provided to the livestock sector/Animal husbandary departent in the 11th five year plan. The year wise financial achievements has been presented in Table 1.7. In addition to the outlay allotted to the department an additional amount of Rs. 5406 crores has also been made and provided under Rashtriya Krishi Vikas Yojana and National mission for protein supplement.

Table 1.7 Year-wise budgetary estimate revised estimate and actual expenditure during 11th five year plan (Rs. crores)

Year	Approved B.E.	Revised estimate R.E.	Actual expenditure	% utilisation w.r.f to R.E	% utilisation w.r.f to B.E
11th Plan 2007-12	**Allocation 8174.00 (Crores)**				
2007 - 08	910.00	810.00	784.00	96.80	86.16
2008 - 09	1000.00	940.00	865.00	92.05	86.53
2009 - 10	1100.00	930.00	873.38	93.91	79.40
2010 - 11	1300.00	1257.00	1104.68	87.88	84.98
2011 - 12	1600.00	1356.52	1243.11	91.64	77.70
Total	5910.00	5293.52	4870.53	92.01	82.41

Source : Annual report of A.H. Department, Govt. of India 2018-2019

The planning commission has in principle allotted an amount of Rs. 14,179 crores to the department of Animal husbandry for the 12th five year plan. This includes an amount of Rs. 7628 for AHD, Rs. 4976 crore for the dairy development and 2483 crores for the fisheries sector. The financial achievements made by the department during the 12th five year plan against amount lay of 14179 crores provided by the Govt. of India has been presented in Table 1.8.

Table 1.8 Year-wise budgetary estimate revised estimate and actual expenditure during 12th five year plan (Rs. crores)

Year	Approved B.E.	Revised estimate R.E.	Actual expenditure	% utilisation w.r.f to R.E	% utilisation w.r.f to B.E
12th Plan 2012-17	**Allocation 14179 (Crores)**				
2012 - 13	1910.00	1800.00	1736.37	96.47	90.91
2013 - 14	2025.00	1800.00	1748.80	97.16	86.36
2014 - 15	2174.00	1800.00	1738.07	96.56	79.94
2015 - 16	1491.14	1491.14	1418.20	95.11	95.11
2016 - 17	1600.00	1748.02	1743.12	99.71	108.94
Total	9200.14	8639.16	8384.56	97.05	91.13

Source : Annual report of A.H. Department, Govt. of India 2018-2019 (Internet)

Even though the expenditure is increased in monetary terms, the allocation to animal husbandry and dairy sector as a percentage allocation to agriculture varied from 7 percent in third plan to 4.6 percent in seventh plan. There was a gradual increase in the percentage allocation on animal husbandry and dairying sector from the first to the third plan periods and from thereafter it showed a decline except in the sixth plan which might be due to the quantum jump in the plan expenditure to the agricultural sector.

An attempt has been made to present the share of agriculture and live stock in the GDP along with the % share in the GDP for the benefit the students particularly to the better understanding by the dairy enterpreneurs in Tables 1.9 and 1.10.

Table 1.9 Share of agriculture and livestock sector in GDP (at current prices in Rs. crores)

Year	GDP Total	GDP Agriculture		GDP Livestock sector	
		Rs.	% share	Rs	% share
2004 - 05	2971.464	476636	16.04	119333	4.02
2005 - 06	3390503	536882	15.83	127518	3.76
2006 - 07	3953276	604672	15.30	142695	3.61
2007 - 08	4582086	716276	15.63	169296	3.69
2008 - 09	5303567	806646	15.21	200.440	3.78
2009 - 10	6108903	928586	15.20	237059	3.88
2010 - 11	7266966	1132048	15.58	276105	3.80
2011 - 12	8353495	1268081	15.18	327838	3.92
2012 - 13	9388876	1417468	15.10	386246	4.11

Source : National accounts statistics 2014. Central statistics organisation Govt. of India.

Table 1.10 Share of agriculture and Allied and livestock sector in GVA (at current prices in Rs. crores)

Year	GVA Total	GVA Agriculture and allied		GVA Livestock sector	
		Amount	% share of Total GVA	Amount	% share of Total GVA
2011 - 12	8106948	1501947	18.5	327334	4.0
2012 - 13	9202692	1675107	18.2	368823	4.0
2013 - 14	10363153	1926372	18.6	422733	4.1
2014 - 15	11504279	2093612	18.2	510411	4.4
2015 - 16	12574498	2227533	17.7	582410	4.8
2016 - 17	13935917	2496358	17.9	672829	4.8

Source : National accounts statistics 2019. Central statistics organisation Govt. of India.

India's milk production was stagnant until 1970 and began to rise crossing the 30 million tonne in 1980, 50 million tonne in 1989, 75 million tonne in 1998 and is proposed to cross 100 million tonne mark in 2007. The trends in average annual growth rate of milk production from 1950-2015 has been given in the Table 1.11.

Table 1.11 Trends in average annual growth rates of milk 1950 - 2015

Year	Average annual milk growth rate (%)
1950 - 51 ----1960 - 61	1.64
1960 - 61 ----1973 - 74	1.15
1973 - 74 ----1980 - 81	4.51
1980 - 81 ----1990 - 91	5.48
1990 - 91 ----2000 - 01	4.10
2000 - 01 ----2010 - 11	4.22
2010 - 11 ---- 2014 - 15	4.68

Source : India Dairy.com

The milk production of the major milk producing countries for the year 2003-05 indicating the percentage of growth rate for the years 05/04 has been given in table 1.12. The projected population of the country, milk production and the per-capita availability from 2005-06 to 2020 has been presented in table 1.13. Whereas the annual milk production and the per-capita availability of milk from 1950-51 to 2009-10 has been given and presented in Table 1.14 as well as graphically in the bar diagram Fig. 1.1 and 1.2.

Table 1.12 Milk production of major producing countries 2003-05

	2003	2004 (provisional)	2005 (forecast)	% growth (05/04)
WORLD	615.5	625.2	640.2	2.4
EU - 25	147.6	146.4	147.4	0.7
India	87.3	91.1	95.4	4.7
U.S.A.	77.3	77.5	80.6	4.0
Russia Fed.	33.4	32.2	32.2	0.0
Pakistan	27.8	28.6	29.5	3.2
Brazil	23.5	23.7	24.6	3.8
China	21.5	27.1	32.5	20.0
New Zealand	14.4	15.0	14.6	-2.7
Ukraine	13.7	13.6	13.6	0.0
Mexico	9.9	10.0	10.1	1.2
Argentina	8.2	9.6	10.1	4.8
Turkey	10.6	10.5	10.5	0.2
Australia	10.3	10.1	10.1	- 0.2
Japan	8.4	8.4	8.3	- 0.6
Canada	7.7	7.9	7.8	- 1.4

Source : India Dairy.com

Table 1.13 India's Projected population, milk production and per capita availability 2005-2020

Year	Population (million)	Production (million tonnes)	Per capita availability (g/day)	(kg/yr)
2005 - 06	1080	94.5	240	87.5
2006 - 07	1094	97.9	243	88.6
2007 - 08	1110	100.0	247	90.1
2008 - 09	1127	103.5	252	91.8
2009 - 10	1144	107.0	256	93.5
2010	1161	111.0	263	96.0
2015	1246	134.5	296	108.0
2020	1331	165.0	340	124.0

Source : India Dairy.com

Table 1.14 Annual Milk production and per capita availability 1950-51 to 2016

Year	Production (million tonnes)	Per capita availability (g/day)	(kg/yr)
1950 - 51	17.0	132	48.1
1955 - 56	19.0	130	47.4
1960 - 61	20.0	127	46.3
1968 - 69	21.2	113	41.2
1973 - 74	23.2	111	40.5
1980 - 81	31.6	128	46.7
1985 - 86	44.0	160	58.4
1990 - 91	53.9	176	64.2
1995 - 96	66.2	197	71.9
1997 - 98	72.1	207	75.6
1998 - 99	75.4	213	77.8
1999 - 2000	78.3	217	79.2
2000 - 01	80.6	220	80.3
2001 - 02	84.4	225	82.1
2002 - 03	86.2	230	83.9
2003 - 04	88.1	231	84.3
2004 - 05	92.5	283	84.7

Table 1.14 *Contd...*

Year	Production (million tonnes)	Per capita availability (g/day)	(kg/yr)
2005 - 06	97.1	241	87.5
2006 - 07	102.6	251	88.6
2007 - 08	107.9	260	90.1
2008 - 09	112.2	266	91.8
2009 - 10	116.4	273	93.5
2010 - 11	121.8	281	102.6
2011 - 12	127.9	290	105.9
2012 - 13	132.4	299	109.1
2013 - 14	137.7	307	112.1
2014 - 15	146.3	310	113.2
2015 - 16	190.0	380	anticipated

Source : Dairy India year book seventh edition

An attempt has been made to present the projected Milk production at different rates of annual growth upto the year 2025 for the better understanding of the students in table 1.15.

Table 1.15 projected Milk production at different rates of annual growth 2014-2025 (Million tonnes)

Year	Rate of growth at percent				
	4.5	5.0	5.5	6.0	6.5
2014-15	146.3	146.3	146.3	146.3	146.3
15-16	152.9	153.6	154.3	155.1	155.8
16-17	159.8	161.3	162.8	164.4	165.9
17-18	167.0	169.4	171.8	174.2	176.7
18-19	174.5	177.8	181.2	184.7	188.2
19-20	182.3	186.7	191.2	195.8	200.4
20-21	190.5	196.1	201.7	207.5	213.5
21-22	199.1	205.9	212.8	220.0	227.3
22-23	208.1	216.2	224.5	233.2	242.1
23-24	217.4	227.0	236.9	247.2	257.9
24-25	227.2	238.3	249.9	262.0	274.6

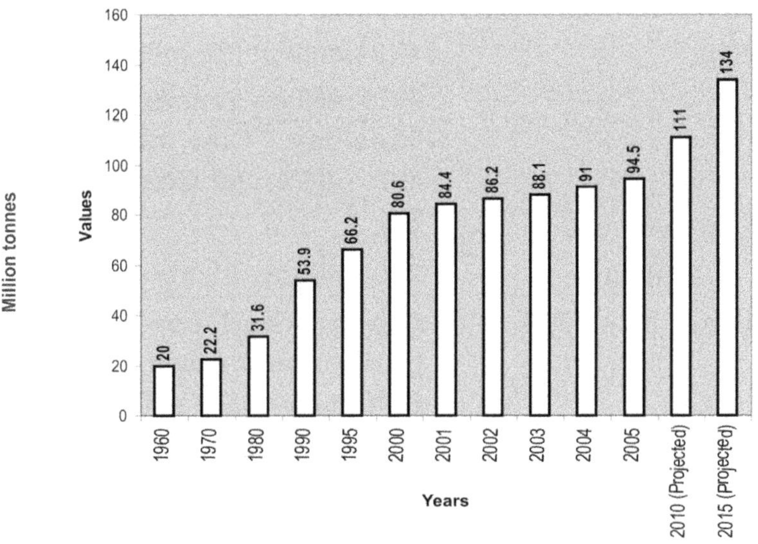

Fig. 1.1 Milk production (million tonnes)

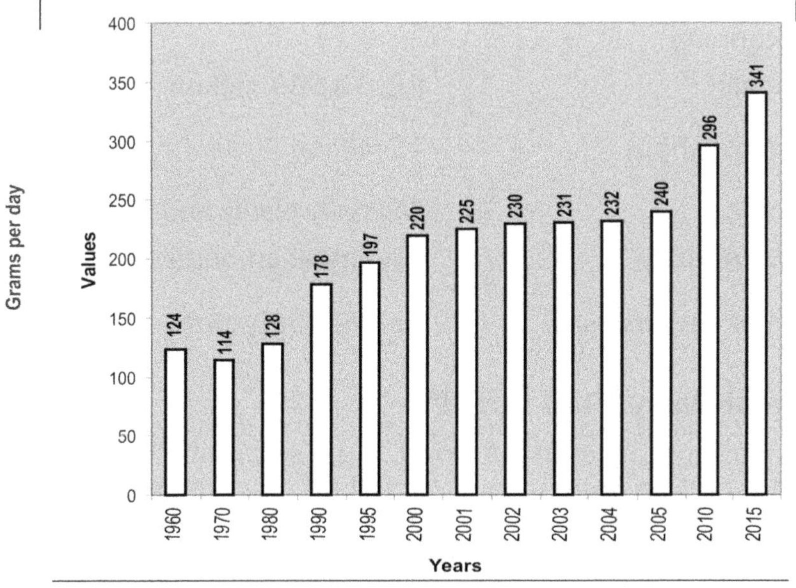

Fig. 1.2 Milk production and availability : 1960 - 2015
(Per Capita availability grams per day)

Source : Dairy India year book 2007

Some indicators of India's dairy development during/up to the year 2005 have been given hereunder :

Human population	-	1094 million (70 million dairy farmers)
Milk production	-	94.0 million tonnes/year 258 million litres/day. Cow milk - 41%, buffalo milk - 55% and Goats etc - 4%
Average annual growth rate	-	4%
Per capita milk availability	-	240 gm/day; 88 kg/year
Bovine population in milk 2003	-	Cross breed cows - 8.2 million, Indigenous cows - 27.6 million, Buffaloes - 33.0 million
Average milk yeild per day	-	Cross breed cows - 6.5 kg
Per animal 2003	-	Indigenous cows - 1.9 kg
		Buffaloes - 4.2 kg
Cattle feed production	-	2.5 million tonnes per year
(Organized sector)		
Market for veterinary		
Pharmaceuticals	-	Rs. 12,000 million
Dairy plants throughput	-	45 million litres/day (18%)
Dairy plants and their installed capacity	-	748 dairy plants and 81.3 million litres/day installed capacity
Value of output of dairy industry	-	Rs. 191,000 crore

1.16 National Dairy Plan - (NDP)

As the largest milk producer in the world, India's contribution is 121.8 million tonnes in 2010-11. The national action plan for dairy development has been envisaged to double the income of the milk producers at farm level by 2021-22 by increasing the access to the organized milk processing sector. According to the planning commission estimates and subsequent corrections on account of consistantly higher growth in G.D.P. the demand for milk is expected to be about 155 million tonnes by the end of 12th five year plan period 2016-17. The milk production has been envisaged to be

254.55 million tonnes by the year 2021-22 from the existing 176.35 million tonnes requiring an annual growth rate of 8.56%. This would lead to increase the per capita availability from the current level of 375 gms/day to 515 gms/day by 2021-22. In order to achieve the desired milk production levels the average in productivity of milk by the animals would be required to grow annually at the rate of 4.7% to 6.14 kg per day by 2021-22 from the existing 4.65 kg per day. It is necessary to maintain the annual growth over 4% in the next 15 years in order to meet the ever growing demand. Therefore it is imperative that a country wide well planned scientific approach / initiative has to be launched to increase the productivity levels in the existing herds, through a programme focusing on breeding and better feeding. The National Dairy Plan (NDP) has been envisaged with a 15 year horizon, considering that 3-5 years of time is required to produce more productive animals and time required to develop and expand systems to increase milk production.

The first phase of the National Dairy Plan is to be financed largely by the world bank which will be implemented over a period of six years with the following objectives.

- To help increase the productivity of milch animals and thereby increase the milk production to meet the rapidly growing the demand for milk and
- To help provide rural milk producers with greater access to the organised milk processing sector.

The estimated project outlay for the phase-I of the NDP is obout 2242 crores of which the share of the world bank is 1584 crores and that of Govt. of India is about 176 crores. The details of phase-I project outlay has been given in Table 1.16.

Table 1.16 Project Outlay Phase-I (NDP)

Component	Activity	Outlay in Rs. crores
Component-A	Breed Improvement	715.0
	Animal Nutrition	425.0
Component-B	Village based milk procurement system	488.0
Component-C	Project Management and Learning	132.0
	Sub-Total #	**1760.0**
	EIA Contribution	282.0
	NDDB's Contribution	200.0
	Grand Total	**2242.0**

Source of funds : World Bank-IDA 1584 crores, Govt. of India 176 crores

Implementation

The Phase-I of the NDP consists of a series of multi pronged approaches or initiatives that are to be implemented over a period of six years commencing from 2012-13.

1. **Increasing the productivity through scientific breading and feeding of the animals.**

   A. Productivity of high genetic merit and disease free cattle and buffaloe bulls using the internationally established practise of Progeny Testing (PT), and Pedigree Selection (PS) and import of Jersey and HF bulls/ embryos or semen.

   B. Strengthening A and B graded semen stations and producing high quality disease free semen.

   C. Setting up of a perfect module for AI delivery service through a professional service provider following Standard Operating Procedures (SOPs).

Milch animals produce milk which will commensurate with their genetic potential when they are fed with a balanced ration/feed leading to a significant decrease in the cost of production, because the feed accounts for about 70% of the total cost of milk production, thereby increasing the farmer's income.

   D. Education of milk producers on ration balancing and nutrients required by their milch animals by providing advisery services at the farmers door step through the trained Local Resorce Persons (LRPs).

   E. Increase fodder yield by making available quality fodder seed of high yielding improved varieties to farmers and demonstrate silage making and fodder enrichment.

2. **Strengthening village based milk procurement systems**

   The milk production is in the hands of about 70 million rural houses holds, the majority being small and marginal farmers and landless labour. The dairy co-operative ensures inclusiveness and livelihood for small holders especially the women.

   (i) Expansion and setting up village based milk procurement systems to collect milk in a fair and a transparent manner ensuring timely payments.

   (ii) Strengthening of the existing dairy co-operatives and promote producer companies of new generation of co-operatives to put in place village level infrastructure for the weighing, testing collection and cooling of milk.

   (iii) Provide support for creating institutional structures and training.

3. Project management and Learning

(a) Implementation of Information and Communication Technology (ICT) based information systems for integration of various activities as well as monitoring and reporting at different levels, to carry out necessary analysis and facilities necessary changes in project implementation.

(b) Carry out base line, mid term and project completion surveys and other special surveys/studies.

(c) Facilitate learning and documentation of learning experiences.

Benefits

In terms of overall benefits the NDP will put in place a scientific approach and systematic process which would take the country on the part to improve the genetics of milk producing animals in a consistent and continuous manner.

(a) It will make much more prudent use of the continous scarce natural resources.

(b) Have an impact on reducing methane emissions.

(c) Improve the quality of milk being marketed.

(d) Helps in strengthening the regulatory and policy measures that will provide an enabling environment for future of dairying in the country and

(e) Contribute for the improvement of the livelihoods of small holder milk producers who are the bed rock of India's milk production system.

1.17 Present Status of Dairy Industry

Adopting the popularly called Amul or Anand Pattern in the milk procurement, processing and marketing, almost all the state governments, have set up their dairy development co-op. federations for the development of the dairy industry in their respective states. The data of which have been summarized in the Table 1.17.

As on March 2009, India's 1,33,349 village dairy co-operative societies federated into 177 milk unions and 15 Federations procured on an average 25.1 million litre of milk per day, 13.9 million farmers are presently members of Village dairy co-op.

India's 3.7 percent annual growth of milk production in between 1999-2000 and 2009-10 surpresses the 1.8 percent growth in population, the net increase in availability is around 2 percent per year.

Table 1.17 Details of various state Dairy Federation

Sl.No	Name of State Dairy Faderation	Address	No.of co-op-Unions	No.of Dairy Plants	Capacity (TLPD)	Brand name	Products
1.	2.	3.	4.	5.	6.	7.	8.
1.	Andaman-Nicobar Islands Integrated Development Corporation Ltd., (ANIIDCO)	Port Blair Pin 744101	N.A.	N.A.	N.A.	N.A.	---
2.	Andhra Pradesh Dairy Development Co-op. Federation Ltd.,	Lalapet, Hyderabad Pin-500 017 Email: apddcf@hd 1. Vsnl.netin Website: apdairy.com	9	12	2437	Vijaya	Table butter, Ghee, Pasteurised Butter UHT Milk, flavoured milk, kulfi SMP, Khoa, Processed Cheese
3.	West Assam Milk Producers Co-op. Union Ltd.,	Guwahati	N.A.	1	60000	Purabi	Double toned milk, toned milk, paneer, Curd, Ghee, skim milk etc.
4.	Bihar State Co-op. milk Producers Fed. Ltd.,	Dairy development complex, Bihar veterinary College Patna-800014. compled@bihnic.in	5	10	780	Sudha	Pure Ghee, Butter Ghee, Paneer, Lassi, Rasagolla, Peda, Ice Cream, Cream Milk.

Table 1.17 *Cond....*

1.	2.	3.	4.	5.	6.	7.	8.
5.	Goa State Co-op. milk Producers Union Ltd.,	Curti-Ponda Pin- 403 401 Goa	--	--	--	Goa Dairy	--
6.	Gujarat Co-op. milk marketing Federation Ltd.,	Amul Dairy Road, Anand – 385 001 Email: gcmmf@amul.com Web: amul.com	12	19	6595	Amul Sagar	Infant milk food, Instant milk mix, ice cream, SMP, Ghee, Dairy whitener, paneer, Swetened, Condensed milk, Malaipeda, Gulab jamun mix, Shrikhand, Pijia cheese, Butter
7.	Haryana Dairy development Co-op. Federation Ltd.,	SCO 127-128 Sector 17-C Chandigarh Pin- 160 017 Email: vita@satyam.net.in web: vitaindia.com	6	5	470	Vita	Ghee, Paneer, table Butter, WMP Dahi, Flavoured Milk, milk cake
8.	Himachal Pradesh State Co-op. milk Producers Federation Ltd.,	Tota, Simla pin 171011	3	3	370	HIM	Paneer, Ghee, Table Butter, Curd
9.	Jammu Co-op. Milk Producers Federation Ltd.,	Jammu Contt. Pin- 180 003	--	--	--	Jamfed	
10.	Karnataka Co-op. milk Producers Fed. Ltd.,	KMF Complex Dr. M.H Marigowda Road D.R. College (PO) Bangalore 560 029 Web.Nandinimilk.com. nandini	13	15	2130	Nandini	SMP, Paneer, Pure Ghee, Badam Berfi, Gulab Jamun, Pesterurized butter, pure milk Khoa, ice cream, Toned milk, Curd.

Table 1.17 ***Cond....***

1.	2.	3.	4.	5.	6.	7.	8.
11.	Kerala State Co-op. milk Marketing Federation Ltd.,	Milma Bhavan Pallam Thiruvananthapuram 695004 Email: milma@md2.Vsnl.net.in Milma.com	3	9	300	Milma	Instant dairy whitener, Ice Cream, flavoured milk, SMP Ghee, Skimmed milk, curd, Peda, Pasteurized Butter
12.	Madhya Pradesh State Co-op. Dairy Fed. Ltd.,	19-B, Zone Maharana pratap Rananagar, Bhopal- 462 011 Email: mpcdf@bomb vsnl.net.in sanchi-milk.com	5	5	1030	Sanchi Sneha	Pure Ghee, Lassi, UHT milk, SMP Pasteurized Butter, flavoured milk, Butter milk
13	Maharashtra Raiya Sahakari Dudh Mahasangh Maryadit	M.K.M. International House 3rd Floor, 178, Back Bay Reclamation B M Chinai marg Mumbai- 400 020 Email: mahafed@ am3vsanl.net.in	23	29	3820	Vikas Mahanand Gokul Dhawal Dudh Panchi Warana Krishna Katnej Rajgans Koyna and Sivamrut	Pasteurized Butter, Pure ghee, Lassi, Srikhand, Butter milk, Flavoured milk, Paneer Peda
14.	Manipur State Co-op. milk processing and marketing Federation	Kesiampet Junction Imphal- 798 001					

Table 1.17 *Cond....*

1.	2.	3.	4.	5.	6.	7.	8.
15.	Kohima Dist. Co-op. milk Producers Union Ltd.,	Milk Chilling Plant Veterinary Compound Burma Camp Dimapur- 797 112				KEVI	
16.	Orissa State Co-op. Milk Producers Fed. Ltd.,	D-2 Sahid Nagar, Bhubaneshwar 751 007	5	5	135	Omfed	Pure Ghee, Butter, Sweet curd
17.	Pondicherry Co-op. milk Producers Union Ltd.,	Vazhudavar Road, Kurumampet, Pondicherry – 605 009.				Ponait	
18.	Punjab state Co-op. milk Producers Fed. Ltd.,	SCO.No. 153-155, New City Centre, Sec. 34-A, Chandigarh Email: milkhod@chd.nic.innic. inmilkfed	11	9	1550	Verka	Ghee, Panner, flavoured milk, ice cream, SMP, WMP, Lassi, Table butter, Sweets, Cheese
19.	Rajasthan Co-op. Dairy Fed. Ltd	Saras Sankal, Jawaharlal Nehru marg, P.O. Box. 1003 Jaipur – 302 017	16	15	1425	Saras	Pure Ghee, Srikhand flavoured milk, tetra pack milk, Cheese, Paneer, Dairy whitener, SMP, WMP, Lassi, Table butter, White Butter, Dahi, Ice Cream, Rasgulla, Chach, Khoa, Mawa, Peda, Kalakand
20.	Sikkim Co-op. Milk Producers Union Ltd.	Tadang (PO) Gangtok 737 102				Sikkim Milk	

Table 1.17 Cond....

1.	2.	3.	4.	5.	6.	7.	8.
21.	Tamil Nadu Co-op. Milk Producers Fed. Led.,	Aavin Illam, Madhavaram Milk Colony, Chennai 600 051 Email: mdaacin@satyam.net. inaacinmilk.com	12	15	2700	Aavin	SMP, Milk, Khoa, Peda, Pure Ghee, Butter, Cheese, Milk butter, flavoured, UHT, Yogurt, Table standarized milk curd, butter milk, Badam Powder, Dates, Khova
22.	Tripura Co-op. Milk Producers Union Ltd.,	Agarthala Dairy, Indra Nagar, Agarthala 799 006				Gomati	--
23.	Pradeshik Co-op. dairy Fed.	29 Park road, Lucknow 226 001 Email: fedf1@satyam.net.in Paragmilk.com	30	16	1680	Parag	Ghee, Milk, Cake, Paneer, White Butter, Table butter, SMP
24.	West Bengal Co-op. milk Producers Fed. Ltd.,	LB-2 Sector III, Salt Lake city Kolkata 700091	14	6	1272	Ben's Bhagiradhi Himul, Mid milk, mother	Table butter, Ghee, Ice Cream, Peda, Lassi, Paneer, curd, Flavoured milk, Yoghurt, mistidoi

In 2009-10 average dairy co-op. milk marketing stood at 2112 lakh litres, Annual growth has averaged about 6.2 percent compounded over the last five (5) years. Dairy Co-operatives now market milk in all metros, metro cities and more than 2000 towns/cities. During 2001-10 the daily milk supply by the co-operatives to each urban consumer has increased from 47.7 to 61.9 kg. per day.

Performance of the Co-op. at a glance

The cooperative profile of India up to 1996-97 has been presented in Table 1.18:

The dairy Co-op. network as on 2010.

(a) Includes 177 milk unions.

(b) Operates in over 346 districts of the country.

(c) Covers 1,40,227 village level co-op. societies.

(d) Owned by about 14 million farmer members of which 4 million are women.

Table 1.18 Co-op. Profile in India upto 1996-97

Item	1994-95	1996-97
Anand Pattern Co-op Milk Societies	67,770	72,740
Milk shed areas	170	170
Membership of Dairy Co-op. Societies	89.88 lacs	93.14 lacs
Milk Procurement	10.2 ml pd	11.5 ml pd

Source : India Dairy.com

There were several innovative programmes for women in dairying. Five states took up specific task of “All Women co-op. dairy societies” formation in selected milk shed areas.

	State	No.of all women co-op.dairies
1.	Andhra Pradesh	210
2.	Bihar	280
3.	Rajasthan	408
4.	Uttar Pradesh	1228
5.	Gujarat	350

1.17.1 Private Sector - Dairy Industries

In addition to the various dairy plants that have been setup under the different state governments with the help of NDDB and dairy co-operative federations yet many dairy plants have come up under private sector competing with the co-operative federations. Some of them have been listed below :

1. ***Nestle India***

   Brands - Nescafe, Everyday, Dairy whitener, Nestle milk maid, Dahi, Neslac, Ghee, Nesvita, Pro Heart milk.

   Nestle, India's first production facility was set up in 1961 at Moga, Punjab, and it was followed by its second plant set up at Choladi, Tamil Nadu in 1967. Consequently Nestle India set up factories in Nanjangud, Karnataka in 1989. Samalkha Haryana in 1993. This was succeeded by Commissioning two more factories at Ponda and Biocholim, Goa in 1995 and 1997 respectively. The seventh factory was set up at Pantnagar in 2006 Uttarakhand. The Nestle Corporate head office is located in Gurgoan, Haryana.

2. ***Hindustan Milk Food Manufacturers Ltd***

   Brand Name - Horlicks, Boost, Maltova, Viva

   The flagship product Horlicks is highly popular and respected brand since 1930 in the Indian market and it underwent a revamp in 2003 to further increase its relevance.

   In 1958-60 the Hindustan milk food manufacturers (Pvt) Limited has built its factory at Nabha, Punjab to make Horlicks from buffaloe milk in India.

   During 1975, the HMM (Pvt) Limited has their second factory at Rajahmundry, AP and expanded its production capacity in 1978. The company was made public in 1979 and constructed third factory at village Kherva Sonepat, Haryana which is fully automated. The full operational capacity of the plant will be 261000 tonnes of Horlicks per annum.

3. ***Glaxo:*** It was founded by Bunny Thorpe Newzealand in 1904. Originally it was a baby food manufacturer, processing local milk into baby food by the same name and the product was sold in the 1930's under the slogan "Glaxo builds bonny babies". It was named as Glaxo Laboratories in 1950. The milk processing plant for the manufacture of infant foods under the brand name "Glaxo' was built in India at Aligarh in the year 1960.

4. ***Modern Dairies***: It is a part of the Modern business group headquartered at Chandigarh established in 1973. It was awarded the prestigious ISO 9001-2000, HACCP (Food safety) and an ISO 1400-2004 for environment and management certification to its credit. The company manufactures a wide range of milk and milk products. They have constructed their second factory near Karnal to meet the ever increasing demand.
5. ***Heritage Foods***: It was founded in the year 1992 and located near Hyderabad is one of the fastest growing private sector enterprises in India, with four business divisions namely Dairy, Retail, Agriculture and Bakery, under the flagship company Heritage foods India Limited. The annual turnover of Heritage foods crossed 900 crores in 2009-10 and 1100 crores in 2010-11. The milk and milk products are being marketed in AP, Karnataka, Kerala, Tamilnadu, Maharashtra and Orissa.
6. ***Dodla Dairy***: This is the first and patent dairy plant of the company established in A.P. The original installed capacity of the plant was 60,000 litres/per day. The capacity of the plant has been increased to 4 lakh liters per day at present. The plant is equipped with excellent infrastructure and the state of art facilities for processing the milk and for the manufacture of milk by-products namely ghee, butter, paneer, flavoured milks, curd, butter milk and doodh peda. There is a separate facility in the same premises for the manufacture of skimmed milk powder with an installed capacity of 15 tons per day.
7. ***Tirumala Milk and Milk Products (Pvt) Ltd.,***: It is a professionally managed company established in 1998 in AP is one of the fastest growing private sector enterprise. The company has the most modern and versatile plants with the state of art technology. The total capacity of the plant is about 7 lakh litres per day, having the market presence in A P, Karnataka and Tamilnadu. The dairy products include:

   sweets, flavoured milks, curd, milk powder, butter, ghee and butter oil both in bulk and consumer packs.
8. ***Durga Dairy***: Although it is a public limited enterprise located in AP. The on going venture came under the mantle of Crane group in Feb. 1995. It has the goodwill of the people of all over India through the commitment to the quality. It has branches almost all over the country and produces 1500 tonnes of ghee annually.

9. ***Gokul Dairy***: The Kolhapur Dist Cooperative Milk Producers Union Ltd was established on 16-03-1963 under the Co-op.act. It made a moderate begining with 500 litres/day from 22 village co-op. societies. The milk receipts capacity at present is about 3.2 lakh litres/day raw milk and 3.8 lakh litres/day chilled milk with a total processing capacity of 7.0 lakh litres/day. The milk powder manufacturing capacity is about 40 mt/day.
10. ***ABIS Dairy Pvt Ltd.,***: It is a subsidary of the Indian Broilers group. The company has its own dairy farm and milk processing industry to produce hygienic and do safe dairy food of with good quality. It is situated at Rajnandgoan in C.G. State. It was established in 1999 and an ISO 9001-2000 certified plant. The products include packaged milk, Khoa, Paneer. The capacity of the plant is about 15 mt per day.

As already stated the milk production in our country is in the hands of the small farmers who are scattered in the remote villages which do not have proper transport facilities. Most of the packed liquid milk segment in India is dominated by the Cooperative federations and the private sector dairies. The liquid milk contribution in the total revenues of the cooperatives ranges from 60-80 percent. Apart from few private organisations many individuals are mainly focussed on milk products other than packed liquid milk. There is a huge potential for processing and value addition in the organized sector particularly in the ethic Indian sweets, and other milk based products, which are largely sold in "unbranded" form in the market. These organisations or individuals either in the organised or un-organised sector also collect the milk for the manufacture of their milk products such as ice-cream, sweets, biscuits and confectionary chocolates etc. either directly or through the contractors/middleman engaged by them. The key differences between the organised and the unorganised sectors concern the level of investment in presenting quality of milk, the technolgy used for processing and the compliance with the legal food standards.

It is a very common sight to find either a halwaii or a sweet shop in almost every street of a town or city being managed by the individuals on a cottage industry basis in the unorganized sector. These people prepare the milk based sweets of different varieties by collecting the milk from the local producers or by the milk vendors. In addition to these people there are some branded sweets being manufactured and sold in the market in an organised manner. A few examples of these organisations have been listed below for information.

Agarwal sweets	-	Delhi, Lucknow etc.
Sri Krishna Sweets	-	Chennai, Pondicherry, Hyderabad
G. Pulla Reddy Sweets	-	Hyderabad
Bikaneri Sweets	-	Rajasthan, Panipat, Delhi.
K.C. Das Sweets	-	Kolkatta, Delhi, Bangalore.

Similarly, milk and cream is used in the manufacture of Ice cream both in the organised and unorganised sector. These people have their own mechanism of collecting the milk for the manufacture of their products. Some of the private sector organisations who are manufacturing the branded ice cream in a large scale, having their own marketing outlets at many places in the country have been mentioned below :

Vadilal Ice cream	-	Ahmedabad
Kwality walls India	-	Delhi, Hyderabad
Hatsun Arun Ice Cream	-	Chennai, Hyderabad
Dinshaw Dairy Foods	-	Pune, Karnataka

In addition to the sweets and ice cream, milk is being used to some extent in the manufacture of some varieties of biscuits and confectionery industry. Some of the leading manufacturers of biscuits and the confectionery industries have been given hereunder.

Biscuit Industry :

Britania Industries Ltd.,	1892	Kolkata
Parle Products Pvt Ltd.,	1929	Mumbai, Bangalore
Surya Food & Agro Ltd.,		
Priya Gold Brand	1992	Noida, Delhi
I.T.C. Ltd., - Sunfeast Brand	2004	Chennai

Confectionary :

Cadbury India Ltd.,	1948	Thane, Mumbai
Nutrine	1952	Chittoor, A.P.

CHAPTER 2

Milk Processing Plant Layout and Management

Milk processing plants converting raw milk into processed milk and/or milk products are usually termed as dairy plants or simply dairies. Number of unit operations are carried out to transfer raw milk into value added finished products employing all sorts of resources like raw ingredients, utilizes (water, air, electricity, refrigeration, steam and waste treatment etc.) plant machinery and human resource. All these resources are housed in one or more buildings located in the same campus. In order to optimise the economy of interactions between these housing and various resources several factors need to be considered in the planning stage. A good planning therefore helps in the smooth flow of resources towards the effective operations, to create hygienic atmosphere during the production at one hand and economy of operations and value addition on the other.

In our country less than 15% of the total milk production is processed in dairy plants of the organised dairy sector. Rest of the milk is either consumed at home or being used in unorganised production centres like sweet makers. Most of the plants are under utilized especially during the lean periods the utilisation capacity further goes down. Operation of unnecessary large size plants and equipment-becomes uneconomical due to more expenses for the operation along with their overheads which depends on the size and capacity of the plant and equipment. In this way in properly designed dairy plants require higher investment and working capital. Therefore various factors that are relevant to design and layout of dairy plant must be looked into scientific layout and planning is the need of the hour because of the fact, the dairy industry in India is growing at a much faster rate. The processing of milk commences at the dairy plant for market milk and various milk products. The subject of dairy plant layout and design includes the designing a layout for the milk factory which includes:

1. Layout of various sections in dairy building
2. Equipment layout
3. Laying of dairy equipment/machines in each section for economical and efficient movement of men and material.

The subject will also deal with all the basic aspects of learning that are needed for planning and layout of dairy factory, such as location selection of site for dairy plant etc., and also deals with the practical need of prospective enterpreneurs aspiring to set up a dairy plant. The milk and milk products plays a fundamental role in human nutrition, therefore, in the interest of the nation, they must be available to every individual at the lowest/cheapest possible rate/price. The margin of profit expected by the dairy enterpreneur is very limited, hence, every possible economy measures in capital outlay on building and plant should be sought.

A good layout design and use of proper materials and techniques make great contribution towards hygiene. Milk being the most suitable medium for the growth of the microorganisms every possible measure should be taken to reduce the contamination especially after processing by maintaining the highest standards of hygiene. The layout of a dairy plant needs a careful thought and planning keeping in view the following aspects namely:

1. Manufacture of the dairy products.
2. Commercial viability of the products including the marketing feasibility and
3. Scope for future expansion without incurring heavy expenditure on capital outlay by way of utilizing the available floor space.

2.1 Location/Selection of Site

Selection of a location for the dairy plant involves a region with a special emphasis of a specific site within that region which involves huge commitments of capital as well. The location of the plant is primarily influenced by the nature of the product which is to be manufactured. The most favourable location is one that attains the lowest unit cost in the production and distribution of the product or service to the consumers. The terms location and site are often confused and misunderstood with each other. It must be clearly understood that, the term "Location" implies to an area, region or territory where as site is the place, where the dairy plant buildings are to be constructed. There are three types of locations namely:

1. City location
2. Sub-urban location and
3. Country location

are suitable for the construction of the dairy plants.

In all the major cities, areas are earmarked or notified for the development of industries and are called “Industrial area”, and generally the land for dairy plants is allotted in that location only. In such circumstances the neighbourhood industries or factories emitting, smoke dust, offensive smell, and gasses poses great risk to the dairy plant. Therefore, it is essential that the location where the milk or dairy plant is going to be set up should be free from all pollution as far as possible, suburban location means a suburban town or the outskirts of the major large city. If the dairy plant is situated in this location it has all the advantages of being closer to the consumption area, with the possibility of availing all the facilities such as electricity, water, waste disposal and labour etc. Similarly, when the dairy plant is located in a country or milk shed area, there will be little competetion from other industries and the cost of the land will be cheaper. The plant will not create any noise problems and it is unlikely that atmospheric pollution will cause any difficulty or problem. There will be no traffic congestion and the milk procurement routes will be short. However for all the dairy plants in whatever region they are situated there are some important points which are common and essential to be considered in determining the location or site.

2.2 Classification of Dairy Plants

The dairy plants can be broadly classified into:

1. Fluid milk plants
2. Milk Product Factory and
3. A combination of the above two i.e., both fluid milk and milk product factory.

Generally the fluid milk factory, where the liquid milk is considered as the main consumer product, it is located as close as possible to the consumer areas in order to have a good network of distribution, at a shorter distance without loosing much time thus reducing the cost of transportation substantially. In the same manner the milk product factory can be located in the milk shed area, because the chief milk/dairy products namely butter, ghee, cheese or milk powder, do not require daily distribution. Further in this case also, the cost of transportation on milk procurement can be reduced substantially, which will be helpful in determining the cost of production of these dairy products.

However, before selecting the location or site for the establishment of the dairy plant the following points must be given due consideration.

1. The location of the dairy plant where it is going to be established should have an easy accessability to railway siding or near highway such that the raw material as well as the finished products are transported easily without incurring heavy expenditure.
2. Transportation of the pasteurized milk in bulk or large quantities should be avoided in order to avoid contamination. In case, if the pasteurised milk is intended for re-use, when it is returned, the plant should be located closely to the consumption area.
3. If the dairy plant is situated in a location close to the proximity of the busy market or on a highway, the expenditure on advertisement can be reduced substantially to a certain extent.
4. The dairy plant buildings must be located in a place where all the necessary services such as electricity, water, waste disposal are available at a reasonable cost.
5. The area in which the dairy plant is going to be established must be large and sufficient enough to meet the requirements in planning building and it should allow freedom for future expansion.
6. The area should be free from atmospheric pollution.
7. Availability of labour must be adequate near the dairy plant where it is located and if necessary housing facilities must be provided to the essential staff.
8. There should not be any congestion of traffic in and around the dairy plant therefore enough space should be provided in order to control the traffic.

In addition to the above described any individual or enterpreneur who is desirous of setting up of the dairy plant, it is advisable to go through some of the following in order to have a check on the accessability or suitability of the location or site which include.

(a) Availability of labour

(b) Influence of local industries on labour

(c) Management potential

(d) Supply of raw material i.e, liquid raw milk

(e) Availability of electricity

(f) Availability of fuel oils, natural gas and coal

(g) Good quality water supply
(h) Water pollution
(i) Transportation facility i.e. rail road with good connectivity
(j) Disposal facilities for the trash and garbage as well as sewage facility
(k) Good planning and zoning including the atmospheric pollution, residential housing, easy approach to the site and
(l) Fire protection aspects.

It is also essential to consider the following points by the enterpreneur before the establishment of the dairy plant.

2.3 Topography

Topography of the location means a detailed description or representation of the natural and artificial features on the map of the place where the dairy building is to be constructed. The study of the topography of the place will help in taking necessary corrective measures in order to ensure the suitability of the site for the construction of the dairy plant.

2.4 Nature of Ground

It is utmost important to have a knowledge of the nature of ground of the selected site that confirms to certain requirements. The upper layers of the soil must allow for easy drainage as is the case with gravel, sand and like soils such that the water quickly disappears from the surface. At the same time the soil should provide a firm foundation for the building structures, roadways and surfaced areas. Locations with loose clay silting sand or high water table should be avoided, because the foundation works on such soils will be expensive and time consuming. In general information on geographical location, climate, milk production, milk utilization, packaging and distribution, services, plant organisation, availability of plant materials, facilities for major repair shops, electric motor maintenance shops, lubricants, availability of sanitary materials, transport services such as local trucks, railway, postal services and professional services etc., is much needed and must be obtained in the preliminary survey for locating the milk plant.

2.5 Planning and Principles of Dairy Layout

Planning a layout for a dairy calls for a careful thinking and designing a dairy plant layout is a joint venture of architects, dairy managers, dairy engineers and administrators because it involves an overall management function. Management policies affect the plant layout as they determine the plant layout objectives and scope of plant activities. A plant layout should be so planned and arranged, that the needed capacity to produce new or related products can be added at low cost with minimum of plant revision and interruption of production schedules. A good layout must improve or facilitate production operations, minimize material handling, maintain flexibility of the operation for alterations and expansion, minimize investment on equipment, make economical use of floor area, promote effective utilization of the labour force and provide for employee convenience and comfort. It is always better to make a modest start or begining and keep provision in the plant for future development or expansion. The form and shape of the dairy building is utmost important because it affects the public reaction as well as the upkeep and initial cost. Before planning the dairy building the following are some of the basic factors which must be given due consideration.

(a) Quantity of milk to be processed everyday and variation if any, which may be expected during the year due to the normal supply and demand changes.
(b) Method of milk collection and optimum rate of intake.
(c) Method of pasteurisation.
(d) Choice of containers
(e) Disposal of the pasteurised milk.
(f) Choice between the cold storage at the dairy and immediate distribution to local depots.
(g) Method of distribution to consumers.
(h) Choice between supply of electricity supply, water supply and waste water disposal.
(i) Choice of fuel for steam raising.
(j) Provision for subsequent future expansion.
(k) Approach to main road.

The scheduled layout of processing is drawn, after a careful consideration of the factors related to the planning mentioned above. Process scheduling means arranging the flow of the products through various operations in the plant in such a manner that maximum utilization is made relating to the labour and equipment which is accomplished in a shortest possible time, with minimum delay in the processing of different products. The operational layout describe the operations which take place in processing or in the manufacture of the items/products. All the operations involved are represented diagramatically in a chronological order or sequence wise like a flow diagram, which can be easily understood even by a lay man. A good dairy layout, in general should incorporate the following principles, which inturn will help in an economical and efficient dairy plant. The principles for a good dairy layout are as under :

1. The milk route should be as short as possible. This will help in minimising the cost of the pipe length and save time in cleaning.
2. Reception and despatch platforms must be arranged in relation to the plant in such a way that congestion of the transport vehicles is avoided.
3. The reception and despatch docks must be separate in order to avoid the rush of the vehicles. Generally the milk reception and despatch of washed cans is on one side and the despatch of finished products on the other side.
4. Where the space is available, a single storey building is most suited with roads on all the sides.
5. The floor level of the milk reception and despatch docks and all the rooms concerned with milk cans and bottles should be at the same height suited the vehicles. However, the weigh tank and raw milk pump should be at a lower level such that to have a convenient tipping height.
6. The raw milk storage tank and pasteurised milk balance tank may be mounted on a staging in order to save floor space and to provide gravity head to fillers.
7. The refrigeration compressor rooms and boiler house floor level must be at par with the ground level.
8. Laboratory should have an easy approach to the processing room, reception dock and filling room.
9. Refrigeration machinery room along with the brine tank should be nearer to the process room and cold store.
10. Separate building should be assinged to the administrative offices.

The planning of operation or otherwise known as process schedule or work plan ensures that all the proposed operations are carried out smoothly without any hinderence and this must be done well in advance so that maximum use is made of men and material with minimum wastage of times. Provision for the storage of the processed milk and other dairy products must be considered at the time of process planning itself. There is a need for cold store maintaining a temperature of 2 °C to 4 °C which is mainly required for the over night storage of pasteurized milk etc.

The water supply for a dairy plant is utmost important and must be given due consideration at the planning stage itself. An adequate supply of pure water is essential for washing of the equipment and also for the purpose of cooling. An ideal water supply is one that is soft and free from all the impurities and such a water is seldom available hence, the best available water must be treated and used to meet the plant requirements. Water can be classified into two types namely soft water and hard water.

2.6 Soft Water

It forms lather easily with ordinary soap, whereas hard water does not easily give a lather with soap, but does so after consuming much soap.

2.7 Hard Water

It contains calcium and magnesium salts in solution as carbonates and sulphates. Hard water is not suitable for washing because it will not give any froth with the soap and is also unsuitable for cooling as well as the production of steam by the boiler in some industries. If the hardness of water is due to the carbonates of lime and magnesium it is called as the temporary hardness and can be removed with the removal of carbondioxide by boiling with the formation of $CaCO_3$ as the precipitate. If the hardness is due to the presence of sulphates and chlorides of calcium or magnesium and is not removed by boiling then it is called as permanent hardness.

Hardness is removed by adding 6 lb or 3 kgs of fresh lime for every 1,00,000 gallons of water per degree of hardness. Permanent hardness is removed by converting the soluble sulphates and chlorides of calcium and magnesium present into insoluble carbonates of calcium and magnesium by the addition of sodium carbonate or washing soda to it. Water for cleaning purposes should meet local or state bacterial standards. The water supply should be free from pollution at the source and should be safe guarded as to prevent the pollution at the well, pump house from cross, connections in the distribution lines or fixtures or from the hoses.

The most common defects of water are (1) dissolved solids (2) excessive hardness and (3) Suspended matter. The United States Geological Survey classifies water as follows on the basis of hardness.

Type of hardness	Parts per million (ppm)	Grains per gallon
Soft	upto 60	0 to 3.5
Moderately hard	60 to 120	3.5 to 7.0
Hard	120 to 180	7.0 to 10.5
Very hard	Over 180	over 10.5

An attempt has been made to present a detailed sketch incorporating all necessary structures which are needed for the establishment of the dairy plant in simple design for the benefit of the students and better understanding.

General Layout Sketch for a Small Dairy Plant (not to Scale)

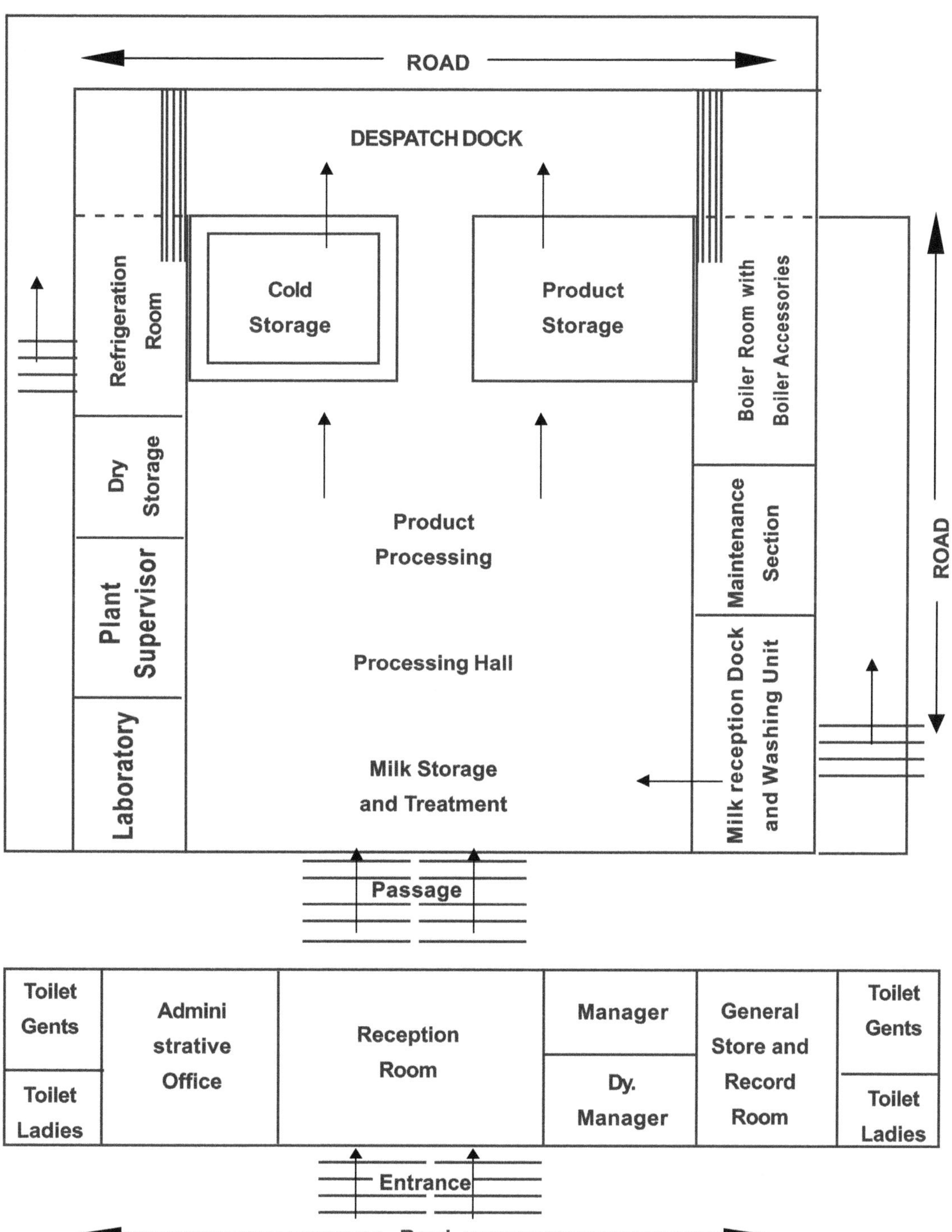

CHAPTER 3

Milk and its Constituents

Among the most important sources of food, milk and its products and its kind considered, as one of its kind for all the civilized nations. It has been observed that the more highly developed and prosperous nations consume greater amounts of milk and milk products. Some the native races belonging to Africa, America and Australia which have never developed beyond the stage of barbarism do not use milk as food. The use of milk as food was made by the primitive races and their descendents of Europe and Western Asia. According to Mc Collum the keeping of dairy animals was the greatest factor in the history of the development of man from a state of barbarism.

Man has been in possession of cattle and has made use of milk and milk products like butter and cheese as per the references made in the Bible. The knowledge of the use of milk and milk products like cheese was spread all over the Europe by the Romans. The art of cheese making was best known and largely developed in the monasteries and the monks for several centuries were the leaders in carrying out the business and in teaching it to others. By fifteenth century, some important cheese markets were in existance in Switzerland, which were very much as they are at present. One type of Swiss cheese introduced by the monks has been made in the same region under the same conditions since about tenth century.

Holland and Switzerland were the two early centres of development of the dairy industry in Europe and the U.K apparently learned the use of milk especially in the form of butter and cheese from the Romans.

The characteristic feature of the mammals is the presence of the mammary glands which are capable of producing milk required for the nourishment of their young ones.

3.1 Definitions of Milk

Milk has been defined in a number of ways by different workers. Some of the important definitions are as under:

Thomas Cogen (1636) in his treatise "The heaven of Health" has stated "Milk not withstanding that it seems to be wholly of one substance yet it is compact or made of three several substances namely Cream, Whey and Curds".

Milk is defined as "the entire product of the complete and un interrupted milking of milch cows, which are properly cared for and are in good health".

Geneva Congress has defined it as "the integral product of entire and un interrupted milking of the milch cows in good health and well nourished and not over worked. It is ought to be collected in proper manner and contains no colostrum".

Federal definition of U.S.A. or according to United States Public Health Services, it has been defined as "Milk is the fresh and clean lacteal secretion practically free from colostrum and obtained by the complete milking of one or more healthy cows, properly fed and maintained and excluding that obtained within fifteen days before and ten days after calving and containing not less than 8.25 percent solid not fat and 3.5 percent fat.

It has also been defined as the "Physiological secretion from the mammary gland of animals". Since the dairy cow is most commonly and extensively used source of milk, the term milk denotes to mean milk of the dairy cows unless otherwise specified.

Kon (1972) has decribed it as" A complex mixture consisting of an emulsion of fat and a colloidal dispersion of proteins, together with the milk sugar (lactose) in the solution. It is obtained in a natural process from the lactating females through their mammary gland only when they give birth to their young ones".

In India the legal definition of milk reads "Milk is the secretion derived from the complete milking of healthy milch animals which shall be free from colostrum".

Milk may be defined as the lacteal secretion particularly free from colostrum obtained by the complete milking of one or more healthy animals. According to PFA rules (1976) cow milk should contain not less than 8.5 percent milk solid not fat and fat not less than 3.5 percent of milk whereas the buffalo milk should contain not less than 9.0 percent solid not fat and not less than 6.0 percent of milk fat.

Milk may be defined as the whole, fresh, clean, lacteal secretion obtained by the complete milking of one or more healthy milch animals, excluding that obtained within 15 days before and 5 days after calving or such periods as may be necessary to render the milk practically colostrum free and containing the minimum prescribed percentages of milk fat, milk solids not fat. In India, the term milk, when unqualified refers to cow or buffallow milk or a combination of the two.

The term "Market Milk" refers to fluid whole milk that is sold to individuals usually for direct consumption. It excludes milk consumed on the farm and that used for the manufacture of dairy products.

According to chemist, milk contains a number of separate chemical compounds as a physical mixture. Fat is present in the form emulsion 'Oil in water type', the proteins and some mineral matter are in colloidal state and lactose with the remainder of mineral matter in true solution. The three distinct physical states are co-existent and to some extent are in a state of poise more particularly to the equilibrium between the colloids and the true solution.

In modern context, milk is described as a mixture of three main constituents, water, Fat and Solids Not Fat (SNF).

The composition of milk deals with the qualitative and quantitative aspects, but on several occassions it became necessary to refer to the variations in concentration shown by certain constituents. These variations make it difficult to assess the quantitative composition of milk. There is only one approach to the problem, and that is to analyse large number of samples taken from different herds of different breeds in various parts of the country(ies) and over different periods of the year. These results, whenever to mean values an "average" composition is obtained. By this means, the factors such as individuality breed, stage of lactation, season etc. which are known to influence the composition of milk are dumped down or practically eliminated.

Inspite of this fact sometimes a question may be raised, what is the usage of this average, when most practical problems are concerned with the milk of localised herds of small groups containing ten to fifty animals.

It is often emphasized that the 'average' man is non-existant. Whether this contention might apply to milk or not is a matter to be considered giving a serious thought. There is no hard and past rule that can be laid down, with regard to the number of samples necessary to enable an average to be calculated, however, most of the investigators have worked on a minimum number of 700 or more. Some have taken individual samples, while others have taken bulk samples with no record of the number of cows in the herds. Both types of investigations will be encountered in a search through the literature of the subject. The student must acquire the habit of critical reading or atleast should be able to execercise descrimination so as to avoid coming to conclusions which the investigators never intended.

Average values alone are of very limited use. However, much more information is required. The question that is generally being asked at the time of investigation is "What is the extent of variation on either side of the mean composition of the genuine milk? or what is the check likelihood of genuine milk falling down the legal presumptive standards? Therefore, the mere statement of average values is not enough. The distribution of the results on each side of the mean values being of equal or perhaps greater importance. Another way of conveying information regarding the extent of variation around the mean values is to calculate the standard variation using the statistical analysis. This value gives the range on each side of the mean into which the majority of the observations may be expected to fall. Thus the statement of mean together with standard deviation is of greater value than the average figures even when the latter is accompanied by the extreme of variations.

3.2 Composition of Milk

The average composition of the cow milk has been given in the Table 3.1 whereas the detailed chemical compsition of milk has been given in Table 3.2.

Table 3.1 Average Composition of Milk (Cow)

Constituent	Percent	
Water	87.54	
Fat	3.71	
Casein	2.63	Protein 3.18
Albumin	0.31	
Globulin	0.11	
Proteose Peptone	0.13	
Other nitrogenous substances	0.11	
Lactose	4.70	
Ash	0.76	
	100.00	

Source : Textbook of Dairy Chemistry - Edgar. R Ling

Table 3.2 Detailed chemical composition of milk

	Constituents/Group of Constituents	Approx Concentration (Weight/Litre of Milk)
(a)	Water	860 - 880 g
(b)	Lipids in emulsion phase Milk Fat (A mixture of mixed triglycerides)	30 - 50 g
	Phospholipids (Lecithins, Cephalins Sphingomyelins etc.)	0.30 g
	Sterols, Caretenoids, Vitamins, A,D,E and K	0.10 g
(c)	Proteins in Colloidal Dispersion	
	Casein (a, b, c fractions)	25 g
	b - Lactoglobulins	3 g

Table 3.2 *Contd...*

Constituents/Group of Constituents	Approx Concentration (Weight/Litre of Milk)
a - Lactalbumin	0.7 g
Albumin, Pseudo globiulin etc.	
Enzymes (Catalase, Peroxidase, Phosphatases, amylases, lipases, proteases etc.)	
(d) Dissolved materials (True solution)	
1. Carbohydrates	
(a) Lactose (α,β)	45.50 g
(b) Glucose	
2. Inorganic and organic ions and salts	50 mg
Calcium	1.25 g
Phosphate (as PO_4)	2.10 g
Citrates (as citric acid)	2.0 g
Chloride	1.0 g
Sodium, Potassium, Magnesium etc.	
3. Water soluble Vitamins	
Thiamine, Riboflavin, Niacin	
Pyridoxine, Pantothenic acid, Biotin, Folic acid	Traces
Vitamin - B_{12} etc.	
Ascorbic acid.	
4. Nitrogenous Materials (non-proteins or Vitamins)	
Ammonia, Amino acids, urea, Creatine or creatinine, Uric acid etc.	Traces
5. Gases (milk exposed to air)	
Carbon dioxide, oxygen, Nitrogen etc.	Traces
6. Trace elements - Those usually present are Copper, Iron etc.	Traces
Partially in Colloidal dispersion.	

Source : Outlines of Dairy Technology - SK De

The average chemical composition of milk of few breeds of Indian cows and Murrah Buffaloe are given in the Table 3.3

Table 3.3 Chemical Composition of Milk of Indian Breeds

Breed Cow/Buffalloe	Percentage composition				
	Water	Fat	Protein	Lactose	Ash
Red Sindhi	86.07	4.90	3.42	4.91	0.70
Gir	86.44	4.73	3.32	4.85	0.66
Tharparker	86.58	4.55	3.36	4.83	0.68
Sahiwal	86.42	4.55	3.33	5.04	0.66
Murrah	83.63	6.56	3.88	5.23	0.70
Cross Bred Cow	86.87	4.50	3.37	4.92	0.67

Source : Outlines of Dairy Technology - SK De

The average chemical composition of milk of different species of animals has been given in Table 3.4.

Table 3.4 Chemical Composition of Milk of Different Species

Name of Species	Percentage composition				
	Water	Fat	Protein	Lactose	Ash
Ass	90.0	1.3	1.7	6.5	0.5
Buffaloe	84.2	6.6	3.9	5.2	0.8
Camel	86.5	3.1	4.0	5.6	0.8
Cat	84.6	3.8	9.1	4.9	0.6
Cow (Foreign)	86.6	4.6	3.4	4.9	0.7
Dog	75.4	9.6	11.2	3.1	0.7
Elephant	67.8	19.6	3.1	8.8	0.7
Ewe	79.4	8.6	6.7	4.3	1.0
Goat	86.5	4.5	3.5	4.7	0.8
Grinea-Rig	82.2	5.5	8.5	2.9	0.9
Human	87.7	3.6	1.8	6.8	0.1
Llama	86.5	3.2	3.9	5.6	0.8
Mare	89.1	1.6	2.7	6.1	0.5
Porpoise	41.1	45.8	11.2	1.3	0.6
Rein Deer	68.2	17.1	10.4	2.3	1.5
Sow	89.6	4.8	1.3	3.4	0.9
Whale	70.1	19.6	9.5	-	1.0

Source : Outlines of Dairy Technology - S.K. De

Table 3.5 Chemical Composition of Milk of some exotic Breeds of Cows

Breeds	Percentage composition				
	Water	Fat	Protein	Lactose	Ash
Holstein Friesian	87.74	3.40	3.22	4.87	0.68
Short Horn	87.19	3.94	3.32	4.99	0.70
Ayrshire	87.10	4.00	3.58	4.67	0.68
Brown Swiss	86.59	4.01	3.61	5.04	0.73
Guernsey	85.39	4.95	3.91	4.93	0.74
Jersey	85.09	5.37	3.92	4.93	0.71

Source : Outlines of Dairy Technology - S.K. De

The average chemical composition of some exotic breeds of cows is given in the Table 3.5.

Efforts have been made to make a comparative study of the human breast milk constituents with those of other animals and presented in Table 3.6. Similarly the different constituents of cows milk has been compared in the with of buffalo, goat, sheep, camel and human and has been given in Table 3.7. Whereas, the Table 3.8 illustrates a comparison of the different constituents between blood plasma and milk.

Accordig to PFA (1976) and as amended by the Food Safety and Standards Authority of India (FSSAI) act (2017) different standards have been fixed for various categories of milk and milk products on all India basis, and also as applicable to certain states, which have been presented in Table 3.9.

Table 3.6 Average composition of breast milk, cow's milk and milk from other animals

	Dry Matter [%]	Fat [%]	Total Protein [%]	Casein [%]	Whey Protein [%]	Lactose [%]	Ash [%]
Human	12.4	3.8	1.0	0.4	0.6	7.0	0.2
Cow	13.0	4.0	3.4	2.8	0.6	4.8	0.7
Goat	13.2	4.5	2.9	2.5	0.4	4.1	0.8
Sheep	19.3	7.3	5.5	4.6	0.9	4.8	1.0
Donkey	8.5	0.6	1.4	0.7	0.7	6.1	0.4
Horse	11.2	1.9	2.5	1.34	1.2	6.2	0.5
Buffaloe	17.2	7.4	3.6	-	-	5.5	0.8
Camel	13.6	4.5	3.6	2.7	0.9	5.0	0.7
Lama	16.2	2.4	7.3	6.2	10.1	6.0	-
Yak	17.3	6.5	5.8	-	-	4.6	0.9
Elk	21.5	10.0	8.4	-	-	3.8	1.5
Reindeer	33.1	16.9	11.5	-	-	2.8	-

Source : Text Book of Dairy Chemistry - M.P. Mathur, D.D. Roy & P. Dinakar

Table 3.7 Constituents of cow, buffalo, goat, sheep, human and camel milk (per cent)

Constituents	Cow	Buffalo	Goat	Sheep	Human	Camel
Moisture	87.6	81.0	85.2	80.71	87.4	87.61
Protein	3.3/3.4	4.3/3.9	3.7/3.5	4.8	1.6	2.98
Fat	3.6/4.9	8.8/6.6	5.6/4.5	6.0	3.75	5.98
Carbohydrates	4.5/4.1	5.0/5.2	4.7/4.6	4.9	6.98	3.26
Mineral Matter (ash)	0.7	0.8/0.7	0.8	0.8	0.21	0.70

Table 3.7 *Contd...*

Constituents	Cow	Buffalo	Goat	Sheep	Human	Camel
Calcium	0.12	0.21	0.17	-	-	-
Phosphorus	0.09	0.13	0.12	-	-	-
Iron	0.2	0.2	0.3	-	-	-
Vitamin A (IU/100g)	180	162	182	189.7	189.9	-
Vitamin B (mg/100g)	0.05	0.04	0.04	-	-	-
Riboflavin (mg /100g)	0.2	0.19	0.15	-	-	-
Vitamin C (mg/100g)	1.6	1.5	1.5	5.5	4.0	-
Nicotinic acid (mg/100g)	0.1	0.1	0.1	-	-	
Calories	67	117	72	-	-	-
Total solids	13.7	17.02	13.5	16.3	12.57	-
Solids-not-fat	8.8	9.2	9.0	10.3	8.80	-

Source : Text Book of Dairy Chemistry - M.P. Mathur, D.D. Roy & P. Dinakar

Table 3.8 Comparative Composition of Blood Plasma and Milk (%)

Constituents	Blood Plasma	Constituent	Milk
Water	91.00	Water	87.00
Glucose	0.05	Lactose	4.80
Serum Albumin	3.20	Casein	2.90
Serum Globulin	4.40	Lactalbumin	0.52
Amino Acids	0.003	Lactoglobulin	0.05
Neutral fat	0.09	Neutral fat	3.80
Phospholipids	0.24	Phospholipids	0.04
Cholesterol ester	0.17	Cholesterol ester	Traces
Calcium	0.009	Calcium	0.12
Phosphorous	0.011	Phosphorous	0.10
Sodium	0.34	Sodium	0.15
Potassium	0.03	Potassium	0.15
Chlorine	0.35	Chlorine	0.11
Citric Acid	Traces	Citric Acid	0.20
		100.00	100.00

Source : Text Book of Dairy Chemistry - M.P. Mathur, D.D. Roy and P. Dinakar

Table 3.9 PFA (1976) standards as amended by FSSAI (2017) act for different classes and designation of milk

Sl. No	Class of milk	Locality or State or Area	Minimum milk fat (%, m/m)	Minimum milk SNF (%. m/m)
1.	Buffalo milk	Assam, Bihar, Chandigarh, Delhi, Gujarat, Haryana, Punjab, Jharkhand, Maharashtra, Sikkim, Meghalaya, Uttar Pradesh, Uttarakhand, West Bengal	6.0	9.0
		Andaman and Nicobar Islands, Andhra Pradesh, Arunachal Pradesh, Chhatisgarh, Dadra and Nagar Haveli, Goa, Daman and Diu, Himachal Prasesh, Jammu and Kashmir, Karnataka, Kerala, Lakshadeep, Madhya Pradesh, Manipur, Mizoram, Nagaland, Odisha, Puducherry, Rajasthan, Tamilnadu, Telangana, Tripura	5.0	9.0
2.	Cow Milk	All India	3.2	8.3
3.	Goat or Sheep milk	Chandigarh, Haryana, Kerala, Madhya Pradesh, Maharashtra Punjab, Uttar Pradesh and Uttarakhand	3.5	9.0
		Andaman Nicobar Isalnds, Andhra Pradesh, Arunachal Pradesh Assam Bihar, Chattisgarh, Delhi, Dadra and Nagar Haveli, Goa, Daman and Diu, Gujarat, Himachal Pradesh, Jammu and Kashamir, Jharkhand, Karnataka, Manipur, Lakshadeep, Meghalaya, Mizoram, Nagaland, Odisha, Puducherry, Rajasthan, Sikkim, Tamilnadu, Telangana, Tripura and West Bengal	3.0	9.0
4.	Camel Milk	All India	2.0	6.0
5.	Mixed Milk	All India	4.5	8.5
6.	Standardised milk	All India	4.5	8.5
7.	Toned Milk	All India	3.0	8.5
8.	Double Toned Millk	All India	1.5	9.0
9.	Skimmed Milk	All India	Not more than 0.5	8.7
10.	Full Cream milk	All India	6.0	9.0

Note: 1. When any class of milk is offered for sale in contravention of the requirements specified above the standards prescribed for mixed milk will apply 2. These standards would only be applicable at the point of sale.

Source: FSSAI act 2017 publised in the Gazette GoI New Delhi

3.1 Flow Diagram Distribution of milk constituents

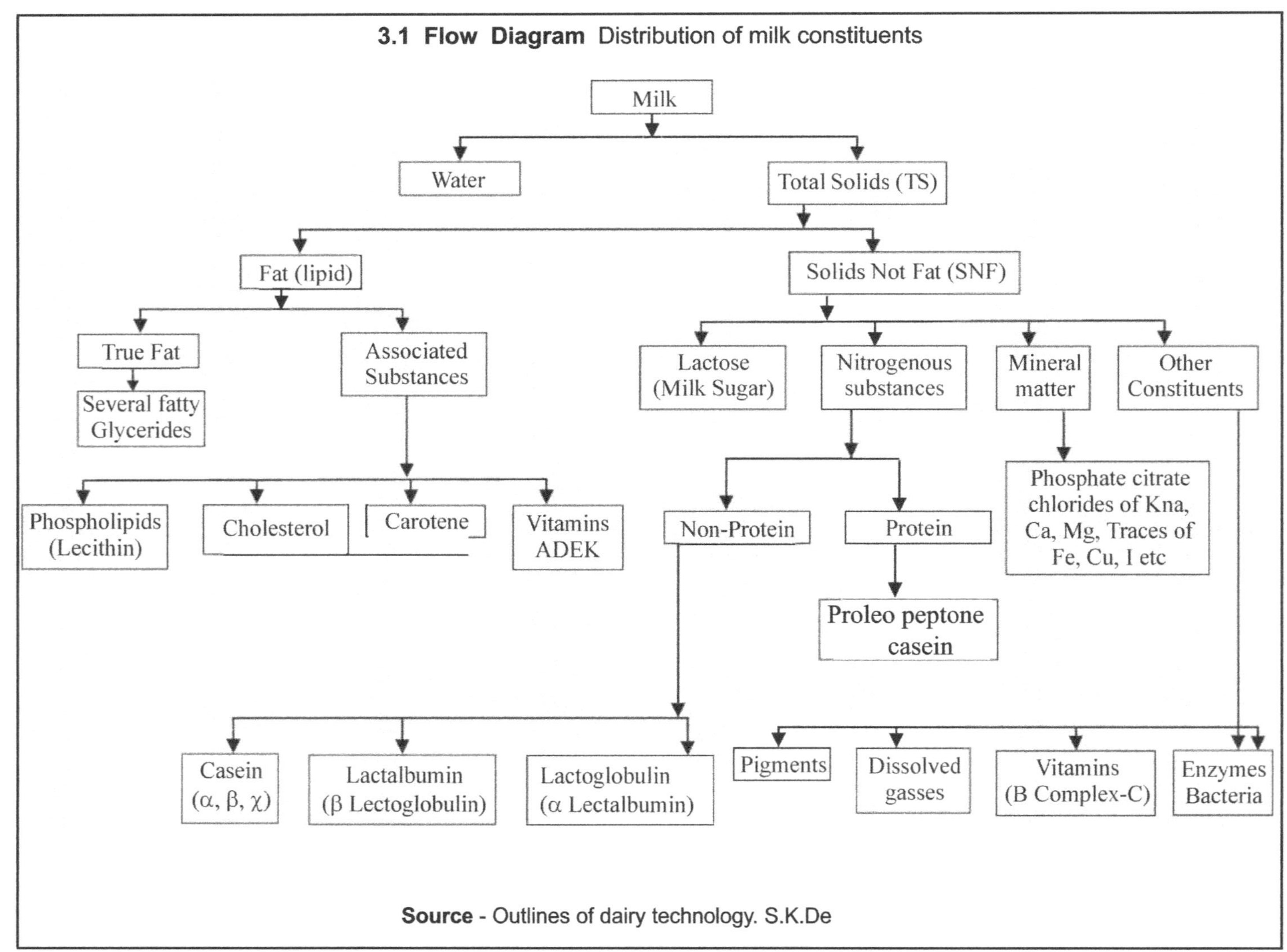

Source - Outlines of dairy technology. S.K.De

3.3 Constituents of Milk

Further examination of the fat and non-fatty solids (SNF) resolves them into several constituents as shown in the flow diagram 3.1

The major constituents are Water, Fat, Protein, Lactose, ash or mineral matter. The minor constituents are Phospholipids, sterols, vitamins, enzymes, pigments etc. The true constituents are milk fat, casein and lactose.

The process of resolving the milk constituents into their simple components is still in progress. For example, the true fat is known to be a mixture, but the precise number and constitution of its components are not yet fully established. Casein, one of the milk proteins, is believed by some workers to be a mixture and it is hoped that further research will establish the identity of these fractions. Some of the constituents which are present in milk are exceedingly minute quantities, but they cannot be ignored on that account.

Considering the fact that composition of the milk is a mixture and that the proportions of its constituents may show considerable variation. Therefore, one should not expect milk to be rigidly constant in its composition because of the fact that variation being a salient feature of all mixtures. Although, this may appear to be a very elementary fact, it is never the less one which is frequently over looked.

Each of the milk constituents will be studied in detail. Milk being a mixture in the sense that it contains a number of separate chemical compounds, milk is also a physical mixture, fat is present in the form of an emulsion, the proteins and some mineral matter are in the colloidal state and lactose together with the remainder of the mineral matter are in solution. Three distinct physical states are co-existent and to some extent are in a state of poise. This applies more particularly to the equilibrium between the colloids and the true solution. Each of the milk constituents will be studied in detail if necessary they must be considered separately, but it is a known fact that in milk they donot enjoy such independence. For example, the protein casein is linked to some of the mineral constituents, while the fat is intimately associated with number of other substances.

3.4 Water

Water constitutes the medium in which the other milk constituents are either dissolved or suspended. Most of it is 'free' and only a very small portion is in the 'bound' form, being finally bound by milk proteins, Phospholipids etc.

3.5 Lactose

Lactose or milk sugar seems to be formed exclusively in the milk of mammals. In cows milk it is present to the extent of about 4.7 percent whereas, human milk contains about 6.3 percent. Therefore, it becomes necessary to add lactose, while modifying the cows milk for infant feeding. It is one of the substance, which is present in true solution of milk. On crystalization from water it forms hard, gritty crystals ($C_{12} H_{22} O_{11} H_2O$). These crystals appear in some milk products under some special circumstances, Even though it is present in sweetened condensed milk they are normally far too small to be detected by the tongue. However, if the conditions of manufacture permit the growth of larger crystals, the product will assume a gritty texture. The crystals may also appear in ice cream when the mix contains too high a proportion of milk solids. Some peculiar properties of dried milk are due to the union of anhydrous lactose $C_{12} H_{22} O_{11}$ with water to form the hydrate $C_{12} H_{22} O_{11} H_2O$.

Lactose is only fairly sweet, being about one sixth the sweetness of cane sugar, therefore it imparts merely a faint sweet taste to milk. Further, it is very interesting to observe that, even though the sweet whey, which contains about the same concentration of lactose, it appears to be sweeter than milk due to the fact of the removal of casein, which masks the flavour in the milk.

Lactose belongs to the di-saccharide group of carbohydrate, and is composed of one molecule each of glucose and galactose, united with the removal of a molecule of water in such a manner as to leave only one reactive aldehyde group in the lactose molecule. By the action of the enzyme lactase which is present in the secretion of the intestinal glands, lactose undergoes hydrolysis into its two components glucose and galactose.

$$C_{12} H_{22} O_{11} + H_2O = \underset{\text{Glucose}}{C_6 H_{12} O_6} + \underset{\text{Galactose}}{C_6 H_{12} O_6}$$

Dilute mineral acids will also bring about this change. Certain varieties of yeast such as those present in Kephir cultures are able to carry out the initial hydrolysis and to proceed with the alcoholic fermentation of the two mono saccharide sugars. Thus fermented milk contains small amounts of alcohol (approximately 0.7 percent) carbon dioxide, together with lactic acid and is occassionally prepared for special dietaries. It is claimed that the casein of this product is in a very digestible form.

Aqueous solutions of lactose reduce Fehling's solution and ammonical silvernitrate solution, but after hydrolysis the reducing power is doubled owing to the production of two active reducing group of sugars. It is insoluble in alcohol or ether but readily dissolves in hot acetic acid. But in slightly acid conditions lactose may be oxidised to formic acid and levulinic acid, whereas oxidation with concentrated nitric acid yields oxalic and carbonic acids.

Lactose, when heated between 110-130 °C the hydrated crystals loose their water of crystalization. Above 150 °C they turn yellow, while at 175 °C they turn to brown considerably forming Lacto Caramel. These temperatures are very interesting because of the general belief that the slight browning and true characteristic cooked flavour of the heated milk are due to the formation of caramel from milk sugar. It is a well known fact that when lactose is gently warmed with nitrogenous bases like ammonia, certain amines or amino acids any of which could be present in heated milk brown coloured complexes are formed. Naturally the extent of these changes will depend upon the severity of heating and the time of holding, while the pasteurised milk may be but slightly affected, and the changes are much more apparent in sterilised and evaporated milk. None of the milk products would be processed at temperature, that are high enough to caramatize pure lactose. This apparent anamoly compels us to recognise that the properties of the individual milk constituents may be different from their behaviour in milk, where they are in contact with so many other compounds. Thus it is possible for lactose to caramalise at much lower temperature as it exists in milk.

Commercially, lactose is a very valuable product being used in the preparation of infant and invalid foods. Its high nutritional value is well recognised. Gortner deprecates the substitution of cane sugar for lactose in the preparation of infant foods, because of the fact that it is the only source of sugar galactose, to the human beings and most of the animals which is present in the brain and nervous tissues. It is also believed that lactose helps in the assimilation of calcium and phosphates from the intestines being a part of the tonic effect on the alimentary canal. The chief source of lactose is whey and the method of its preparation will be described in latter chapter.

3.6 Lactic Fermentation

Lactose is an easy pray for microorganisms, which find their way into milk, even under most careful methods of production and handling. The faintly sweet taste of raw milk gradually changes to a more stringent flavour with simultaneous production of the characteristic odour of souring of milk under certain favourable conditions of temperatures. The gradual progress of this change can be demonstrated by incubating the milk at 37 °C. The acidity tests are carried out at regular intervals of time and the results are plotted on a graph. The points lie on a smooth curve from which it is observed that initially the development of acidity is slow, which is followed by one of vigorous development until the final stage is reached, showing a marked reduction or retardation at an acidity level corresponding with about 1 percent lactic acid. It is possible to detect the sour taste at an acidity of 0.3 to 0.4 percent, whereas the sample of milk will curdle at an acidity of 0.6 to 0.7 percent under ordinary room temperatures. The exact time taken for these changes to occur will depend upon the number and nature of the microorganisms present and the temperature of storing which may be varied from merely a few hours to days.

These changes which occures in the milk are mainly due to the bacterial conversion of lactose to lactic acid.

$$C_{12} H_{22} O_{11} + H_2O = 2C_6 H_{12} O_6$$

$$2C_6 H_{12} O_6 = 4CH_3 \text{ CHOH COOH (Lactic Acid)}$$

The raw milk which generally contains the usual mixed bacterial population, there is a possibility to check the further acid production approximately at 1 percent level of lactic acid, not because that there is no lactose present, but because of the fact that this concentration is unfavourable for further active production of the acid by the microorganisms. Neutralization of milk when it reaches this stage, results in a renewed burst of activity with the production of more acid. A close glance of the equation will clearly show that 1 percent of lactic acid is formed from approximately 1 percent of lactose, thereby about 75 percent initial lactose is still available. When this retarding effect is established, pure cultures of different bacteria show pronounced differences in their toleration of accumulation of acid.

Lactic acid in its pure form has neither the odour nor the flavour of sour milk. The souring of milk cannot be represented by one equation, like most of the biological reactions. There are numerous by-products which are being produced includes corbon-dioxide gas, acetyl methyl carbinol diacetyl and butyric acid, the latter two possessing powerful odours. The nature and quantity of the by products is determined by the types of bacteria and the conditions of storage of milk sample even though the main reaction that generally occurs is certainly as shown in the equation. This explains the case with which the cream starter cultures are propagated and stored. Only the correct cultures will produce the required flavours and aroma in quantities that is sufficient to give its desired characteristics to the butter. It is observed that out of every 100 parts of lactose that is being attacked during the souring of milk about 95 parts goes into lactic acid, while the remainder is converted into different by products.

These changes are of great value in the manufacture of butter and cheese, when under control, otherwise they cause severe losses to the dairy industry causing considerable annoyance to the consumer. The modern methods of clean milk production which aims at the reduction of the bacterial contamination at the source and also by keeping the milk below 50 °F or at as low a temperature as possible to avoid such kind of wastage. At such low temperatures, although it seems that the bacterial action is very considerably retarded it does not cease entirely, thereby it becomes necessary to excercise care in the bacteriological cleanliness of storage vessels, when the milk is kept.

The development of lactic acid effects both casein and the mineral matter. Casein is normally united with calcium however, as the concentration of lactic acid increases more and more of the calcium is withdrawn from casein to form calcium lactate. Further, some of the colloidal calcium phosphate is converted to soluble hydrogen phosphate. The development of lactic acid thus alters the physical equilibrium between the true solution and the colloidal state. The former increases at the expense of the latter, and the net result is an increase of osmatic pressure and a corresponding decrease in the freezing point of the sample. This effect on the freezing point assumes great practical significance in the detection of added water by the cryoscopic method.

Lactic acid can be produced by controlled fermentation of lactose in whey because of its valuable importance as a commercial product. The initial product is calcium lactate, from which the pure lactic acid, metallic lactates or lactic esters are prepared. These substances are used in a variety of industries including plastics and resins, textiles, fine chemicals etc. for which, whey may be regarded as a possible raw material. Further, as lactic acid may be fermented to butyric acid according to the following equation, it offers another possible utilization of surplus whey.

$$2C_3\ CHOH\ CHOH = \quad C_3\ H_7\ COOH + 2CO_2 + 2H_2$$

Lactic acid Butyric acid + Corbon di-oxide + Hydrogen

3.7 Milk Fat

Milk fat consists almost entirely of triacyl glycerides or triglycerides which constitute nearly 97-98 percentage of total lipid. Small quantities of di and mono acyl glycerols are also present in traces. These glyceroides are formed by the esterification of the hydroxyl groups of glycerol with the fatty acids. The nature of these fatty acids varies vastly and has a significant influence on the chemical, physical and organoleptic properties of the fat.

As many as 437 different fatty acids have been identified and compiled from the bovine milk. Inspite of the fact that such a large variety of fatty acids have been identified only a handful, limited number of fatty acids are present in significant quantities which are of nutritional, chemical and physical importance. Only five fatty acids make up nearly 80 percent of the total fatty acids between them, narrowing the field further. These are oleic, palmitic, butyric, stearic and myristic acids. The remaining fatty acids are present in small quantities. Nearly 50 of the 80 commonly encountered fatty acids together make up hardly 1% of the total. Some of them have been isolated and studied while others have been identified only on the basis of analytical data.

The average composition of the fatty acids present in the milk fat has been presented in the Tables **3.10, 3.11 and 3.12.**

Table 3.10 Component Fatty Acids of Milk Fat

Formula	Name	Approx. % weight	At ordinary temperatures	Volatility in Steam	Water Solubility
A. Saturated compounds of the type $C_nH_{(2n+1)}$ COOH					
C_3H_7 COOH	Butyric	3.5	Liquid	Volatile	Soluble
C_5H_{11} COOH	Caproic	2.0	Liquid	Volatile	Slightly soluble
C_7H_{15} COOH	Caproic	1.0	Liquid	Volatile	Very slightly soluble
C_9H_{19}COOH	Caproic	2.0	Solid	Volatile	Very slightly soluble
$C_{11}H_{23}$COOH	Lauric	2.5	Solid	Slightly Volatile	Almost insoluble
$C_{13}H_{27}$COOH	Myristic	10.0	Solid	Very slightly Volatile	Insoluble
$C_{15}H_{31}$COOH	Palmitic	25.0	Solid	Non-volatile	Insoluble
$C_{17}H_{35}$COOH	Stearic	10.5	Solid	Non-volatile	Almost insoluble
$C_{19}H_{39}$COOH	Arachidic	0.5	Solid	Non-volatile	Insoluble
B. Unsaturated compounds with one double bond C_nH_{2n-1}COOH type					
C_9H_{17}COOH $C_{11}H_{21}$COOH $C_{13}H_{25}$COOH	Decenoic Dodecenoic Tetra Decenoic	5.0	Liquid	Non-volatile	Insoluble
	Oleic	33.0	Liquid	Non-volatile	Insoluble
C. Unsaturated acids of the C_nH_{2n-3} COOH type (Two double bonds)					
$C_{17}H_{33}$COOH	Linoleic	4.0	Liquid	Non-volatile	Insoluble
D. Other more unsaturated acids containing 20 to 22 carbon atoms					
		1.0	Liquid	Non-volatile	Insoluble

Source : Text Book of Dairy Chemistry - Edgar. R Ling

Note: In giving the above proportions of the fatty acids, it is well recognised that various factors may cause variation in the composition of milk fat. These will be discussed in a later chapter. In spite of these variations the balance of steam volatile/ non-volatile acids, saturated/unsaturated acids, etc., renders milk fat unique and thus serves to distinguish it from animal or vegetable fats.

Table 3.11 Lipids of Fresh Milk

			Fatty acid residues					Percentage of the lipid		
Lipid class	Alcohol residue	Other constituent	MW	Number	x	y	Percentage in milk fat (w/w)	Core of globule	Globule membrane	Milk plasma
Neutral Glycerides :							98.7			
Triglycerides	Glycerol		728	3	14.4	0.35	98.3	–100	+	
Diglycerides	Glycerol		536	2	14.9	0.38	0.3	90?	10?	?
Monoglycerides	Glycerol		314	1	15.0	0.36	0.03	+	+	+
Free fatty acids	--		253		15.8	0.36	0.1	60	10?	30
Phospholipids[2] **:**		Phospho group					0.8	--	65	35
Lecithin	Glycerol	Choline	764	2	17.2	0.6	0.26			
Ph : ethanolamine[3]	Glycerol	Ethanolamine	742	2	17.9	1.0	0.28			
Ph. serine[3]	Glycerol	Serine	784	2	17.8	0.8	0.03			
Ph. inositide[4]	Glycerol	Inositol	855	2			0.04			
Plasmalogens	Glycerol	Choline[6]	-700	1[7]			0.02			
Sphingomyelin[5]	Sphingosine	Choline	770	1	19	0.2	0.16			
Cerebrosides[4,5]	Sphingosine	Hexose	770	1	20	0.2	0.1	--	70	30
Gangliosides[4,5]	Sphingosine	Hexose[8]	-1600	1		0.01		-	70?	30?
Sterols :							0.32	80	10	10
Cholesterol			387				0.30			
Cholesteryl esters	Cholesterol	642	1	16	0.4	0.02?				
Carotenoids + Vitamin A	--						0.002	95?	5?	+

1 - Approximate Average values ; not complete ;
x = average number of carbon atoms ;
y = average number of double bonds.
2 - Approximately 1% is present as lysophosphatides
3 - Phosphatidylethanolamine + Ph.serine = cephalin
4 - Glycolipids
5 - Sphingolipids
6 - Orethanolamine
7 - Also a fatty aldehyde residue
8 - Also neuraminic acid

Source : Dairy Science & Technology Pieter Walstra et.al

Table 3.12 Fatty Acids in Milk Fat

Acid	Notation[2]		Melting Point (°C)	Solubility[3] (g/L)	Composition (in mol %) of			Percentage in 3 position
	x	y			Neutral glycerides	Phospholipids	Free fatty acids[5]	
Saturated:					69 (57-80)	45	72	97
Butyric	4	0	–8	Miscible	8.5 (7-14)	--	14.5	84
Caproic	6	0	-4	174	4.0 (2-7)	--	4.5	45
Caprylic	8	0	16	58	1.8 (1-3.5)	0.2	2	33
Capric	10	0	31	17	3.0 (1.5-5)	0.2	2	26
Lauric	12	0	44	5.6	3.6 (2.5-7)	0.5	2	17
Myristic	14	0	54	1.6	10.5 (8-15)	3	9	12
Palmitic	16	0	63	0.49	23.5 (20-32)	19	21	22
Stearic	18	0	70	0.14	10.0 (6-13)	12	13	7
Odd-numbered					2.5 (1.5-3.5)	4.5	2.5	
Branched					1.1 (0.7-1.8)	0.7	1	
Other					0.7 (0.3-2)	5		
Monoene:					27 (18-36)	41	23	
Palmitoleic	16	1Δ9[6]			1.4	?	1?	23
Oleic	18	1Δ9[6]	16	0.42	21 (13-28)	38	20	32
Other					5.5	3	3.5	
Diene:					2.5 (1-4.3)	8		
Linoleic	18	2Δ9,12[6]	-5		1.8	8	2.3	20
Other					0.7	0.2		
Polyene:					0.8 (0.4 - 2)	4		29
σ-Linolenic	18	3Δ9,12,	15[6]		–12	0.4	2	
Other						0.4	2	
Keto						0.3	?	
Hydroxy						0.3	?	
Fatty alcohol						0.01	0.15	
Fatty aldehyde						0.02	0.01	
Unclassified							2	1

1 - Properties, approximate average fatty acid composition of some lipid classes, and average percentage of each fatty acid residue esterified in the 3 position of the triglycerides.

2 x = number of C atoms ; y = number of double bonds ; D refers to the position in the carbon chain : D9, 12, for instance, indicates that the two double bonds occur at the ninth and twelfth bonds, counting from the carboxyl group.

3 - Critical micelle concentration in the presence of Na+ ions.

4 - In parentheses is the approximate range.

5 - Free fatty acids liberated by the action of milk lipase.

6 - All cis.

Source : Dairy Science and Technology Pieter Walstra et.al

An attempt has also been made to present the fatty acid profile of the milk fat of different species of animal in the Table 3.13.

Table 3.13 Fatty Acid profile of milk fats of different species (Mole %)

Component mole %	Fatty acid	Cow	Buffalo	Goat	Sheep	Camel
Butyric	4:0	8.8	11.4	7.6	8.4	5.9
Caproic	6:0	3.5	3.1	4.5	5.4	1.9
Caprylic	8:0	2.0	1.0	6.2	5.8	1.1
Capric	10:0	3.0	1.6	11.1	10.1	2.1
Lauric	12:0	3.8	2.6	5.1	6.0	5.7
Myristic	14:0	9.9	10.6	11.2	11.8	7.9
Palmitic	16:0	26.1	30.3	21.5	20.4	28.3
Stearic	18:0	9.1	10.5	7.3	5.4	9.7
Higher Saturation	20-26.0	1.0	0.7	0.1	1.3	-
Unsaturated (Mono)	10:1-14:1	1.8	1.0	-	-	-
Lower	16:1	3.6	3.6	-	-	-
Unsaturated (poly)	18:1	26.2	21.6	24.2	22.2	34.1
		3.5	2.0	1.2	3.2	3.3

Source : Text Book of Dairy Chemistry - M.P. Mathur, D.D. Roy & P. Dinakar

Microscopic examination of milk reveals that the bulk of the fat exists in the form of small globules, which average approximately 2 to 5 microns in size (range 0.1 to 22 microns). This is an oil in water type emulsion. The surface of these fat globules is coated with an adsorbed layer of material commonly known as the fat globule membrane. In any one sample of milk there is a great diversity of globule size but there is a definite increase in the proportion of smaller globules with advancing lactation. There are also differences between the breeds. Generally those giving milk of greatest fat content also give more of the larger globules whereas those giving least fat are characterized by a much larger proportion of smaller globules.

The size and distribution of the milk fat globule has been given in table 3.14.

Table 3.14 Size Distribution of Milk Lipid Globules

Class	Diameter (mm)	Proportion of the total globule Population (%)	Fraction of total milk lipid (%)
Small	Below 2	70 - 90	< 5
Intermediate	3 - 5	10 - 30	90
Large	8 - 10	0.01	1 - 4

Source : Text Book of Dairy Chemistry - M.P. Mathur, D.D. Roy & P. Dinakar

Cooled raw milk sample shows the individual fat globules will group together resulting in clusters when it has not been subjected to violent agitation. This tendency to form clusters is of considerable great importance in the separation of cream by centrifugal force or by the force of gravity. Therefore, it appears that each fat globule enjyos an independent existance through the good offices of a protective layer. This layer is 'tacky' by nature or may be rendered so by some of the milk constituents or by suitable physical conditions. In an attempt to elucidate the nature of this protective film (fat globule membrane) much work has been done by several workers and there are still many gaps in our knowledge.

The research work carried out on this protective layer has shown that it contains a variety of substances such as protein, phospholipid, phosphatase, an un-identified ether soluble substance and a complex of riboflavin phosphoric acid-protein, which is known to possess the properties of 'Schardinger Enzyme' (i.e. it stimulates the simultaneous reduction of methylene blue and the oxidation of formalin when both of these are added to milk). The average composition fat globule membranes of the buffalo as well as that of the exotic cow has been presented in table 3.15 for a comparative study.

Table 3.15 Estimated average composition of membranes of milk fat globule

Component	Buffaloe	Exotic Cow
Lipid Content	38.5	35.7
Neutral Lipids	74.4	65.7
Phospho lipids	19.7	18.6
Triglycerides	52.7	45.5
Diglycerides	7.4	5.5
Sphingomyelin	4.3	3.2
Phosphatadyl Serine	3.1	2.4
C18.0 Stearic acid	34.4	24.1
C 16.0	22.4	238
C 18.1	14.5	14.5
C 14.0	12.6	14.4
Protein (g/100 g/fat globule)	1.36	1.6

Source : Text Book of Dairy Chemistry - Edgar. R Ling

The physical condition of the fat determines to some extent the precise composition of the protective layer. Some references were made to the very large surface area of the fat globules. These large areas associated with small particles create considerable surface forces (tension) that are responsible for the phenomenon of adsorption. This phenomenon is common to all systems in which small particles of matter are in contact with liquids or gases, the main condition being that they should be small enough to present a great surface area per unit weight. This condition is fulfilled by the particles of colloid size such as casein and the fat globules in milk. The modern view is that certain milk constituents are able to prevent the merging of these fat globules to concentrate at their surfaces into a continous fat phase.

The peculiar properties of the adsorbed layer of the fat globule are able to stimulate or bring about the oxidative changes in the fat phase, when metals like copper are dissolved in minute quantity by the milk. The water soluble metal is brought into direct contact with the fat which otherwise would remain inaccessible. The dissolved metal forms a metallic proteinate by direct union with the surface protein of the globule or it unites with the plasma proteins and is then adsorbed by the globules. In this respect the globule surface material functions in a manner which is far from the protective nature. The conception that the fat globule is neatly covered or wrapped by an inert membrane is removed by the present modern view. Indeed the adsorbed layer is at times is the major attraction centre of various chemical actions.

The physical properties of pure fat extracted from milk, cream or butter, reveals it to be a mixture the colour of which varies considerably. If we consider that the fat is just one compound, it would melt at one definite temperature and solidify at the same point. There are many compounds which normally comprise milk fat and each has its own melting point and other distinctive physical properties. On warming the solid mixture, it is the constituents of the lowest melting point which liquify in the first instance and the remaining solids will dissolve in the liquid, liquifying at a temperature well below their normal melting points. More and more of these solids will dissolve until finally the whole of the fat is in liquid condition, as the temperature is increased. The temperature at which this occurs is not strictly the melting point, rather it is the temperature at which the least soluble fat fractions will dissolve in the liquid obtained at lower temperatures. Similarly, the setting point or solidification state is really the temperature at which the various fat fractions crystalize from the mixed solutions. Invariably there is a time and temperature lag in the crystalization, hence the setting point do not correspond exactly with the melting points. The first crystals, would encourage others to appear and the setting would take place more rapidly and over a small range of temperature, when the fat is considered to be in a continuous phase, when the milk is subjected to alterations in temperature. It is not the case and the consequence is that several hours must elapse before the fat globules are adjusted. The heterogenous nature of milk fat is furnished by the evidence of the variations in its specific gravity 0.936 - 0.946 at 15°C and refractive index 1.459-1.462 at 15°C. It is soluble in fat solvents petroleum and ether, and slightly soluble in ethyl alcohol and is readily soluble in hot amylalcohol. Fat is not soluble in water, however, fat is able to dissolve very small quantities of water. The physical properties of milk fat are being determined by the nature and the proportions of its constituents. Chemically it consists of a mixture of true fats (i.e. triglycerides of fatty acids) which hold in solution or by adsorption a number of substances including lecithin, cholesterol carotene and the fat soluble vitamins.

The physical properties of the milk fat varies/differs at different temperatures as shown in the table 3.16:

Table 3.16 Physical Properties of Milk Fat

Temperature (°c)	Density (kg. m^{-3})	Refractive index at $\lambda = 0.589\ \mu m$	Viscosity (mPa - s)	Solubility of water (% w / w)
10	922	(1.465)		(0.11)
20	915	1.462	70.8	(0.14)
30	909	1.458	45.7	0.17
40	902	1.454	30.9	0.20
50	895	1.451	22.1	0.24
60	889	(1.444)	16.6	0.27
70	882	(1.444)	12.5	0.32
80	876	(1.440)	9.8	0.36
90	(869)	(1.436)	7.6	0.41
100	(863)	(1.433)	6.2	0.46

Source : Dairy Science and Technology Pieter Walstra et.al

True Fats

It has already been stated that the physical properties of milk fat are being determined by the nature and proportions of its constituents. Glycerol being a tri-hydroxy alcohol, combines with three molecules of the same or different fatty acids, producing fat, which may be resolved into its components by hydrolysis according to the following equation.

$$\begin{array}{ll} CH_2-OOC\,C_{17}H_{35} & CH_2OH \\ | & | \\ CH-OOC\,C_{17}H_{35}+3H_2O & =CHOH+3C_{17}H_{35}COOH \\ | & | \\ CH_2-OOC\,C_{17}H_{35} & CH_2OH \\ \text{Stearin} & \text{Glycerol + Stearic Acid} \end{array}$$

The enzyme lipase is the natural hydrolytic agent of the fats. The decomposition of the fat may be accomplished either by treating with superheated steam or with alcoholic sodium hydroxide, where it is saponified to the sodium salts of the constituent fatty acid (soaps) and glycerol. This solution on acidification with sulphuric acid liberates the fatty acid.

$$
\begin{array}{ll}
CH_2-OOC\,C_{17}H_{35} & CH_2OH \\
| & | \\
CH-OOC\,C_{17}H_{35} + 3NaOH = & CHOH + 3C_{17}H_{35}\,COONa \\
| & | \\
CH_2-OOC-C_{17}H_{35} & CH_2OH \\
\text{Stearin} & \text{Glycerol + Sodium acetate}
\end{array}
$$

$$2\,C_{17}H_{35}\,COO\,Na + H_2SO_4 = 2C_{17}H_{35}\,COOH + Na_2SO_4$$

All the true fats which are formed or produced in the similar manner, will react in the same manner, however, they differ in the number and relative proportions of the fatty acids which are combined with the glycerol. Milk fat when treated in the same manner, it would produce/furnish a mixture of at least fifteen (15) fatty acids of which some are soluble in water, however, majority of them are insoluble hence appear as a white precipitate. The physical and chemical properties of fats which are dominated by their constituent fatty acids have been the subject and attracting more researchers to carryout further research.

In association with many other natural fats, the oleic and palmitic acids constitute the greatest quantity, representing one third (1/3) and one fourth (1/4) of the total respectively, followed by the stearic and myristic acids each representing approximately one tenth (1/10) of the total. However, the minor components are responsible to distinguish the milk fat from others and the most important consideration is that about 43 percent of the total acids are unsaturated, which accounts for the milk fat being chemically active under the right circumstances/ conditions.

There would be about 16 individual fatty acids will be constituted, in milk fat when each of them are allowed to combine with the glycerol molecule. The evidence available till date indicates that the glycerides, which are actually present in milk fat are the compounds of glycerol with two or even three different fatty acids, hence the total number of possible compounds are indeed very large. The problem of discovering the precise constitution of these molecules is obviously very complex, however, the evidence available to date suggests that the following conclusions can be arrived.

1. The acid in greatest quantity, oleic is mainly distributed between four types of glycerides.

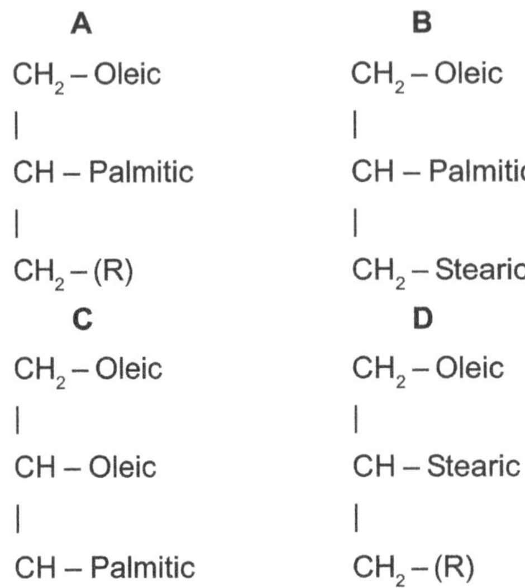

where R - denotes one of the lower fatty acids. Out of the four, the group A is the most common fatty acid present in the milk fat.

2. Majority of the molecules are fully saturated.
3. The palmitic radical is present in about 75 percent of the molecules.

The actual percentage of each acid is subjected to some variations, hence, there is a possibility that their distribution in the various glycerides is expected to show similar variation. The modern theory of milk fat synthesis was conceived due to the extensive research carried out on the chemical constitution of the glycerides. The former theory, that the phospholipids present in the blood were the precursors of the milk fat is no longer valid, with the simpler theory that the mammary gland elaborates the milk glycerides from the oleo glyerides of the blood is more acceptable, keeping in view with the results of the research available at present. Hydrolysis of the fat is brought out by the enzyme lipase and the change resulting in the liberation of free fatty acids which causes the taint commonly known as rancidity.

3.8 Milk Fat Constants/Properties

In order to characterise the more common fats certain well known physical and chemical constants have been formulated. These constants serve as an indication of the type of the component fatty acids present in the fats. They also enable the detection of fat adulteration qualitatively and in some instances quantitatively.

The physico-chemical characteristics of buffaloe and cow milk fat have been given in the Table 3.17.

Table 3.17 Physico-chemical Constants of Buffaloe and Cow milk fat

Physico Chemical Constants	Buffaloe Milk Fat	Cow milk Fat
Softening Point range	34.3 - 36.3°C	33.5 - 35.9°C
Melting Point Range	33.4 - 46.4°C	31.5 - 35.2°C
Acid value	0.17 - 0.352	0.26
Refractive index	1.4515 - 1.4533	1.4498 - 1.4530
B.R. Reading	41.00 - 43.50	41.05 - 42.40
Saponification number	218.23 - 236.10	221.0 - 238.0
Iodine Value	27.00 - 33.90	27.70 - 37.32
R.M. Value	27.83 - 35.50	24.60 - 29.70
Polenske Value	0.70 - 1.60	1.30 - 1.80
Density	0.905 - 917 g/ml	0.888 - 0.911 g/ml
Grain Size	0.20 - 0.41 mm	0.098 - 0.190 mm
Fat Globule size	4.15 - 4.60 mm	3.36 - 4.15 mm
Unsaponifiable Matter	392 - 398 mg/10 ml	414 - 450 mg/100ml

Source : Text Book of Dairy Chemistry - M.P. Mathur, D.D. Roy and P. Dinakar

Refractive Index

The basic principle in determining refractive index concerning fat is the degree of binding of light waves passing through a liquid or transparent solid will be a characteristic for the particular liquid or solid. It is the ratio of velocity of light in vaccum to the velocity of light in the sample medium. More generally, it is expressed as the ratio between the sine of the angle of incidence to the sine of angle of refraction, when a ray of light of a definite wave length (usually 589.3 μ the mean of the D lines of soidium) passes from air into fat.

In the case of milk fat the constant may be determined readily with Abbe refractometer. The reading is normally made at 40°C. The refractive index of milk fat generally ranges between 1.4527 to 1.4566. This value is low in comparison to that for other fats and oils mainly because of the presence of greater number of saturated glycerides and short chain acids in the milk fat. The refractive index of a fat is influenced by both the molecular weight and the degree of saturation of the component fatty acids. Its determination may be useful as an indication of adulteration but would seldom, in itself be conclusive.

Saponification Number

The saponification number is defined as the number of milligrams of decinormal potassium hydroxide required to sapoinfy 1 gm of fat. This value may range from

210-233 for milk fat and more often falls within the range of 225-230. Thus this constant is an indication of the average molecular weight of the fatty acids present. With the exception of coconut and palm kernel oil, this constant for milk fat is well above those for other fats and oil, such as liquid paraffins, in ghee as they are not acted upon by alkali and such a sample does not form a homogenous solution on saponification. The saponification value is related to the molecular weight of ghee from which can be calculated the saponification equivalent, which is the amount of ghee saponified by one gram equivalent of potassium hydroxide and is equal to 56100 divided by the saponification value.

Iodine Number

The iodine number is defined as the number of grams of iodine absorbed by 100 g fat under specified conditions. This constant is a measure of the unsaturated linkages present in the fat. The iodine number for milk fat falls within the range of 26-35 which is low in comparison to most other fats. The absorption of 1 Br. (Hanus method) and 1 Cl. (wijs method) gives a measure of the unsaturated acids present in a fat. One molecule of halogen compound is absorbed by each unsaturated linkage and the absorption is expressed as the equivalent number of grams of iodine absorbed by 100 g fat.

Reichert Meissel Number - (R.M. Value)

It is defined as the number of millilitres of decinormel (N/10) sodium hydroxide or aqueous alkali solution required to neutralize the steam volatile water soluble fatty acids distilled from 5 grams of fat under the precise conditions specified in the method. This constant for milk fat is quite significant since it is primarily a measure of butyric acid and caproic acid. The value for milk fat ranges between 17 to 35 which is well above that for all other fats and oils. Milk fat contains more of these acids than does any of the fats from which it might be desirable to distinguish it.

Polenske Value/Number

The polenske value is the number of 0.1N aqueous solution or decinormal sodium hydroxide required to neutralise the steam volatile water in soluble fatty acids distilled from 5 g of fat under the precise conditions specified in the method. Caprylic acid, Capric acid, which are somewhat steam volatile but largely insoluble in water are indicated mainly by the polenske number. The polenske value or number for milk fat ranges from 12 to 24. As most of the volatile acids obtained from milk fat are soluble, this test helps to distingush between this fat and one of high caprylic acid content (eg. coconut oil).

Kirschner value

The Kirschner value is the number of 0.1 aqueous alkali solution or decinormal alkali required to neutralise the steam volatile, water soluble fatty acids

(R.M. Value distillate) which form water soluble silver salts distilled from 5 g fat under the precise conditions specified in the method.

Melting Point

Melting point ranging from 30°C to 41°C have been reported in the literature for milk fat.

Rancidity

Commercially milk fat is the most valuable component of milk and at the same time it is most susceptible to oxidative spoilage or rancidity, which in turn rapidly decreases the commercial value of the products. With the onset of rancidity, the normal pleasant appetising flavour and aroma of the product is replaced by undesirable and unpleasant flavours. This is due to the formation of volatile compounds which can impart noticeable off flavours even when present in very small quantities. There are three types of rancidity development in milk fat viz, the hydrolytic rancidity, the ketonic rancidity and the oxidative rancidity.

A. Hydrolytic Rancidity

This change as given to understand is never complete due to the fact/reason that the accumulation of water soluble butyric acid inhibits the further reaction of the enzyme. The pronounced rancid order and flavour will become evident even if there is a slight degree of reaction due to the fact that the butyric acid has a powerful odour.

In this type of rancidity, the fatty acids which are esterified with glycerol to form the triglycerides and are broken off from the glycerol and released as free fatty acids. This leaves behind a diglyceride, or mono glyceride depending on the extent of hydrolysis. Fresh butter or ghee contains very small amount of free fatty acids (FFA), but as a result of hydrolysis, the free fatty acid's level increases. This type of hydrolysis requires the presence of moisture and lypolytic enzymes, commonly known as lipase. Both these agents are commonly found in butter but to a much lesser extent in the ghee.

The nature of the free fatty acids released on lypolysis influence the rancidity. It was generally thought that butyric acid is responsible for off flavours, but it is now accepted that all the lower free fatty acids from C_4 to C_{12} i.e. butyric, caproic, capric, caprylic acids have powerful bad odours, which are responsible for the rancidity. Milk fat, characteristically contains an abundance of these short chain saturated fatty acids and is, therefore, readily susceptible to this type of rancidity. The long chain fatty acids whether saturated or unsaturated do not possess any distinct odour or aroma of their own, and by themselves do not significantly contribute to rancidity. They are however, subject to transformation, which in turn leads to 'off' flavoured products.

The reasearch carried out by many workers on the occurence and behaviour of the lipase of milk and dairy products have emerged a number of facts which are of great importance. It is still impossible to connect all these observations to a satisfactory theory due to the fact that some of these observations/findings of the researchers are apparently contradictory. The most important and potent disturbing factor is the secretion of lipase by the microorganisms and the consequent lypolysis being governed by different conditions those pertaining to the natural lipase of milk. The measurement of either water soluble fatty acids or the acidity of fat as the criteria of lypolysis, has become another source of confusion because of the fact that the reaction liberates both type of acids. The complete picture is possible only when both serum and fat acidity are measured.

The sailent features of some of Research findings are given below :

1. Lypolysis may occur at relatively low temparatures (i.e. near freezing point of water). There exists a negative temperature coefficient in the Milk, which means, that whenever the lypolysis occurs it is more extensive at low temparature than the higher.
2. The natural lipase functions with positive temperature coefficient in a significant manner whenever steps are taken to suppress the microorganisms.
3. The extent of lipolysis is greatly reduced by rapid cooling when the milk is stored at 5 °C (41 °F) for 48 hours.
4. Inactivation of lipase was first noticed when the raw cream is heated to 110°F and held there for about two and half hours. Momentary heating at 155°F also prevents lipolysis.
5. Under normal conditions there is little lypolysis of raw milk owing to the protective action of the adsorbed globule surface.
6. It is difficult to discriminate between lipase activity and surface permeability because of the fact that lipolysis is largely controlled by the physical condition of the fat and the nature of the globule surface.
7. Reduction in surface tension results in the liberation of fatty acids which in turn causes an increased foaming and difficulty in churning of cream.
8. Lipolysis varies considerably in the milk of different cows, which apparently has no relation with breed, season or yeild.

   There is no established clear relationship between stage of lactation and lipase activity. There are some reports which state that some of the cows in advanced stage of lactation giving milk a bitter flavour due to lipolysis. It has also been reported that feeding of cows with green feeds tend to inhibit the lipase activity in late lactation milk.

9. Homogenisation, agitation, particularly when the globules are liquid or semi-liquid, or warming pre cooled samples to 30°C (80°F) and cooling to 10°C (50°F) or below, tends to encourage or increase lipolysis due to the fact that the normal globule surface is altered in all these conditions.
10. Oxidation adversely affects the lipolysis, particularly when the dissolved oxygen is accompanied by the presence of traces of copper or other heavy metals. Lactic acidity is also considered to be an influencing factor in restricting the lipase activity.
11. Homogenisation produces milk fat globules with a newly absorbed surface which infact does not afford the same protection as the original fat globular surface and with positive temperature coefficient, the reconstituted samples show an increased lipolysis.

B. Ketonic Rancidity

Free fatty acids are present in milk fat either inherently or due to hydrolysis of glycerides. These free fatty acids are more prone to degradation through the β-oxidation. This deterioration is brought about by the enzymes elaborated molds like pencillium, asperigillus etc or by bacteria like micrococci bacillus etc. In β-oxidation the carbon atom at a β position with respect to the carboxyl group is activated and undergoes transformation. The methylene group loses hydrogen to form a double bond (unsaturation) which then adds on a molecule of water before loosing a molecule of hydrogen to give keto acids. The keto acid further loses a molecule of CO_2 to yield a ketone which is responsible for imparting the off flavours. In this case the lower fatty acids are broken down to ketones of the type $C_5H_{11}COCH_3$ or $C_9H_{19}COCH_3$ which possess a heavy odour.

The type of off flavour imparted depends on the aldehyde or ketone produced, Glycerides on deterioration yield either saturated or mono unsaturated aldehyde or short chain length which impart tallowing or oily flavours. Phosholipids on the other hand produce larger aldehyde with one or two double bonds and are responsible for metallic or fishy flavours. Phospholipids are much more vulnerable to β-oxidation than the glycerides.

C. Oxidative Rancidity

The most common type of rancidity which is generally encountered in the deterioration of fats is oxidative rancidity. Oxygen is implicated as the main causative factor in this process. The oxygen could be present either in the air in contact with the fat or in the dissolved air within the fat. The unsaturated fatty acids in the lipid are the main targets of oxidation leading to the formation of products which impart unpleasant off flavours and also cause a loss of nutritive and commercial value of the product.

Oxidation of fat is often called auto-oxidation due to the fact that once the process is initiated its reaction rate increases rapidly. This phenomenon proceeds through a free radical chain mechanism and can be classified into three distinct stages.

1. Initiation or formation of free radicals.
2. Propagation of free radical chain reaction and
3. Termination through the formation of non-radical compounds.

Initiation

In this stage of initiation an unsaturated hydrocarbon loses a hydrogen atom to form a radical and oxygen adds on to the adjacent double bond forming a di-radical. It has been postulated that oxygen in the singlet state can directly attach between a liable hydrogen atom to form a hydro peroxide directly. This phenomenon is not very common and may not necessarily be a free radical chain machanism.

Propagation

The hydro peroxides radical thus formed propogates by attracting hydrogen alone from an adjacent molecule and thus forming/creating a new free radical. The new radical thus formed in turn is very active and results in the formation of more peroxy radicals, hydro peroxides and new hydrocarbon radicals. The new radical formed reacts very readily with oxygen.

Termination

Termination could occur when two free radicals react, when no more free radicals are available for further reaction with oxygen it is necessary for a new initiation reaction to occur before oxidation can continue.

Secondary Reaction

The oxidation process does not end with the formation of hydro peroxides. In fact, up to the formation of hydro peroxides there is no loss of flavour or orgenoleptically perceptible deterioration of the fat. This is because the hydroperoxides are odourless substances and are quite stable. However, these hydroperoxides are readily decomposed or altered and the resultant products contribute to the development of the 'off' flavours.

Reference had already been made to the important fact that about 40 percent of the fatty acids are unsaturated and present in the milk fat while dealing with its composition. These unsaturated fatty acid molecules are the centres of the oxidative reactions, which inturn results in the production of taints or flavours which have been described as 'oily', 'card board', 'metallic' or 'tallow' in accordance with their concentration or intensity. They are more often correctly described as 'oxidised' taints or flavours due to the fact that they are the products of oxidation. Undoubtedly, the double bonds of the un saturated fatty acids are the centres of these changes.

The fat absorbs oxygen to produce peroxide compounds which may rupture at the peroxide linkage to give the aldehydic substances, or may undergo molecular rearrangement to give rise to hydroxy ketones, after a period of induction of varying duration. The peroxides or decomposition products such as epihydrinaldehyde may produce the peculiar flavours and odours of oxidative taints.

It is suggested that epihydrinaldehyde is derived from linolic acid, which reacts with phloroglucinol in the 'kreis' test in the determination of the oxidised fat. The presence of peroxide compounds may be detected with their ability to liberate iodine from acid potassium iodide solutions. Acidity of fat, whether it may be lipolysis or by absorption of lactic acid is considered to be favourable to these oxidative reactions. Undoubtedly, the double bond of the unsaturated fatty acids, is the centre of attraction for all these changes, it must not be assumed that the saturated fats are incapable of oxidation. It is considered that in certain favourable conditions, ketones may be produced under the influence of peroxidase, however, such a possibility is a matter of doubt under the conditions prevailing normally in the milk.

It is a matter of consideration and most important to recognize that 'Lecithin' contains unsaturated fatty acid molecules that are susceptible to oxidation. It is also considered by many of the research workers that the oily taint originates in lecithin which is not only absorbed on the fat globules, but also present in the aqueous phase. These claims are further strengthened by the observation that the measurement of iodine value of the fat of the milk during the process of oxidation and that the separated milk has been known to develop this taint. It is reasonable only to expect the Iodine value of fat to decrease as the double bonds are broken, because of the fact that the absorption of iodine is a measure of the degree of unsaturation. However, the researches carried into this aspect of the problem have failed to establish this anticipation/theory. It is very significant to observe that there were marked decreases in the iodine values for the phospholipid (lecithin containing) fraction, however, the fat did not show any substantial changes in its iodine value during the development of an oily taint.

It is an established fact that the taint originates from the double bonds of the unsaturated fatty acids and some researchers believe and insists that the true fat is involved whereas others still maintain that lecithin is the reacting substance. Lecithin is responsible for the initial development of the oily taint in the early stages of oxidative process and similar reactions are able to be stimulated on the surface of the fat globules within the fat itself. Therefore, the powerful tallowy taint due to the advanced oxidation process, would therefore, represent the climax of the chain reactions that are commenced with the attack of the lecithin.

The oxidation of fat differs from lipolysis not only in the chemical reactions involved, but also the different conditions which governs the change. Obviously, oxygen is an essential requirement, whether it may be atmospheric or in the dissolved condition.

A stimulating effect is exercised by the light more particularly in the Ultra Violet wave lengths. Milk when it is exposed to direct sunlight, which is a very common cause of the development of an oxidised taint, wherein the dissolved oxygen is rendered much more active. The development of bacteria considerably reduces the amount or quantity of dissolved oxygen in milk, which explains the apparent anomaly that the taint is seldom developes in the milk of poor hygenic quality.

The occurrence of the taint is more frequent in winter than in summer and in the milk kept in cold store. Exposure of milk to sunlight is known to cause a rapid removal of dissolved oxygen due to its activation and absorption by the unsaturated fats. It is estimated that when the milk is de-aerated to a level of 0.4 to 0.5 mg oxygen per litre, there is no oxidation of fat or loss of ascorbic acid (Vitamin C) Evidence available to date indicates that, traces of the heavy metals, particularly copper exercises a powerful catalytic effect on these changes. The presence of protein favours the solution of the copper by the formation of copper proteinates, which are absorbed by the fat globules thus bringing the metal into contact with the fat. Iron, cobalt, nickel, chromium and manganese are also effective but to a much lesser degree than copper. Tin and alluminium have little or no effect. Copper is normally present in milk to the extent of 0.12 parts per million, however, if the concentration raises to 1.5 ppm an oily taint may result.

Many biological reactions are there, which demonstrate the existence of the inhibitory substances and oxidation of fat is one of them. A great deal of attention has been devoted in recent years to substances which prevent or delay the fat oxidation, which are designated as anti-oxidants, which are able to break the chain of reactions that occur during the fat oxidation. Thus in this process they themselves gets oxidized, thereby loosing their anti-oxidative powers, which occurs during the induction period, the duration of which is dependant on the nature and quantity of the anti-oxidants present. Therefore, they cannot be considered as negative catalysts owing to their inability to survive the reaction which they influence.

Vitamin-E, hydroquinone, and substances of the poly-hydroxy benzene type carotene, ascorbic acid (Vitamin C), lipase pancreatic enzymes, sulphites, potassium iodide, hydrogen-sulphide, and its derivatives, cereal flours and their extracts and cereal germ oils are some of the anti-oxidants, which are known to occur. Some of them are the normal milk constituents namely carotene, ascorbic acid, lipase, and hydrogen-sulphide and its derivatives which may be formed in heated milk. Some of these anti-oxidants are specific in their action, protecting one kind of fat only.

Reference has already been made to the more frequent occurrance of oily taint in the milk during the winter months due to the lack of supply green fooder in sufficient quantity assumes an important factor. The evidence is conflicting, however, it leads to the conclusion that the ration which provides an adequate supply of good quality green fodders is likely to furnish milk which is rich in carotene and ascorbic-acid, thereby protected to the extent against oxidation. It appears that the anti-oxidant properties of carotene rich milk are not due to the carotene but to some other compound closely associated with it.

The formation of sulphydril (-SH) compounds by the decomposition of serum or fat globule proteins is a matter of great concern, which is explained by the fact that when the milk is heated to 80^oC did not develop the oxidised taint. The greater resistance offered by the condensed milk might be explained on the similar lines. Addition of anti oxidants, particularly the extracts of oat flour and by de-aeration of milk in order to prevent the oxidation are partially successful and met with varying degree of success.

The mechanism of the development of the oxidized taint in copper contaminated milk has furnished vulnerable information pertaining to the protective nature of the ascorbic acid. The oxidised taint is developed in the milk which has been contaminated with copper, when the ascorbic acid is present in lower concentrations only, whereas, the higher concentrations were found to be protective in nature. The oxidised taint which is produced during the copper induced oxidation of ascorbic acid with the production of hydrogen peroxide, which was confirmed by the fact that addition of 0.1 percent hydrogen peroxide to the copper contaminated milk samples resulted in the appearance of the oxidized taint.

The mechanism of the development of the copper-induced oxidized taints can be explained as stated here under :

1. Reduction of cupric copper to the cuprous form with the simultaneous oxidation of the ascorbic acid.
2. Oxidation of the cuprous copper by the atmospheric oxygen with the simultaneous production of hydrogen peroxide.
3. Oxidation of the phospholipids by the hydrogen peroxide produces the oxidised taint.

Potassium iodide in concentrations greater than 0.1 percent results in the inhibition of the cycle of changes. The development of copper-induced oxidative taints in milk might also be produced due to the action of the tri-peptide - glutathione which enters the same cycle of reactions. However, the other active oxidation reduction substances, which are normal constituents such as thiamine, pyridoxine, and riboflavin do not enter into these reactions.

3.9 Substances Associated with Milk Fat

Lecithin: The biological significance of this compound, which is a member of the most important group known as the phospholipids - phosphoric acid + Fatty acids is better understood from the fact that they are present in all the living cells, more particularly the nervous and brain tissues as well as egg yolk. The cell protoplasm is a complex mixture of phospholipids, sterols and protein. Phospholipids in general are the compounds of glycerol, fatty acids both saturated and unsaturated and phosphoric acid which is attached to the nitrogenous base choline as a common member. Hence lecithin may be written as :

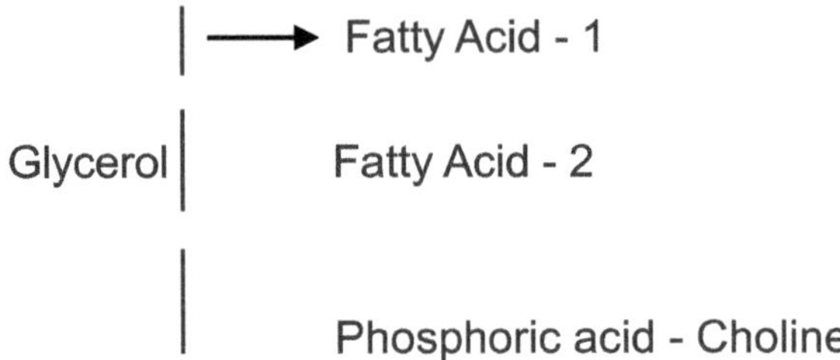

or more particularly to understand the chemical structure/nature it may be explained.

$$
\begin{array}{l}
CH_2\,(COOC\,R_1) \\
| \\
CH\,(COOC\,R_2) \\
| \\
CH_2 - O - \underset{\underset{O\quad OH}{/\!/\ \backslash}}{P} - O - CH_2 - CH_2 - N - (CH_3)_3\,OH
\end{array}
$$

When lecithin is subjected to complete hydrolysis it is broken down to produce all the four constituents namely glycerol, fatty acids, phosphoric acid and choline.

CH_2OH | $CH\,OH$ | CH_2OH Glycerol	$R_1\,COOH$ $R_2\,COOH$ Fatty Acids	$HO - \underset{\underset{O\quad OH}{/\!/\ \backslash}}{P} - OH$ Phosphoric acid	$CH_2\,OH\,CH_2\,N(CH_3)_3\,OH$ Choline

$CH_2OH\ CH_2N(CH_3)_3OH$ - Choline

Phospholipids said to possess some exceptional properties. They are intensely hydrophilic in nature having strong affinity to water, even though they are fatty in nature and soluble in fat and fat solvents. They imbibe large quantities of water and swell in a peculiar budding manner, when they came in contact with water, thus the lecithin and its compounds forms a link between the fat and the aqueous phase in the milk and cream. Lecithin is able to combine with both acids and bases because of the fact that it contains a basic nitrogenous group in addition to the free acid group of the phosphoric acid.

Three types of phospholipids are present in milk namely - Lecithin, cephalin and sphingomyelin approximately in the ratio of 9 : 5 : 1. An attempt has been made to summerise the concentration of the phospholipid fraction in the milk of cow and buffaloe in table 3.18.

Table 3.18 Concentration of Phospholipid Fractions in Buffaloe and Cow milk (mg/100 ml)

Phospholipid fraction	Concentration	
	Buffaloe Milk	**Zebu Cow Milk**
Lecithin	7.29	10.01
Cephalin	9.12	15.57
Sphingomyelin	4.63	8.13

Source : Outlines of Dairy Technology - S.K. De

Lecithin and sphinogomyelin contain choline radical, however, the sphingomyelin possess an additional nitrogenous base in its structure. Whereas cephelin contains an amino ethyl alcohol $CH_2OH\ CH_2\ NH_2$ which is united to the phosphoric acid instead of the choline as the nitrogenous base. Inspite of the fact that there exists some differences amongst them, it will be more convinient to group all the phospholipids of milk under the single heading 'Lecithin', because of the fact that it constitutes about 60 percent of the total. They are excellent emulsifying agents, and no doubt serve to stabilize the milk fat in an emulsion state even though fat soluble and at the same time hydrophilic in nature. The intense water binding capacity of these phospholipids assumes considerable importance in the moisture relation of dried milks. They combine with the protein of milk forming lecitho-protein complexes which in turn serve to bring the metals into contact with the milk fat, thereby initiating the process of fat oxidation. The lecithin content of the milk even though is very small, yet it assumes an important status due to their reactivity and outstanding versatality of the physical properties. It is evenly distributed between fat and aqueous phase as is observed from the table 3.19 given below.

Table 3.19 The Average Lecithin Content of milk and some Milk Products

Products	Percent
Milk	0.035 -0.036
Seperated Milk	0.015 - 0.018
Cream (41 percent)	0.182
Butter Milk	0.126 - 0.187
Butter	0.182 - 0.212
Whole Milk	0.24

Source : Outlines of Dairy Technology - S.K. De

Attention has been focused on the study of fatty acid constituents of the Lecithin. Contrary to the milk fat, there are no acids lower than myristic which the saturated fatty acids of the type Cn H_2n + 1 COOH are represented by myristic, palmitic, stearic and arachidic acids. Oleic acid predominates over the four or five unsaturated fatty acids that are present. Obviously has a provision for only Lecithin two fatty acid radicals there must be a number of these compounds just as there are many individual glycerides in milk fat. Earlier lecithin was considered to be heat sensitive but this view is no longer valid. Lecithin is considered to be partly responsible for the rich flavour of the butter milk and other dairy products.

Lipase more precisely lecithinases bring about the hydrolysis of lecithin which take place at any glyceryl ester bonds and at the choline phosphoric acid link. The products which are obtained on hydrolysis do not impart any taint unless and until the occurrence of oxidation which results in the production of two possible taints. Oxidation of choline produces a fishy taint, whereas an oily taint is developed when the double bonds of the unsaturated fatty acids are attacked. The fishy taint due to the oxidation of choline is assumed to be due to the substance named trimethylamine [$(CH_3)_3N$] which is known to possess a very powerful odour. However, this assumption is being questioned by several researchers owing to the fact that the slightly acid reaction of milk is not conducive to the existence of any free bases. Evidence available also indicates that the causative substance, is trimethylamine oxide, when combines with the unsaturated fatty acids produces compounds of fishy nature and these substances are characteristically more fishy than the trimethylamine itself.

Feeding of sugar beet by products such as molasses pulp results in the production of trimethylamine which is derived from betaine, a normal constituent causing fishyness as the base is diffused into the milk from blood. The factors which are known to favour the oily taint will also stimulate the fishyness due to the oxidative reactions that are taking place. Butter and milk powders, the two dairy products are more susceptible to this oily taint. Contamination with/sources of copper is known to act as a most powerful stimulant and all the factors are cumulative as in the case of oily taint.

Cholesterol : (C_{27} H_{45}OH)

Sterols - otherwise termed as solid alcohols - a group of substances in which the compound Cholesterol is an important member distributed widely in animal and plant cells similar to phospholipids, and present in the bile, brain, nervous tissues, blood and egg yolk. The chemical structure of the cholesterol is closely related to sex harmones, Vitamin D and saponins, containing the (1) Phenanthrene nucleus attached to a cyclopentane ring (2) Which inturn combines with the structural foundation using the chemical formula as represented.

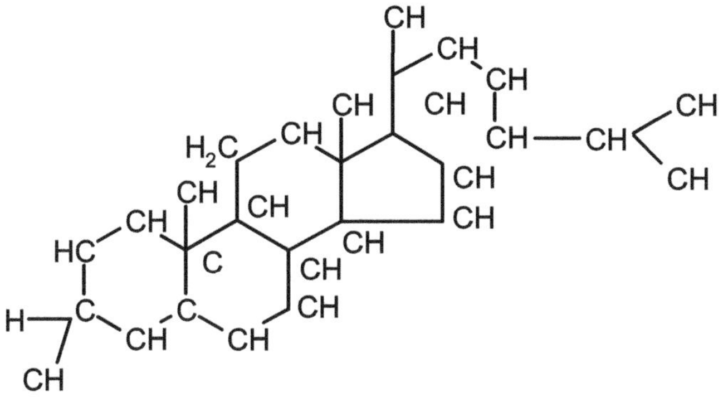

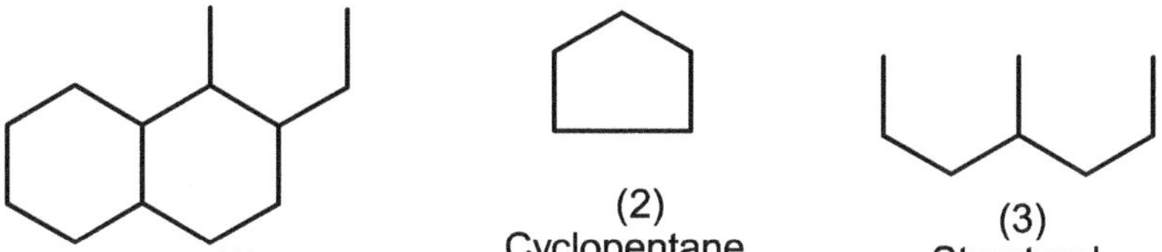

(1) Phenanthrene Nucleus

(2) Cyclopentane ring

(3) Structural Foundation

Generally all the sterols are having the same configuration, but differs in the number and position of their double bonds. In fact they are the outstanding examples, wherein, even a slight alteration in the ornamentation of the original structural design may lead to physiological differences which are of very significant in nature. Cholesterol is absorbed from the digestive tract, whereas, the sterols of vegetable origin are not. Yeast contains a sterol known as ergosterol, do not possesses any vitaminic properties, however, exposure to ultra-violet-light brings about a rearrangement of the molecule in which the second phenanthrenering foundation is opened by creating an extra double bond. Thus the resultant compounds are having the strong anti-rachitic properties which in fact one of the Vitamin-D (Calciferol).

They are essentially unsaturated monohydroxy alcohols capable of forming esters with the acids, even though, they are very complicated in their structure. Sterols are not affected by their saponification, so that the extraction of the residue with ether furnishes a method of isolating the sterols.

Lecithin is the name which is used to denote all the phospholipids in milk, similarly there are many vegetable sterols which are generally grouped together and are called phytosterols. Generally the ration of the cow contains vegetable sterols which do not pass into the milk, however, the differences in the physical properties between the cholesterol and the vegetable sterols is used as a measure to detect the presence of vegetable fats in butter. Cholesterol is insoluble in water, however the presence of lecithin assists it to be in solution. It is assumed that cholesterol is able to exercise some control over the intense hydrophilic action/nature of lecithin, thus the close association of these two substances is very interesting hence, some of the peculiarities of lipolysis are attributed to the fact that all these effects are very closely related.

Cholesterol, in combination with lecithin, undoubtedly plays an important role in maintaining the stability of the fat emulsion. Milk fat contains about 0.32 percent of Cholesterol which corresponds to 0.01 percent in milk having 3.2 percent fat and more than one sterol is present. When the milk is exposed to ultra violet rays, there is an increase in the anti rachitic properties, because of the closely related structure of cholesterol and vitamin D. It is assumed that the presence of $C_{27}H_{43}OH$ the dihydro cholesterol on irradiation is probably transformed into vitamin-D. Ergosterol which might be present $C_{28}H_{43}OH$ passes into calciferol by the same process of molecular rearrangement, and these transformations are shown in the formulae as given under.

(A) Dihydro Cholesterol : $C_{27}H_{43}OH$

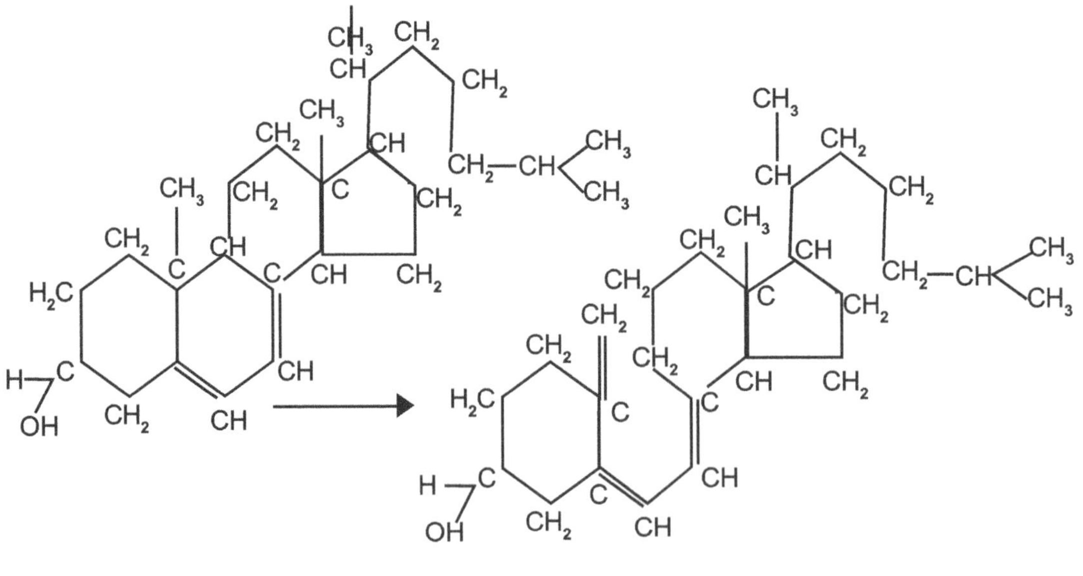

(B) Vitamin - D : $C_{27}H_{43}OH$

(C) Ergosterol : $C_{28}H_{43}OH$

(D) Calciferol : $C_{28}H_{43}OH$

Carotene

Cows milk which is very rich in fat soluble pigment, carotene and to some extent xanthophyll imparts a yellowish taint whose intensity is dependent on various factors such as breed, feeds, fat globules percentage of fat in milk etc. Certain breeds of cows imparts a deeper yellow taint to their milk than others. The greater the intake of green feeds the deeper yellow the colour of cows milk. The larger the fat globule and the higher the fat percentage the greater the intensity of the yellow colour. Carotene is one of the pigments present in the green leaves and the pure carotene possesses a red brown colour, whereas in dilute fat solutions it gives a rich yellow colour. It is insoluble in water, therefore, is able to colour only the fat phase of milk. Along with xanthophyll it is a member of an important family of compounds, carotenoids, having a wide distinction in the vegetable kingdom just like lecithin and cholesterol. Carotene and xanthophyll are the yellow pigments of the green leaves.

Carotene has an emperical formula $C_{40}H_{56}$ which in its structure consists of two six membered carbon rings connected by a long hydro carbon chain $C_{22}H_{26}$ having nine double bonds. Each terminal ring contains a double bond and it is evident that the compound is highly unsaturated. There are many possibilities of isomerism in view of the fact that the structural formula reveals the fact that the compound is identically symmetrical on either side and the formula shown is that of β-carotene.

It has been suggested that phytol a green pigment which is the active agent in photo synthesis and a component of the chlorophyll molecule is said to have similar structure as that of carotene and the milk fat contains about 0.3 mg percent. The

C-CH:CH-C:CH:CH:CH:C:CH:CH:CH:CH-C-CH:CH:CH-C-CH:CH-C

β-Carotene C_{40} H_{56}

β-Carotene

C -- CH:CH-C:CH:CH:CH:C:CH:CH2OH

Vitamin - A (2 molecules)

α-Carotene C_{40} H_{56}

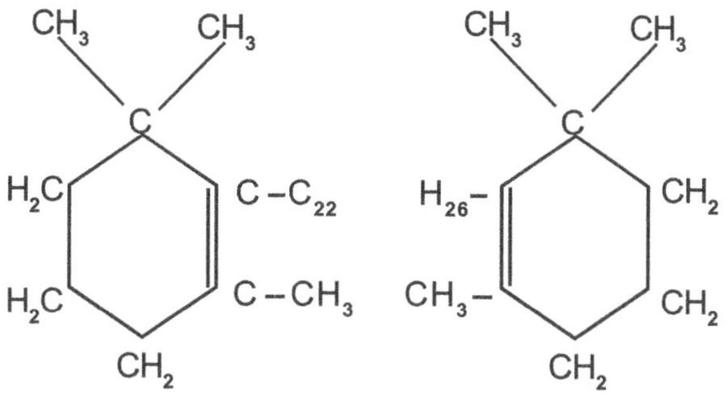

CH_3 CH_3 CH_3 CH_3
C C
H_2C $C-C_{22}$ $H_{26}-$ CH_2
H_2C $C-CH_3$ CH_3- CH_2
CH_2 CH_2

Xanthophyll C_{40} H_{54} $(OH)_2$

extensive fluctuations in the carotene content of milk are due to the variation in the content of the ration. Small amounts of xanthophyll accompany carotene in milk, however, the latter predominates in its content. The importance of carotene in milk fat is mainly due to the fact that it is known to be the precursor of Vitamin-A but not to its colouring properties or its possible role as an antioxidant. It is evident from the perusal of the structural formulas of β-carotene, which gives two molecules of Vitamin-A, while only one molecule is produced from α-carotene, whereas the closely allied xanthophyll pigment did not yeild any molecule.

Vitamin-A exercises a great deal of importance in human nutrition. The retina of the eye contains carotenoid protein and it may be expected that lack of carotene or Vitamin-A in animal or human nutrition will have a serious effect on the eye sight.

3.10 Milk Proteins

Nitrogenous Compounds of Milk

Without any exaggeration proteins are the most complex substances and their precise constitution still remains to be established inspite of the fact that there are many modern methods are available for investigation. A general understanding of this important group to which they belong will help to place them in their true perspective, before commencing a detailed study. Proteins in general contain the elements Carbon, Hydrogen, Nitrogen, Sulphur and Phosphorus. The approximate composition of proteins is as under.

Element		Percentage
Carbon	-	52%
Hydrogen	-	7%
Oxygen	-	22%
Nitrogen	-	16%
Sulphur	-	2%
Phosphorus	-	1%
Total		100%

Proteins are vital for all the living organisms, because they constitute an indispensable part of the individual body cell. Proteins are composed of a large number of amino-acids some of which are essential while the others are termed as non-essential. Normally milk contains approximately 0.5 percent of nitrogen which corresponds to an average of 3.3 percent (range 3.2 - 3.4) protein which is distributed within a number of compounds as shown in the flow diagram 3.2.

Flow Diagram 3.2

Nitrogenous Compounds (100 Parts)

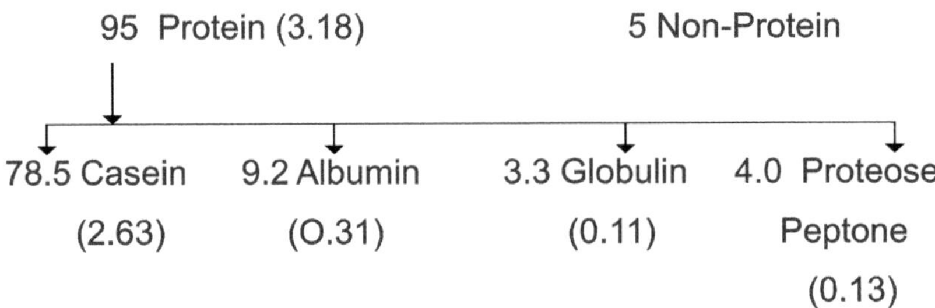

Proteins on hydrolysis yields a mixture of amino acids of which glycine (amino-acetic acid) may be considered as the simplest example. The general formula RNH_2COOH may be considered to represent the group provided that it is realised the variations occur in the numbers of amino (NH_3) and acid (COOH) groups. These amino acids are able to combine with the elimination of water molecule, in the similar manner as the simple carbohydrates unite to form the complex starches for example.

$$nC_6H_{12}O_6 \text{ -- } nH_2O == (C_6H_{10}O_5)_n$$

Two molecules of the same or different amino acids combine with the elimination of water from the COOH group one and the NH_2 group of the other as explained by the example.

$$NH_2\text{--}CH_2\text{--}CO\text{-}OH + H\text{-}NH\text{--}CH_2\text{--}COOH == NH_2CH_2CONHCH_2COOH + H_2O$$

Glycine Glycel Glycine

It is very interesting to observe that the resulting di-peptide still retains the antagonistic groups (NH_2) and (COOH), so that the building process of more complex molecules may proceed in the same manner. The fact that the NH_2 group of amino acids possesses basic properties, whereas the COOH is acid by nature hence these are able to combine with both acids and bases. This amphoteric reaction is shared by the proteins, however complex they may be, some of these reactive groups are exposed, thus allowing the proteins to manifest the amino acid properties.

The proteins behave as colloids and shows the phenomenon of adsorption due to their enormous molecular dimensions. Proteins are solid by nature but they are able to imbibe considerable quantities of water thereby the particles are in the fluid state, thus holding the water in bound form. Ultimately, the particles of colloidal protein solutions are very concentrated proteins/water mixtures possessing gelatinous or fluid properties and such colloidal solutions are termed as emulsoids.

Proteins exhibit a characteristic feature due to their ability to unite with metals, and the precise mode of combination is still obscure, however, it is assumed that the methods are simply united to one or more of the COOH groups or undoubtedly they may be adsorbed by the protein surface. Solubility of majority of the metals in water is considerably increased by the presence of protein due to the fact that the metal proteinate disturbs the equilibrium, thereby more metal goes into solution in an endeavour to restore the balance. Proteins function as milk buffer compounds due to their amphoteric nature and are able to resist to some extent the change in the hydrogen ion concentration (pH) which results from the addition of either acids, or bases to equivalent quantities of water. Proteins do manifest a very slight degree of ionisation with the exposed outer groups even though they are in colloidal state in the solution. Every protein has a certain pH where the ionisation is at minimum and the protein is least soluble, hence its separation from the solution is most complete and this particular pH is known as the iso-electric point. The important nutritive function of the proteins is to supply a number of amino acids which are in turn transformed by the aminal into its body or milk proteins. Therefore in the consideration of the milk proteins to their high nutritive value lies in the fact that they supply the essential amino acids and are able to rectify the deficiencies that are common in the human and animal food.

Casein

It is the name of the product which is obtained by the precipitation of milk at pH 4.6. British workers have used the term 'caseinogen' to denote milk protein and casein which is obtained in the production of cheese curd protein, whereas the Americans have used the name casein for the milk protein and paracasein for the casein of the cheese curd protein. In order to remove the confusion over the nomenclature of the casein used by both the British and the American systems it is proposed to follow the American system only in the text book in view of the fact that there has been a general drift towards this system. The elementary composition of casein, along with albumin and globulin has been presented in the table 3.21, which reveals that the casein belongs to the phospho proteins. Reference has been made to the composition of the proteins in general and these are in confirmity with them.

Table 3.20 Composition of the Milk Proteins (in Percentage)

Name of the Protein	Carbon	Hydrogen	Oxygen	Nitrogen	Sulphur	Phosperous
Casein	53.50	7.13	22.14	15.80	0.72	0.71
Albumin	52.51	7.10	23.04	15.43	1.92	--
Globulin	51.88	6.96	24.62	15.44	0.86	0 .24

Source : Text Book of Dairy Chemistry - M.P. Mathur, D.D. Roy & P. Dinakar

The concentration of various proteins in the Cow and Buffaloe milk has been given in the table 3.21.

Table 3.21 Concentration of various proteins in buffaloe and Cow milk (g/100 ml)

Poly Peptide	Concentration in milk		
	Buffaloe	Zebucow	Exotic cow
α S_1 Casein	1.44 - 1.8	1.04 - 1.3	1.2 - 1.5
α S_2 Casein	0.22 - 0.26	0.26 - 0.34	0.3 - 0.4
β Casein	1.26 - 1.58	0.94	0.9 - 1.1
γ Casein	0.16	0.15	0.1 - 0.2
κ Casein	0.43 - 0.54	0.3 - 0.37	0.3 - 0.4
β Lactoglobulin	0.39	0.31- 0.38	0.2 - 0.4
α Lactolbumin	0.14	0.71- 0.1	0.1 - 0.15
Proteose - Peptone	0.33	0.24	0.6 - 1.8
Serumalbumins	0.029	--	0.1 - 0.4
Lactoferrin	0.032	0.005	--

Source : Text Book of Dairy Chemistry - M.P. Mathur, D.D. Roy and P. Dinakar

The valuable nutritive properties are believed to be attributed to its exceptional acid qualities due to the inclusion of the phospheric acid in its molecule and its ability to form salts with both acids and bases of which the acid nature predominates in milk. Casein milk is united to calcium partially nullifying the acid properties and the resulting compounds being an acid calcium caseinate, more correctly calcium, hydrogen caseinate.

The acidity of the milk is determined by titrating with sodium hydroxide and in doing so some of the solution is being used to neutralise casein irrespective of the presence or absence of lactic acid. Therefore the addition of sodium hydroxide to milk neutralises calcium hydrogen caseinate in a similar manner as its addition to a solution of acid sodium phosphate will result in the formation of saturated trisodium phosphate.

Calcium acid caseinate + NaOH === Calcium -- Sodium -- Caseinate + Water

Approximately 0.32 ml of 1N NaOH per gram of casein is required to neutralise the acidity of casein and on this basis it appears that about one third of the acid groups of casein are normally free in milk. The development of lactic acid results in a gradual reduction of calcium from the colloidal to the more soluble condition, which continues until the milk curdles. The following equation will explain the process of removal of calcium from calcium hydrogen caseinate which inturn represents the state of affairs at the iso-electric point at pH 4.6.

$$\underbrace{\text{Calcium hydrogen caseinate + Lactic acid}}_{\text{Colloidal}} == \underbrace{\text{Acid Casein}}_{\text{Precipitate}} + \underbrace{\text{Calcium lactate}}_{\text{Soluble}}$$

There is a progressive removal of the phosphate from the surface of casein, due to the development of lactic acid as explained by the following equation :

$$Ca_3\,P_2\,0_8 + 4CH_3\,CHOH\,COOH == \underset{\text{(Soluble)}}{Ca\,H_4\,P_2\,0_8} + \underset{\text{(Soluble)}}{2(CH_3CHOHCOO)_2\,Ca}$$

The soluble phase is enriched at the expense of colloidal state and on this occassion the mono calcium phosphate enters the soluble state along with the calcium lactate. It is not appropriate at this stage to discuss about the development of lactic acid, and it may be advantageous to summerise the effects briefly as under:

1. There is a decrease/reduction in the lactose concentration.
2. Progressive conversion of colloidal tricalcium phosphate attached to casein into soluble mono calcium phosphate and calcium lactate.
3. Gradual removal of calcium from calcium hydrogen caseinate to form calcium lactate and more acid casein.
4. There is an increase in the osmotic pressure, consequently resulting in the decrease of freezing point due to the effects mentioned above.
5. There is an increase in the hydrogen ion concentration i.e. the pH figures show a decrease due to the development of acidity.
6. Precipitation of casein occurs at pH 4.6.

The different types of casein and their manufacture and uses will be discussed separately later while studying the by-products of milk. However, it is appropriate to have a brief knowledge about the edible casein. Edible casein may be defined as casein, which has been isolated from skim milk by taking special precautions to ensure its suitability for use in patented foods and pharmaceutical preparations.

The main and important precautions which are concerned are:

1. Strict control on the quality of the raw material (i.e. skim milk)
2. Use of standard equipment and correct technique for production
3. Maintenance strict hygienic conditions on production and
4. Usage of good packaging material and storage under recommended conditions.

The B.I.S. specifications for the edible quality of the casein is given in the table 3.22.

Table 3.22 B.I.S. Specifications for casein (edible quality) (IS : 1167 : 1965)

Characteristics	Requirement	
Moisture (% wt) Max.	..	10.0
Total ash (% wt) Max. (on dry basis)	..	2.5
Acid insoluble ash Max. (on dry basis)	..	0.1
Fat (% wt) Max (on dry basis)	..	1.5
Nitrogen (% wt) Max. (on dry basis)	..	14.5
Total acidity - interms of 0.1 NaOH (NaoH Per g)	..	6.14
Free acidity - interms of 0.1 NaOH (per 10 g) Max	..	5.6
Bacterial Count (per g) Max	..	50,000
Coliform Count (per g) Max	..	10
Mould count (per g) Max	..	50
Rate of Solubility	to confirm to test	

Source : Outlines of Dairy Technology - S.K. De

Preparation

Adopting the special precautions mentioned above, edible casein is prepared on the similar lines to those followed for acid casein - preferably grain curd - hydrochloric. These include:

1. Precipitation of curd at pH 4.1-4.3
2. at least a minimum of 3 separate washings of curd in water of proper pH with a contact period of 15-20 minutes each time
3. last but not the least washing with hot water at a temperature of 71-77°C (160 - 170°F) to give an effect of pasteurization of curd for reducing the bacterial count and
4. the last washing with neutral water at 41°C (105°F) etc.

Uses

Edible casein in its original form or as sodium caseinate is used in various food products such as:

(1) Ice cream (2) Coffee Whiteners (3) Imitation milk (4) Whipping powders (5) Instant breakfast (6) Water binders in sausages and (7) Protein hydrolysates.

Lactalbumin and Globulin

The chemical composition of these two proteins has been the subject of discussion, in order to establish the identical nature with the corresponding albumin and globulin of blood and colostrum. The globulin of milk and blood are identical whereas the respective albumins even though similar in their composition they are not identical. The mammary gland appears to allow small quantities of globulin to filter into the milk from blood stream, however, the lactalbumin is obliged to elaborate. Both albumin and globulin present in the colostrum are the same as those present in milk. However the quantity of globulin is higher in the colostrum. The concentration of globulin is as low as 0.1 percent, whereas it may be as high as 6 percent which is high enough to cause visible coagulation on boiling and this property is lost as the globulin content declines gradually to the normal level.

These two proteins constitutes about 1/8th of the total, and generally described as soluble proteins of milk which is considered as the superficial distinction only. Even though, both are dissolved in milk, they differ in that the albumin is soluble in distilled water whereas the globulin is insoluble in water but soluble in dilute salt solutions. Globulin appears to contain phosphorous belonging to a phospholipid. Both albumin and globulin are coagulated by heat, which is governed by time of holding, temperature, salt concentration and pH of the solution. Coagulation commences at 65 °C and increases with extent as the temperature and time of holding are increased at a pH of 4.5 which is considered as the most favourable reaction for the change. Some of the heat coagulated protein re-dissolves as the temperature exceeds 90 °C thus imparting a typical cooked flavour by the heated milk. Both the proteins possesses a high nutritive value, supplying all the essential amino acids. The low sulphur content of the casein is rectified to some extent by the albumin. The nutritive value of the combined milk proteins is higher than the one measured separately and are complimentary to each other.

Other Proteins in Milk

Casein, albumin and globulin, when removed from milk the filtrate still contains a nitrogenous substances consisting some proteose - peptone substances, which are considered as the normal constituents of milk and are precipitated by saturation with ammonium sulphate. The presence of these substances in the milk is debatable since they are the products produced by the degradation of protein. Always there existed a doubt whether these compounds occurs naturally in normal milk or as a result of the analytical separation of the proteins. It is now generally agreed that they are the normal constituents naturally present in the milk.

It has been generally accepted that milk contains some traces of other proteins. The study of fat globular surface has revealed the presence of a riboflavin protein phosphoric acid complex. Possibly there are more than one flavo protein which are reddish brown in colour, but colorless in the reduced condition. The main reactivity of these compounds in oxidation/reduction changes makes them the centre of attention. It is also interesting to assume that they are likely to participate in the chemical reactions that are associated with fat oxidation. Enzymes which are present in milk are the other substances of protein in nature, undoubtedly their total percentage weight is very small, their importance cannot be ignored, because being catalysts, they are able to bring about important reactions even in very low concentration.

The behaviour of these three proteins to some common laboratory reagents is presented in the table 3.23.

Table 3.23 Behaviour of Milk proteins to Laboratory Reagents

Reagent	Casein	Albumin	Globulin
Rennet	Coagulated	Not coagulated	Not coagulated
Heat	Coagulated above 100 °C	Coagulated below 100 °C	Coagulated below 100 °C
Cold dilute acids at pH 4.6	Precipitated	Not Precipitated	Not Precipitated
Excess of ammonium Sulphate	Precipitated	Precipitated	Precipitated
Equal volume of saturated ammonium sulphate	Precipitated	Not Precipitated	Precipitated
Excess of Sodium Chloride	Precipitated	Not Precipitated	Precipitated
Excess of Magnesium Sulphate	Precipitated	Not Precipitated (unless acid)	Precipitated

Source : Text book of Dairy Chemistry - Edgar R. Ling

Non-Protein Nitrogenous Substances

A variety of substances which are included in this group such as, amino acids, certain vitamins, urea, uric acid, ammonia, creatine and creatinine etc and constitute about 5 percent of the total nitrogen content of milk, which is equal to about 0.025 percent nitrogen suggesting that the quantities of each member is very minute. In normal circumstances they are filtered into milk from blood along with globulin. Physiological disturbances might cause a higher permeability of the membranes, thereby increasing the content of soluble nitrogen in a considerable manner and thus disturbing the normal value of the casein number which is explained as

$$\frac{\text{Casein N}}{\text{Total N}} \times 100$$

The concentration of same of the non-protein nitrogenous substances has been detailed in table 3.24.

Table 3.24 Concentration of some minor constituents in Buffaloe and Cow milk

Constituent	Concentration in milk		
	Buffaloe	Zebucow	Exotic cow
Choline (mg / 100 ml)	-	-	13.7
Folacin (mg / 100 ml)	-	-	0.0059
Myo-Inositol (mg / 100 ml)	-	-	11.0
Non-protein Nitrogen (mg /100 ml)	-	-	22.9 - 30.8
Urea Nitrogen (mg /100 ml)	-	-	8.4 - 13.4
Creatine Nitrogen (mg /100 ml)	2.5	1.6	0.6 - 2.0
Uric Acid Nitrogen (mg /100 ml)	24	37	0.5 - 0.8
Taurin (moles/100 ml)	5.9	4.1	4.1
Orotic Acid Nitrogen (mg /100 ml)	19.5	52.6	1.2 - 1.3
N acetyle neuraminic acid (mg /100 ml)	--	-	12 - 27
Lactic acid (mg /100 ml)	--	--	3.4 - 10.4
Acetic acid (mg /100 ml)	--	--	0.3 - 5.0
Formic acid (mg /100 ml)	--	--	1.0 - 8.5

Source : Text Book of Dairy Chemistry - M.P. Mathur, D.D. Roy and P. Dinakar

3.11 Mineral Matter of Milk

Evaporation of milk to dryness followed by its ignition, a grey ash is obtained, which on analysis reveals the presence of phosphates, chlorides, potassium, calcium and sodium in appreciable amounts, with smaller amounts of magnesium sulphate, carbonate, and traces of iron. Very minute traces of copper, alluminium, boran, zinc, manganese, and silicon have also been reported to be present in the ash content of milk. However, it is necessary to study the precise relationship between the ash and the milk from which it is obtained. The process of ashing has distinctly altered the original mineral balance making it as basic in nature because of the fact that milk is normally but slightly acidic. The alterations which might be occurred during the process of ignition of the ash content of milk are also follows :

1. Phosphates of the ash have been augmented by the P_2O_5 derived from organic phosphorous.

2. Sulphates have been derived from the burning of the proteins.
3. Carbonates are obtained by the union of CO_2 with the bases.
4. Carbonates are also obtained by the conversion of the citrates present in milk on ignition.
5. Loss of alkaline chlorides might occur due to volatilisation, unless the temperature of ignition is kept below dull red heat.
6. Traces of iodine which is normally present in milk is lost in ashing.
7. Normal milk on an average gives about 0.76 percent ash.

Based on the very few complete mineral analyses that have been made by different investigators, a composite picture must serve in presenting the complete mineral picture of the ash content of the milk obtained on ignition as given in tables 3.25 and 3.26.

Table 3.25 Concentration of Major Minerals in Buffaloe and Cow milk

Elements	Concentration mg/100 ml		
	Buffaloe	Zebucow	Exotic cow
Major Minerals			
Calcium	183.9	114.2	123
Magnisium	19.02	11.0	12
Sodium	44.75	50.0	58
Potasium	101.6	148.0	141
Phosphate	88.74	84.0	96
Citrate	177.6	166	160
Chloride	63.82	106	119

Source : Text Book of Dairy Chemistry - M.P. Mathur, D.D. Roy and P. Dinakar

Table 3.26 Concentration of Trace elements in Buffaloe and Cow milk

Elements	Concentration mg/100 ml	
	Buffaloe/Zebucow	Exotic cow
Trace elements		
Boron	52 - 14.5	27
Cobalt	0.69 - 1.61	0.6
Copper	7 - 21	13
Iron	42 - 152	45
Manganese	38.2 - 65.8	22
Sulphur	15700 - 31400	30000
Flouride ppm	12.3 - 18.5	20 - 180
Iodine ppm	8.5 - 19.4	5 - 50
Zinc	147 - 728	390

Source : Text Book of Dairy Chemistry - M.P. Mathur, D.D. Roy & P. Dinakar

It may not be possible to construct the complete salt balance, a more practical approach to problem may be made by the consideration of the distribution of the major constituents between the soluble and colloidal phases. The chlorides of potassium and sodium are in true solution while the phosphates of calcium and magnessium are partly colloidal and partly soluble. As already stated, that a considerable amount of total calcium and phosphate are associated with the protein casein, and thus appear to be in colloidal phase. However, the distribution of calcium and phosphorous in milk is summarised as a flow chart 3.3 for better understanding.

Flow Chart 3.3 Distribution of Calcium and Phosphorous in Milk

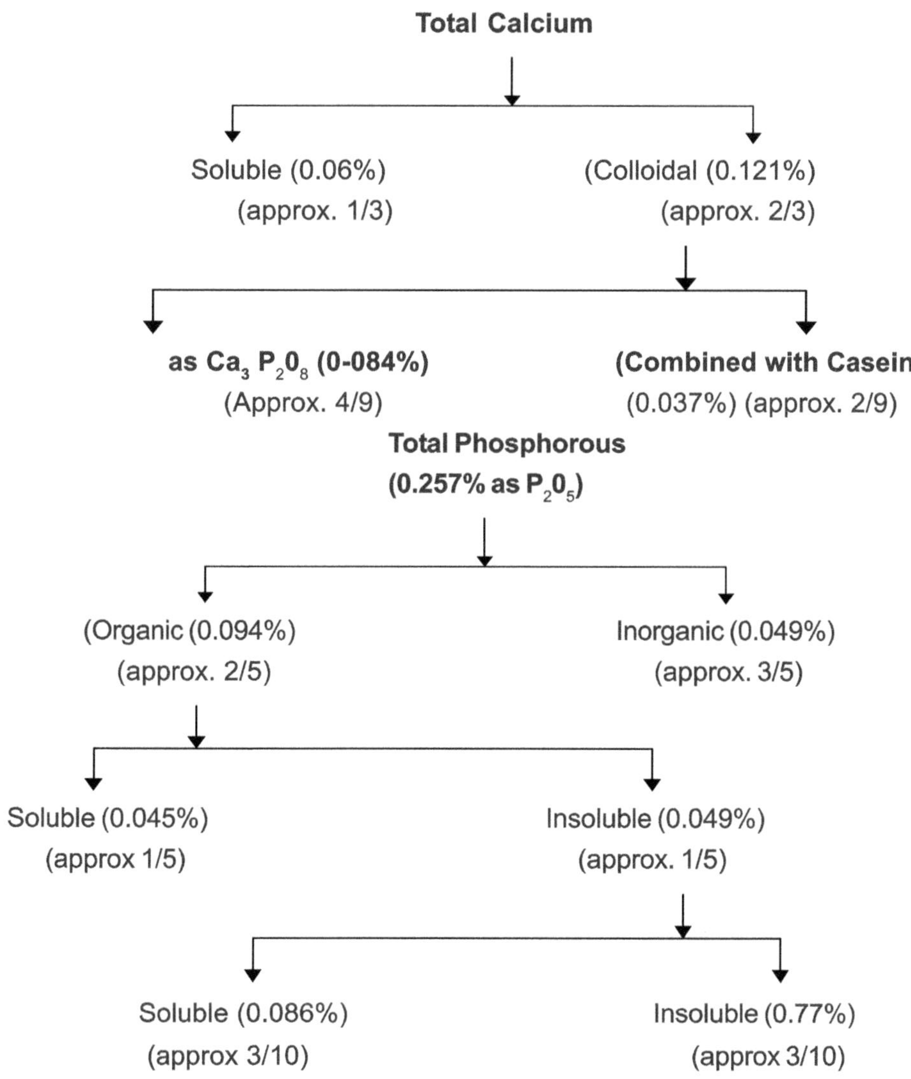

The presence of soluble organic phosphorous is accounted for by the phospholipid fraction, and by the unidentified esters of the phosphoric acid and to some extent by the water soluble vitamins even though the amount is extremely very small.

The rich nutritive properties of milk are attributed to the mineral constituents. All the mineral elements of milk are essential for nutrition, however, it is necessary to emphasize that some minerals such as iron and iodine are not present in quantities which would justify the extravagent claim that milk is a perfect food.

3.12 Vitamins of Milk

The pioneer research and the experiments that have been carried out in isolating the various factors which have been directed to the problem in studying their physiological functions and chemical nature, and the possibilities of their synthesis either in the laboratory or on a commercial scale. The chemical nature of those vital factors, when obscure were called as "fat soluble A" and "Water soluble B" etc. These substances are now termed as vitamins and their chemical structure is known and is hoped that they will be called by their chemical names because of the fact that the alphabetical system is not the same universally.

The vitamins that are present in the milk are:

Fat soluble - A,D,E, and K.

Water soluble - B complex consisting of B_1, B_2 Nicotinic acid, Pantothenic acid, Pyridoxine etc. and C (Ascorbic acid).

It is necessary and essential to distinguish, between those which are merely present and those present in sufficient quantity to make the milk a valuable source of supply, while considering these Vitamins. Milk is regarded as an important supply of vitamin A and the B groups at least as far as the human nutrition is concerned even though all the above mentioned vitamins are present. An attempt has been made to summarise the availability, level, significance, sensitivity and the daily requirement of the various important vitamins that are present in the milk and milk products for the humans has been presented in the table 3.27.

Vitamin-A

A great deal of interest has been generated, due to the recognition of the fact that the yellow pigment carotene is the precursor of the vitamin - A. This vitamin is necessary to maintain the tone and health of the mucous membranes of the nose, throat, lungs, ears and genital organs and its absence brings the changes by the process of 'hardening' or the "keratinisation" rendering them prone to infection. During the discussion of carotene a reference has already been made that the retina contains a carotenoid protein and lack of vitamin - A or carotene results either directly or indirectly visual disorders such as night blindness and xeropthalmia.

The need for vitamin - A is more and greatest during the growth of the young ones, however, it is required to maintain good health in the adults.

Table 3.27 Important Vitamins in Milk

Type	Available in [mg/l or mg/kg]	Level	Significance	Sensitivity	Daily Require-ment for humans (mg)
Fat soluble					
Vitamin A Retinol Provitamin Carotene	Cow's milk Summer butter Winter butter cheese	0.02 - 0.2 1-20 0.2 - 5	Results in yellow colour of butter, if deficient causes skin and eye damage (night blindness)	To UV rays and oxygen, sur-vives braising	1-2
Vitamin D Cholecalciferol Provitamin Ergosterol	Cow's milk Butter	0.0002 - 0.0004 0.0004 - 0.002	Promotes tooth growth and bone reconstitution when deficient, causes rachitis and infections	Not sensitive to oxygen and heat	0.01
Water soluble					
Vitamin B2 Lactoflavin Riboflavin	Cow's milk Whey Milk powder Whey powder Kefir powder	0.2 0.1 - 0.27 1 - 3 2 - 6	Discoloration of whey (green-yellowish), important for growth, esp.for lactobacilli when deficient, intestinal illness	Not sensitive to oxygen and heat	2 - 4
Vitamin B6 Ademin	Cow's milk Whey	0.01-0.15 0.1	when deficient, nervos-ness and insomnia	Not sensitive to heat and acid, light sensitive	2 - 4
Vitamin C Ascorbic acid	Cow's milk Whey	0.5 - 2.8 1.0	when deficient, apathy during spring, infection sensitive, tooth decay	Sensitive to heat and light	30 - 100

Source : Dairy Science and Technology Pieter Walstra et.al

The Vitamin is purely vegetable in its origin and the animals cannot synthesize it and excess vitamin is stored in the liver. The sources of vitamin - A are liver oil, milk fat, egg yolk, green foods, and carrots. vitamin - A is fairly heat stable, neither pasteurization nor sterilization does not cause any appreciable reduction in its quantity. Similarly condensation and drying have no effect on vitamin-A, however, both carotene and vitamin-A are susceptible to oxidation due to the fact that both are highly unsaturated.

Vitamin of the B Group

In the earlier days, it was considered that the water soluble B was found to be one substance, however, the later research has resolved the issue by stating that it consits of two groups one the heat labile (B_1) and the second heat stable (B_2). Consisting of a mixture of number of Vitamins. The flow chart 3.4 gives the details of the vitamin -B complex group.

Flow Chart 3.4 Distribution of 'B' Vitamins

Water Soluble - B

- Heat labile (B1 Aneurin or Thiamine)
- Heat Stable Original B2
  - Riboflavin Vit-B2
  - Nicotinic acid (Niacin)
  - Pantothenic acid
  - Pyridoxine
  - Vitamin B12
  - Others

Vitamin B1 - (Aneurin or Thiamine)

This Vitamin is formerly known as anti-beriberi factor, in the absence of which causes a kind of paralysis beri-beri in humans and polyneuritis in the poultry and animals. The important function of this vitamin is growth stimulation, maintenance of the tone of the aliminetary canal, and is known to participate actively in the carbohydrate metabolism. In milk it is present partly in diffusible state either freely or combined with phosphoric acid and partly fixed as compounds of proteins in either of these forms as stated below.

N═C—NH_2 HCl | CH_3—C, C—CH_2—N, C═C—CH_3, CH_2, OH, CH—S | N—CH

The rich sources of this vitamins are the whole cereal grains, eggs and yeast. In the grains it is situated in the germ or adjacent layers which is removed from the grains during the process of milling.

Riboflavin

Vitamin B2 - This Vitamin is known to exist as a phosphoric acid riboflovin protein and is identified with the green-yellow pigment of whey. Colostrum contains higher concentration (0.4 mg percent) which declines very rapidly to the normal level in the milk. It is present in liver, Kidney, eggs, and green foods. The growth-promoting function is identified with the oxidation-reduction changes, where its role is either catalyst or co-catalyst. Riboflavin is known to act as a sensitizer, in the destruction of Vitamin-C by oxidation under the influence of light. The structural formula of Riboflavin is as under.

$CH_2-CHO-CHOH-CHOH-CH_2OH$

CH C N CO

CH_3 C

CH_3 C NH

C C

CO

Riboflavin Vitamin B2

Nicotinic acid and its Amide

This was identified as the substance responsible to prevent the disease known as pellegra. This disease is manifested with serious skin disorders, gastro intestinal symptoms and diarrhoea. Liver, kidney, milk, eggs and whey are good sources of nicotinic acid or Niacin. Nicotinic acid and/or its amides are significantly associated with riboflavin in a number of oxidation-reduction enzymes and the structural formula of the nicotinic acid and its amide are as under.

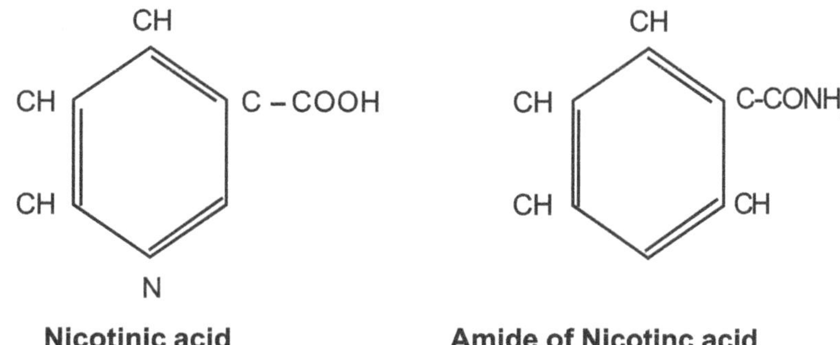

Nicotinic acid **Amide of Nicotinc acid**

Pyridoxine

This substance is known to prevent the pellegra in rats. It is a member of the B complex group and is present in milk. The structural formula is as under.

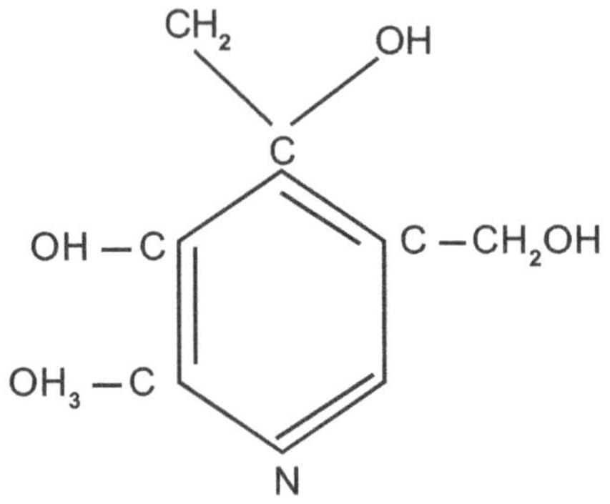

CH_2 OH

C

OH – C C – CH_2OH

OH_3 – C

N

Pyridoxine

Pantothenic acid

It is identified as the anti-chick dermatitis factor and is present in the milk in small quantities.

The B complex group of water soluble vitamins to some extent are synthesized in the rumen/intestines by its microorganisms. However, the amounts of these substances that are synthesized are not able to meet their own requirement hence they need to be supplemented through the feed. The B complex as a group are much more heat stable than thiamine or vitamin B1 and the others are unaltered by such treatment or by condensing and drying.

Vitamin-C (Ascorbic acid)

The amount of ascorbic acid present in the milk is approximately 1-2 mg percent which is further reduced by oxidation and exposure to light. Vegetables and fruits supply the majority of the requirements of ascorbic acid in human beings. 'Scurvy' a disease which is known in the earlier times is due to the lack of ascorbic acid or vitamin-C in the diets of marieners or explorers, who were compelled to subsist on highly artificial diets for longer periods. The symptoms of acute 'scurey' are weekness of the blood capillaries with a tendency to rupture resulting in haemorrhage or bleeding.

It undergoes oxidation, very easily hence the processing of milk bound to cause losses, pasteurization causes 20 percent, sterilization 50 percent, drying 20 to 30 percent losses where as, in the preparation of evaporated milk the losses are up to 60 percent, however, the lossess are much less severe in the making of sweetened condensed milk (15 percent). Presence of traces of copper and visible light accelarates the removal of ascorbic acid from milk. Ultra violet light along with the

Riboflavin is a potent accelerator and these factors are apparently ineffective without dissolved oxygen. The chemical structure of vitamin C is presented here under.

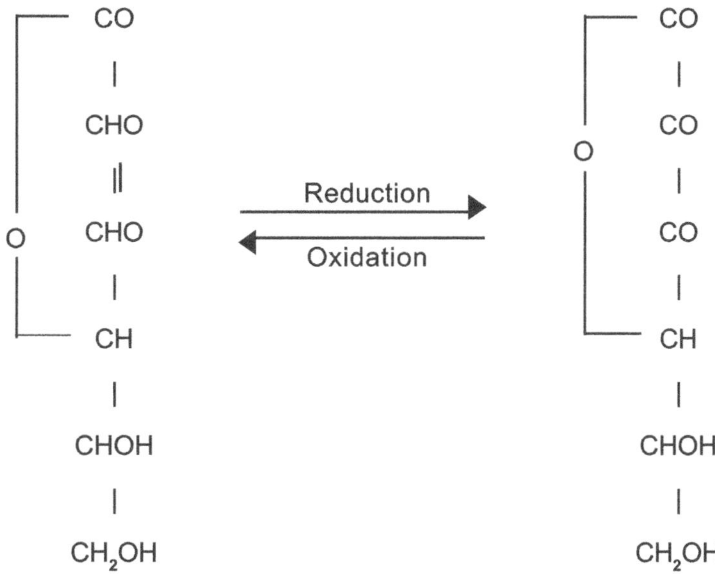

The known reversibility of the reaction and the case with which the ascorbic acid suffers the oxidation suggests that like the B-Complex group, one of the biological function of this compound is to assist in the oxidation-reduction reactions.

Vitamin-D

The body needs the supply of these elements not only for the purpose of bone formation but also for the general maintenance purposes also. Reference has already been made to the relationship between the vitamin-D and the sterols. Deficiency of vitamin-D causes rickets, but moderate deficiencies are considered to have an ill effect on health and resistance to desease. Exposure to Ultra-Violet rays may result in the production of vitamin-D by its action on the sterols present in the epidermal cells of animals and thereby augments the supply of this vitamin. Good sources of vitamin-D are fish liver oils, egg yolk, and synthetic preparations of irradiated sterols such as dihydrocholesterol and ergosterol. Preparations of vitamin-A and D are extensively used in the infant feeding and in the fortification of infant feeds.

Production of irradiated milk needs careful control because exposure to longer periods results in the production of 'off' flavours in milk. Pasteurization, sterilization, condensing or drying, did not adversely affect the vitamin-D content of milk.

Vitamin-E or Tocopherol

It is a fat soluble vitamin, which prevents the sterility in rats. Hence it is also called as the anti-sterility vitamin. The rich sources of this vitamin are, wheat germ oil and green leaves, milk and meat. Table 3.28 summarises the distribution of the common vitamins in milk and various milk products.

Table 3.28 Distribution of the common vitamins in milk and various dairy products

Name of Product	A	B1	B2	C	D
Whole milk Raw	Fair-good	Very fair	Good	Very fair	Very fair
Pasteurised milk	Fair-good	Slight	Good	Slight	Very fair
Sterilized milk	Fair-good	Very slight	Good	Slight	Very fair
Whole milk Dried	Good	Slight	Good	Slight	Fair
Whole milk evaporated	Good	Very Slight	Good	Very Slight	Fair
Condensed Sweetened					
whole milk	Good	Slight	Good	Slight	Fair
Separated milk					
Butter milk	Absent	Very fair	Good	very fair	Absent
Whey					
Cream	Good	Slight	Slight	Slight	Fair
Butter	Good	--	--	--	Fair
Cheese	Good	Slight	Very fair	--	Fair

Source : Text Book of Dairy Chemistry - Edgar. R Ling

3.13 Gases Present in Milk

Milk when freshly drawn from the udder contains about 8 percent of dissolved gases by volume among which the carbondioxide predominates. The milk when comes in contact or exposed to atmosphere, this amount falls to 6% approxiately, thereby bringing a significant change altering the composition as the milk gases reach an equilibrium. There is a loss of carbon dioxide and a gain of oxygen and nitrogen as shown in the table 3.29.

Table 3.29 Distribution of Gases present in Milk

Description	Oxygen (%)	Nitrogen (%)	Carbon Dioxide (%)	Total (%)
Fresh milk from Udder	0.092	1.184	6.577	7.853
Milk from commercial source	0.47	1.29	4.45	6.21

Source : Text book of Dairy Chemistry - Edgar R. Ling

The equilibrium concentration of oxygen and nitrogen is of the same order as the solubility of these gases in distilled water. The quantity of oxygen in the milk may be reduced to negligible proportions by a combination of heat and vaccum treatment in view of the importance of dissolved oxygen in fat oxidation and removal of ascorbic acid. With this kind of process the ascorbic acid is able to withstand the normal oxidative influences. Due to the decomposition of protein, sulphuretted hydrogen and its derivatives might be present in the heated milk.

3.14 Enzymes in Milk

Several Enzymes are present naturally in the milk. An attempt has been made to present these enzymes their occurence/origin function and significant role they play in the milk and milk products in the table 3.30.

Table 3.30 The most important enzymes in milk

Enzyme	Origin and function	Significance
Peroxidase	Splitting of atomic oxygen from peroxides transfers easily to other compounds $2H_2O \rightarrow 2H_2O + O_2$ at T > 85 °C t > 10 °C inactivation	Proof of sufficient heat treatment of milk based on UHT process
Catalase	Splitting of oxygen in molecular form from hydrogen peroxide $2H_2O \rightarrow 2H_2O + O_2$ The level is high in colostral milk, from old cows or cows with mastitis, Inactivation during Pasteurization.	Natural levels in milk are absent to low, they indicate the hygienic status of the abnormality of milk (catalase reaction)
Phosphatase	Important enzyme group of the hydrolases. Catalyzing the hydrolysis from phosphoric acid esters.	
Alkaline phosphatase	The optimum reaction is found at pH 9.0 - 9.5 and 37 °C; inactivation at 70-75 °C	Proof of sufficient heat treatment for milk in short-time-treatment processes
Acidic phosphatase	The optimum is at pH 4.0 - 4.5, inactivation at 100 °C	Survives the pasteurization process, is one of the heat-stable enzymes.
Xanthinoxidase	Adsorbed into the fat globule membranes, also called Schardinger enzyme	Oxidizes xanthine with methylene blue and water into uric acid.

Table 3.30 ***Contd...***

Enzyme	Origin and function	Significance
Lipase	Lipase is an esterase which hydrolyzes the ester bonds of fats. Found at low levels in milk as a natural enzyme. Bacterial lipases indicates a strong contamination or recontamination with microorganisms. Optimum is pH 8.5 to 9.1 and 38 - 40 °C; inactivation during high temperature treatment at T > 85 °C	Splits milk fat from defective fat globule membranes, develops noticeable taste and flavor modifications in milk and dairy products (rancidity). Bacterial lipase in cream and butter indicates recontamination.
Protease Proteinase	Denatures proteins and their subgroups while cutting the peptide bonds. Milk contains low levels of natural protease, bound to casein. They survive the pasteurization. Bacterial protease are formed in large quantities from desirable and nondesirable microorganisms.	Bacterial protease split proteins into products of desirable quality e.g., during cheese ripening, but can cause spoilage and taste defects. Releases important amino acids, which are necessary for the growth of microorganisms.
Carbohydrase Amylase Diastase	Splits the glycoside bond in carbohydrates. Natural carbohydrase in milk, low levels,modifies starch into maltose. Optimum condition at pH 7.4 and 44 °C. Inactivation during pasteurization.	
Lactase	Bacterial carbohydrase, formation from the metabolism of lactobacillus and yeast. Splits lactose into glucose and galactose.	Base for lactic fermentation and alcoholic fermentation in acidified dairy products.

Source : Dairy Science and Technology Pieter Walstra et.al

3.15 Colostrum Milk

The advent of parturition is concomittant and simultaneous with bio synthesis and secretion of milk, a biological phenomenon triggered by the lactogenic harmones. For the first few days after parturition mammary glands secrete, a fluid known as "Colostrum" which has a strong odour, a bitter taste, very slimy viscous acidic and slightly reddish yellow in colour, containing a high percentage of immunoglobulins. It is a rich source of all milk constituents except for lactose, potassium and pantothenic acid. The colostrum gradually changes to normal milk after about five days after parturition. The gradual transition of colostrum to normal milk is given in the table 3.31.

Table 3.31 Transition from Colostrum to Milk (Percent)

Time after coloring (Hr)	Total Protein	Casein Solids	Albumin	Fat	Lactose	Ash	Total
0	17.5	5.08	11.34	5.10	2.19	1.01	26.20
6	10.0	3.51	6.30	6.85	2.71	0.90	20.46
12	6.05	3.00	2.96	3.71	3.71	0.89	14.53
24	4.52	2.76	1.48	3.98	3.98	0.89	12.77
30	4.01	2.56	1.20	3.40	4.27	0.86	13.63
36	3.98	2.77	1.03	3.55	3.97	0.83	12.22
48	3.74	2.63	0.99	2.80	3.97	0.84	11.46
72	3.86	2.70	0.97	3.10	4.37	0.83	11.86
96	3.76	2.68	0.82	2.80	4.72	0.84	11.85
120	3.86	2.68	0.87	3.75	4.76	0.85	12.67
168	3.31	2.42	0.69	3.45	4.96	0.84	12.13

Source : Text Book of Dairy Chemistry - Edgar R. Ling

The specific gravity, acidity, casein, albumin and globulin contents of normal milk as well as that of colostrum of murrah buffaloe and Red sindhi cow has been given in the table 3.32 for comparison.

Table 3.32 Composition of Colostrum and Normal Milk of Murrah Buffaloe and Sindhi Cow (Percent)

Constituents	Colostrum		Normal Milk	
	Murrah Cow	Red Sindhi Cow	Murrah	Red Sindhi
Specific gravity	1.0747	1.063	1.033	1.031
Acidity	0.35	0.42	0.1 to	0.16
Water	71.1	81.8	81.74	85.28
Total solids	28.9	18.2	18.26	14.72
Milk fat	3.9	2.0	8.11	5.67
Solid not fat	25.0	16.2	10.15	9.05
Milk sugar (lactose)	2.2	2.2	5.0	4.69
Ash	1.4	1.0	0.82	0.76
Protein	21.4	12.8	4.33	3.60
Casein	6.8	4.0		
Albumin	21.4	4.6	12.8	2.6
Globulin	10.8	6.2		
Total	100.0	100.0	100.0	100.0

Source : Text Book of Dairy Chemistry - M.P. Mathur, D.D. Roy & P. Dinakar

Further the average components of Colostrum of various species of animals has been presented in the table 3.33.

Table 3.33 Components of Colostrum of various species of animals (percent)

Components	Cow	Ewe	Goat	Sow	Mare
Water	73.3	74.8	79.2	77.5	87.1
Lipid	3.0	1.8	0.8	0.7	0.4
Lactose	2.2	2.2	3.4	2.4	4.7
Protein (Casein)	20.6	20.2	15.7	18.8	7.2
Ash	1.0	1.0	0.9	06	0.6
Total	100.0	100.0	100.0	100.0	100.0

Source : Text Book of Dairy Chemistry - M.P. Mathur, D.D. Roy and P. Dinakar

The globulin (lactoglobulin) and albumin (lactalbumin) of the colostrum are identical with the serum globulin of the blood. Colostral fat shows a different trend from milk fat containing a greater proportion of caprylic and capric acids. The fat of colostrum contains nine times carotene, eight times the vitamin-A and twice the vitamin-D than the normal butter fat. Colostrum is responsible for passing over natural antibodies to the new born thus giving passive immunity. Bacterial agglutinins also pass over from colostrum to the new born and rapidly get mixed up with the blood. The cow is unable to pass the anti bodies to the developing foetus in the uterus because of the placental barrier. The fat globules of colostrum are large and irregular in their shape whereas those of normal milk are round in shape.

The globulins of colostrum are α, β, γ and κ types and these globules are absorbed by the intestinal cell walls of the newly born calves within first 24-36 hr of their life. Immune bodies are being manufactured in the body by help of these globulins, for the safe-guard of the calves against different bacterial and viral infections. It is therefore advisable, and worth considerable, that most of the colostrum must be fed to the calves within this period for providing the immune bodies to them. The colostrum contents did not show any difference between the udder halves except the casein content which was higher from posterior quarter as compared to the anterior quarter.

Functions of Colostrum

1. It helps providing the passive immunity to the new born calf against several diseases through the globulins and immuno globulins.
2. Colostrum contains high quantity of vitamin-A, in comparison to vitamin-D found in normal milk. This vitamin provides a high rate of resistance power in calves against various diseases and to eye infections in particular.
3. Protein casein provides 'heat' and 'vitality' to the developing calf at a much faster rate than the human infant.

4. The high mineral content is helpful in the formation of bones and teeth to the calf thus giving a better growth response.
5. It is laxative in action, and helps in the passage of muconeum, the first foecal matter, which was already produced in the intestines of the calf, while it is developing in the uterus.

3.16 Nutritive Value of Milk

Milk is generally considered as nearly/almost complete food having a high nutritive value. Food is required by all living beings for their growth, supply of energy, maintenance, reproduction and during sickness or convalescence. These requirements are met from varied sources all of which contribute to the diet depending on their nutritive value. In this context milk plays an indispensable role in the nutrition of human beings. Milk is the only food consumed by the infant and has to provide all the nutrients in an assimilable form. Milk supplies body building proteins, bone forming minerals, energy giving fat and carbohydrates and health giving Vitamins. Besides supplying certain essential fatty acids, it contains all the above nutrients in an easily digestible and assimilable form. Milk continues to hold its place as an important diet of human beings either as fluid milk in the form of variety of milk products because of its properties, for pregnant mothers, growing children, adolescents, invalids, patients and convalescents in the similar manner. The reasons for this exalted projection of milk and milk products can be analysed in terms of the role played by each of the major constituents that are present in the milk.

The Nutritive value of various animal products has been presented in the table 3.34 for a comparative study and better understanding.

Table 3.34 Nutritive value of various Animal Products (Percent)

Product Milk from	cal/ 100/g	Moisture	Protein	Fat	Carbo-hydrates	Ash	Cal	Phosp	Iron
Cow	65	87.6	3.3	3.6	4.8	0.7	0.12	0.09	0.2
Buffaloe	117	81.0	4.3	8.8	5.1	0.8	0.21	0.13	0.2
Goat	84	85.2	3.7	5.6	4.7	0.8	0.17	0.12	0.3
Human	67	88.2	1.0	3.9	7.0	0.1	0.02	0.01	0.2
Butter	15	98.0	0.8	1.1	0.5	0.1	0.03	0.03	0.8
Curd	51	90.3	2.9	2.9	3.3	0.6	0.12	0.09	0.3
Skim milk	29	92.5	2.5	0.1	4.6	0.7	0.12	0.09	0.2
Khoa	21	30.6	14.6	31.2	20.5	3.1	0.65	0.42	5.8
Cheese	348	40.3	24.1	25.1	6.3	4.2	0.80	0.50	2.1
Mutton	194	71.5	18.5	13.3	--	1.3	0.15	0.15	2.5
Pork	114	77.4	18.7	4.4	--	1.0	0.03	0.20	2.3

Source : Dairy Science & Technology Pieter Walstra et.al

An attempt has been made to present some data relating to the composition and the nutritive value of milk and milk products in table 3.35.

Table 3.35 Quantities of certain nutritive components in milk and milk products (100 g)

Product	Calories	Water	Protein	Fat	Ash	Total carbo-hydrate	Calcium	Phos-phorus	Iron
		g	g	g	g	g	mg	mg	mg
Whole cow milk	68	87.0	3.5	3.9	0.7	4.9	118	93	0.1
Skim cow milk	36	90.5	3.5	0.4	0.8	5.1	123	97	0.1
Evaporated milk	138	73.7	7.0	7.9	1.5	9.9	243	195	0.2
Condensed milk	320	27.0	8.1	8.4	1.7	54.8	273	228	0.2
Skim milk (dry)	362	3.53	5.6	1.0	7.9	52.0	1300	1030	0.6
Butter	716	15.5	0.6	31.0	2.5	0.4	20	16	0
Butter milk	36	90.5	3.5	0.1	0.8	5.1	118	93	0.1
Cheddar cheese	398	37	25.0	32.2	3.7	2.1	725	495	1.0
Cheddar cheese proposed	370	40	23.2	29.9	4.9	2.0	673	787	0.9
Cottage cheese	95	76.5	19.5	0.5	1.5	2.0	96	189	0.3
Cream	204	72.5	2.9	20.0	0.6	4.0	97	77	0.1
Ice Cream	207	62.1	4.0	12.5	0.8	20.6	123	9	0.1
Whey dried	344	6.2	12.5	1.2	7.7	72.4	679	576	--

Product	Sodium mg	Pota-ssium mg	Vitamin A mg	Vit B_1 mg	Vit B_2 mg	Nico-tinic acid	Vitamin C mg	Niacin mg
Whole cow milk	50	140	160	0.04	0.17	0.1	1.0	0.1
Skim cow milk	52	150	Traces	0.04	0.18	0.1	1.0	0.1
Evaporated milk	100	270	400	0.05	0.36	0.2	1.0	0.2
Condensed milk	--	--	430	0.05	0.39	--	1.0	0.2
Skim milk (dry)	77	1130	40	0.35	1.96	1.1	7.0	1.1
Butter	980	23	2300	Traces	0.01	0.1	0	0.1
Butter milk	130	140	Traces	0.04	0.18	0.1	1.0	--
Cheddar Cheese	700	92	1400	0.02	0.42	0.42	0	Traces
Cheddar cheese Processed	1500	80	1300	0.02	0.41	0.41	0	--
Cottage cheese	290	72	20	0.02	0.31	0.31	0	--
Cream	--	--	830	0.03	0.14	0.14	1	--
Ice Cream	100	90	520	0.04	0.19	0.19	1	0.1
Whey dried	--	--	50	0.49	2.5	2.5	-	--

Source : Dairy Science and Technology Pieter Walstra et.al

Milk Proteins

The proteins supply the amino acids, the building blocks from which the tissue proteins are synthesized thus helps in the growth of children, making it possible to replace the proteins, that are used in the metabolism or lost by injury. The human body synthesizes ten (10) amino acids from the available sources of proteins out of the twenty amino acids required for the synthesis of the tissue proteins. Therefore, the amino acids which cannot be synthesized need to be supplied through the diet which are known as esential amino acids and the milk is a rich and ready source for these essential amino acids compared with the reference protein as recommended by FAO/WHO has been presented which are presented in the table 3.36.

Table 3.36 Percentage of essential amino acids in the ideal reference protein and in milk protein

Essential amino acids	Reference Protein (FAO / WHO)	Grams in 100 g of protein in milk protein
Isoleucine	4.0	6.3
Leucine	7.0	10.0
Lysine	5.5	8.1
Methinine + Crystine	3.5	3.5
Phenylalamine + Tyrosine	6.0	10.3
Threonine	4.0	4.9
Tryptophane	1.0	1.4
Valine	5.0	6.9
Total	**36.0**	**51.4**

Milk protein compares very well with the standard reference protein proposed by FAO/WHO and even surpasses to a great extent. Thus the milk protein provides all the essential amino acids in the proper proportion and concentration as required by the human beings for their optimum utilization. The milk protein has an excellent supplementary value for blending with other foods in order to increase the nutritive value and has an excellent buffering effect as a very good remedy in controling the excessive stomach acidity in the human beings to some extent. Dairy products, are therefore used in the dietetic therapy of patients suffering from gastritis and ulcers of the stomach. In some circumstances the milk proteins are known to produce allergic responses adversely in sensitive individuals. This reaction is usually species specific and it is best to avoid the milk of such species from the diet.

Milk is a perfect food, all mammals secrete milk for a certain period because the milk of each species is disignated for the rapid growth of their young ones. The milk is nutritionally very rich hence it can also be called as "nutritional bank". The nutritive value of protein is evaluated by:

1. Estimation of amino acid content of the Protein which are specific for that protein.
2. Utilization of protein by the body after its digestion and absorption.
3. Synthesizing of protein in the body.
4. Non-protein nitrogen-Nutritional constituents and their availability in the body.

Animal proteins whether from milk, meat, eggs or fish are very good sources to supply these essential amino acids to man.

Probably the milk is the only source of animal protein for vegetarians who form a significant portion of the Indian population. Other sources of protein cannot and adequately substitute the role played by the milk and milk products, and the biological value of milk protein is 90 (cows milk) and higher biological value can be obtained when combined with potato and rice or maize protein.

Further, an attempt has also been made to present the protein and essential amino acids composition of some important foods for a comparative study for a better understanding by the students of dairy science and presented in table 3.37.

Table 3.37 Protein and Essential Amino Acids Composition of Some Foods

Source	Nitrogen g / 100 g	Protein Food	Iso-leucine	Relative mg /g	Lysine	Sulphur containing amino acids	Trypto-phane	Valine	Total amino acids
Human (raw)	0.19	1.2	252	548	428	185	105	284	5530
Cow (raw)	0.55	3.5	295	596	487	208	88	362	6463
Buffaloe (raw)	0.63	4.0	339	631	489	257	92	380	--
Soya milk	0.56	3.2	305	497	348	195	85	294	6010
Groundnut	3.10	19.4	281	480	413	150	50	390	6229
Til (sesamme)	3.42	18.1	226	420	170	290	84	288	5916
Sun flower seed	2.38	12.6	267	401	225	212	85	317	5871
Fish	12.00	75.0	269	452	484	248	60	318	5859
Wheat	1.90	10.9	228	440	130	250	67	258	6216
Rice	1.13	6.7	258	509	224	225	84	353	5904
Maize	1.52	9.5	230	783	167	217	44	303	6093

Source : Dairy Science and Technology Pieter Walstra et.al

The term "biological value" was first used in 1909 by Thomas and later it was modified by Mitchell and presently known as Thomas Michell Biological value and is used to indicate the nutritive value of proteins. The basis of the calculation is the amount of nitrogen in protein absorbed by the human body and the source of nitrogen is from protein only which the body can utilize. Always there is a balance in our body between the anabolism (bio synthesis) and catabolism (break down of Protein). Biological value (B.V.) may be defined as the quantity of nitrogen or protein in grams absorbed from 100 grams of nitrogen or protein that may be considered as having participated in the tissue construction or removal of protein or the percentage of nitrogen absorbed that is made use of by the body.

$$\text{B.V.} = \frac{\text{Intake } N_2 \text{ - Urinary } N_2 \text{ Output + Faecal } N_2 \text{ Output (d)}}{\text{Intake } N_2 - \text{Faecal } N_2 \text{ Output}} \times 100$$

$$\text{B.V.} = \frac{\text{Nitrogen retained}}{\text{nitrogen absorbed from food}} \times 100$$

Based on this method the B.V. of the egg protein is 96, cow milk 90, goat meat 76 and raw soya 57. When the milk protein is supplemented with cereals the B.V. of milk protein increases and this is known as "the supplementary value" of milk protein.

Another method for evaluating the protein value is by calculating the Net Protein Utilization (N.P.U).

The food protein ingested cannot be digested completely by the body, therefore, assuming that 96 percent of the cow milk protein is digested, the N.P.U. can be calculated by the following formula.

$$\text{B.V.} \times \text{Digestibility} = \text{N.P.U.}$$

$$90 \times 0.96 = 86.4 \text{ or } 86$$

The respective N.P.U. values for different milk proteins are

Cow milk - 86

Goat Milk - 76

Raw Soya - 48

Another method for estimation of protein value is the protein efficiency ratio (PER). This ratio gives the increase in weight of the body to the grams of protein consumed.

$$\text{PER} = \frac{\text{Increase in weight (grams)}}{\text{grams of protein consumed}} \times 100$$

The body weight gain is not solely due to the protein but substances such as lipid water and other sources also contribute. Therefore, when the gain in weight is calculated on the basis of newly synthesized tissue which has protein only and 80 percent water, then

$$PER = \frac{80 + 20}{20} \times 50$$

However, practically this high value is never achieved. The PER value for cow milk is 3.2 and that of egg protein is 3.8.

Milk Lipids

The milk fat is a very rich source of energy yeilding 9.0 kilocalories per gram. Fat gets deposited in the adipose tissue after conversion as body fat when consumed in excess of the body requirement, which is not desirable and hence there is a trend towards the consumption of low fat milk and dairy products. The stored body fat is broken down and is utilized in meeting energy needs of the body during the periods of extreme starvations acting as a store house of energy in emergency conditions. Milk fat acts as a carrier for the fat soluble vitamins viz. A, D, E and K and this important function is underscored by the fortification of these vitamins in the dairy products by the usage of milk fat as the medium.

The body requires certain essential fatty acids which cannot be synthesized by the body similar to the case of essential amino acids. Milk fat which contains the essential fatty acids linoleic and arachidonic acids in significant quantities is in a position to meet the requirements of the body. Triglycerides constitute nearly 99 percent of the milk fat which includes a good amount of esterified lower chain of fatty acids particularly butyric acid, significantly contributing to the pleasant and acceptable rich flavour of milk fat and is not duplicated by any other type of fat. Phospholipids particularly lacithin, cephalin and sphingo myelin form up to 1 percent of total milk fat and possesses emulsifying properties. The presence of up to 0.40 percent of sterols, mainly cholesterol is a discouraging factor in the consumption of excessive milk fat, which is implicated as the causative agent of atherosclerosis and this awareness has led people to avoid the fats containing cholesterol. Milk fat apart from directly contributing to the nutrition, also plays a secondary role which is no less important. The pleasant flavour imparted to the products containing milk fat greately enhances their palatability and acceptance leading to an increased consumption. Therefore, milk fat is widely used in a variety of foods as a frying medium, hence it is the most highly priced constituent of milk.

Carbohydrate

Lactose is the predominent milk sugar, which is present normally to the extent of about 5 percent and acts as a ready source to provide energy. In addition to this lactose adds to the taste of milk. The sweetness greately enhances its palatability of the milk. Generally, carbohydrates are considered to provide heat and energy to the body and when ingested they are broken down to their elementary monosaccharide structure.

Lactose takes a much longer time compared to other sugars for its digestion which is beneficial, because it favours the growth of lactobacillus species in the stomach. Thus a favourable microflora is maintained by preventing the growth of undesirable putrifying organisms. Lactose is essential for the absorption of calcium from the gastro intestinal tract by chelating the mineral thereby delaying its precipitation during transport. Lactose is broken down to its constituent monosaccharide glucose and galactose in the small intestine by the enzyme β galactoside which is commonly known as lactase. A congenital absence of this enzyme occurs in some individuals and it could be destroyed either by illness or surgery etc., hence such persons are not able to utilise lactose in their diet. Lactose intolerance need not necessarily mean intolerance to milk, because other foods, which are ingested along with milk containing a small amount of lactose consumed may not cause any adverse effects. In the intestine the putrifactive bacteria acts on the intact lactose and produces gas, alcohol and other undesirable fermentation and such a condition is known as lactose intolerance where the affected individuals are not able to consume lactose in their diet.

Human milk is very unique in its nature because it contains oligosaccharides including sialic and or N-acetyl-neuraminic acid. The intestinal flora of the breast fed babies, significantly differs by the presence of *"Lactobacillus bifidus"* and the "bifidus" factor causing this difference is traced to the oligosaccharides of human milk. The growth of favourable micro flora like lactobacillus prevents the growth of other harmful pathogens in the intestinal tract.

Minerals

Various inorganic minerals like calcium, phosphorous, iron, copper, manganese, sodium, potassium, chloride, iodine, cobalt, zinc and so on are required by the human body for its proper growth and functioning. Some of these minerals are required in significant quantities, while others are required only in traces. Milk and Milk products are a good source for majority of these minerals especially for calcium and phosphorous present in a ratio of 1.3 : 1 for their optimum utilization and are important for the formation of teeth and bones.

In this ideal ratio the minerals i.e calcium and phosphorous present in the milk can meet almost all the requirements of a growing child and for most other age groups. Milk is said to be dificient in iron because it cannot meet the requirements adequately. Milk also contains a plethora of trace elements present in ppm or μg quantities in addition to the major minerals which are present relatively in high proportion. The various major and trace minerals that are present in the milk have been presented in tables 3.38, 3.39, 3.40, and 3.41.

Nutritionally these trace elements are essential in micro quantities and the role played by these elements should not be under estimated. Cobalt forms the centre of vitamin B_{12} molecule and is essential for its synthesis. Iron is essential for the formation of haemoglobin and its deficiency may lead to anaemia. Iodine is a constituent of hormone thyroxine which plays an important physiological role in the metabolism of the individual and its deficiency causes enlargement of the thyroid gland resulting in goiter. As already stated Iron is a constituent of haemoglobin and some of the respiratory enzymes. Milk is a poor source of Iron hence, it needs to be fortified with an external source of iron. Selenium is required in traces, however high concentrations are toxic. Milk is a good source for zinc, nickel, molybdenum etc. which are essential for certain specific enzyme activities. Even though milk contains almost all the minerals required by the human being, it is considered as a poor source for iron and manganese.

Table 3.38 Mineral (Salt) Composition of Milk

Constituent	Average content ml/100 g	
	Buffalo milk	Cow Milk
Sodium	44.63	58.0
Potassium	103.00	140.00
Calcium	175.50	123.30
Magnesium	19.25	11.10
Phosphorous Total	97.6	95.1
Phosphorus (linegainic)	--	--
Chloride	62.84	104.5
Sulfate	--	10.0
Carbonate as CO_2	--	20.0
Citrate (as citric acid)	163.6	176.0

Source : Text Book of Dairy Chemistry - M.P. Mathur, D.D. Roy and P. Dinakar

Table 3.39 Trace elements in milk

Elements	Cow (μg/litre)	Buffaloe	Goat
Boron	100 - 400	0.82 (ppm)	--
Zinc	1000 - 6000	6.50 (μg/litre)	3.84 (ppm)
Iron	About - 300	2404 (μg/litre)	--
Copper	0.2 - 1.4	212 (μg/litre)	--
Sulphur	--	239.4 (μg/litre)	__

Source : Text Book of Dairy Chemistry - M.P. Mathur, D.D. Roy and P. Dinakar

Table 3.40 Average Value of some trace elements in milk

Elements	Average value
Chlorine	119 mg/100 ml
Sulphur	30 mg/100 ml
Iron	3 ppm
Zinc	3 ppm
Silicon	2 ppm
Copper	0.3 ppm
Florine	0.15 ppm
Aluminium	Traces
Manganese	Traces
Iodine	Traces
Boron	Traces

Source : Text Book of Dairy Chemistry - M.P. Mathur, D.D. Roy and P. Dinakar

Table 3.41 The partition of milk salts

Elements	Concentration (mm)		
	Skim milk	Diffusible	Colloidal
Ca	30.3	10.0	20.3
Mg	5.2	3.3	1.9
Cit	9.5	9.1	1.0
Pi	21.4	11.9	9.5
Na	22.8	21.8	1.0
K	39.7	37.5	2.2
Cl	31.5	31.5	0.0

Source : Text Book of Dairy Chemistry - M.P. Mathur, D.D. Roy and P. Dinakar

Vitamins

Milk contains almost all the vitamins that are required by the body in various quantities. It is considered as a good source for vitamins like Vitamin-A, Riboflavin. Thiamine etc whereas the milk is a poor source for the other Vitamins like Vitamin-D, Niacin, Vitamin-C etc. Deficiency of the vitamins leads to various diseases and disabilities such as night blindness (Vitamin-A), Beri-beri (Thiamin) scurvy (Vitamin-C), Rickets (Vitamin-D) and so on. Efforts are made to fortify the Vitamin-D in order to enhance the nutritive value of the milk. The fat soluble vitamins A, D, E and K can be introduced by blending in the milk fat. Some of the vitamins are destroyed on heat processing or storage, whereas others are more stable and resist deterioration, hence the amount of vitamins available to the consumers could be significantly different from those estimated in the raw milk. The various vitamins that are present in the milk has been given in table 3.42.

Table 3.42 Vitamins in Milk

Vitamins	Concentration (Per litre)
Vitamin - A	1590 I U
Vitamin - D	22.11 I U
Vitamin - E	1.0 mg
Vitamin - K	0.04 mg
Vitamin - B_1	0.4 mg
Vitamin - B_1	1.5 mg
Pan to thermic acid	3.0 mg
Biotin	50 μg
Niacin	0.2 - 1.2 mg
Pyridoxine B_6	0.7 μg
Folic acid	1.0 μg
Vitamin B_{12}	7.0 μg
Victamin C	20.0 mg
Inositol	150.0 mg
Choline	--

Source : Text Book of Dairy Chemistry - M.P. Mathur, D.D. Roy and P. Dinakar

Energy Value

The energy giving milk constituents and their individual contributions are as follows:

Milk fat	..	9.3 C/g
Milk Protein	..	4.1 C/g
Milk sugar	..	4.1 C/g

where1C (Kilocalorie) = 1000 (small calore). The energy value of milk varies with its composition. On an average cow milk furnishes 75 C/100 g and buffaloe milk 100C/100 g.

3.17 Effect of Processing

1. Pasteurization carried out with reasonable care has no effect on vitamin-A, carotene, riboflavin and a number of remaining vitamins of B-Complex and vitamin-D, whereas a 10% and 20% loss of Thiamine and ascorbic acid (vitamin-C) respectively may be expected due to pasteurization.
2. Sterilization increases the losses of thiamine and ascorbic acid to a level of 30-50 percent and 50% respectively, although the remaining vitamins are affected in a negligible manner.

A balanced diet is essential for proper health and growth therefore, the role of milk and milk products in providing the nutrients required for a balanced diet have been presented in the table 3.43.

Table 3.43 Role of milk and milk products in a balanced diet

Nutrients	Purpose	Sources
Proteins	Essential for muscle building and repair; give the body energy and heat	Meat, poultry, fish, milk, cheese, beans, peas, nuts
Carbohydrates	Body energy and heat	Bread, cereals, pastry, sugar, vegetables, fruit.
Fats	Body energy and heat	Butter, ghee, oils
Minerals Vitamin A	Bone, teeth, body cells Growth, health of the eyes, structure and function of the skin and mucous membrane.	Dairy products, fruit, vegetables. Fat-rich dairy products, eggs, spinach,carrots, tomatoes, fish liver oils.
Vitamin B_1 (Thiamine)	Growth, aids appetite, prevents beri-beri, function of the nervous system.	Whole grains, eggs, green vegetables, yeast, liver, kidney.
Vitamin B_2 (Riboflavin)	Growth, health of the skin and mouth, functioning of the eyes.	Milk, cabbage, carrots, spinach, liver, eggs, yeast, lean meat, prones
Niacin	Functioning of the stomach, intestines and nervous system	Meat, heart, kidney, liver, eggs, fish, milk, peanuts, yeast
Vitamin C	Aids bones and teeth, prevents scurvy	Citrus fruit, maize, tomatoes, lettuce, cabbage
Vitamin D	Aids in calcium-absorption which strengthens bones, prevents rickets	Eggs, milk, fish liver oils

Source : Dairy Science and Technology Pieter Walstra et.al

In order to make the milk more suitable to certain special nutritional purposes they are modified according to the need. They are :

3.18 Soft Curd Milk

The formation of curd either from cow or buffalo milk in the stomach due to the peptic action may be hard as compared to the mothers milk. The curd tension of milk can be decreased to bring it closer to that of human milk thereby facilitating better utilization by the base exchange process. Milk acidified with 0.3 percent citric acid is percolated through a Zeolitebed (Sodium aluminium silicate containing potassium). During this process up to 20 percent of calcium and phosphorous is replaced by sodium and potassium. It is possible to maintain the pH and the calcium: Phosphorous ratio by proper regulation while reducing the curd tension appreciably. Homogenisation, high heat treatment, ultrasonic sound and use of pancreatic protease to partially digest the protein are some of the other methods that can be used to reduce the curd tension of milk.

3.19 Humanisation of Buffaloe Milk

Very often it so happens that infants, have to be fed with other milks even before weaning because of the fact that the mothers milk is insufficient to meet the requirements. In such circumstances the alternative source of milk namely the buffaloe milk or cow milk may not be suitable or acceptable to the infant since it differs with the human milk in its composition very widely. The buffaloe milk can be modified or humanised to make it more acceptable in order to overcome these difficulties, which can be brought about by the following changes or modifications.

1. Removal of α S1 Casein (by controlling the Tryptic Proteolysis)
2. Replacement of milk fat with vegetable fat.
3. Addition of lactose/polysaccharide.
4. Enrichment of vitamins.
5. Removal of calcium by electrodialysis.

The product thus produced can be spray dried and used for infant feeding when required.

3.20 Vitamin-D Fortified Milk

Already, it has been stated that milk is a poor source of vitamin-D and it is deemed fit to fortify the milk in order to increase its nutritive value. vitamin-D regulates the absorption and metabolisms of calcium and phosphorous in the process of bone building. Consequently lack of vitamin-D could lead to Rickets. The milk can be enriched with vitamin-D by any one of the following process.

1. Addition of vitamin-D concentrate to the milk.
2. Irradiation of milk with ultra violet light to convert the precursor 7- dehydrocholesterol to vitamin-D.
3. Feeding of irradiated yeast to the animal.

3.21 Low Sodium Milk

People suffering from heart and kidney troubles are advised to take a low sodium diet. Milk contains approximately 500 milli grams sodium per litre, therefore it is necessary to remove some sodium in order to make it acceptable as low sodium diet. This can be done by ion exchange resins, where sodium is replaced by potassium from the resin. The concentration of sodium is reduced drastically, while that of potassium is increased considerably.

3.22 Lactose Hydrolysed Milk

In some individuals the enzyme lactase or β galactosidase is lacking, which is essential for the break down of lactose, the principal milk sugar into its component glucose and galactose moieties. Such people are not able to digest lactose and are called lactose intolerant. Intolerance to lactose need not necessarily mean an intolerance to milk, because the concentration of lactose in milk may not be sufficient to cause allergic reactions. Moreover, along with other foods, when milk is usually consumed the effect of lactose may not be very drastic. However, in the case of infants dependant on milk, the lactose intolerance could be a great problem and can be overcome by hydrolysing the lactose enzymatically. Commercially, the lactose enzyme is immobilised on a column through which the milk passes. The outgoing milk contains hydrolysed lactose and is acceptable by intolerant individuals. Alternatively the β galactosidase enzyme is available in small sachets, which can be added to the milk before its consumption. Hydrolysed lactose milk is available commercially to fulfil this specific need.

3.23 Cow Milk Vs. Buffaloe Milk

Out of the total milk production in the world 84% derives from the milch cows, 12% from the buffalo and 2% each from sheep and goat. On the contrary in India buffaloes contribute 54%, 43% from the cow and 3% from sheep and goat of the total milk production. Most of the milk produced and processed in India is buffalo milk due to its qualities, which makes it more favourable as compared to the cow milk. However, the usage of buffalo milk has created, some problems, during the manufacture of the products and storage due to some basic differences in the physico-chemical properties. It has been reported that the low heat stability is due to the high content of calcium which can be improved by the partial replacement of calcium by sodium and potassium.

The study relating to the chemical composition and the physico-chemical properties of the cow and buffalo milk reveals that there exists some differences between them. In view of this efforts have been made to discuss these differences briefly hereunder in order to re-capitulate them for the benefit of the students.

1. **Low cholesterol content** - buffalo milk contains 0.65 mg/g as compared to 3.14 mg/g in the cow's milk.
2. **More protein** - Buffalo milk - 2.74 PER
   - Cow milk - 2.49 PER
3. **More minerals:** Buffalo milk contains calcium 92%, Fe 37.7%, phosphorus - 118% higher as compared to the cow milk. The calcium : phosphorous ratio is 2.26 : 1.96 respectively for cow and buffalo milk. High salt balance ratio in buffalo milk has low heat stability.
4. **More Vitamins:** Vitamin-A in the buffalo milk, because the buffalo metabolizes all the carotene and converts it to vitamin-A.

5. There is no difference in the nutritive value and digestibility of milk and milk products between the two milks.
6. **Better whey protein:** Proteins of buffalo milk, more particularly the whey proteins are more heat resistant to heat denaturation as compared to cows milk protein.
7. **Better health foods:** Higher levels of various bio-protective factors such as immunoglobulins, lactoferin, lysozyme, lactoperoxidase as well as bifidogenic factors renders the buffalo milk more suitable than cow milk.
8. More viable commercially than cow milk for the manufacture of fat based and SNF based milk products because of low water content, high fat and white colour of buffalo milk.

Differences in milk fat

1. Buffalo milk fat has a higher Richert-Meissel (RM) value, saponification value, Kirschner value and melting point, with a lower polenski value, iodine number, and butyro-refractometer index as compared to cow milk.
2. Buffalo milk fat is higher than cow milk in respect of butyric, palmitic stearic and oleic acids but lower in caproic, caprylic capric, lauric and myristic acids as shown in the table 3.44.

Table 3.44 Fatty acid composition of Cow and Buffalo milk

Sl.No.	Fatty acid	Buffalo milk fat	Cow milk fat
1.	Butyric	4.4	3.2
2.	Caproic	1.5	2.1
3.	Caprylic	0.8	1.2
4.	Capric	1.3	2.6
5.	Decinoic	Trace	0.3
6.	Lauric	1.8	2.8
7.	Myristic	10.8	119
8.	Palmitic	33.1	30.6
9.	Stearic	12.0	10.1
10.	Oleic	27.2	27.4
11.	Linoleic	1.5	1.5
12.	Linolenic	0.5	0.6

Source : Hand book of Milk Micro Biology - M.K. Shrivastav

The differences in the physico-chemical properties of cow and buffalo milk has been presented in the table 3.45. Buffalo milk normally has a higher pH, acidity buffer value, density, viscosity, and fat globule size than cow milk.

Table 3.45 Differences in physico-chemical properties of buffalo and cow milk

Sl.No.	Physico chemical properties	Cow milk		Buffalo milk	
		Mean	Range	Mean	Range
1.	Density at 20°C	1.0287	1.028 - 1.030	1.0305	1.030- 1.032
2.	SP. gravity at 20°C	1.0330	--	1.0323	---
3.	Viscosity at 20°C (Centi Poises)	1.86	1.5-2.0	2.245	2.086- 2.538
4.	Curd Tension (g)	23.3	28-54	402	32-85
5.	Surface tension at 20°C (dynes / cm)	55-9	40-60	55.4	45.5-48.2
6.	Freezing Point	0.540	0.525- 0.565	0.544	0.5454- 0.5443
7.	pH at 20°C	6.64	6.5-6.7	6.73	6.63- 6.87
8.	Titrable acidity (as % lactic acid)	0.15	0.14- 0.18	0.16	0.10- 0.28
9.	O.R. Potential (+ volts)	0.273	0.18- 0.38	0.310	0.129- 0.469
10.	Electrical Conductivity (mmho/cm) at 30 °C	4.67	4.14- 5.03	3.60	3.28- 3.81
11.	Thermal Conductivity Kcal / h / m (degree celsius)	0.4572 at 36.6°C	-- --	0.5689 -- at 42.5 °c	0.5487- 0.5937
12.	Refractive Index	--	0.3449- 1.3448	1.0	--
13.	Heat capacity at 40°C (cal/g/degree celcius)	0.930	--	0.8191	0.7402- 0.9201
14.	Fat globule size (mm)	3.85	--	5.01	--
15.	Phosphatase (units)	83	--	28	--
16.	Ultra violet flourescence	Pale Bluish		Greenish Yellowish	

Source : Hand book of Milk Micro Biology - M.K. Shrivastav

Some of the problems that are encountered in the processing and storage of buffalo milk products have been presented briefly hereunder :

1. **In Cheese Making**

   Slow ripening, faster rennet action, excessive synerises, slow cheddaring and curing of cheese, slightly bitter taste, hard body and texture in cured cheese.

2. **In Condensed Milk Preparation**

   There is a great scope and livelihood in the formation of desired gel during the production of both condensed and evaporated milks. Greater incidence of age thickening during storage of condensed and evaporated milk. There is a possibility of sandiness effect in sweetened condensed milk. Browning, discolouration, and cooked flavour are likely to occur/develop to a greater extent during the storage of these products.

3. **In milk Powder Making**

   Low heat stability low stability of the resultant dried whole or skim milk powder. Similar problems are also encountered in the manufacture of Instant food.

4. **In Butter Making**

   The milk fat of buffalo contain large amounts of fatty acids, with a high melting points, hence it is harder than the cow milk fat. Buffalo butter has been found to appear much harder, when compared with the butter prepared from the cow milk which appear to be soft when prepared under identical conditions.

5. **In Ghee Making**

   The buffalo milk fat contains fatty acids of high melting point hence responsible for the larger grains as compared to the ghee prepared from cow milk. The fat in the solid state is hydrolyzed with a slow rate of hydrolysis, because of which the buffalo ghee is less vulnerable to the development of rancid flavour than the cow ghee. The higher content of highly unsaturated tetraenoic and penta enoic fatty acids, in buffalo milk fat, etc is less stable to oxidation as compared to that of the cow milk fat.

6. **In Making Channa**

   The channa prepared from buffalo milk is normally greasy, hard in body with coarse in texture and does not suitable for the best quality sweets viz Rasgolla and sandesh due to high level of fat higher calcium and protein in the milk as well as the miscellar differences in milk protein.

7. **In making Khoa**

   The khoa prepared from buffalo milk is not inferior to the khoa prepared from cow milk.

CHAPTER 4

Factors Affecting the Composition of Milk

4.1 Variations in the Composition of Milk

Various factors are known to influence the composition of milk and before considering them, it is better to recognise that some,constituents as well as some physical properties show much less variation than others. Reference has already been made that the milk varies widely in its composition containing the same kind of constituents, but in varying amounts. Milk from the individual cows/buffaloes shows greater variation than the mixed milk of the herd. The day-to-day variations are always greater in small herds than in large herds. It is essential, to examine the extent of these day-to-day variations should be examined strictly without any impartiality because fat being the most variable major constituent which shows the greatest daily variation. In general, other constituents which shows greatest daily variation in addition to fat are protein followed by ash and sugar.

The various factors that are responsible for affecting the composition of milk are.

4.1.1 Species

Each species gives milk of a characteristic composition, which has been presented in the table 3.4. in chapter 3 while discussing the composition of the milk.

It is difficult to determine accurately the relationship between the specific factor and its effect on the composition of milk because the variation in milk composition is dependent on many factors in different species. Generally the composition of milk of each species is designed to provide adequate nutrition for the natural rate of growth of the new born, which is an inherited trait for that species. The faster the rate of growth the more concentrated are the milk components needed for this growth which is presented in the table 4.1.

Table 4.1 Growth rate of young animals in relation to Protein content of their mother milk

Species	Days required to double birth weight	Protein percentage of milk
Man	180	1.60
Cow	47	3.50
Sheep	15	4.80
Pig	14	5.21
Cat	9.5	7.00
Rabbit	6	10.38

Source : MP Mathur, DD Roy, P. Dinakar.

4.1.2 The Individuality of the Cow

Generally it has been observed that there are some individual cows, which constantly give milk showing some variation in its composition, even though they are kept under the same ideal management and feeding conditions of the herd. These variations may be consistantly high or low percentage of fat or protein etc., which cannot attributed to any of the known factors.

4.1.3 Breed

It is a well known fact that the breed of the animal bears a distinct relation to the quantity of milk produced and especially to the fat content. The composition of the milk is strongly influenced by breed and the variation in milk yeild is far greater than that of the fat percentage. Milk produced by the individual cows within the breed varies much less in composition when compared to their milk yeild than that of the fat percentage which is generally fairly typical of the breed. Breed also exerts its effect on the non-fatty solids of milk. It is advantageous to calculate the non-fatty solid content on fat free basis is

$$\frac{\text{Percentage SNF in milk}}{100 - \text{Percentage fat}} \times 100$$

The variations in the percentage of fat are very common and frequent. Generally there is a tendency for high fat content to be associated with a somewhat higher percentage of SNF and vice versa when the milk of one breed is compared with the milk of another breed. It may also be argued that the breeds which are very high producers of milk may give a low fat and SNF content, which is very much true and such milk may fall below the presumptive standards laid down under the law. The composition of milk of some exotic European breeds and Indian breeds of cows and buffaloes have been presented in tables 4.2, 4.3, 4.4 and 4.5 respectively.

Table 4.2 Milk Composition of some European breeds of Cows (Percent)

Breeds	Water	Total solids	Fat	Protein	Lactose	Ash	SNF
Holstein Friesian	88.0	12.0	3.6	3.2	4.6	0.7	8.4
Jersey	85.4	14.6	5.1	3.8	4.9	0.8	905
Shorthorn	87.2	12.8	3.8	3.3	5.0	0.7	9.0
Guernsey	85.4	14.6	4.9	3.9	5.0	0.8	9.5
Ayrshire	87.1	12.9	3.9	3.3	5.0	0.7	9.0

Source - M.P. Mathur, D.D. Roy, P. Dinakar

Table 4.3 Milk Composition of European breeds of cows in relation to Total Solids (Percent)

Breed	Fat	Protein	Sugar	Ash
Holstein Friesian	29.0	26.4	39.0	5.7
Jersey	35.3	26.0	33.4	5.2
Shorthorn	29.6	26.0	39.0	5.4

Source - Milk and its properties S.M.Srivastava

Table 4.4 Milk Composition of Indian Breeds of cows and buffaloes (Percent)

Constituents	Cow	Buffalo	Sahiwal	Sindhi	Gir	Tharparkar	Cross bred
Water	87.2	82.8	86.4	86.0	86.5	86.6	86.87
Total Solids	12.8	17.2	13.6	13.9	13.5	13.4	3.20
Fat	3.8	7.2-7.8	4.6	4.8	4.5	4.6	4.5
SNF	8.9	9.3	9.0	8.4	9.4	9.6	8.7
Protein	3.5	3.8	3.3	3.4	3.2	3.3	3.4
Lactose	4.8	4.8	5.0	4.9	4.9	4.8	4.9
Ash	0.7	0.8	0.7	0.7	0.7	0.8	0.7
Specific Gravity	1.031	1.035	1.031	1.031	1.031	1.031	1.032
Acidity	0.51	0.76	0.16	0.15	0.16	0.16	0.16

Source - Milk and its properties - S.M. Srivastava

Table 4.5 Milk Composition of Indian Breeds of Cows in respect of Fat and SNF (Percent)

Breed	Fat	SNF
Murrah	7.70	9.68
Hariana	4.60	9.25
Sindhi	4.84	8.42
Ongole	5.05	8.05
Gir	4.54	9.15
Tharparkar	4.60	9.63
Sahiwal	4.90	9.10
Karan Swiss	4.70	9.10
Jaffarbadi	7.30	10.1
Surti	8.40	10.3

Source : Milk and its properties S.M. Srivastava

4.1.4 Size of the Herd within Breed

Generally the dairy farmers who are primarily interested in the sale of milk and its products prefer the cows of medium to large size of animals of the breed. Large animals of breed have produced high milk and fat yeild without any exception. It is understood that stimulation to give more quantities of milk may be inherited independent of the size of the animal.

Even though a small cow has a great stimulation and potentiality to give milk is limited in her capacity to digest food which is required to meet the requirement of milk production. Therefore, it cannot compete with a large size animal which has the potentiality and stimulation to the same degree and also has the capacity and strength to handle the food necessary for high milk production. There is no relationship between the size of the animal and the richness of the milk within the breed.

4.1.5 Intervals between Milking

The composition of the milk generally varies to a little extent, when the cows are milked at twelve hours interval twice daily though the fat content of the evening milk shows a higher tendency. The general experience is that the larger the milking interval the greater the milk yield, correspondingly containing smaller percentage of fat, whereas the shorter the interval for the milk drawn the richer the fat with smaller quantity of milk. Attempts have been made to estimate the precise relationship between the milking hours and the (fat percentage) milk composition however, the results did not show any convincing degree of agreement. There is no doubt that the effect will be of different magnitude depending upon the individuality of the cow etc. The variation in the fat percentage may be from 0.5 to 1.0 percent under the conditions as stated above. The disturbing effect of the long night intervals, coincides with other factors such as large proportion of cows in early lactation in the herd, or the change order from stall feeding to grazing on pastures, then there is a serious risk of fat percentage falling below the normal standards that have been laid down under the law. The variations in the composition of milk at distinct intervals of time has been presented in tables 4.6 and 4.7.

It is recognised that many considerations are involved in fixing the milking hours for any particular herd, it is possible to avoid the large disputes by shortening the night intervals, during those periods, when low fat percentages are anticipated in the morning milk.

Table 4.6 Composition of morning and evening milk (%)

Breed	Fat		SNF		Difference and Percent	
	Morning	Evening	Morning	Evening		
Tharparkar	4.4	4.8	9.14	8.92	0.22	2.46
Hariana	4.4	4.8	9.20	8.96	0.24	2.68

Source : Milk and its properties S.M. Srivastava

Table 4.7 Three times a day milking efficiency

After Interval : Hours	Fat%	Yield	Butter fat per milking
Night :12.5 hrs (0 hour)	2.6	54 kg	1.4 kg
Morning : 5.5 hrs (5:30 AM)	4.8	40 kg	1.8 kg
Afternoon : 5.0 hrs (170 PM)	4.9	30 kg	1.4 kg

Source : Milk and its properties S.M. Srivastava

4.1.6 Variation from Day to Day

The milk which is drawn from the same cows shows greater variations in the fat percentage from day to day. Even though the cause of such variations, sometimes may be recognised but our knowledge of milk secretion is still incomplete to arrive at a satisfactory explanation in majority of cases. Perusel of table 4.8 reveals the variation in the fat percentage at different stages of milk during the day between the cows.

Table 4.8 Variation in fat content during various milkings (Percent of fat)

Portion	Cow A	Cow B	Cow C
First	0.9	1.6	1.6
Second	2.6	3.2	3.2
Third	5.3	4.1	5.0
Stripping	9.8	8.1	8.3

Source : Milk and its properties S.M. Srivastava

4.1.7 Efficiency of Milking

This is an important factor which is concerned with the yeild and fat percentage per milking. The percentage of fat increases by about 1 to 2 percentage in the first drawn or fore milk to 6 or more in the strippings as the process of milking proceeds. The percentage of non-fatty acids will show a decrease as the fat content increases, however, when calculated on fat free basis practically, there is no change or alteration. Therefore, the necessity is apparent for the complete milking and also for the maintenance of the milk yeild and the fat content. Further it has been observed that when the quarters are milked separately, there is a decrease in the percentage of fat in the milk drawn from the last quarter. This effect has been attributed to the reduction of pressure within the udder. Therefore, in order to ensure a satisfactory, fat percentage milking must be carried out quickly and thouroughly. It is also a common observation, that the employment of un-skilled milkers might be detrimental to both yeild and the quality of milk. Incomplete milking is also detrimental, because, when it is practiced continually, it has tendency to decline in lactation. The higher fat content of the strippings can be explained in part by the possibility that certain gland cells produce fat and that release of fat from these is greately increased as the pressure of the milk in the udder declines as the milk is withdrawn. However, the most logical approach for this difference in the fat percentage of the fore milk and the stippings can be explained on the basis of the differences in the specific gravity of the different constituents of the milk. Reference has already made that the specific gravity of the fat is 0.93 and it is lighter than other constituents of the milk. It is also necessary to consider here the bowl or funnel shape of the udder. The fat being lighter tends to float on the surface of the solution. Therefore it can be argued that the fore milk is therefore contains a lower fat as compared to the strippings or the last milk which is very rich in the fat content. The variation of the percentage of fat in different portions of the milking is given in the table 4.9.

Table 4.9 Variation in the percentage of fat in different portions during milking

Portions	Fat Percentage	SNF Percent
1	1.04	8.64
2	1.92	8.63
3	3.02	8.51
4	4.40	8.37
5	5.32	8.15
6	7.63	7.77

Source : Text Book of Dairy Chemistry - M.P. Mathur, D.D. Roy & P. Dinakar

4.1.8 Different Quarters of the Udder

Anatomically and physiologically all the four quarters of the Udder are separate units. Because of this structural difference the quarters milked first yeild milk having the highest percentage of fat than the milk from the quarter which is milked last. However, if a milking machine is used and all the four quarters are milked simultaneously the milk from the different quarters will be very similar in the fat percentage as shown in the table 4.10.

Table 4.10 Variation of Fat Percentage from Different Quarters of Udder

Animal	Left Fore	Left Hind	Average Left side	Right Fore	Right Hind	Average Right side
1	4.36	4.42	4.39	4.33	4.20	4.26
2	4.42	4.31	4.37	4.68	4.38	4.53
3	4.19	4.32	4.26	4.12	4.27	4.19

Source : M.P. Mathur, D.D. Roy - P. Dinakar

4.1.9 First and Last Milk Drawn

It is a well established fact that the first milk or fore milk drawn from the udder contains a very low fat percentage and that of the last milk or strippings as termed ordinarily contains very rich fat percentage and the difference is confined only to the fat content thus showing a striking difference. The variation in the fat content between the fore milk and the strippings is usually greatest with the heavy milk producing cows. In order to secure the maximum amount of the high fat testing milk, the cow should be stripped more effectively and efficiently.

Table 4.11 Composition of First and Last Milk drawn

	Total Protein %	Sugar %	Fat %	Ash %	Total %
First Milk	3.58	5.30	1.87	0.75	10.67
Strippings	3.38	5.33	6.28	0.70	14.86

Source : Milk and Milk Products Eckles et.al

4.1.10 Stage of Lactation and Milk Yeild

The fat content of milk is high after parturition, which decreases to a low level during the first or second month of lactation and then gradually increases with a change in the composition during the remainder of the lactation period. The fat content of milk variers widely during the lactation period, however, the general trend of the fat percentage is similar to the trend of the solids not fat. Towards the end of lactation there is a slight decrease in the percentage of lactose. Colostrum the first secretion after parturition is very different from the normal milk in its composition and general properties. The chief characteristics are its viscous nature, high percentage of globulin, low lactose content and high chlorides. Due to the high percentage of soluble protein, colostrum coagulates on heating, but this property is lost as the concentration falls to normal with a gradual change and the comprehensive figures are shown in the table 4.12.

Table 4.12 Change of colostrum to normal milk

	%T.S	%Fat	% Protein (diff.)	Lactose	S.G	%As h	%CaO	%MgO	%K_2O	%Na_2O	%P_2O_2	%Cl
At parturition	25.38	5.4	14.70	4.12	59.4	1.16	.288	.056	.150	.106	.439	.122
After 6 hr.	21.15	4.85	11.36	3.91	4.57	1.03	.216	.020	.182	.67	.282	.117
12hrs.	20.32	8.46	7.00	3.95	3.89	.91	.199	.031	.168	.097	.325	.121
18 hrs.	14.84	4.90	4.44	4.67	3.42	.83	.196	.022	.183	.092	.298	–
24 hrs.	13.70	4.43	–	–	3.43	.79	.174	.022	.185	.088	.295	.177
32 hrs.	11.65	2.31	4.09	4.38	3.46	.87	.186	.022	.214	.070	.302	.105
40 hrs.	13.81	3.92	4.17	4.81	3.45	.91	.202	.023	.217	.071	.343	.091
48 hrs.	12.22	3.14	4.06	4.16	3.22	.86	.192	.23	.205	.078	.134	.100
56 hrs.	12.52	3.23	3.73	4.74	3.32	.82	.176	.020	.179	.073	.275	.102
64 hrs.	12.98	4.25	3.83	4.08	3.21	.82	.178	0.23	.193	.070	.284	.094
72 hrs.	12.87	4.00	3.86	4.16	3.24	.85	.183	.025	.196	.076	.286	.096
10 days	13.56	3.73	3.17	5.91	36.1	.75	.168	.018	.183	.063	.252	.068

Note:TS Total Solids. Figures in the SG Column represents the Lacto meter reading i.e. 1000 (SG - 1).

Source : Text Book of Dairy Chemistry - Edgar. R Ling

The yeild of milk increases to the maximum during the second or third month of lactation and then declines at a steady rate to the end of the period. Further it has been observed that there is a reciprocal relationship between yeild and percentage of fat and non-fatty acids and the latter declines to the minimum value corresponding to the peak of production. The extent of this decline varies considerably with different individuals, however, the observations recorded indicate that for fat 0.2-0.4 percent and for non-fatty solids 0.2 - 0.3 percent may be acceptable. Breed, individuality condition of the cow at the time of calving and subsequent management are all known to modify the lactation effects.

Stage of lactation brings about some changes in fat composition. Particularly the proportion of steam volatile soluble fatty acids decreases regularly throughout the lactation period with a simultaneous reduction in the size of the fat globules. Lactose shows a gradual decrease of about 0.5 percent from the beginning of the lactation to the end of lactation.

Table 4.13 Influence of Stage of Lactations on milk quality

No.of lactations	**1**	**2**	**3**	**4**	**5**	**6**	**7**	**8**	**9**
No. of cows examined	67	67	49	34	21	12	9	4	3
Relative Yields (Lbs)	100.0	116.3	133.4	143.7	151.8	151.9	156.3	141.4	111.6
Fat (percent)	3.95	3.90	3.90	3.84	3.79	3.7	3.74	3.67	3.7
SNF (percent) on Fat Free Milk	9.40	9.32	9.26	9.31	9.10	9.38	0.03	8.94	8.9

Source : Milk and its properties S.M. Srivastava

Table 4.14 Stage of lactation in relation to milk yield and fat (Percent)

Month of lactation	Daily Milk yield in terms of percent	Fat (percent)
1.	99.6	4.0
2.	100.0	3.9
3.	92.0	4.0
4.	86.0	4.0
5.	82.0	4.1
6.	75.0	4.1
7.	71.0	4.2
8.	69.0	4.2
9.	64.0	4.2
10.	52.0	4.5
11.	34.0	4.6
12.	11.5	4.7

Source : Milk and its properties S.M. Srivastava

Table 4.14(a) Milk Constituent of Cross breed Cows during 1 to 3 lactations (Percent)

Milk Constituents	Jersey × Hariana	Brown Swiss × Hariana	Holstein Friesian × Hariana
Milk Fat	5.14+0.04	4.58+0.02	4.39+0.02
Total Solids	15.03+0.06	13.65+0.03	3.39+0.03
SNF	9.16+0.02	9.10+0.07	9.02+0.02

Source : Milk and its properties S.M. Srivastava

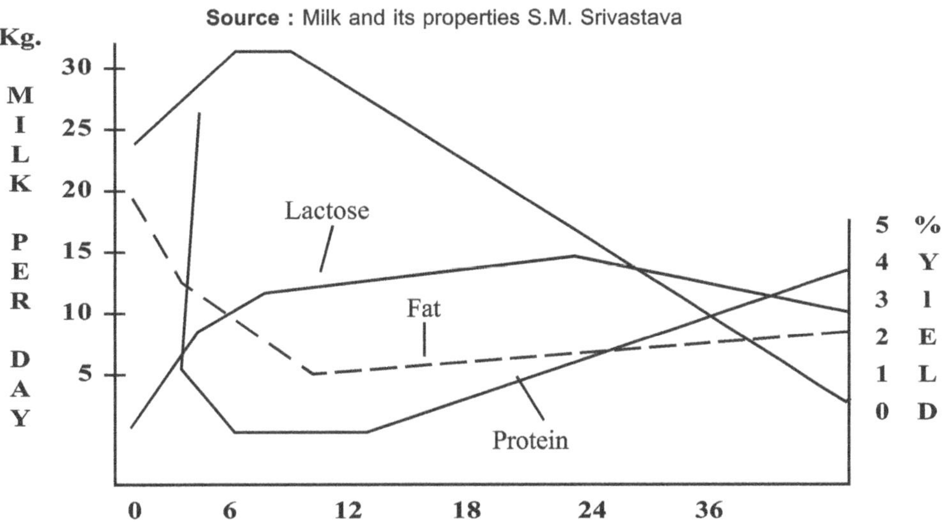

Fig. 4.1 Normal Lactation Curve

Source : Milk and its properties S.M. Srivastava

Whereas the chloride content increases with the advancement of the lactation showing an inverse relationship between the lactose and chloride content of the milk. Natural acidity commences from a high level and decreases to the normal value during the first three months and remains fairly steady until the last month when it declines rapidly. The two extremes may differ by as much as 0.05 to 0.06 percent calculated as lactic acid.

4.1.11 Season

Season of the year is the factor which causes variations in the composition of milk, when many other disturbing factors are eliminated. Both fat and solids not fat shows a slight but well defined variation during the course of the year. Variation in the fat percentage due to season shows maximum values in May and minimum in November, whereas solids not fat shows highest value in October for the cows of Red sindhi Breed and the values have been presented in Table 4.15.

Table 4.15 Effect of season on yield and composition of Red Sindhi cows milk in particular region (Percent)

Month	Season	Yield	Total	Fat	S.N.F Solid	Lactose	Ash	Protein
January	Winter	5.12	13.830	5.14	8.652	4.656	0.784	3.349
February		4.92	13.439	4.86	8.573	4.691	0.746	3.143
March	Spring	4.34	13.574	5.03	8.541	4.668	0.757	3.186
April		4.82	13.312	4.82	8.453	4.738	0.743	3.176
April		4.82	13.312	4.82	8.453	4.738	0.743	3.176
May	Summer	4.01	13.545	5.19	8.453	4.517	0.748	3.254
June		4.00	13.237	4.84	8.360	4.480	0.751	3.139
July		3.34	13.358	5.11	8.355	4.232	0.749	3.243
August	Rainy	4.80	13.614	5.18	8.240	4.449	0.756	3.205
September		5.38	13.141	4.85	8.438	4.602	0.731	3.085
October		6.16	13.443	4.82	8.291	4.609	0.774	3.237
November	Winter	6.01	13.332	4.72	8.683	4.700	0.747	3.231
December		5.01	13.459	4.84	8.618	4.575	0.753	3.265

Source : Text Book of Dairy Chemistry - M.P. Mathur, D.D. Roy & P. Dinakar

The lower values for solid not fat occur in the months of July and September. A number of separate factors such as temperature, hours of sunshine, length of day, exercise, rainfall, draught, stall feeding or pasture feeding, in flux of newly calved cows, thunder storms etc., are embraced the season with different magnitude, so far as the milk production is concerned.

Therefore, very often, the observations, regarding the effect of these complicated association of factors composition and milk appear as contradictory. However, the following points are clear with good degree of agreement.

1. An increase in the temperature favours slight decrease in fat percentage and vice-versa.
2. The amount of Vitamin-D is directly related to the exposure of the cows to sunshine.
3. Carotene and riboflavin reach a maximum level when green feeds are available in plenty during summer.
4. Summer drought depresses the solid not fat percentage, however, there is no effect on the fat percentage.

Lactose is the least variable of the gross components on percentage basis, but because of its osmotic relationship in milk, it increases slightly, as the other solids decrease, and decreases as the other solids increase.

The effect of season is apparently the result of weather conditions, especially the temperature. The percentage of fat is depressed during the period of hot weather, whereas, it increases, when the conditions are reversed. Depending upon the locality, the intensity of the effect of season exerted through the conditions of weather differs/ varies considerably. An effort has been made to present the effect of season or location on the protein in relation to the total solids as well as the different constituents in buffaloe and cow have been presented in the tables 4.16 and 4.17.

Table 4.16 Ratio of protein to total solids (percent) in milk of Indian and exotic dairy cattle

Species	Breed	Location	Protein	% in total solids in milk
Cow	Red Sindhi	India	3.34	24.00
	Gir	India	3.36	24.00
	Crossbreds (Red Sindhi × Brown Swiss)	India	3.31	24.42
	Holstein Fresian	U.S.A	3.32	27.03
	Jersey	U.S.A	3.78	26.02
	Guerensey	U.S.A	3.90	26.62
	Ayrshire	U.S.A	3.51	26.94
	Murrah buffalo	India	3.91	23.84

Source : Milk and its properties S.M. Srivastava

Table 4.17 Seasonal Variation in protein fractions of cow buffalo milk

Season of the year	Casein		Albumin		Globulin	
	Cow	Buffalo	Cow	Buffalo	Cow	Buffalo
Winter	2.55	3.20	0.26	0.36	0.20	0.17
Summer	2.54	2.88	0.23	0.30	0.21	0.17
Monsoon	2.58	3.22	0.28	0.36	0.16	0.16

Source : Milk and its properties S.M. Srivastava

During the dry months, the total milk yeild tends to decrease together with a decrease in the solid not fat content with an increase in the percentage of fat. Depending on the level of nutrition, a decreasing tendency in the percentage of both solid not fat and fat, with or without an increase in the yield of milk during the wet months. However, controlled humidity as opposed to natural weather conditions, does not appear to influence the milk composition when the temperature is kept below 24 °C.

The increase in fat yeild results from the higher rate of milk production that occurs during the winter months viz October to February because of lush green pasture and the availability of nutritive forage crops. The reduced milk fat yield has been observed during the stress summer months (March to June).

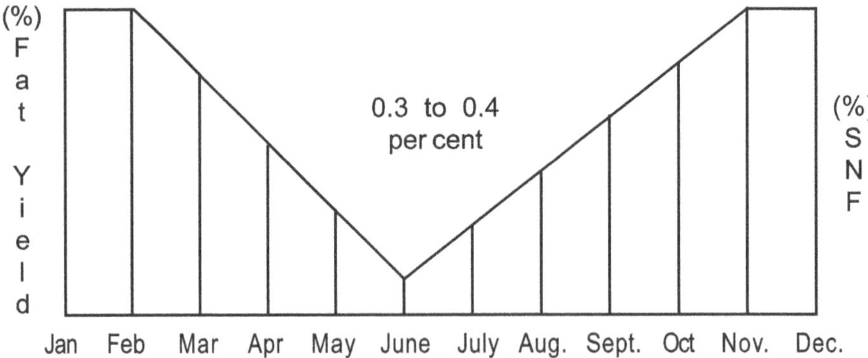

Fig. 4.2 Effect of season of year on milk fat yield ("V" shaped curve)

Source : Milk and its properties S.M. Srivastava

When the forage and range land pastures are poor with the rise in the ambient temperatures which is high and above the comfort zone. Due to the reduction in the total milk yield, the fat yield in reduced steadily during the lean summer months, yet the fat content is not affected adversely. The extent of variation is 0.3 to 0.4 percent only.

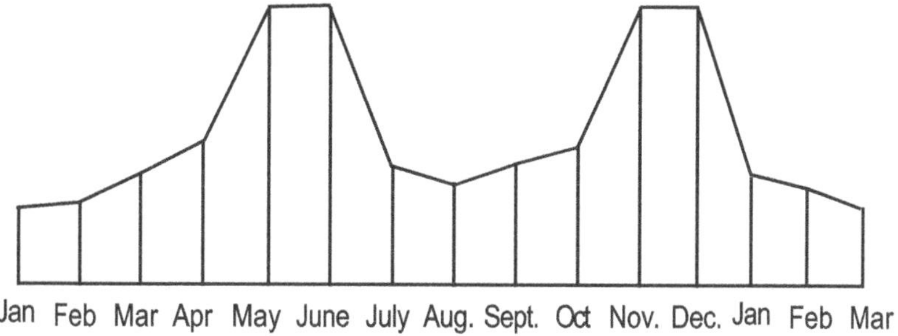

Fig. 4.3 Variation in SNF due to season (W shaped curve)

Source : Milk and its properties S.M. Srivastava

The milk protein and the Non-Protein Nitrogen (NPN) contents rise to a peak during summer and again during September month with the lowest value during February to April (Fig 4.2). Lactose content tends to be high from February to June (Summer months) however, the variation is not large. Total solids and SNF contents (Fig.4.3) are at a minimum during March and April with SNF at peak during May and June and the total solids are at a peak during October and November months.

4.1.12 Age of the Animal/Number of Lactations

With the advancement of age and the lactation numbers, it appears that there is a decrease in the average percentage of fat, along with a gradual though irregular decrease in solid not fat content. The decline in solid-not fat percentage with age is approximately twice the magnitude of the decline in fat percentage lactose and casein are the solids not fat components that are most affected by the ageing of the cow. The decline in solids not fat is ranging from 0.21 to 0.45 percent among the first seven (7) years of lactation according to some reports. Maximum milk yields may be obtained from fifth to the Ninth lactation. Age factor is most potent in advance lactation. The fat percentage or the richness of the milk is a matter of inheritance and is fixed even before the animal is born and nothing can be done to change it. Selection of the breed of breeding stock from high testing families and selection of individuals are the only means to be used in securing a high testing herd.

Table 4.18 Influence of age of cow on fat content of milk(%)

Age (Years)	Fat (%)	Age (Years)	Fat (%)
2.	4.35	11.	4.15
3.	4.33	12.	3.19
4.	4.31	13.	3.92
5.	4.26	14.	3.91

Table 4.18 ***Contd...***

Age (Years)	Fat (%)	Age (Years)	Fat (%)
6.	4.23	15.	3.90
7.	4.20	16.	3.90
8.	4.19	17.	3.94
9.	4.18	18.	3.30
10.	4.16		

Source : Milk and its properties S.M. Srivastava

Table 4.19 Relation of Age to Fat Percentage

Lactation Period	Jersey's Per Cent	Holstein's Per Cent
1	4.99	3.30
2	5.04	3.23
3	5.00	3.24
4	4.89	3.29
5	4.97	3.19
6	4.85	3.34
7	4.76	3.18
8	4.63	3.21
9	4.54	3.16
10	4.40	3.10
11	4.22	
12	4.00	

Source : Milk and Milk products Eckles et.al

Table 4.20 Relation of Age to Fat Percentage (Results from Official Tests)

Age (Years)	Holstein (Percent)	Ayrshire (Percent)	Guernsey (Percent)	Jersey swiss (Percent)	Brown (Percent)
2 - 3	3 - 42	4 - 05	4 - 98	5 - 37	4 - 01
3 - 4	3 - 39	4 - 00	5 - 04	5 - 45	4 - 06
4 - 5	3 - 39	3 - 95	4 - 01	5 - 36	4 - 02
Over 5	3 - 42	3 - 91	5 - 00	5 - 27	3 - 93

Source : Milk and Milk products Eckles et.al

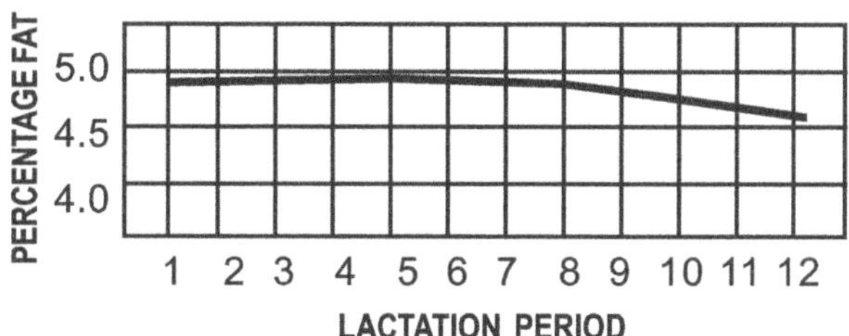

Fig. 4.4 Influence of Age on Fat content of Milk

Source : Milk and Milk products Eckles et.al

4.1.13 Effect of Gestation

It is a matter of common observation that a milking cow which is not bred or which fails to become pregnant holds up in milk production better than the one that is carrying foetus. Gestation can affect the changes in the composition of milk in an indirect manner by hastening the end of lactation, which has been widely accepted. An increase in milk solids especially solids not fat starts about fourth/fifth month of pregnancy and continues to increase till the end of the lactation. The decline in milk production after the animal reaches about six months in gestation is due to the effect of harmones produced by the ovaries, which exert an inhibiting effect on the secretion of the lactogenic harmones produced by the pituitory gland.

4.1.14 Level of Nutrition - Influence of Feeding

The amount of milk, a cow produces depends upon two things. First there are some internal factors, which are the results of inheritance and the second are the factors of environment, that is feed and management. It should be borne in mind that the feed alone cannot cause a cow to give milk beyond the limits of her inherited ability. Feeding more than the normal requirements by way of additional concentrate mixtures, has been reported to increase the solids not fat slightly. Feeding excess protein, does not alter the protein content of the milk appreciably however the non-protein nitrogen content and at times the fat content may be enhanced. The popular view is that a fall in both fat and non-fatty solid percentages may be anticipated, when the animals are shifted to grazing on pasture, and it is important to note that there is ample evidence for and against this belief. Some studies suggests the possibility that the new grass has a depressing effect on the fat percentage of the milk of the spring calved cows, whereas it may have little effect on that of the autum calvers. The effect of feeding on the composition of milk has been presented in tables 4.21, 4.22 and 4.23.

Table 4.21 Protein constituents of milk of cows and buffaloes (%)

Constituents	Cow milk	Buffaloe milk
Crude Protein	3.178 (940)	3.748 (804)
True protein	2.798 (192)	3.386 (177)
Casein	2.480 (205)	3.003 (170)
Albumin	0.406 (196)	0.383 (164)
Globulin	0.116 (199)	0.310 (174)
Proteose peptone	0.13	
Non-protein nitrogen	0.05	0.05
Total	**3.178**	**3.748**

Source : Milk and its properties S.M. Srivastava

Table 4.22 Proportion of Proteins in milk of various mammals (percent)

Species	Casein	Whey proteins
Cow	2.51	0.6
Buffaloe	3.0	0.5
Cat	3.8	3.2
Dog	4.8	2.6
Reindeer	8.4	1.5
Human	0.9	1.2

Source : Milk and its properties S.M. Srivastava

Table 4.23 Ratio of protein to total solid(%) in milk of various mammals

Species	In milk	In total solid
Cow	3.60	26.76
Buffalo	3.60	20.88
Goat	3.53	27.10
Sheep	5.23	27.11
Camel	2.98	24.00
Mare	2.70	24.54
Ass	2.01	18.32
Reindeer	10.30	28.06
Human	1.63	12.79
Llama	3.90	29.00
Egyptian Buffalo	4.16	23.23

Source : Milk and its properties S.M. Srivastava

Some intersting experiments have been conducted to ascertain the effect of feeding certain oils, such as palm oil, butter fat, lard and coconut oils which resulted in the increase of fat percentage, whereas cod liver oil or herring oil produced an opposite effect. It must be emphasized, that the results are variable not only between the individuals but also with the same animal at different times. The unsaturated fats have a depressing action whereas the saturated fat increase the fat percentage. The specific effect of cod liver oil is largely removed either by hydrogenation, which renders the fat less un-saturated, by eliminating some of the double bonds of the constituent fatty acids or by feeding the fat in the form in which it is protected from degradation and bio-hydrogenation in rumen. Food fats are able to modify to a limited extent the composition of milk fat. An increase in the proportion of certain groups of fatty acids in the food fat will lead to similar but slight modifications in the composition of milk fat. This is particularly true with the unsaturated fatty acids of the oleic type which are known to produce soft fats. It has been widely accepted that feeding of cows on pasture grass or the circumstances associated with it, bring about an increase in the unsaturation of milk fat and the reason for this change in the composition are still obscure. However, the technology to prevent rumen degradation and bio-hydrogenation of fat has enabled to modify the fatty acid profile tailored to our requirements, thereby increasing the content of essential fatty acids by about 35 percent or so.

4.1.15 Excitement

Both yeild and composition are liable to transient fluctuation during the periods of sexual excitement, storms, worrying by dogs or insects, but it is not possible to generalize any one of these factors. Some cows are more sensitive to unusual conditions than others. In such cases some animals are known to produce milk with low solid not fat content without affecting the fat content.

4.1.16 Heat or Oestrum

It is recognised that any unusual excitement or excercise may change the composition of the milk especially with regard to the fat content. However, the oestrum is a perfectly normal function of the animal and should hardly be classed, as un natural or abnormal and it is doubtful if any harmful effect would be produced upon the milk. The effect of oestrum on milk composition is real but inconsistent. There is usually much variation in the fat test, together with a slight decrease in yeild during this period. These changes have been attributed to increased excilability and nervousness causing the cow to either hold up some of the milk or to secrete less milk. The chemical analyses of milk showed that it is practically normal and fit for consumption.

4.1.17 Condition of Flesh at Calving

At the time of freshening, the cow must be in a strong, and vigorous condition as a good practice also should have considerable surplus flesh as a reserve. The milk yeild especially in the early period of the lactation, depends to a considerable extent upon this point. The fat percentage is not influenced as much by the physical condition of the cow as in the milk yeild. However, when the conditions are rather extreme, a pronounced effect upon the fat percentage results. A cow which calves in a decidedly fat condition may for the first twenty to thirty days have a fat test much above the normal level for the year, and decidedly more when compared that if the same cow was found to be in a thin condition. In some cases the effect may continue for as long as three months.

4.1.18 Relation between Daily and Yearly Production

The only means of knowing what a cow will produce in a year is to keep systamatic records. Estimates of yearly production based upon the results of a few days are certain to have a limited value. Very often the average can be determined, that will represent reasonably well the results from a group. Individuals, however, will vary widely from the average. However, there is fairly a definite relation between the highest point of the lactation period and the total production in a herd, where uniform conditions of feeding and management are maintained. The official test records are the results of a combination of a high class cow and of expert care and management. Because of this reason they are subject to wide variations in the relation between the daily production and yield for the year. This is especially true in the groups showing the exceptionally high daily production hence these results should be looked upon as an indication of the relation between daily and yearly production in a general way.

Table 4.24 Relation of Daily Milk Yield to Yearly Production, Twice-a-Day Milking, Calving Within Twelve Months

Average Daily yield	Average yield for Lactation period				Probable yearly milk production
Lbs.	Holstein Lbs.	Ayrshire Lbs.	Guernsey Lbs.	Jersey Lbs.	Lbs.
15 - 20	3,787	3,985	3,914	3,600	3,000 - 4,500
20 - 25	5,480	5,158	4,614	5,498	4,500 - 5,500
25 - 30	6,864	5,577	5,996	6,172	5,500 - 6,500
30 - 35	7,079	6,367	6,548	7,133	6,500 - 7,500
35 - 40	7,880	6,913	8,052	8,010	7,500 - 8,500
40 - 45	9,598	8,557	8,900	9,385	8,500 - 9,700
45 - 50	10,190	10,800			9,700 - 11,000
50 - 55	11,435				11,000 - 12,000
55 - 60	12,693				12,000 - 13,000
60 - 65	13,118				13,000 - 14,000
65 - 70	14,851				14,000 - 15,000

Source : Milk and Milk Products Eckles et.al

Table 4.25 Relation of Daily Milk Yield to Yearly Production, (Official Test Conditions)

Average Daily yield	Average Milk yield			Probable yearly milk production
	Holstein	Guernsey	Jersey	
Lbs.	Lbs.	Lbs.	Lbs.	Lbs.
20 - 25			6,112	6,000 - 6,500
25 - 30		8,030	6,770	6,500 - 8,000
30 - 35		9,394	8,800	8,000 - 9,500
35 - 40	10,745	10,935	10,505	9,500 -11,000
40 - 45	12,722	12,035	11,513	11,000 - 13,000
45 - 50	13,993	12,883	12,515	12,000 - 14,000
50 - 55	15,675	14,362	14,254	13,500 - 16,000
55 - 60	17,092	16,440	15,337	15,000 - 17,500
60 - 70	18,354	17,599		16,500 - 19,000
70 - 80	20,096			18,500 - 21,000
80 - 90	21,063			20,500 - 23,000
90 - 100	21,811			22,000 - 25,000
100+	28,821			24,000 - 29,000

Source : Milk and Milk Products Eckles et.al

4.1.19 Official Vs. Ordinary Milk Production Records

In recent years, official testing of pure breed dairy cows has become more common and the records of milk and fat production have been given wide publicity and are used as the basis for breeding.

A pertinent question that arises at this stage is to what extent the records of the recent years are the result of better breeding and to what extent they are to be attributed to more skill in the feeding and management. Another important question which also arise in connection with the records from the dairy herd improvement associations and their relation to records from official test conditions because of the fact that the records of these associations are decidedly below those made under conditions of advanced register testing. In view of this, the question that arises is as to how far it is fair to compare them, and if the wide range in records between the two is due to the quality of the animals or to the difference in the feed and management.

Under typical farming conditions, where most dairy herd improvement testing is done the milking is done twice daily and the cow is expected to drop another calf within twelve months. The feed and care vary widely, but the cow at best receives far less attention than under official test conditions. However, it should be borne in mind that the cow under test for advanced registration is first of all prepared for the test by liberal feeding preceeding the time of calving. When the test is underway she is milked three or four times daily, breeding is generally delayed, the best possible ration is provided and she receives the most careful attention that a skilled herdsman can give. The production under ordinary conditions, it should be noted, was excellent, indicating that cows that make good records under official test conditions are undoubtedly superior animals and may be relied upon to give a liberal production under ordinary conditions. Sometimes it is also argued that at times there may be a tendency to over rate the cows with official records when compared with the cows tested under the cow test conditions. The large records that are made under official test conditions are the result of a cow with outstanding ability as a milk producer combined with the most favourable environment.

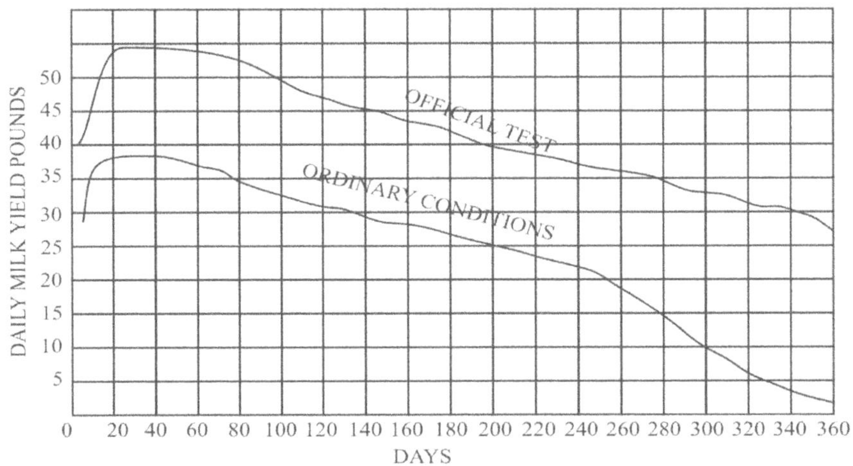

Fig. 4.5 Comparison of Milk Production under farm conditions Vs. Official testing

Source : Milk and Milk products Eckles et.al

4.1.20 Hormonal Influence

The various ductless glands of the body secrete certain chemical compounds, which are carried by the blood to other glands or directly to the target organs which they stimulate to activate and these compounds are generally known as hormones. Sometimes these hormones are known to suppress rather than to activate. Some hormones are produced by the sex organs especially the ovaries in the cow. The pituitary gland which is situated at the base of the brain is considered as the master gland because it acts as a central control measure of these substances which are regarded as chemical messengers produced from various ductless glands. These messages may be relayed by other glands or they may operate directly from the pituitary.

Prolactin, one of the harmone secreted by the pituitary gland is known to be responsible for the maintenance of lactation. When this substance is administered to cows which have crossed their peak of production, it stimulates a higher level of production with an increase in the percentage of the fat have been observed. Occassionally an increase in lactose concentration is also seen when the cows are treated with the pituitary extracts. Thyroxine an iodine containing compound secreted by the thyroid gland is known to be essential for milk production since the harmone controls the metabolic rate of the body. Feeding small doses of thyroprotein, a substance which stimulates the thyroid gland thereby releasing the harmone thyroxine to the cows which have crossed the peak production level led to an increase of milk yield and fat percentage. The same harmone also excersises a depressing effect, when administered at the peak of production or when the dose is too large.

Oestrogen, the harmone secreted by the ovaries of the female sex organs exercises two effects on lactation. One a stimulating effect and the other depressing effect. In the former both fat and non-fatty solids are increased although the yeild is diminished to some extent. There is an increase in the percentage of protein but the casein number remains normal. The stimulating effect is continued for some weeks after the administration of the oestrogen is discontinued. The depressing effect is exercised by administration of increased doses of the compound. There is a possibility that the harmones may pass into the milk, which may have some undesirable effects on consumer, when they are administered either synthetic or natural oestrogen preparations. Therefore, as it is necessary and important to establish beyond doubt that harmone treatment has no detrimental effect upon the animal. Undoubtedly it can be concluded that oestrogen stimulates the lactation harmones of the pituitary gland.

4.1.21 Effect of Disease

Pronounced changes in the composition of milk are associated with an inflammed condition of the udder. There is a marked reduction on the volume of the milk secreted with the onset of Foot and Mouth disease, whether the udder is inflammed or not with a moderate increase in the percentages of fat, protein and Ash content, whereas the percentage of lactose decreases.

Mastitis

When the cow is infected with mastitis, the disease in which the udder is involved, the composition of milk shows the following characteristics.

(a) Low lactose and casein percentages.

(b) Increases in chloride content.

(c) Increases in the percentage of soluble nitrogen and a consequent reduction of the casein number.

(d) Some reduction in natural acidity. These changes occurs in such a manner as to maintain the normal osmotic pressure of milk having normal freezing point with an invariably low percentage of non-fatty solids. The decrease in osmotic pressure is due to the reduction of lactose, which is counter balanced with an increase of chlorides thus resulting in the lower percentage of non-fatty solids.

Other Diseases

It is considered that diseases which causes a marked reduction in the milk yeild may be expected to modify the composition of milk. Ketosis a metabolic disorder that occurs in the milch cattle reduces the milk yeild depending on the severity of the disease.

4.1.22 Influence of Drugs

The interest in the possible usage of drugs as a means to stimulate milk production arises largely in connection with the phenominal milk and fat records which have characterised official testing records. Most of the experimental evidence available indicates a depressing effect due to the usage of drugs rather than an increasing effect. Some of the changes has depressing effect on milk secretion whereas others do not exhibit any such effect.

4.1.23 Miscellaneous Causes

(a) Methods of milking

Milking is generally considered as a simple operation, that any common labourer is supposed to be able to milk. However, there is an immense difference between the milkers and the greatest difficulty in the farm operations is that of securing competent men for this work that is the milking operation.

(b) Change of milkers

It has been observed that frequent change of milkers may lead to less milk yeild. One milker may be able to get 20 percent more milk than another. One may dry the cow within a few months, while another may keep her in milk the entire year. The milker should not be allowed to excite or worry the cows by loud talking or cruelty or absue of any kind. Milking of the cow must be carried out quietly and quickly.

(c) Hard milking cows

Some cows cause considerable annoyance to the milker because they milk unusually hard. This difficulty is caused by a strong sphinctor muscle, which closes the opening of the teat tighter than it should be. Teats which become dry and harsh also tend to produce this condition.

(d) Leaking Teats

Some quantity of milk is lost by the leaking teats from the udder between the milkings due to the weak contraction of the sphinctor muscle.

(e) Bloody milk

Blood in milk is more common than generally understood. Often it may be noticed in separator slime, when its presence is not suspected in the milk. It is not an indication of disease or any unhealthy condition in the cow. It is caused due to the rupture of a small blood vessel that allows blood to escape into the milk cistern or the milk ducts.

(f) Chapped teats

Sore teats may be caused by cold weather milking with wet hands or other causes of local irritation. When so affected the cow does not stand quietly for milking on account of the pain.

(g) Warts on the teats

Warts on the teats are often troublesome.

(h) Kicking cows

Due to faulty management the cows are given to few vices of which the most common vice is that of kicking when being milked.

(i) Self-sucking cows

This vice is not very common, but it is annoying when begun and difficult to stop.

4.2 Inter Relationship between the Milk Constituents

The major constituents of milk are independent of each other and the concentration of these constituents are influenced by a large variety of factors that have been described earlier. Physiologically milk has to maintain Isotonic equilibrium with blood. Consequently it is seen that the milk constituents which contribute towards the osmotic pressure, namely the lactose, chloride sodium etc., bear a relationship with each other. Fat is completely independent of any other constituents and is influenced only by the hereditary and environmental factos like breed, type of feed, interval of milking stage of lactation and so on.

Fat and Solids not Fat

Generally it is seen that an increase in the fat content is concomitant with an increase in the solid not fat content. This relation is not quantitated in specific terms but is only empirical in nature. The relation between fat and solid not fat is fairly understandable on the basis that protein and fat content are positively correlated whereas the lactose content is fairly constant. This correlation is significant only from the point of view of fixing legal standards and pricing policy of milk. This correlation also gives a broad indication of adulterated milk with water particularly in India where attempts to dilute buffaloe milk are quite common.

Vieths Ratio

This like the nitrate test is valuable in furnishing confirmatory evidence to the freezing point test and other analytical data. In normal milk the ratio of percentage of lactose. Protein ash bears a definite relationship. Vieth estimated this as 13 : 9 : 2. For Indian milks this ratio may not hold good and has been estimated as 12.4 : 8.6 : 2 for cow milk and 12.4 : 9.7 : 2 for buffaloe milk. This ratio varies for abnormal milks, where the udder is infected etc. If the milk is diluted with water all the three constituents decrease proportionately maintaining the ratio without any change however, the freezing point would change confirming the adulteration with water.

Lactose-Chloride Number

The most important factors that are responsible in maintaining the osmotic pressure of milk are lactose and chloride. Hence, a decrease, or increase in one of these is matched by a proportionate increase or decrease in the other. This is particularly true in the case of mastitis infected animals. Where it is seen that the lactose concentration decreases with a simultaneous proportional increase in the chloride content is observed. This relationship has been quantified as the chloride lactose number or the Koestler number.

$$100 \times \frac{\text{Percent chloride}}{\text{Percent lactose}}$$

This number varies between 1.5 and 3.0 for normal milks. This relationship between chloride and lactose arises from the need to maintain isotonic equilibrium with the blood.

The Nitrate Test for Added Water

Natural water supplies sometimes contain nitrates, whereas milk contains no appreciable traces, therefore, it is clear that the presence of nitrates in milk may be taken as evidence of watering. The test has many limitations, and their absence cannot be taken as a criterion of genuineness because of the fact that some public water supplies are free from nitrates. Another limitation lies in the fact that nitrates in milk are slowly reduced and as a consequence an acid sample which originally would have given a positive result may escape detection.

Refractive Index of Milk Serum

The addition of a prescribed copper sulphate solution to milk, the colloidal matter is precipitated and on filteration a clear serum is obtained, which may be examined with a dipping refractometer. The most serious limitation lies in its inability to discriminate between watered milk and abnormal samples. Regarding other forms of adulteration it is impossible at present to detect addition of separated milk or removal of fat so that reliance must be placed on the estimation of fat, when it is below 3 percent raises a presumption of adulteration.

Chapter 5

Physico-chemical Properties

5.1 Physical State of Milk

The constituents of milk are either in dissolved or suspended state, where water is the continuous phase. Lactose and a small portion of the mineral salts are found in solution, proteins and the remaining minerals in colloidal suspension, whereas the fat is present as an emulsion.

The physico-chemical properties of the cow and buffalo milk have been summerised and presented in a tabular form (Table 5.1) .

Table 5.1 Physical properties of buffaloe and cow milk

Physical Property	Buffaloe	Cow
Specific gravity at 20 °C	1.0323	1.0317
Viscosity at 27 °C in centipoise	2.245	1.450
Curd tension (g)	32.85	28.54
Surface Tension (dynes per cm at 20 °C)	45.50	42.50
Freezing point (centigrade)	–0.5454	–0.530
Titrable acidity (as % Lactic acid)	0.16	0.15
pH (at 20 °C)	6.7	6.6
Redox Potential (Volts)	+ 0.310	+0.258
Electrical Conductivity (mmho)	3.22-6.67	4.00 - 5.50
Thermal conductivity (Kcal / hm C at 37 °C)	0.569	0.46
Heat capacity (cal / g C at 30 °C)	0.852	0.93 0.95
Colour	White	Creamy

Source : Text Book of Dairy Chemistry - M.P. Mathur, D.D. Roy & P. Dinakar

5.1.1 Colour of Milk

Milk has a characterstic white opalescence due to the scattering of light by the colloidal particles. Actually the colour is a blend of the individual effects produced by:

(a) The colloidal casein particles and the dispersed fat globules, together scatters the light and

(b) The carotene and to some extent xanthophyll which gives a yellowish tint to the milk.

The colour of the milk ranges from a yellowish creamy white which is a characteristic colour of the cow's milk to a creamy white in the case of buffalo milk. The intensity of the yellow colour in the case of cow's milk depends on various factors such as breed, feed, size of fat globule, percentage of fat in milk etc. A deeper yellow tint/colour of milk is seen in certain breeds of cows than in others. Greater intake of green feeds also imparts a deeper yellow colour to the cow's milk. Larger the size of the fat globules and higher the fat percentage of the milk, the greater the intensity of the yellow colour. Skim milk has a bluish whereas the whey has a greenish yellow colour, which in milk has been masked by the other constituents present in the milk.

5.1.2 Flavour and Taste

This is a property which is difficult to define, however, there is no doubt that taste and smell are used in its assessment. The flavour of the milk is a blend of the sweet taste of lactose which is balanced against the salty taste of the minerals and is damped down by the proteins. The phospholipids, fatty acids and fat of milk also contribute to the flavour. The normal balance of flavour of milk might be disturbed, when some flavouring essences present in such plants as mint, caromile and garlic finds a passage into the milk. Freshly drawn milk has a certain 'cowy' odour, which passes off by the time it reaches the consumer. As lactic acid develops the flavour of the milk changes in a very definite manner due to the resultant production of various by-products such as butyric acid, diacetyl etc. Changes with the flavour of milk also occur due to the nature of the feed, season, stage of lactation, condition of the udder, sanitary condition prevailing in the milk barn and subsequent handling of milk during storage. The sulfhydryl compounds significantly contribute to the cooked flavour of heated milks. Any pronounced flavour other than the normal is considered as abnormal and the various sources of abnormal flavours may be :

1. Bacterial growth
2. Feed
3. Absorbed
4. Chemical composition
5. Processing and handling
6. Chemical changes and
7. Addition of foreign material.

5.1.3 pH of Milk

pH is generally expressed in number which is equal to the negative logarithm of hydrogen ion concentration in the system and any solution is said to be neutral when it shows pH of 7.0. Since the pH is a logarithmic function, any change in one limit of pH indicates a ten fold difference in the H^+ ion concentration of the solution. pH can be measured either colorimetrically using indicators or buffer solutions or electronically with pH meter either with a pair or a combined electrode. The use of pH meter is more suitable for measuring the pH of milk. The average pH of the buffalo milk at 10 °C is 6.79 and decreased to 6.26 at 70 °C which is significantly but negatively correlated to fat content of the milk.

The hydrogen ion concentration of milk is approximately $10^{-6.6}$ gm per litre and corresponds to the pH of 6.6 in the freshly drawn milk. This concentration lies slightly on the acidic side of the neutral pH 7.0. The normal pH value of normal, fresh, sweet milk of cows varies from 6.4 to 6.6 and 6.7 to 6.8 for buffalo milk. Higher pH values for fresh milk indicates or suggestive of infections of udder, whereas the lower values might be due to bacterial action on milk. The pH of the fresh milk may vary with

1. Species
2. Breed
3. Individuality
4. Stage of lactation and
5. Health of the animals etc.

5.1.4 Specific Gravity and Density

The density of a substance is defined as its mass (weight) per unit volume i.e., M/V = D or W/V = D, whereas the specific gravity is the ratio of density of the substances to the density of standard substance i.e., water. The unit of volume normally used is the milli litre and the unit of weight is gram. One milli litre of water at 4 °C to be more precise 3.98 °C weighs 1 gram and the specific gravity water is exactly 1 at 4 °C. Water expands at temperatures higher or lower than 4 °C, hence the water has a specific gravity relative to 4 °C of less than 1.

It is necessary to specify the temperatures, while reporting the densities or specific gravities of any substance because the density of the substance varies with the temperature at which it is determined. The specific gravity of a substance when referred to water at 4 degree celsius is numerically equal to the density of that substance in the metric system. Generally the specific gravity of milk is expressed at 60 degree F or 15.6 degree celsius ranging between 1.028-1.034 and on an average it is found to be 1.032 but is most variable in its properties. The specific gravities of both fat and solids not fat are subjected to slight variations, due to the fact that they themselves are mixtures of varying composition, and these variations merely reflect the fluctuating proportions of water, fat and solids not fat. Further there are some differences between the specific gravity of solid fat and liquid fat even though, both are at the same temperature.

The density or specific gravity of milk may be determined by either determining the weight of a known volume or the volume of a known weight. The weight of a known volume may be determined either with pycnometer or with a hydrostatic balance, whereas the volume of a known weight is determined by using lactometers the scale of which is calibrated not in terms of volume but as a function of either density or specific gravity. The density of milk can also be measured by means of the lactometer which is a special hydrometer calibrated over the range 1.025–1.035(25.0-35.0 as lactometer degree) or 0.000-040(0-40 as lactometer degree). Lactometers are calibrated at different temperatures. After recording the lactometer reading at the calibrated temperature of lactometer or corrected lactometer reading at the temperature of milk and after applying the correction factor the specific gravity can be calculated as follows.

$$\text{Specific gravity} = 1 + \text{CLR} / 1000$$

The common types of lactometers are Zeal, ISI or BIS, Quevenne etc. Milk is heavier than water and the average specific gravity ranges from 1.028 to 1.030 for cow's milk, 1.030 to 1.037 for skim milk. The specific gravity of different milk constituents approximately as follows :

Water 1.000, fat 0.93, proteins 1.346, lactose 1.666, salts 4.12 and solids not fat 1.616. One must be able to appreciate the limitations of commonly used formulae connecting the specific gravity of milk with its chemical composition. One such formula, proposed by 'Richmond' has proved to be most convenient out of the several formulae calculated from time to time.

$$T.S = L.R/4 + 1.2F + 0.14$$

where, T.S = Total solids in percentage

L.R = Lactometer reading at 60°F i.e., 1000 (S G^{-1})

F = fat percentage

The Richmond's formula has been revised and recalculated for use with the modern improved density hydrometer is as follows:

$$T.S = D.H/4 + 1.21F + 0.66$$

where, T.S = Total solids in percentage

F = Fat percentage

D.H = Density hydrometer reading at 20°C (i.e., 1000 × density^{-1})

In our country, the Bereau of Indian Standards has proposed the following formula for the calculation of total solids percentage or solids not fat contents in milk as per IS : 1183.1965 (Revised)..

$$\% TS = 0.25D + 1.22F + 0.72$$

$$\% SNF = 0.25D + 0.22F + 0.72$$

where $D = 1000\ (d - 1)$

d = Density of sample of milk at 20 °C

F = Fat percentage of the sample

These are very convenient, being used very commonly on the assumption that the sample contains the normal constituents of average specific gravity in normal proportions. Addition of water lowers the specific gravity of the milk, however milk with high fat content also has a low density, whereas on the contrary, addition of (addition of cream or by increasing temperature) skim milk removal of fat or reduction of temperature increases the specific gravity. The specific gravity of fresh drawn milk is lower than is indicated an hour or later. This difference is of the order of 0.001 which corresponds to one degree of lactometer. This phenomenon has discovered by 'Recknagel' and is known by his name. Further, he found that the time required for the completion of this change could be shortened by keeping the sample cold and concluded that these changes are attributable to casein. The changes in the physical condition of the fat are also partly responsible for the 'Recknagel phenomenon'. In view of this phenomenon which occurs in the milk the specific gravity of the milk should not be determined for at least one hour after it is drawn from the udder of the animals, otherwise a lower value than the normal will be obtained.

5.1.5 Acidity of Milk

The milk which is freshly drawn from the udder possess a certain amount of acidity which is termed as the 'natural acidity' and is amphoteric in its reaction to litmus i.e., it turns the red litmus to blue and the blue litmus to red. The acidity which is present in the milk is determined by the titration with an alkali (sodium hydroxide N/9) in the presence of an indicator, phenolphthalein which is called as 'Titrable' or 'natural' or 'apparent' acidity which in turn is caused due to the presence of casein, acid-phosphates and citrates and to a lesser degree due to albumin, globulin and carbon dioxide in the milk. This is the natural property of the milk. The natural acidity of the milk of individual cows varies considerably depending on species, breed, stage of lactation, physiological state or condition of the udder etc, however the natural acidity of fresh herd milk is more uniform in nature. The titrable acidity varies on an average from 0.13 to 0.14 percent in cows, and 0.14 to 0.15 percent in buffalo milk. The higher the solid not fat (SNF) content of the milk the higher the natural acidity and vice versa. Developed or real acidity is due to the lactic acid, which is produced as a result of bacterial action on the lactose present in the milk. Therefore titrable acidity of stored milk is equal to the sum of natural acidity and developed acidity and it is generally expressed as the percentage of lactic acid.

Titrable acidity value of milk = Natural acidity + Developed acidity

(TA = NA + DA)

The use of titrable acidity as a rapid platform test, in rejecting the milks, which are unsuitable for converting into evaporated milk because of the fact that the developed acidity has a powerful effect/influence in lowering the temperature of heat coagulation of casein, which may create trouble on this account. In the absence of any convenient platform test to estimate the lactic acid in the milk, we are obliged to rely upon the organo-leptic observation and taste coupled with titrable acidity. Therefore, an arbitrary maximum of 0.18 to 0.19 percent is taken as the safe margin in dividing the satisfactory and unsatisfactory milk deliveries.

5.1.6 Viscosity

It is defined as the tangential force on unit area of one of two parallel planes at unit distance apart with the space between the planes filled with fluid and one of the planes moving relative to the other with unit velocity in its place. The unit of viscosity (μ) is the poise (centipoise).

Milk is considerably more viscous than water, mainly due to the fat emulsion and colloidal particles. Any alterations/changes in the physical nature of the fat and changes in the hydration of proteins might have some effect on viscosity. Homogenization, where the fat globules are sub divided, causing alterations in the state of division increases viscosity. This property of milk assumes greater importance in considering both the formation and properties of cream. The viscosity of the surrounding liquid resist the movement of the fat globules, whether the fat is separated either mechanically or by the force of gravity because of the fact that the viscous cream creates an impression of "richness" i.e., very rich in fat content.

Dynamic Viscosity (n)

Kinematic Viscosity(r)

This is defined by the equation $\gamma = \eta/Q$

where Q = is the density of fluid.

Kinematic viscosity = Absolute viscosity / density.

The unit of kinematics viscosity is the stoke (or centistoke) corresponding to the absolute unit of viscosity the poise (centipoise).

Poise

Poise is defined as the force in dynes/cm required to maintain a relative velocity of 1 cm per second between two parallel planes 1 cm a part. Centipoise is one hundredth of a Poise.

The viscosity of milk is measured at 20 °C and it is decreased with an increase in the temperature. The relative viscosity of different milks are as under

Cow milk – 2.000 cp
Buffaloe milk – 1.800 cp
Skim milk – 1.500 cp
Whey – 1.200 cp
Absolute viscosity of water – 1.005 cp

The viscosity of the milk of the indian animals appears to be in the order: buffalo, cow, goat, and ass respectively. Viscosity of milk is influenced by state and concentration of protein, fat, temperature and age of the milk. The viscosity of pasteurized milk, both homogenized and un homogenized increases with age. The viscosity may be measured in three ways.

1. The time of flow under fixed pressure such as with 'Ostwald viscometer'. The viscosity of the liquid is directly proportional to the time taken for a certain volume of a liquid to flow out under a certain pressure gradient and also is proportional to the density of the liquid when flows through a capillary tube. Viscosity of milk is compared with that of standard liquid distilled water and absolute viscosity of milk can be obtained by knowing the viscosity of water.
2. By measuring the force required to move two layers of liquid past each other as with MacMichael viscometer.
3. By measuring the fall of a ball through a column of liquid as it is accomplished with the falling ball viscometer.

Apparent viscosity and plasticity are the two terms used in addition to the viscosity of milk and its products. Viscosity of the fluid is a measure of the resistance to flow and the internal friction forces decreases the rate of flow. Viscosity is related to the flow properties of fluids which in turn control the manufacturing operations of many milk products hence it is considered as an important physical property which needs more attention.

Theories of Viscosity

There are three main theories of viscosity which will help to visualize the phenomenon of viscosity at molecular level with the molecular forces. They are:

Molecular Theory

The intra molecular affinity/attraction between the molecules in a liquid brings them together or couple to each other. Thus, when two adjacent layers of a fluid move with different velocities will tend to drag each other due to the inter molecular attraction, which will dissipate the velocity gradient within the liquid resulting with deformity of the molecular structure of the fluid. The configuration of the molecules is affected and becomes more ellipsoidal than the usual radial distribution due to the dragging force on the surrounding molecules. The principle axis of this ellipsoidal distribution determined by the local velocity gradient and the degree of deformation of the molecular structure determines the coefficient of viscosity.

Rate Process Theory

The molecules which are present in the liquid has to pass over a potential barrier in order to flow is the most important factor of this theory. On collision, two molecules forms a pair instantaneously and this double molecule passes through its neighboring by rotation, when there is a free space in the adjacent layer. This free space in the adjacent layer is created by the difference in local density due to the fluctuation of temperature. The forward and backward motion would be equally frequent when no external stress is applied and the potential barrier will be deformed in order to favour high forward move, when an external stress is applied. Thus it appears that a molecule in order to move into an unoccupied site, passes through potential energy bottleneck which is lowered when shear force is applied.

Diffusion Theory

According to this theory, self diffusion of molecules increases in the direction of shear field. Under application of shear force and naturally the displacement of molecule decreases with opposite direction of the shear. In case the molecules are considered to have overcome the activation energy for diffusion, in self diffusion, and in general the diffusion theory model in essence is similar to the rate process theory model.

5.1.7 Specific Heat

The specific heat of the milk can be calculated from the knowledge of the specific heat of the milk constituents and their percentages. Considering the specific heat of fat as 0.5, lactose 0.3, protein 0.5, ash 0.7 and for water 1.0, and the resultant value is 0.93 for milk of normal composition. The observed specific heat of warm milk, in which the fat is liquid, is somewhat lower than the values obtained below 19 °C, the reason being some of the heat supplied to milk at a temperature near the melting point of the fat is used to supply the latter with its latent heat of fusion. Due to these reasons, the specific heat curves either for milk or cream show 'humps' being maximum at 19 °C, whereas whey or separated milk did not show such maximum. The following specific heats are observed at 60 °F for :

milk - 0.94,

skim milk - 0.94,

whey - 0.98,

butter - 0.53;

cream with 30 percent fat - 0.98, 60 percent fat cream -1.05 and cheese 0.64.

Whereas at 40 degree celsius the specific heat values are:

milk 0.93, 30 percent fat cream 0.85 and 0.72 for the cream with 60 percent fat.

It is defined as the amount of heat required to raise the temperature of one gram of a substance through 1°C. Generally it is measured at a constant volume or at a constant pressure even though the difference between the two values is negligible in the case of solids and liquids, whereas in the case of gases it is appreciable and high ratio of the two values gives us valuable information about the atomicity of the gas molecule. The specific heat value of normal milk is 0.93 and is approximately the same for the warm milk in which the fat is in liquid condition.

5.1.8 Surface Tension

It is defined as the force (in dynes) acting upon a line of 1 cm length on the surface of the liquid. There are several methods of measuring surface tension:

(a) The force required to pull a metal ring from the surface of the liquid.
(b) The number of drops formed when a given amount of liquid is allowed to fall from a pipette and
(c) The pressure required to force a bubble through a nozzle immersed in the liquid.

The surface tension of milk is determined by an apparatus called 'Torsion' balance, based on the method of force required to pull a metal ring from the surface of a liquid. The average value ranges between 40-60 dynes/cm at 20 °C. The surface tension of cow and buffalo milk at 20 °C are 51.84-51.93 and 50.40 dynes/cm and at 27 °C 51.09-51.19 and 49.64 dynes/cm respectively.

5.1.9 Bound Water

The surface tension of milk is low when compared with water, the actual ratio being 5:2 and to some extent higher values are shown by the separated milk, whereas cream has a lower surface tension. It is considered that the colloidal constituents, particularly the proteins are responsible for the lowering of the surface tension. The strong tendency for the substances which lower the surface tension is to be concentrated at the liquid/air interface is the characteristic feature of such colloidal solutions. Calcium caseinate gathers at the liquid/air surface along with small quantities of albumin, globulin and fat constituting the familiar skin when the milk is warmed. Similar precipitation occurs around the air bubbles resulting in the phenomenon of frothing when the milk is agitated. Reduction of surface tension, when the milk is warmed/heated above 30 degree celsius, thereby resulting in a marked tendency to foam,on account of the more active concentration of protein around the air bubbles. To a certain extent the stability to the form is attributed to the protein films, being somewhat elastic in nature, which ultimately collapses, leaving the disrupted films on the surface of liquid. Milk which is cooled below 20 °C shows an increase with the foaming properties, whereas the minimum foaming are observed to be in the ranges 20-30 °C temperature. The effects of surface tension might partly explain the foaming qualities of milk certainly they are not the only factors involved.

The percentage of water present in milk on an average is about 87.5 and all is not in the free form. Some amount of water is firmly bound by the milk proteins, phospholipids and some by the substances adsorbed on the fat globules. The quantities of water bound in milk and dairy products have been given in the table 5.2.

Table 5.2 Distribution of water in milk and milk products

Product	total water %	Water bound per 100 gm of product	Age in hours
Raw milk	86.75	3.18 gm	24
Skim milk	90.56	2.13 gm	24
Cream	70.92	2.50 gm	8
Butter milk	91.75	1.75 gm	-
Colostrum	80.83	4.65 gm	24

Source : Text Book of Dairy Chemistry - M.P. Mathur, D.D. Roy and P. Dinakar

Out of 3.18 gm of bound water present in 100 gm of milk, 50 pecent is held by casein, 30 percent by albumin and globulin and about 15 percent by the fat globular membrane. It is observed that when milk is stored at 40°F the bound water slowly increases until equilibrium is established. Bound water $\rightleftarrows$ Free water. The equilibrium thus attained may be disturbed by heating above 160°F, when after cooling and ageing the samples were found to contain less bound water.

5.1.10 Freezing Point

A solution freezes at a lower temperature than the pure solvent and the depression of freezing point depending upon the concentration of the dissolved substances. Milk freezes at a temperature which is slightly lower than water due to the presence of soluble constituents such as lactose, soluble salts etc, thus lowering or depressing the freezing point. The depression of the freezing point is considered to be the function of the osmotic pressure of the solution. It is reasonable to believe that the osmotic pressure of the milk is approximately the same as that of blood, which is fairly constant when considered on physiological grounds. In fact, the freezing point of milk is the most constant property and addition of water is bound to alter the normal value hence, the estimation of freezing point is valuable in detecting this kind of adulteration. The average freezing point depression of Indian cow milk may be taken as – 0.547 °C (31.02 °F) and buffalo milk – 0.549 °C (31.01°F). Majority of the bulk milk samples have a freezing point depression – 0.530 °C (31.05 °F). Any depression lower than this value indicates added water. Souring of milk results in the lowering of the freezing point due to an increase in the amount of soluble molecules. Therefore, the freezing point should be determined on unsoured samples for greatest accuracy. Boiling and sterilization increases the value of freezing point depression, however pasteurization has no effect. The fat and proteins of milk have no direct effect on the freezing point of milk. The drawbacks of the freezing point test are:

1. It does not detect the addition of skim milk or removal of fat from the milk sample and
2. Watered milk, which has subsequently soured may pass the test.

5.1.11 Boiling Point

A solution boils at a higher temperature than the pure solvent according to the concentration of the dissolved substances. The constituents of milk which are in true solution are mainly responsible for the elevation of the boiling point above 100 °C. The boiling point of the milk has been reported to be 100.15 °C, but the actual observation proves to be underestimated by about 0.40 °C. The reason for this discrepancy is undoubtedly an alteration of the normal ionic molecular colloidal equilibrium as a result of heating. Addition of water lowers the concentration of the dissolved substances, thus responsible for the elevation of the boiling point. Hence its determination might be helpful to detect this particular form of adulteration.

5.1.12 Refractive Index (RI)

It is directly related to the soluble constituents of milk. Milk has a refractive index of about 1.35 and that of water being 1.33. Considerable variations has been observed with values for genuine milk. It is not possible to use this property as a criterion for the genuineness of milk samples, as the addition of water would therefore lowers the refractive index of milk.

Refractometer, the instrument gives a reading, which is a measure of refractive index, and the test has been used as a rapid means of sorting the suspected samples for further examination. Indeed, this test has the advantage over the routine estimations of fat and non fatty solids.

Light when passes through at an oblique angle from a less dense medium like air into a more dense medium such as water it is bent or refracted. The magnitude of the bending is expressed as a ratio of the 'sines' of the angles of incidence (i) and the refraction of light (r) is known as the refractive index (R.I) and is designated by letter 'n'.

$$n = \sin(i)/\sin(r)$$

The refractive index of a pure liquid is a characteristic constant for that liquid under specified conditions of temperature and wavelength of light. The refractive index of water at 20 °C with the D line of sodium spectrum (589.3 mμ) is nd^{20}=1.33299 whereas for milk it ranges from nd^{20}=1.348d. Refractive index of milk itself to certain extent is difficult to determine because of its opacity, however , by using 'Refractometer' such as 'Abbe Refractometer', which employs a thin layer, it is possible to make a satisfactory measurement of milk. The values of refractive index (R.I) for milk at 20 °C is 1.4600 whereas at 40 °C the values are 1.3460 (1.3450-1.3472) for cow milk and 1.3477 (1.3460-1.3490) for buffalo milk. The individual constituents of milk make the following contributions to the refractive index at 20°C with the D line of sodium spectrum are as under

Casein	= 0.0049 - 0.0060
Serum protein	= 0.0021 - 0.0035
Lactose	= 0.0063 - 0.0067
Miscellanious constituents	= 0.0013 - 0.0022
Fat	= nothing

5.2 Butero-Refractometer Reading (B.R)

Milk fat does not contribute to the refractive index value however the ghee has a constant refractive index value, which depends upon the composition and the constituent fatty acids. The butero-refractometer measures the index of refraction between the air and the liquid fat, which varies with the nature of the fat. The butero-refractometer reading is usually taken at 40 °C because the ghee generally exists either as semi solid or solid at ordinary temperature. Generally ordinary day light is used by compensating arrangement built into the refractometer, and the value thus obtained is corresponding to the D line wavelength of sodium (589 mμ). At times the instrument used is calibrated in butero refractometer degrees instead of the usual absolute refractive indices, and the two values are being inter convertible as

B.R in degrees = 42.0 + factor (observed refractive index – 1.4538)

Refractive index = 1.4538 + factor (observed B.R – 42.0)

Refractive index and B.R vary inversely with temperature correction from

B.R reading –0.54 BR degrees/°C

Correction Refractive index= 0.00038/ °C

Ghee exhibits a B.R at 40 °C of about 43.0 which corresponds to the refractive index value of 1.4545. An increase in the B.R value suggests the adulteration either with vegetable oil or animal body fat in the ghee. PFA (1976) has given B.R standard for ghee at 40 °C which is 40.0-43.0 °C in all areas of India except for the cotton tract areas of Maharashtra, Madhya Pradesh and some areas of Saurashtra and kutch where the value is 41.5-45.0 °C.

5.3 Oxidation-Reduction Potential

Oxidation is defined as the loss of electrons whereas reduction as the gain in electrons. Every oxidation is followed by a reduction and vice-versa. Compounds, which loose electrons or which gains electrons are characterized as oxidizing or reducing agents respectively. The potential difference created on a platinum electrode in a solution of an oxidant or reductant is measured by completing a circuit through a calomel half cell and a potentiometer. The voltage measured under these conditions reflects the oxidizing or reducing capacity of the solution. This potential is called the oxidation-reduction potential or redox potential, which is designated by the symbol 'Eh'. There are two methods for measuring the 'Eh' of a system and these are electrometric and colorimetric of which the electrometric method is much more reliable.

Fresh milk has + 0.20 to 0.30 volts O-R potential, which is mainly due to the dissolved oxygen present. Heating removes oxygen from the milk and also results in the production of sulfhydryl (-SH) groups from β-lacto globulin resulting in the decrease of O-R potential. Within the normal range of pH of milk Eh varied to the extent of + 0.03 v for every unit decrease in pH of milk. Fresh zebu cows milk at 30°C the Eh value is + 0.258 v and fresh buffalo milk Eh value is + 0.312 v and the Eh value of milk drawn anaerobically is + 0.13 v and quickly decolorizes methylene blue. Reduction of methylene blue at Eh value of + 0.12 to 0.03 v and that of resazurin occurs at + 0.18 to 0.09 v.

5.4 Electrical Conductivity

The ability of electrolyte to conduct a current is measured in terms of specific resistance, which is defined as the resistance in ohm's of a column of solution 1 cm long and 1 sq.cm in cross section. The specific conductance is the reciprocal of specific resistance which is expressed in ohm's. The specific conductance is expressed as

$$\text{Specific conductance} = K / R$$

where, K = "cell constant" i.e., a proportionality factor previously determined for the given cell with solution of known specific conductance and

R = measured resistance ohms

The specific conductance of whole milk is lower than that of skim milk. There is a relationship between percentage of chloride and conductivity with the increase of electrical conductivity the chloride percentage also increases.

Factors affecting the Conductivity

The factors which affects the electrical conductivity of milk are:

1. **Temperature:** Increase in temperature results in an increase of electrical conductivity due to an increase in the dissociation of electrolytes and decrease in viscosity.
2. **Acidity:** Increase in the acidity of milk brings an increased electrical conductivity because of the fact that the colloidal minerals are brought into the soluble state.
3. **Fat:** Skim milk has higher electrical conductivity than the whole milk. An increase in the fat percentage of milk decreases the electrical conductivity due to the reason the fat globules restricts the mobility of the ions since the fat phase constitute an inert dilutant.
4. **Dilution:** Dilution of milk decreases the electrical conductivity due to the decrease in electrolyte concentration which is slightly compensated by the increased dissociation of electrolytes.
5. **Concentration of milk:** The concentrations of solids are increased up to 30 percent due to the increased concentration of electrolytes by the electrical conductivity whereas the electrical conductivity is decreased, when the milk solid concentration goes beyond 30 percent due to the increase in viscosity.
6. **Neutralization of milk:** Increase in the concentration of electrolytes results in the increase of the electrical conductivity due to the addition of neutralizers to the milk.

5.5 Buffering of Milk

Fresh milk acts as a complex buffer because of its content of carbon dioxide, protein, phosphate, citrate and a number of minor constituents. Bacterial action introduces lactate and other organic anions as additional buffers. Milk behaves as buffer over the pH range from 4.8 (curdling point) to 8.3. Titration of fresh milk in the pH range of 6.6 to 8.3 requires 13 to 20 ml of 0.1N NaOH per 100 ml (1.3 to 2.0 meq per 10 ml). Most of the fresh milk samples fall in the range of 1.5 to 1.8 meq/10 ml.

CHAPTER 6

Microbiology of Milk

Among all the foods, milk is the most perishable food because of the fact that it is an excellent medium for the growth and development of bacteria, which may be pathogenic and or nonpathogenic to human beings. The milk that has been produced, processed and distributed under rigid sanitary conditions having relatively low bacterial count, with good flavour, and appearance possesses good keeping quality. Under normal conditions the milk is sterile, until it reaches the milk ducts in the udder of the cow therefore, the first contamination of the milk most likely occurs due to the milk pails, dairy equipment, barn dust and the personnel who are handling the milk etc. Dried dirt and filth is picked up by air movements and carried about as dust in the atmosphere, consequently contaminating the milking area.

The microorganisms which are mostly responsible and involved in the spoilage of milk are psychrotropic in nature and are destroyed by the pasteurization temperature. However, some organisms like *Pseudomonas fluorescens* clate and *pseudomonas fragi* can produce the extra cellular enzymes like porteolytic and lipolytic, which are heat stable and capable of causing spoilage. Pasteurization has decreased the threat of milk borne diseases such as tuberculosis, brucellosis and typhoid fever. In the past, where the milk was not pasteurized properly or handled in poor conditions has resulted in a number of food borne diseases resulting in post processing contamination due to the ingestion of such milk or dairy products.

Milk secreted by the mammary glands (udders) of the healthy cows is sterile, but it frequently gets contaminated at the time, when it is withdrawn from the animal. The number of microorganisms present in the milk at this stage varies from less than 1000 to 50000 under controlled methods of production. The types of microorganisms found in milk vary considerably and are dependent upon the specific conditions associated with the batch of the milk. The principal sources that contribute to the contamination by the microorganisms found in milk are the interior of the udder, the exterior of the teat and its immediate surroundings where the milk is produced i.e., the milking area and the milk handling equipments. The most important microorganisms in the dairying are bacteria, mould, yeast etc of which the bacteria is the predominating, which are visible only with the aid of the microscope only. A few microorganisms are desirable, while most of them causes undesirable changes and relatively small number are disease producing types and are called as 'pathogens'. In the dairy industry efforts are being made on large scale in different laboratories and research organisations to control the micro-organisms which are responsible for the spoilage of milk. The greater the bacterial count or the presence of bacteria in large number per ml of milk, the lower is the bacteriological quality.

Freshly drawn milk contains substances which are capable of exerting a bactericidal action, because of the fact that the bacteria does not multiply/grow for sometime in milk. Lactenin 1 and 2 are the substances that have been identified which exerts the bactericidal action in the milk and are present in combination with the fat globules and are destroyed by higher temperature. The bactericidal stage of milk is dependent on various factors such as:

1. Temperature of storage.
2. Initial bacterial contamination and
3. The species of the animal.

Bacteriophages are abnoxious for the starter cultures because they kill the bacteria or the starter culture thus stopping the fermentation of the milk. The distribution and the presence of microorganisms either in milk or milk products depends upon several conditions such as:

1. Bio-chemical activities and nutritional requirements of the microorganisms.
2. Temperature response and
3. Capacity to cause infection and disease.

Bacteriophages are viruses that are very small and cannot be seen under an ordinary microscope, which attack and destroy the bacteria. The bacteriophages require a host cell for their multiplication or reproduction and are very strain specific, due to this fact, culture rotation and resistence are used as control mechanisms. The bacteriophages can enter the dairy plant, through the raw milk supplies and some culture strains may act as carriers. The most important key to prevent the bacteriophage contamination is the rigid and stringent handling of culture and plant sanitization procedures.

6.1 Importance of Study of Microbiology

The following are the few important points of significance to be considered in the study of milk micro flora.

1. To judge the sanitary quality and conditions of milk and milk products.
2. Microorganisms of milk, under suitable conditions degrade milk components like proteins, fat and carbohydrates, thus making the milk unpalatable.
3. Many milk borne epidemics caused by the organisms are present in the milk.
4. Some microorganisms produce desirable changes resulting in the production of butter, cheese etc., the undesirable effects of milk microbes are "blue milk" *(pseudomonas cyanogens)* 'red milk' *(serratia marcescens)* or 'ropiness' *(Alkali genes viscosus)* or sometimes milk appears dirty or stale and at sometimes 'off' flavours also develops in the milk.

6.2 Classification of Microorganisms of Milk

Microbes are able to bring about various biochemical changes in the raw milk, which may be lipolytic, proteolytic or acid producers etc. In the course of normal or neutral fermentation a characteristic sour flavour develops in the milk due to the conversion of the lactose to lactic acid by *streptococcus lactis* and *streptococcus cremoris*. The bacteria which produces only one type of acid are homo fermentative e.g. *S.lactis, S.cremoris* and some other lacto bacilli, the micro bacterium, coli forms also ferments lactose to lactic acid and other products.

6.3 Types of Microorganisms in Milk

As already stated a number of microbes are present in the milk depending on the specific conditions/circumstances and batch of milk viz. bacteria, fungi, moulds, yeasts, bacteriophages and at times even protozoans, which are seldom observed.

6.3.1 Bacteria

The most common bacteria that are present in the milk are, cocci usually gram positive, gram positive non spore forming rods and gram positive spore forming rods as well as gram negative non spore forming rods. There are many examples *Strepto coccus lactis*, *Entero cocci, Pseudomonas aeuroginosa* and other species like *Salmonella, micrococcus, clostridium, Corynebacterium pyogenus* etc. The pathogenic bacteria like *myco bacterium, bovis* and *myco bacterium tuberculosis* may also present.

These are microscopic unicellular organisms which occur principally in the form of spherical, cylindrical or spiral cells, which reproduce by transverse fission. These bacteria belong to four main groups namely :

1. Cocci
2. Gram positive non spore forming rods
3. Gram positive spore forming rods and
4. Gram negative non sporing rods.

The spherical and cylindrical forms are the most predominant forms that are present in the milk and its products, which vary from 1 to 5 microns in their size. The individual bacterial cells are invisible to the naked eye, but the 'colonies' consisting of large number of individual cells formed by them are visible.

These bacteria are found in large numbers in the soil, sewage, decaying plants or animals and are also present in air, water etc. Under favourable conditions, bacteria multiply very rapidly and may double their numbers in about 15 minutes or less. Some bacteria also form 'spores' which are tough resistant bodies which cover the bacterial cell wall and when the environmental conditions are favorable they transform into new vegetative cells and actively multiply their growth. The spore forming bacteria cause trouble in the dairy industry because of their resistance to pasteurization and sanitization procedures.

6.3.2 Yeast

Very commonly and most frequently encountered in milk. They act upon lactose producing lactic acid and carbon dioxide. e.g. torula species generally found on raw cream in hot weather however prevails throughout the year.

These are unicellular, but are somewhat larger in size than bacteria and acts on lactose to produce acid and CO_2. Spores of yeasts are readily destroyed during pasteurization.

6.3.3 Moulds

These grow in large numbers and appear as fluffy or fuzzy growth on the surface of the milk products like butter, cream, khoa or cheese etc and produces undesirable flavours and odours, however, certain type of moulds are essential in the production of a particular typical type of cheese.

These are multicellular organisms differing greatly in many aspects from the bacteria. The individual cells are not visible to the naked eye, when matured they can be seen as 'mycelium' by the naked eye. They are found in soil, feeds, manure and poorly washed utensils. They are black, gray, green, blue or white. Most spores of moulds are destroyed by pasteurization. They are of considerable importance in cheese making and are responsible for some defects in butter and other milk products.

6.3.4 Viruses

Viruses include all the ultra-microscopic forms of life. In the dairy industry only these viruses, which are parasitic on the lactic acid bacteria and are known as starter. Bacteriophage or simply phage are of special importance. The viruses range in size from 0.22 to 0.23 microns. The lactic phages are usually not destroyed by normal pasteurization of milk employed for cheese and cultured butter milk, but they can be destroyed by higher heat treatment.

6.3.5 Bacteriophages

These are particularly obnoxious (i.e., harmful) and are seen in starter cultures of cultured milk, butter and cheese, which kills the bacterial cultures thus preventing the process of fermentation products formation.

6.3.6 Hetero Fermentative Organisms

These organisms produce acids like propionic acid, acetic acid, gases like $H_2 CO_2$, alcohol etc. Lactic acid continues to be produced until it becomes a limiting factor for the bacteria itself. As the acidic pH reaches 4.6 precipitation of casein takes place and the organisms that are able to metabolize lactic acid like yeast and moulds develop diminishing the acidity, and producing the products of protein degradation like amines ammonia etc by species of *bacillus, clostridium and pseudomonas etc.* Sometimes the changes which are not deleterious to health are also brought about by the microbes e.g. ropiness a sliminess giving a thread like appearance by *Alkali genes visco lactis*, a tepid or stormy fermentation of lactose by *Clostridium perfringes*. Whereas some others may impart same brilliant colors to milk *Pseudomonas syscynea* (blue colour) *Pseudomonas synxantha* (yellow colour) and *Seriatia marcerceris* (red colour) are seen in milk.

A. On the basis of Bio-chemical Activity

Based on the biochemical activity the microorganisms are classified into five major types/groups.

TYPE-I : Acid producers : The representative microbes are:

1. **Strepto cocci**

   Like *Strepto coccus lactis* and *Strepto coccus cremoris*

   - The source of these microbes is dairy utensils, manure, equipments, udder etc.
   - The biochemical activity involves souring of milk, conversion of lactose to lactic acid other products. They may be either homo or hetero fermentative, also produces acetic acid, propionic acid etc and alcohol along with CO_2.

2. **Lacto bacilli**

   Lacto bacillus casei, Lacto bacillus planterum, L. brevis and

   L. fermentum etc.

   Source: Feeds, silage and manure.

   Biochemical activity: Fermentation of lactose up to 1-2% lactic acid. Some of the microbes are homo while the others are hetero fermentative.

3. **Micro bacterium :** *Micro bacterium lactium.*

   Source: Manure, dirty utensils and dairy products.

   Activity: Lactic acid is produced by conversion of lactose and also produces other products. Lactic acid is less produced as compared to strepto cocci or lacto bacilli. Some of these micro bacteria can survive even at a temperature of 80-85 ^{o}C for 10 minutes.

4. **Coliform Group**

   Example: *Escherichia coli, Entero bactor aero genus* etc.

   Source: Manure, Polluted water soil and plants.

   Activity: These are hetero fermentative. Produces other products like gases, alcohols etc., along with lactic acid. Some neutral products are also produced and the number of Coliform bacteria present in the milk is an indicator regarding the sanitary quality conditions under which the milk is produced.

5. **Micro Cocci**

   Example: *Micro coccus luteus. M.varians* and *M.freudenreichii* etc.

   Source: Teat canal ducts to mammary gland of the cow, dairy utensils etc.

Activity: Very small quantity lactic acid is produced, weakly fermentative and proteolytic moderately heat resistant. Some of the strains are capable of surviving at 63 °C for 30 minutes.

B. Biochemical TYPE II: Gas producers

Coli form bacteria like *Clostridium butysicum* and *Torula cremoris.*

Source: Soil, manure, water and feed.

Activity: Fermentation of lactose with accumulation of gas which may be a mixture of carbon dioxide and hydrogen or only carbon dioxide in case of yeast fermentation. The bulk containers of milk may hear the sounds of their lids being lifted by the gas pressure when the contamination is very high.

C. Biochemical TYPE III : Ropy or stingy fermentation

Example: *Alkali genes visco lactis, Enterobactor acrogens,*

Strepto coccus cremoris etc.

Source : Soil, water, plants and feed.

Activity: These microorganisms develop a viscous poly saccharide critical to heat, forming a slimy layer or capsule on the cell. Milk favours the formation of capsules resulting in the formation of long thread like appearance while it is poured into another vessel or into the weighing tank.

D. Biochemical TYPE IV: proteolytic

Example : Bacillus sp like *B.sublies, B.cercus. Pseudomonas sp.Proteus sp* and *Streptococcus liquefacicus.*

Source : Soil, water and utensils

Activity: These organisms breakdown the casein to peptides which may be further disintegrated to amino acids. Proteolysis may be proceeded by coagulation of casein by the enzyme rennin. The end products of proteolysis may impart abnormal flavour or odour to the milk.

E. Biochemical TYPE V : Lipolytic Organisms

Example : *Pseudomonas flavour, Achromo bactor lipoliticum, Candidi polytica* and *penicillium* sp.

Source : Soil, water and utensils.

Activity: These microorganisms hydrolyze the milk fat to glycerol and fatty acids. Some of these fatty acids impart a rancid odour and flavour to the milk.

The microbial population which is present in the milk can also be classified according to their temperature characteristics i.e., temperature response growth and biochemical activities of these bacteria in milk largely depends on the temperature at which the milk is being handled. The microorganisms require an optimum temperature for their growth and activity. Low temperatures are generally being used to prevent the activity and any change by the microbes, while the high temperatures either reduce their population to an acceptable limit or kill these bacteria including the pathogenic bacteria. Based on the optimum growth temperatures they are classified into:

1. **Psychrophilic or Psychrotropic**

   These microorganisms grows at a temperature above freezing and refrigeration temperature i.e., 4-5 °C and produces spoilage effect resulting in 'off' flavours and odour, even though the low temperatures are recommended for the prevention of spoilage in milk. e.g. *Pseudomonas achromobactor*, *vibri,* and *flavo bacterium* and *Alcoligenes genera* : they are killed by pasteurization but occasionally found even in the pasteurized milk also. The contamination of milk after pasteurization is by equipment and bottles etc.

2. **Mesophilic**

   These organisms grow at a moderate temperature of 37-39 °C e.g. : *Strepto cocci, Lacto bacilli; Coli form* and produce acid, gas and other off flavours. These are killed by pasteurization process.

3. **Thermophilic Bacteria**

   The optimum temperature for these organisms to grow is 62 °C a temperature slightly below the pasteurization process. However when the milk is pasteurized at this temperature holding it for a period of 30 minutes these organisms are destroyed e.g.: *Bacillus* and *clostridium,* however *Bacillus* and *Stero thermophilis* grow even at temperature of 65 °C and above.

4. **Thermoduir**

   These microorganisms survive the pasteurization process at 62 °C, but do not grow at this temperature. These are extremely troublesome because they are not destroyed by pasteurization. They contaminate and accumulate as a result of improper cleaning of equipment hence the subsequent batches of milk is heavily contaminated.

e.g. *Micro bacterium lactium.*

Micro coccus luteus

Strepto coccus and *Bacillus subtils* etc.

6.4 Growth of Micro Organisms

In microbiology, growth means to an increase or multiplication in the number of organisms. Fresh milk which was drawn from a healthy animal contains some microorganisms and their numbers increase/multiply during handling depending upon the cleanliness of these production operations. Subsequently the number of these organisms may increase substantially still further depending on the condition of storage. The changes that take place in the physical-chemical properties of milk are the result of the activities of the individual microbial cells during their period of development and multiplication or due to the substances produced during such activity.

6.4.1 Stages of Growth

Normally the growth of the microorganisms takes place in the following stages.

(a) Initial stationary phase

(b) Lag Phase or Phase of Adjustment

(c) Accelerated Growth Phase or Log Phase

(d) Maximum Stationary Phase and

(e) Phase of Accelerated death

Factors Influencing Growth

The following factors are responsible in influencing the growth of the microorganisms.

(a) Food Supply

Milk and majority of the dairy products provides the necessary medium and all the requirements for the multiplication of the microorganisms.

(b) Moisture

Milk contains adequate moisture, whereas, the dry products due to their low moisture content are able to keep longer or has long shelf life.

(c) AIS

It supplies oxygen to the aerobic type of bacteria and moulds in general.

(d) Acidity or pH

Majority of the common type of microorganisms prefer a pH ranging from 5.6 to 7.5 for their active multiplication.

(e) Preservatives

These are having a controlling effect or may destroy the growth depending on their concentration.

(f) Light

It is more or less harmful to the growth or multiplication of the microorganisms.

(g) Concentration

The presence of high sucrose or salt content in the milk product checks the growth of microorganisms and

(h) Temperature

It is an importance means to control the growth either to check or by destroying the microorganisms. Every species of the microorganisms have their optimum maximum and minimum temperatures of growth.

An Attempt has been made to consolidate some of the microorganisms and groups of the organisms that are essential for milk have been presented in the table 6.1 for the benefit of the students.

Table 6.1 Survey of some Microorganisms and Groups of Organisms that are essential for Milk

Name	Source	Growth in raw milk	Heat resistance	Pathogenicity	Spoilage
Bacillus anthracis	Diseased cow, soil	–	+	Anthrax	No
Bacillus cereus	Feed, dung, soil, dust	++	+	Food poisoning	Sweet curdling, "bitty cream" in pasteurized milk and cream
Bacillus subtilis and B.stearo-thermophilus	Feed, dung, soil, dust	++	+	Probably not	spoils sterilized milk
Brucella abortus	Diseased cow	–	–	Contagious abortion (cow), Malta fever? (humans)	No

Table 6.1 *Contd....*

Name	Source	Growth in raw milk	Heat resistence	Pathogenicity	Spoilage
Campylobacter jejuni	Dung, water	–	–	Intestinal disorder	No
Clostridium perfringens	Soil, contaminated water	-	+	Botulism	No
Clostridium perfringens	Soil, dung, contaminated water	++	+	Intestinal disorder (infant formulas)	No.
Clostridium tyrobutyricum	Soil, silage, dung	–	+	Not	"Late blowing" in cheese
Coliforms	Feces, milking utensils, contaminated water	++	–	Mastitis, intestinal disorder	Spoil milk and cheese
Corynebacterium bovis	Teat canal	+	–	Not	No
Corynebacterium Pyogenes	Interior udder, flies	+	– (?)	Mastitis	Hardly
Coxiella burnetii	Infected cattle, dung	–	–	Q fever (humans)	No
Lactobacillus spp.	Milking utensils, milk in churns, milking parlor	++	–	Not	Sour milk
Leptospira (hardjo)	Infected cattle, urine, contaminated water, surface water		–	Leptospirosis	No
Listeria monocytogenes	Soil, feed, dung	+	–	Meningitis	No
Microbacterium lacticum	Milking utensils	+	+	Not	Grow in pasteurized products
Micrococcus spp. (1)	Teat canal, skin, milking parlor	+	–	Probably	Hardly
Microccus spp. (2)	Milking utensils	+	+	Not	Not Grow in pasteurized products
Molds	Dust, dirty surfaces, feed	+ / -	–	Some produce toxins	Spoil cheese, butter, sweetened condensed milk
Mycobacterium tuberculosis	Diseased cow or milker	–	–	Mastitis, tuberculosis	No
Mycobacterium paratuberculosis	Cow	–	–	Weak	No
Psychrotrophs, e.g., Pseudomonas	Milking utensils, cold-stored milk, polluted water	++	–	Occasionally?	Hydrolyze protein and fat in cold-stored milk

Table 6.1 ***Contd....***

Name	Source	Growth in raw milk	Heat resistence	Pathogenicity	Spoilage
Salmonella, Shigella	Dung, polluted water	+	–	Intestinal disorder, mastitis	No
Staphylococcus aureus	Teat canal, interior udder skin, diseased milker	++	–	Food poisoning, mastitis, ulcers	Hardly
Staphylococcus epidermidis	Teat canal, skin, milking parlor	++	–	Probably not	Hardly
Streptococcus agalac. tiae, S.dysgalactiae, S. Pyogenes, S. uberis	Interior udder, milking parlor	++	–(?)	Mastitis	Sour milk
Streptococcus thermophilus	Milking utensils, milk in churns, milking parlor	++	+	Not	Sour milk
Vibrio cholerae	Polluted water, diseased ger	–	–(?)	Cholera	No
Viruses	Cow, humans, milking parlor	–	+	Many are, e.g., foot and mouth disease virus	No
Yeasts	Dust, milking utensils	+ / -	–	Not	Spoil cheese, butter, sweetened condensed milk
Mycobacterium tuberculosis	Diseased cow or milker	–	–	Mastitis, tuberculosis	No

Source : Text book of dairy micro biology-Parihar & Parihar

6.5 Milk and Public Health

As already stated milk is an excellent source or medium for majority of the micro-organisms both pathogenic and non-pathogenic, which not only remain viable for a considerable period but also capable of their multiplication. Therefore, several diseases are carried through the milk and milk products and some of the milk borne diseases have been presented in the table 6.2 and table 6.3.

Table 6.2 Milk borne diseases associated with drinking raw milk

Sl. No.	Disease	Organism	Symptoms and Complications
1.	Campylobacteriosis	Campylobacter sp.	bloody diarrhoea
2.	Salmonellosis	Salmonella sp.	bloody diarrhoea
3.	Haemolytic Uremic Syndrome	E.coli 0157: H7	diarrhoea, Kidney failure, death

Table 6.2 ***Contd....***

Sl. No.	Disease	Organism	Symptoms and Complications
4.	Yersiniosis	Yersinsia entero colitica	diarrhoea
5.	Listeriosis	Listeria monocytogenes	meningitis,blood infections
6.	Tuberculosis	mycobacterium tuberculosis	Tuberculosis, pneumonia
7.	Brucellosis	Brucella sp.	Blood infections, heart infections
8.	Cryptosporidiosis	Cryptosporidium parrum	diarrhoea
9.	Stephalo cocci enterotoxin poisoning	Stephalo coccus aureus	vomitting
10.	Q.fever	Coxiella burnetti	high fever,severe headache, muscle aches (can infect the liver and or heart)

Source : Text book of dairy micro biology Parihar and Parihar

Table 6.3 Human microbial pathogens possibly occuring in milk and milk products

Sl. No.	Organism	Disease
1.	**Enterobacterioaceae**	
	(a) *Escherichia coli*	Gastroenteritis
	(b) *Salmonella*	Gastroenteritis, typhoid fever
	(c) *Shigella*	Gastroenteritis
	(d) *Yersinia entero colitica*	Gastroenteritis
2.	**Other gram negative bacteria**	
	(a) *Aeromonas hydrophila*	Gastroenteritis
	(b) *Brucella abortus*	Brucellosis (abortion)
	(c) *Campylobacter jejuni*	Gastroenteritis
3.	**Gram-positive spore formers**	
	(a) *Bacillus caereus*	Intestinal intoxication
	(b) *Bacillus anthracis*	Anthrax
	(c) *Clostridium Perfringens*	Gastroenteritis
	(d) *Clostridium botulinum*	Botutism

Table 6.3 ***Contd....***

Sl. No.	Organism	Disease
4.	**Gram Positive Cocci**	
	(a) *Staphalo coccus aureus*	Emetic intoxication
	(b) *Strepto coccus agalactiae*	Sore throat
	(c) *Strepto coccus pyogenes*	Scarlet fever, sore throat
5.	**Miscellaneous Gram Positive Bacteria**	
	(a) *Mycobacterium tuberculosis*	Tuberculosis
	(b) *Micobacterium bovis*	Tuberculosis
	(c) *Mycobacterium Paratuberculosis*	Jhones disease
	(only in ruminants)	
	(d) *Corynebacterium Sp.*	Diphtheria
	(e) *Listeriamonocytogenes*	Listeriosis
6.	**Spirochetes**	
	Heptospira interrogans	Heptospirosis
7.	**Rickettsia**	
	Coxiella Burnetti	Q.Fever
8.	**Viruses**	
	Enterovirus, rota virus	Enteric infection
9.	**Fungi**	
	Molds	Mycotoxicosis
10.	**Protozoa**	
	(a) *Entamoeba histolytica*	Amoebiasis
	(b) *Crypto sporidium muris*	Cryptosporidiosis
	(c) *Taxoplasma gondii*	Taxoplasmosis

Source : Text book of dairy micro biology-Parihar & Parihar

It is well established that milk acts as a potential carrier and source of disease producing organisms. In the past many milk borne epidemics have occurred throughout the world. Unless proper precautions are taken, such outbreaks of milk borne diseases can occur anywhere and at anytime especially when the raw milk is consumed. Diseases which are known to be transmissible through the milk are listed below together with the manner in which they may gain entrance into the milk.

1. **Infection of Milk directly from the Cow**

   These diseases are essentially of bovine origin and the causative organisms gains their entry into the milk, through the mannary glands or through the faecal contamination and may cause/produce disease conditions in human beings who consume the milk. *Example*: Bovine tuberculosis undulant fever or Malta fever etc.

2. **Infection from man to Cow and then to Milk**

   These diseases are essentially in human in nature, but can become established in the udder of the animals.

   Examples : Septic sore throat, Scarlet fever, Diphtheria etc.

3. **Direct contamination of milk by human beings**

   These diseases may be transmitted to the milk by direct contamination through human contact, either by carrier or patients, the milkers who are suffering from such diseases as: Septic sore throat, scarlet fever, typhoid, Para typhoid fever, dysentery, gastroenteritis and diphtheria etc.

4. **Indirect Contamination of Milk by human beings**

   These are human diseases caused by the pathogenic organisms which find their way into the milk through contaminated bottles or other utensils, water supplies, insects and dust.

   Examples: Typhoid or paratyphoid fever, dysentery or diarrhoea etc.

6.5.1 Source of Contamination

There are several important sources of contamination of milk, from the time milk leaves the udder, until it is dispensed in containers. Everything with which it comes in contact is a potential source for the microorganisms. When milking is carried out under hygienic condition, will reduce the entry of microorganisms in the milk. The various sources of microorganisms are :

6.5.2 Producing Animal

Most of the times, the animal itself is the responsible for the contamination of the milk because, there are many chances for the bacteria to gain its entry through the teat opening and from these into the interior of the udder, unless the external parts of the udder teats including the flanks are specially cared and kept clean. Milk which was drawn under aseptic conditions contains low or very few number of microorganism and the common microbes that present are micro cocci, streptococci and *corynebacterium bovis* etc. The bacteria cannot increase, in case the animal is infected by the disease of the udder called "mastitis", which is the major cause of economic loss in dairy industry. The important microbes that are responsible for mastitis infection in the animals are *Staphylococcus aures E.coli, streptococcus, agalactiae*, *Pseudomonas aeroginosa etc.*

Therefore, washing and massage of udder with warm, detergent sanitizer solution is the most important measure in the prevention of the spread of this disease. The hair, dust and dirt present on the animal body harbours a number of microorganisms and are unknowingly gains entry into the milk through the equipment like milking pails, milking machines, etc. Hence, periodic examination of milk is very essential.

6.5.3 Milking Area

The microbial content of air in the milk barn is greatly affected by many conditions of the environment and management practices. Dust and filth is picked up by the hot wind movements and are carried in the atmosphere, which becomes a major source almost for every kind of contamination. Therefore the environment must be kept clean and hygienic as there is a possibility of contamination of the milk from the bedding dung, feeds etc. In order to have a control on the bacterial count there should be a proper monitoring by taking necessary precautions.

6.5.4 Handling of Milk-Utensils and Equipment

Unhygienic equipments and utensils like teat cups, milking pails, buckets, cans and storage tanks are the most important and principal sources of contamination of milk by the microorganisms more particularly the interior surfaces of these equipments acts as the main source of contamination and account for lakhs to billions of the microbes per ml of milk, when they are not cleaned properly by taking proper care. Milk being a nutritious medium, when these equipments are not cleaned properly nourishes the microbes on their interior surfaces and continues grow when fresh batches of milk are added or handled by them. Therefore, proper sanitization and treatments should be applied and if the cleaning is neglected for a longer periods, the hydrophilic mineral rich deposit known as "milk stone" develops on the surface of the interior of these equipments. These milk stones thus developed are resistant to the sanitizers and are the evidence or 'sanitizer proof' and allows growth of the microbes. The containers having cracks, dents etc are the major habitats for these microbes and the milk stones to develop. Therefore suitable methods for washing these equipments should be applied with proper care which will be discussed later while dealing with the cleaning and sanitization of the milk equipment.

6.5.5 Personnel Sources of Contamination

The persons involved in the milking or the milkers should be healthy and must be aware of the personnel hygiene as well as the sanitary conditions of the milking area. Before the commencement of the milking of the milch animals, they should wash their hands, cleanly and rinsed in the bactericidal solution, dry them with a clean towel. The nails of the fingers should neatly trimmed and wear a neat and clean uniform and should possess a clean hand kerchief or cloth in order to prevent the direct contamination either by cough or sneezing and the use of a surgical mask also acts as an additional measure.

6.5.6 Water as a Source of Contamination

Water is continuously used for various purposes in the dairy industry hence a special care/attention should be given for the supply of clean water. It is used for the cleaning of utensils, washing of hands and the cows as well as the milk barn. Therefore, the quality of the water should be maintained which is dependent on the source of supply. The surface water is always contaminated with many contaminants, hence; the quality should be monitored continuously. The quality of the water can be determined by the "presumptive coli form test" to check the potability. The presence of coliform group of bacteria like *Ecoli* is considered as the guideline for the treatment of water by chlorination, which is commonly practiced for potability. Presently the milk after it is withdrawn from the udder or collected from various milk producers is chilled immediately, before its transportation for further processing.

6.6 Types of Microorganisms in Milk

As already stated a number of microbes are present in the milk depending on the specific conditions/circumstances and batch of milk viz. bacteria, fungi, moulds, yeasts, bacteriophages and at times even protozoans, which are seldom observed.

6.6.1 Bacteriophages

These are particularly obnoxious (i.e., harmful) and are seen in starter cultures of cultured milk, butter and cheese, which kills the bacterial cultures thus preventing the process of fermentation products formation.

6.6.2 Spoilage of Milk

Milk is an excellent source of culture medium for many kinds of microorganisms, being high in moisture with a nearly neutral pH and rich in nutrients which are very conducive for them to multiply. A plentiful supply of food energy is present in the form of milk sugar, butter fat, citrate and nitrogenous compounds, in many forms besides acting as accessory food and minerals required by the microorganisms. The freshly drawn milk possesses some inhibitory substances, which are soon became ineffective. An acid fermentation by bacteria occurs most likely under ordinary conditions because of fermentable sugar present in the raw milk; however other changes may take place when the conditions are unfavorable to the acid producing microorganisms or if they are absent. When the milk sours it is usually considered as spoiled especially if it curdles even though the lactic acid fermentation of milk is utilized in the manufacturing of fermented milks and cheese. In raw milk at a temperatures ranging from 10-37°C, *Streptococcus lactis* is most likely to cause the souring possibly with some growth of coliform bacteria.

The pasteurization of milk destroys killing the more active acid forming bacteria, but permits the serviving of the heat resistant lactis, which will cause lactic acid formation at higher temperature. The various ways of milk spoilage are:

6.6.3 Gas Production

The production of gas by the bacteria is usually accompanied by the formation of acid in the milk. The chief gas producing organisms are coli form bacteria *clostridium sp* and *bacilli* that yield both hydrogen and CO_2, the production of gas in milk is evidenced by foam at the top. The likelihood of gas formation and type of micro-organisms causing it depend upon the pre treatment of the milk and the temperature of holding. In the raw milk the coliform bacteria are most apt to be the main gas formers. Gas forming clostridium and bacillus do not compete well with acid formers at higher temperature, but may function if the acid formers are absent.

6.6.4 Proteolysis

The hydrolysis of milk proteins by microorganisms usually is accompanied by the production of bitter flavour caused by the release of some peptides. Proteolysis is favored by storage at low temperatures. The type of changes produced by the proteolytic microorganisms includes:

1. Acid proteolysis in which acid production and proteolysis occur together.
2. Proteolysis with little acidity or even with alkalinity.
3. Sweet curdling which is caused by rennin like enzymes of the bacteria at an early stage of proteolysis.
4. Slow proteolysis and
5. Residual proteolysis

Acid proteolysis causes the production of shrunken curd. Actively proteolytic bacteria are found among the species of *micro coccus, pseudomonas, flavobacterium etc.*

6.6.5 Ropiness

Ropiness or sliminess or Taetta occur in milk due to :

1. Stringiness caused by mastitis.
2. Sliminess resulting from the thickness of cream.
3. Stringiness due to thin film of casein.

Bacterial ropiness is caused by slimy capsular material from the cells. These are two main types of bacterial ropiness; one of which milk is most ropy at the top and other in which milk becomes ropy throughout. Surface ropiness is caused by *Alcoligenes visco lactis* and *Strepto coccus liquifascience*.

Changes in milk fat

Milk fat may be decomposed by various bacteria, yeast and moulds e.g. *Pseudomonas, Proteus, Bacillus etc.*

6.6.6 Microbiological Flavour

Milk is supposed to be a good medium for the growth of most of the microbiological agents like bacteria, mould and yeasts. The flavour may be fruity, rancid, stale or putrid depending on the nature and chemical changes brought about by the bacteria. Some of these are shown in Table 6.4.

Table 6.4 Flavour changes in milk due to microbial agents

Flavour Type	Organism Responsible
Acidic flavour (clean)	Streptococcus lactis
Acidic flavour (aromatic)	S. lactics and Leuconostoc species
Acidic flavour (sharp)	S. lactic and Leuconostoc species
	Clostridium species, Coliforms
Bitter flavour	Micrococcus species, Alcaligenes species, Pseudomonas species, Proteus species, Achromobacter species, Flavobacterium species, Serratia species, Bacillus species, Clostridium species.
Bitter/musty flavour	Actinomycosis species, certain type of cocci
Burnt or caramel flavour	Streptococcus lactis var maltigenes
Barny flavour	Aerobacter oxytocus
Soapy flavour	Pseudomonas sapolactica
Turnip-like flavour	Eschericia coli, Pseudomonas fluoresceris
Malty flavour	Yellow micrococci
Potato flavour	P-grayeolens and P-mucidolens
Fishy flavour	P-icthyosmia
	Fruity flavour Yeasts
Musty flavour	Actinomycosis species

Source : Text book of dairy micro biology- Parihar & Parihar

Sometimes the colour of milk may change due to excess of organism. These are shown in Table 6.5.

Table 6.5 Colour change in milk due to bacterial proliferation

Type of change	Organism Responsible
Bluish-grey or brownish milk	Pseudomonas syncyariea
Brown milk	Pseudomonas fluorescens
Deep blue milk	Pseudomonas syricyanes and Streptococcus lactis
Red milk	Serratia marcescens, Brevibacterium erythrogenes, Micrococcus roseus, Torula glutinis
Yellow milk flavobacterium	Pseudomonas synxantha

Source : Text book of dairy micro biology-Parihar & Parihar

6.7 Alkali Production

The alkali producing groups include bacteria which cause an alkaline reaction in milk without any evidence of proteolysis. Alkali formers are pseudomonas fluorescens.

Flavour Change

***Sour Flavour*:** Acidity may be produced by *Strepto coccus lactis* and Leuconostoc.

***Bitter Flavour*:** Bitterness usually results from proteolysis but may follow lypolysis or even fermentation of lactose. Milk from cows in late lactation period sometimes slightly bitter in taste and flavours.

***Burnt Flavour*:** Certain strains of *Strepto coccus lactis* produces which resembles the over heated milk.

***Colour changes*:** The colour changes are:

***Blue Milk*:** *Pseudomonas syncyanea* produces a bluish gray to brownish colour in milk, in pure culture but when growing with an acid former like *Strepto coccus lactis* may cause a deep blue colour.

***Yellow milk*:** *Pseudomonas synxantha* may cause a yellow colour in the cream layer of the milk.

Red milk: It is usually produced by the species of *serratia.*

Brown milk: A brown colour may result from *Pseudomonas Putre faciens.*

6.8 Microbial Examination of Milk

Milk, when it is allowed to stand for sometime at room temperature, a large number of microbes develop from the initial state present in the fresh milk drawn from the udder. The number of microorganisms present at the time of consumption is determined by the following factors.

1. The temperature at which the milk is kept.
2. The time interval between the production and consumption of milk.
3. Number of microbes population present initially.
4. Weather the milk is pasteurized or not.

Various microbiological and chemical tests are made to ascertain whether the milk is handled, treated properly under sanitory conditions and a high microbial count does not mean that it contains pathogenic bacteria but indicates excessive contamination. The total mesophilic plate counts are widely used as indicators of milk and its products quality. The examination is also necessary to test the presence of pathogenic organisms and also the spoilage of milk. Recently some indicator organisms have been developed in order to monitor the quality of milk and other food products. A good quality indicator organism is always present in enough numbers. The examination of milk and its products also depends on the nature and type of laboratories and deal with different aspects in their examination.

For example:

Government laboratories: These laboratories carry out the tests to ascertain the public health standards.

The food/plant laboratories: These labs carry out quality control tests of raw products and ingredients as a warning of possible troubles in the manufacture of the products.

Private agencies: These agencies carry out the tests to enable the recommendation or acceptance of typical brand of food or milk products.

Methods of Milk Examination

There are two types of microbial examination of milk, they are:

1. Quantitative and
2. Qualitative.

The quantitative methods of microbial examination are:

(a) Standard plate count (SPC)

(b) Direct microscopic count (DMC)

6.8.1 DMC Method

The examination of milk and other food products, the possibility of detecting the microbes directly under the microscope should not be missed. The microbial content of milk is determined by direct examination under a microscope by preparing a smear and the procedure is as follows.

The smear is prepared by spreading exactly 0.01 ml of milk on an area of 1 cm^2 by means of a sterile pipette or syringe. The smear is air dried and stained with the solution of metheylene blue and is examined under oil immersion objective (100x) for the estimation of the density of the microbes. The final count is carried out as under.

The diameter of the most of the oil immersed lenses is 0.16 mm; hence the area seen under this objective lens is calculated as πr^2 i.e., $3.14(0.85)^2 = 0.02$ mm^2 and the smear is prepared on an area of 1 cm^2 or 100 mm^2. Therefore the smear contains 100/0.02 = 500 fields. The average count of one field is multiplied by the total no. of fields, which will give the total number of microbes present in 0.01 ml of the milk sample. The numerous large clumps of bacteria indicate uncleaned utensils and presence of puss cells indicates the infection of workers. Presence of strepto cocci indicates the possibility of mastitis. It is a common practice to report the results in terms of dump counts and it is used/applied to the unpasteurised milks only as heating causes many bacteria to loose their staining properties.

Advantage of DMC

1. The method is more rapid and takes only about 15 minutes. Thus a large number of samples can be examined.
2. Limited equipments are only required.
3. The slides can be kept on a permanent record and can be re examined whenever required.
4. Different morphological types can be easily distinguished in slide preparations. This is of great value in determining the source and nature of contamination.

Disadvantages

1. The count is not applicable if the number of microbes present in milk is low. Since a large factor is used for converting the number of organisms per field to the number per ml of milk.

2. The method is not applicable to differentiate between viable (living) and non living microbes.
3. This test cannot be used for the pasteurized milk because most of the organisms are dead.

The Direct Epiflourascent Filter Technique (DEFT) is a microscopic technique, which was developed for estimating the bacterial counts in raw milk. It involves concentration of large number of microbes from a higher volume of sample by filtering through the poly carbonate membrane filter. The retained bacteria on the membrane are stained with a cridine orange and then counted directly under epiflourascence microscope. It may be necessary to protect the sample and to allow filtration, the milk can be pre treated with detergent and pretcure enzyme. It is also necessary that bacteria should be trapped in a single focal plane because of limited depths of focus of the microscope at the magnifications required.

A cridine orange is Meta chromatic flouro chromo flugrasing either green or orange when bond to double standard DNA it flourescent green and when bond to a single standard RNA it flouresent orange. Generally it is assumed that the cells which flourescent yellow are variable whereas those flourescent green are non variable. Which is always not true, the actual color of individual cells largely depends upon the concentration of the dye within the cell. In many microbes in which the cell maintains its integrating, the dye doesn't pass through to the cell and flouresce orange and many dead cells from which the dye leaks in flouresce green. This method has also found application in the examination of meat, fish and their products, beverages and vital microbes in water samples.

SPC

It is an enumeration method for the examination of microbes in the milk. In this method agar plates are commonly used for the estimation of number of bacteria in the milk. This is the most sensitive method detecting the presence of a variable bacterium and allowing it to multiply itself to form a visible colony. It forms the basis of the traditional pour plate, spread plate or miler and misra drop plate still widely used in microbiological laboratories.

The normal procedure of this method is to prepare a number of sufficient serial dilutions of milk. One ml of diluted milk sample is mixed with melted agar at 45 °C and the inoculated medium is then poured into the petri dish plates. These plates are then incubated at 32-35 °C for 48 hours for psychrophiles or at 55°C for 48 hours for the enumeration thermophili and for routine work at 7°C for 7-10 days.

The number of visible colonies of the microbes is counted and the numbers of colonies per ml of milk are reported as SPC of given sample. Sometimes at 45 °C the agar destroys the psychrophilic. Spread plate is beneficial because it provides a good aerobic environment but the sample size should be small (0.1 ml).

In a thoroughly mixed sample of milk the number of microbes probable of forming colonies are expected to have a distribution property of which is that the variance is equal to mean i.e.

$$X = VAR = S^2$$

As a consequence of this is that limiting procession of a colony count is dependent on number of colonies counted. The 95% confidence limits (cl) can be estimated as approximately.

$$2S = 2\,X.\sqrt{x}$$

(If X is count on single plate and has to be on estimate of mean). Thus for a plate only with 16 colonies the 95% cl would be approximately 50% (i.e., we have the count between 8-24 with the 95% confidence level).

For plate count the plates must not be over crowded and 30-300 colonies are considered as appropriate. A commonly used dilutant contains 0.1% peptone and 0.85% sodium chloride. However the traditional plate counts are very expensive and recently some new variety of plates have been developed only a small volume of each dilution (20) is considered. A number of dilutions can be grown on a simple plate. It saves about 2/3 of material and time required to develop the colonies after incubation which are counted in a special grid. In addition to the usual agar media that is being used in the plate count various other selective media are also used for specific microbes.

Advantage of Plate Count Method

The method is very easy and can be used as a routine examination method for the estimation of number of bacteria in milk and also for its quality.

Disadvantage

1. For proper plate count each microbial cell should be well separated, but in milk, a number of microbes may be in the form of dumps or clusters. Thus a single colony produced may misguide about the count.
2. Some rapidly growing microbes inhibits the growth of other microorganisms. Thus SPC is only an 'estimation' and not the total count of bacteria in milk.

Numerical relationship between Direct and Plate Counts

The direct count is usually 5 to 10 or sometimes higher than the plate count because the former enumerates the individuals even those in dumps and also dead bacteria. Whereas the plate count enumerates only live bacteria that are capable of developing in the media at a given temperature. Each dump count only as a single bacterium since each dump contain several dozen of individual cells yet a dump usually yield only one colony.

Qualitative Analysis of Milk and Milk Products

Reductase test

Freshly drawn milk from udder of the animal comes in contact with air and acquires an oxidation-reduction potential (O/R) of + 30 mv (milli volts). When the milk is contaminated by the microorganisms which begins to grow results in the lowering of O/R potential. The rate of shift of O/R potential depends upon the number and kind of organisms present and their metabolic rate. This forms the basis of the 'Reductase test' for the examination of milk. This shift of O/R potential can be determined by various dyes like metheylene blue, resazurin, which indicates by their colour change on oxidation. Many redox reactions involve the transfer of oxygen (O_2). One reactant may act as electron (e^+) donor and the other as (e^-) receptor. Thus the rate at which the substrate (milk) is oxidized by the living organisms is a rapid method of examination for manufacturing of milk and its products and also cost effective.

Methylene Blue Reductase Test (M.B.R.T)

It is based on the principle that the colour of the dye changes to colourless when it is reduced. For this test 10 ml of milk sample is pipetted into a sterile test tube and one ml of standard methylene blue solution (1 : 3,00.000) is added. The test tube is then closed with a rubber stopper and inverted three to four times for thourough mixing. The tube is inverted at every one hour for the colour test and then examined at 30 minutes interval until decolourization occurs. The result is reported in terms of time required for the reduction of the methylene blue. The reduction time is inversely proportional to the initial microbes content. When the number of microbes is more the O/R potential is more and the dye reduced to colourless due to rapid consumption of O_2 by the microbes.

Methylene blue Reductase Test Time and the Quality of Milk

MBRT time	Quality of milk
30 minutes	Very poor
30 minutes- 2 hours	Poor
2 hr - 5 hr	Fair
5 hr - 6 hr	Good
More than 8 hr	Excellent

Resazurin Reductase Test

Another dye named resazurin is being used now-a-days by the quality control laboratories for the reductase test as it undergoes a series of colour changes depending upon the O/R potential thus this dye permits readings of degree of reduction at short intervals than does the methylene blue. It undergoes a serial change from state blue to pink to colourless. The test is performed by mixing 1:20000 dilution of 1 ml of resazunin to 10 ml of milk sample in a sterile test tube and kept in a water both at 35-37 °C. The colour change sequence is as below.

Slate blue pink blue pink colourless.

There are two methods employed for this reduction test.

(a) One Hour Method

One ml of dye is added to 10 ml of milk in sterile tube and stand for 1 hour at 37 °C temperature. New colour of milk is compared to already prepared standard solution of different colours mentioned earlier.

	Colour change	Quality
Gr - I	Slate blue - pink blue	– Good, less no. of microbes
Gr - II	Slate blue - marvel	– Higher microbial content.
Gr - III	Slatable pink - pink	-- growth by abnormal
Gr - IV	Pink - colourless	– Highly contaminated.

Gr - IV, The milk may contain many puss cells and is probably teaming with mastitis organisms.

(b) Three Hour Test

The same principle is employed and the readings are taken three times at hourly interval. High grade milk requires at least 3 hours for decolourization. The reading is reported as resazurin reduction time (RRT). Changes in technology especially refrigeration has made this test less reliable because it shows little correlation with psychrophilic bacteria. As reduction of resazurin takes place in two stages of colours, there is wider range of colour that can be scored by using standard computer disc within less time thus.

Resazunin		Resofurin	Dihydro
(blue)	→	(pink)	resofurin (leuco)
oxidized		intermediate	reduced.

6.9 Destruction of Microorganisms

The destruction of the microorganisms present in the milk are carried out by any one of these means or in combination.

1. **Heat:** Most widely used different types of heat treatments temperature time combinations are employed such as pasteurization, sterilization etc.
2. **Ionizing radiation:** Such as ultra violet rays etc.
3. **High frequency sound waves:** Such as super sonic and ultra sonic.
4. **Electricity:** In this case the microorganisms are destroyed actually by the generation heat.
5. **Pressure:** Should be about 600 times greater than atmospheric pressure and
6. **Chemicals:** These include acids, alkalies hydrogen peroxide and halogens etc.

6.10 Bactofugation

It is defined as a special form of separation in which specific types of microorgnisms more specifically spore forming bacilli are removed from milk and inactivated thereafter.

In general all active/vegetative microorganisms are inactivated by pasteurisation to such an extent meeting the requirements of both hygienic and technological aspects. However, the highly heat resistant spores are surviving the pasteurisation process as precursors of spore forming bacilli and clostridia. Spores can form new vegetative cells by growing under favourable conditions and then propagating very quickly/rapidly. They can lead to significant quality defects in hard cheese, semihard cheese, or long life products due to proteolysis, lipolysis and gas formation. Therefore, bactofugation is used mainly in the manufacture of these product groups.

Bactofuges are special nozzle clarifying separators with high separation precision that can remove microorganisms from milk based on their higher density. This is applicable especially for spore forming cells or spores, as the density difference from milk is enhanced by the concentrated cell plasma of the spore and it permits complete separation of this group of microorganisms by a bactofugation. The various parts of the hermetic bactofuge are:

1. Nozzles in the bowl wall
2. Sludge discharge bend
3. Cyclone discharge
4. Discharge for bactofugate
5. Over flow for germ enriched air
6. Bactofugate

Bactofuge-Operating Principle

The microorganisms which have been separated are concentrated on the periphery of the bowl, are transported in milk highly enriched with bacteria (bactofugate) to the out side through the nozzles and are then collected in the sludge discharge pipe. The bactofugate, which still corresponds to 2-3% of the treated milk, is transferred into a cyclone, deaerated and transferred towards the discharge. The air which is highly contaminated with microorganisms, goes into the overflow avoiding the contamination of the environmental air. The relatively high quantity of bactofugate must be treated in such a way that all the microorganisms are inactivated such that the milk must be used in a suitable manner.

Bactotherm Process

The clarified and standardised milk is pumped into a plate heat exchanger, where it is heated up to the bactofugation temperature of 60-75 °C and into the bactofuge. Then it is pasteurised in a plate heat exchanger and rechilled for further processing. The bactofugate is transferred by means of a pump into another preheating section of the plate heat exchanger then into the vacuum deaerator, which is then transferred by a pump into the steam injector. The bactofugate, enters the injector head through a perforated pipe and is mixed with steam at 130 - 140 °C for a period of 3-4 seconds. This time temperature profile is sufficient to inactivate all the spores. The sterilized bactofugate is rechilled in the plate heat exchanger and can be added back into deaeration milk or discharged separately for other suitable applications.

CHAPTER 7

Dairy Operations (Practices)

7.1 Pricing of Milk

Milk purchase rates have to be decided from time to time taking into account the production season (lean/flush) as per the pricing policy, the availability of milk and the competitive prices offered by the contractors/competetors etc. The milk payment charts are prepared in the form of a ready rock over giving slabs of fat percentage on the horizontal axis and the quantity of milk on the vertical axis. The prices per litre a part in decimal and in multiple of litre can be read by correlating the volume and the actual fat percentage.

Although the pricing is based on both fat and S.N.F in actual practice the rates per litre are worked out on the basis of assuming 8.5% S.N.F. in cows and 9.0% S.N.F for buffalo milk which is quite practical because the S.N.F content is not variable widely. Secondly the calculation of S.N.F. is highly combursome in each case unless the milk collection centre (MCC) has any specialised instrument for the calculation of SNF instantly. In contrast the testing of fat is easier and is demonstrable. The pricing policy and milk payment chart also consider the commission payable to the MCC on per litre basis or per kg fat basis.

The importance of pricing in any sectorial growth policy cannot be over emphasized. This is more so in the case of agriculture and its allied sciences/ activities. Price policy assumes a vital role in:

(a) determining the farmer's choice of economic activities,

(b) influencing the farmer's decisions on resource allocations,

(c) accelerating the production and productivity, and

(d) securing adequate increase in the marketed supply.

A positive price policy acts as a catalytic agent in providing impetus to these growth parameters. A number of economic and non-economic factors determine the size of milk production. Among these there are six vital factors that are often considered are:

1. The price farmer's should receive for milk.
2. Profitability of competing on farm enterprises.
3. Input costs of milk production.
4. The condition of the general economy.
5. Individual preferences and skill for dairying.
6. The price-consumption responses to market.

Milk prices vary from region to region. In rural areas milk is priced lower than in the urban markets where the demand is high. Judicious payment to milk producers has been attracting the dairy manager and technocrats from time to time. Previously milk is used to be priced on the basis of volume without any consideration to its quality and this has not provided any incentive to the milk producer to produce milk with better quality. The vendors used to dilute or skim the milk unscrupulously in order to get a better price as that of the whole milk. Farmers, who are the producers of the milk have realised the fact that if they dilute the milk with water or remove the fat by skimming the fat test would show it and proportionately get less price for their milk. This system of payment of milk on the basis of fat alone did not go for long as the knowledge of the nutritive value of the non-fat solids (SNF) of milk has changed the fundamental approach towards the milk. A remunerative pricing structure is aimed at providing sufficient incentives to farmers to enhance milk production.

The sensitivity involved in milk pricing mechanism introduces an element of intricacy and uncertainity in its determination and needs a constant monitoring mechanism for fair pricing based on scientific measurements. A critical factor that has since been considered in the price fixation is the composition of quality milk. This implies an equitable weightage to both ingredients of milk fat and solids not fat (SNF).

A third ingredient which needs equal attention is organo-leptic quality. To overcome the ill effects of faulty pricing system a two axis approach to milk pricing has been evolved. In any developed countries, the milk having fat over and above 3.7 percent receives no extra incentive of payment. There the emphasis to day is on protein rather than fat since a product like cheese has assumed a higher relative importance and the market for low fat products is showing an increased growth.

Taking into consideration the point of view of the milk processors/milk product manufacturers, the price fixation of milk must take into account the following.

1. The stage of operation of the plant.
2. The cost of transportation.
3. The cost of processing/manufacturing and distribution.
4. The capacity of the plant, so as to utilize the surplus milk in the milk shed area and
5. The market objective of the plant including its diversification programme of the products.

Similarly, the consumer's price of the milk must be fixed taking into consideration the size of the population, which is to be covered by the milk scheme and the distribution of people in different occupational and income groups namely the lower, middle and higher, which are to be served. The total cost of transportation, processing/manufacturing and distribution, which should be covered adequately besides leaving a reasonable margin of profit to the milk plant. Therefore, it is necessary to strike a balance between the producers price, which is to serve as a production incentive and the consumers price, which should be within the purchasing power of the average consumer throughout the year.

The following basic considerations have to be borne in mind while fixing the price of fluid milk to be purchased from the producers.

1. It should be related to the cost of milk production.
2. It must ensure a fair margin of profit.
3. It should take into account the seasonal variations in milk production i.e. supply and demand and
4. It should be linked with the consumer price index by taking into account the general market trends.

Various methods for the purchase of milk are employed individually or in combination as given below.

(a) Payment according to Weight or Volume also known as Flat rate.

Advantages

1. It saves time.
2. Simple to calculate and make payment.

Disadvantage

1. Encourages adulteration either by watering or by removal of fat i.e. skimming. Payment by weight is preferred to payment by volume because the former is not affected either by foam or specific gravity. This method is popular in the unorganised sectors, where the milk vendors used to purchase the milk in the rural areas from the milk producers who are still unaware due to their illeteracy.

(b) **Payment according to the fat content of milk:** Among others it includes the straight fat method of payment.

Advantages

1. It is very much practical in its nature.
2. Discourages adulteration with water.

Disadvantages

1. It does not prevent the removal of skim milk.
2. Does not take into account the solids not fat content of milk.

This method has been practiced by majority of the dairies.

(c) **Payment according to the use of the milk for the manufacture of products:** This is the practice that is being followed in the preparation of the milk products. Majority of the private manufactures of the milk use this kind of procedure in order to get the returns for their investment.

(d) **Payment of Premiums:** This method is usually confined to the purchase of market milk, which is based on the quality of milk as determined by:

1. Sediment test
2. Flavour score
3. Bacterial count or Methylene Blue reduction time.

This method encourages the production of high grade quality milk by the producers.

(e) **Payment according to the cost of production:** Even though this method is very rational in its nature because the producers will be benefited in getting a maximum return to their investment however it is much complicated.

(f) **Quality Linked Payment to Milk Producers:** There is a growing concern for the quality of raw-milk produced by the farmer. A new system of payment scheme for raw-milk according to the quality has been introduced in South

Korea during the year 1993 according to this method, the quality of milk has been classified based on the Somatic Cell Count (SCC) in milk. Initially the milk quality was classified into three categories according to the SCC namely:

(a) First class - milk with less than 200,000 scc/ml

(b) Second class - milk with less than 500,000 scc/ml

(c) Third class - milk with more than 500,000 scc/ml

As the consciousness for milk quality increased the standards were revised to grade the milk into five classes such as:

(a) First class - milk with less than 200,000 scc/ml

(b) Second class - milk with less than 350,000 scc/ml

(c) Third class - milk with more than 500,000 scc/ml

(d) Fourth class - milk with less than 750,000 scc/ml

(e) Fifth class - milk with above 750,000 scc/ml

The indian farmer also needs to be given initiative for supplying better quality milk may be based on the korean pattern.

(g) Two axis pricing system: India has been producing large quantities of buffalo milk compared to any other country in the world. The buffalo milk being rich in fat content always attracted a fair price in comparison to cow milk. The fat portion being visible which gives thickness, separable yeilding cream and can be measured as percentage thus making it easier to decide the price of milk.

A **Kilo fat system:** A system based on "Kilo fat" became a practice for the purchase of buffalo milk. Under this system an amount in rupees per kilo fat means an amount payable on that quantity of milk which would yeild one kilo of fat. For example assume that the rate per kilo fat is Rs 550 which means that the said amount will be paid for 16.66 litres of buffalo milk with 6% fat (minimum standard) as shown below.

1 kilo (1000 gm fat / 60 gm / litre (6%) = 16.66 litres on this basis the price per litre works out total 550 / 16.66 = Rs 33.02/ litre.

B. **Double Axis Formula:** Similarly if cow milk with 3.5% fat (minimum standard) has to be purchased under kilo fat system it is going to fetch Rs. 19.25 per litre as shown below.

1000 gm/35 gm/litre (3.5%) = 28.57 litres

Rs 550/28.57 = Rs 19.25/litre

This works out to be 60% of the rate paid for buffalo milk. Thus it amounts to gross injustice to the cow milk producer. There fore with a view to pay for buffalo milk and cow milk on the rationale of their two components viz fat (F) and solids not fat (SNF) a system was devised called as double axis milk pricing. This system considers as if milk is paid for the fat component and also for S.N.F. component. The purchase rate for fat and S.N.F area determined based on previous experience or ruling market prices/consumer appreciation for buffalo milk fat (white ghee) via-a-vis cow milk S.N.F (SMP). Accordingly the difference between prices paid for buffalo milk and cow milk is reduced on scientific basis. Assuming that the rate of Rs 550 per kilo fat (which can neither be purchased nor the selling rate for ghee normally) is translated into Rs 210 per kilo fat and Rs 175/- per kilo S.N.F. which the dairy industry considers as the basic purchase price, then the purchase of buffalo milk and cow milk can be calculated and determined as shown below.

Milk	Fat	SNF	Rate/litre
Buffalo milk	6.0%	9.0%	
Rate/kg Rs	210	175	
Amount Rs	12.60	15.75	28.35
Calculation[x]	(210×60)	(175 × 90)	
Cow milk	3.5%	8.5%	
Rate kg Rs	210	175	
Amount Rs	7.35 (210 × 35)	14.88 (175 × 85)	22.23

*Calculation gms in a litre of milk × paise per gram of component.

In this way the cow milk is paid approximately to the extent of about 80% of the rate for buffalo milk. This also matches approximately 82% with total solids in cow milk as compared to buffalo milk.

A ready Reckoner can also be prepared depending on actual rates decided from sees on to season. For every 0.1 percent increase in fat and SNF the value per litre can be worked out for buffalo milk/cow milk.

Most of the milk federations in various states of the country, are purchasing the milk on the basis of the two-axis pricing system in which both fat and S.N.F. have been taken into consideration. While fixing the price of the milk per litre. (Table 7.1 and 7.2). Though this system of pricing has worked well, it appears to be unjustified because the two major constituents of

SNF are proteins, ash and sugar. They receive the same payment, despite the fact that the present knowledge of nutrition will demand a preferential price for the milk proteins. According to the two axis pricing of milk a unit of SNF consisting primarily of milk sugar and protein, receives only 2/3rds of the payment of a unit of milk fat. There is some justification to pay less for the milk sugar than for milk fat but absolutely there is no justification to pay less for the milk protein as compared to milk fat. Therefore, it appears the more justified approach would to make the payment for milk on the basis of fat and protein content of the milk and even more justified on nutritional grounds also.

An arbitrary system of milk pricing has been developed/evolved by the milk plant or the milk product factories, which seems to have no relation to the actual cost of milk production. Based on the fat and SNF content of milk, the liquid milk plants have a differential pricing system for the flush and lean months, with provisions for the payment of premium for a higher fat and SNF content than the specified standard. Whereas the milk product factories pruchase milk on per kilo (kg) fat basis at different prices for different seasons. It is necessary to adopt a uniform pricing policy which would provide an adequate incentive in order to encourage milk production through the high yeilding indegenous and cross bread cows. In this context, the National Dairy development Board (NDDB) has suggested the two axis milk pricing policy, which ensures the payment for milk on its compositional quality evaluated rationally on its fat and solids not fat components. This would discourage adulteration of buffalo milk and at the same time ensure a common pricing approach to cow and buffalo milk. The PFA (1976) rules has prescribed the minimum standards for cow milk are 3.0 to 4.0 percent fat and 8.5 to 9.0 percent SNF, whereas for buffalo milk are 5.0 to 6.0 percent fat and 9.0 percent solids not fat throughout the country.

There is an urgent need for a sound pricing policy for milk and milk products in this country which has been realized by the dairy planners. No doubt, the problem is very complicated, because it has to be acceptable to the producers, processors and manufacturers, distributors and finally the consumers. Further, the approach must be a long-range significance to the dairy development. The milk pricing policy must be rational and should ensure a guaranteed price and market to the producers throughout the year, which will act as an incentive for them to produce more milk. There must be a regular supply of wholesome milk at reasonable rates to the consumers, and an attractive margin of profit to the milk processors and product manufacturers.

Table 7.1 Cost per litre of Buffalo Milk and Cow Milk being paid by different Federations and Milk Unions (Based on Chothani 1989)

State	Cost per litre buffalo SNF milk testing 7% fat		Cost per litre of cow milk, 4% fat, 8.5% two-axis pricing system Actual			
	Maximum	Minimum	Maximum	Minimum	Maximum	Minimum
Gujarat	6.56	4.90	4.86	3.38	4.88	3.64
Uttar Pradesh	5.36	4.26	3.98	3.19	3.99	3.17
Punjab	4.88	4.55	3.66	3.38	3.63	3.38
Rajasthan	4.76	4.33	3.54	3.24	3.54	3.22
Maharashtra	4.43	4.43	3.46	3.46	3.29	3.29
Haryana	4.54	4.54	3.34	3.34	3.38	3.38
Andhra Pradesh	4.22	3.86	3.25	3.09	3.14	2.87
Tamil Nadu	4.18	4.18	3.60	3.06	3.10	3.1

Source : M. Perumal, P.S. Mohan, M. Suresh

Table 7.2 Cost of Cow Milk and buffalo milk being paid by milk federations and unions and the cost of buffalo milk based on protein + fat basis

State	Cost per litre cow milk testing 4% fat and 8.5% SNF		Cost per litre buffalo milk testing 7% fat and 9% SNF being actually paid		Cost of litre buffalo milk calculated on protein + fat basis	
	Maximum	Minimum	Maximum	Minimum	Maximum	Minimum
Gujarat	4.86	3.38	6.56	4.90	7.26	5.05
Uttar Pradesh	3.98	3.19	5.36	4.29	5.94	4.76
Punjab	3.66	3.38	4.88	4.55	5.46	5.05
Rajasthan	3.54	3.24	4.76	4.33	5.29	4.84
Maharashtra	3.46	3.46	4.43	4.43	5.17	5.17
Haryana	3.34	3.34	4.54	4.54	4.99	4.99
Andhra Pradesh	3.25	3.09	4.22	3.86	4.85	4.61
Tamil Nadu	3.60	3.60	4.18	4.18	5.37	5.37

Source : M. Perumal, P.S. Mohan, M. Suresh

The main difficulty, which is of most significant importance that can be fore seen in this system is the accurate testing of milk at source for its solids not fat content. The estimation of solids not fat in milk is a time consuming process, research would have to be undertaken to develop a rapid test which could be carried out under field conditions. The overall pricing policy for the milk products should encourage, efficiency of production, minimization of costs, quality of the product etc. The management of the dairy factories should try to create confidence in the minds of the milk producers that the price of the milk produced by them is not predetermined unilaterally at a fixed rate but varies with the season and quality.

7.2 Cooling of Milk History

The seasonal harvesting of snow and ice is an ancient practice estimated to have begun earlier than 1000 BC. A Chinese collection of lyrics around this period describes religious ceremonies for filling and emptying ice-cellars, however, little is known about the construction of these ice-cellars and the purpose for which the ice was used. Other ancient cultures such as the Greeks and Romans dug large snow pits insulated with grass or chaft as cold storage but did not use ice and snow to preserve the food, but primarily as a means to cool the beverages. The persians stored ice in a pot called a "Yakhchal" and might have been the first people to use cold storage to preserve food.

Few Americans used ice to refrigerate food due to lack of ice store houses and iceboxes before 1830 and as these two things become more widely available then the individuals started using axes and saws to harvest ice for their storehouses. The history of artificial refrigeration began when William Cullen, a Scottish Professor designed a small refrigerating machine in 1755. Benjamin Franklin and John Hadley collaborated in 1758 on a project investigating the principle of evaporation as a means to cool rapidly an object and confirmed that the evaporation of highly volatile liquids such as alcohol and ether could be used to drive down the temperature of an object past the freezing point of water.

Breweries have become the largest users of harvested ice by 1870. The ice-harvesting industry even though had grown immensely by the turn of 20th century, pollution and sewage had begun to creep into the natural ice making posing a serious problem and the breweries began to complain of tainted ice with the growing knowledge and rise of 'germ' theory in the early 1900's public concern began to increase for the purity of water from which the ice is being produced/formed. Refrigeration units were designed for installation on trucks and Lorries by the middle of 20th century. Although commercial refrigeration quickly progressed it had its limitations which prevented it from moving into the house-hold due to the following reasons namely: 1. Most of the refrigerators were far too large 2. Commercial refrigerators were more expensive for their production, purchase and as well as to maintain 3. These refrigerators were unsafe because it was not uncommon for these refrigerators to catch fire, explode or leak toxic gases. Because of these reasons refrigeration did not become a house-hold until the development of suitable technology to overcome these challenges/problems.

7.2.1 Refrigeration

In the 19th century when the refrigeration was invented and adopted the equipment was initially used to cool cans of milk which were filled by hand milking. These cans were placed into a cooled water bath to remove the heat and keep them cool until they were able to be transported to a collection point/facility. With the advancement and development of more and more methods for harvesting the milk resulting in the replacement of hand milking and the milk can with the bulk milk coolers. "Icebanks" were the first type of bulk milk coolers which was having a double walled vessel with evaporator coils and water located between the walls at the bottom and the sides of the tank. A small refrigeration compressor was used to remove the heat from the evaporator coils. Eventually the ice builds up around the coil until it reaches a thickness of about three inches surrounding each pipe and the cooling system shuts off. When the milking operation starts only the milk agitator and the water circulation pump which flows water across the ice and the steel walls of the tank are needed to reduce the temperature of the incoming milk to about 5 °C degrees or below.

This cooling method was well suited and worked satisfactorily with the small dairies, however it was fairly inefficient and was unable to cope up with the increasingly higher cooling demand of larger milk parlors/dairies. In the mid 1950's direct expansion refrigeration was first applied directly to the bulk milk cooler. This type of cooling utilizes an evaporator built directly into the inner wall of the storage tanks to remove the heat from the milk. Direct expansion is able to cool the milk at a much faster rate than the earlier ice bank type coolers which is the primary method for the bulk tank cooling today on small to medium sized dairy operations.

Another device which brought a drastic and dramatic revolution in the cooling of the raw milk instantaneously almost to 5 °C and thus contributed significantly to milk quality is the plate heat exchanger (PHE). This device utilizes a small number of specially designed stainless steel plates with small/narrow spaces between them and occupies a small space in the dairy plant. Milk is passed between every other set of plates with water being passed between the balance of the plates in order to remove the heat from the milk. This method of cooling can remove large amounts of heat from the milk in a very short time thus drastically slowing the bacterial growth and thereby improving the quality of milk. This results in the reduction of the refrigeration load on the bulk milk cooler as well as the operational costs to the dairy processors and thereby increasing the supply of milk production.

7.2.2 Basic Principles of Refrigeration

The invention of refrigeration system in the 19th century has made rapid strides in its development and with the introduction of direct expansion system in 1950's was made it possible to adopt the same by the dairy plants in the cooling of milk thus enhancing its keeping quality with the reduction of the microbial growth. In order to be familiar with the refrigeration the processors as well as the students should have knowledge and must be familiar with the terminology associated with it for better understanding.

Temperature: It is a measure of the degree of heat contained in a substance. Almost all the countries in the world except the united states of America follows and uses the Celsius (°C) temperature scale. In the United States Fahrenheit (°F) system is still the official scale of temperature however one can see both (°F) and (°C) are given together. Sometimes the student may come across and hear references to the 'Ramkine' and the 'Kelvin' temperature scales. Therefore it is essential to mention them here and all you need to know about them is that their use is generally restricted to scientific and engineering works only. Both these scales have "Absolute Zero" as their minimum point. Absolute zero is the theoretical temperature at which a substance would contain no heat and the molecular particles in the substance are not moving. In theory that is as low a temperature as can be reached. The Kelvin Scale uses the same graduations as that of Celsius scale whereas Ramkine scale uses the Fahrenheit scale graduations.

Ambient Temperature: Generally we may came across and hear the term ambient air or ambient temperature which means that the temperature surrounding the object one equipment and should not exceed 40 °C or 104 °F.

Humidity: It is the amount of moisture in the air whereas the "Relative humidity" (rh) is the percent of moisture in the air as compared to 100 percent of the moisture that air that temperature can hold. The relative humidity is as important as temperature in creating the comfort zone.

Heat: Every substance contain some amount of heat above the absolute zero even the substances that may feel cold. Still contain certain amount of heat. The unit of heat used in the United States is the British Thermal Unit (BTU). The Unit of heat in the metric system is the Joule or the kilo Joule.

- A BTU is the heat that will rise the temperature of one pound of water by 1 °F.
- A Kilo Joule (Kj) is the rough equivalent to one BTU more exactly one BTU is equipment to 1.055 KJ.

The capacity of refrigeration units is often rated in terms of 'tons' which is also used to refer the cooling load and it is equals 12000 BTU per hour.

Heat flow means that it always flows from warmer to a cooler place whereas the heat transfer is the method by which heat flows which is the basic principle of refrigeration. The evaporator transfers heat into the refrigerant and the refrigerant transfers this heat into the condensor whereas the condensor transfers the heat to a cooling medium such as water or outside air. Heat is transferred by 1. Conduction 2. Convection and 3. Radiation.

Conduction: It is the flow of heat through a substance by contact of particles. Heat from hot refrigerant transfers by conduction through the metal of a condensor coil to the outside of the coil.

Convection: It is basically the transfer of heat by the movement of gas or a liquid. The heat which travels through the walls of a condensor coil by conduction is carried into the air by convection.

Radiation: It is the transfer of heat by electro-magnet energy.

Specific heat: It is the amount of heat required to rise one pound of the substance by 1 °F. The specific heat of water is 1 which means that 1 BTU will rise the temperature of one pound of water by 1 °F.

In metric system the specific heat is stated as the amount of heat that is required to rise the temperature by 1 °KJ. The specific heat of water is 4.187 kJ/kg.

Specific Volume: The specific volume of a gas is the volume of a gas in cubic milli meters of 1 kg of the gas at standard temperature conditions.

Specific gravity: It is the ratio of the weight of a liquid or/solid as compared to an equal volume of water. The specific gravity of water is 1 and any solid or liquid that has the specific gravity of less than 1 will float on water.

Pressure: In order to understand the refrigeration cycle the student must be familiar with the pressure and the terms used to measure the same

rsi = pounds per square inch

Inches Hg = inches of mercury

Inches w_g = inches water guage

Latent Heat: There are three states of a substance 1. Solid 2. Liquid and 3. Gas. For example, water exists in all the three forms as stated above – ice – in the form of solid water – as liquid and steam or vapour = as gas. The term vapour also means the same as gas. It also means a gas with droplets of a liquid suspended in it which is also called as wet vapour. The term vapour is used with the refrigerants in the refrigeration cycle. Latent heat is the heat released or absorbed by a substance when it changes its state with no change in temperature which are termed as latent heat of vaporization or condensation which is the basis of refrigeration because it means that relatively small changes in temperature.

Chiller: A Chiller is simply a device that used to remove heat from a process or substance and in order to understand one should have a fundamental knowledge of the basic principles of refrigeration.

(a) **Heat:** It is a form of energy transferred by virtue of a difference in temperature. Heat exists everywhere to a greater or lesser degree. As a form of energy it can be neither created nor destroyed although other forms of energy may be converted into heat and vice-versa. It is important to remember that heat energy travels in only one direction from warmer to a cooler object/substance or area.

(b) **Cold:** It is a relative term which refers to the lack of heat in an object, substance or area. Another definition describes it as absence of heat no process yet has been devised of achieving "Absolute zero" the state in which all heat has been removed from any object, substance or area.

(c) **Refrigeration or Cooling Process:** It is removal of unwanted heat from a selected object/substance or space and its transfer to another object/substance or space. Removal of heat lowers the temperature and may be accomplished by the use of ice, snow, chilled water/brine or mechanical refrigeration.

(d) **Mechanical Refrigeration:** It is the utilization of the mechanical components that have arranged in a "Refrigeration system" for the purpose of transfer of heat.

(e) **Refrigerants:** These are chemical compounds that are alternately compressed and condensed into a liquid and then permitted to expand into a vapor or gas as they are pumped through the mechanical refrigeration system to cycle. The refrigeration cycle is based on the long known physical principle that a liquid expanding into a gas extracts heat from the surrounding substance or area. Refrigerants evaporate or boil at much lower temperatures than water which permits them to extract heat at a more rapid rate than the water on your finger thus helps in the removal of the unwanted heat from one place and discharging the same in another area. In order to accomplish this the refrigerant is pumped through a closed refrigeration system. If the system is not a closed one it would be using up the refrigerant by dissipating the same into the surrounding area/medium. It is a closed system the same refrigerant is used again and again as it passes through the cycle removing the heat and discharging the same. The closed system of refrigeration cycle serves other purpose as well because it keeps the refrigerant from becoming contaminated and controls its flow for it is a liquid in some parts of the cycle and gas or vapor in other phases/places.

7.3 Collection of Milk

The scenario of milk production is confined only to the rural areas, in all most all the developed dairying countries, whereas the demand is mostly urban in nature. Therefore, the milk has to be collected and transported from the production points in the "milk shed" areas to the processing and the distribution points in the cities.

A "Milk shed" is defined as a geographical area from which the milk or milk products factory receives its fluid milk supply for processing and distribution. The allocation of definite milk sheds to individual dairies for the purpose of developing the same is now under consideration in our country.

The common systems for collection of milk from the various points are as follows:

1. ***By milk collection centre or point*** : In this method the management of the milk plant may appoint a person as its agent to collect the milk on their behalf and makes the payment to the milk producers directly through him.
2. ***By cooperative organizations***: The milk coop societies are formed by a group of milk producers under the Coop. Societies Act. The members of the society themselves collect the milk and in turn send it to the milk plant. The society receives the payment from the management of the milk plant, which is distributed among the members, which is best suited because no profit making middle men are not involved in this process.
3. ***By contractors***: These are persons who collects milk from the milk producers in the rural areas and then sells the milk to the milk plant regularly on contract basis. In this way the milk producers will receive less payment from the contractor as he is acting as a middle man between the actual producer and the management of the milk plant.
4. ***Individual Producers***: These are the persons who are practising the dairying as commercial basis by keeping large number of dairy animals, thereby producing good amount of milk. It is possible for these farmers to sell the milk directly to the milk plant, with their own arrangement for transporting the milk. However, it is possible, when these dairies are nearer to the milk plant in order to avoid the extra expenditure on the transportation of the milk.

7.3.1 Milk Collection Chilling Centres/Depots

Since the electricity is now available in almost all the villages, these milk chilling centres have a great scope in our country, largely and predominately due to the small and scattered milk production. The number of organised dairy farms are rather very small and consequently these centres can provide the only alternative solution to the collection of raw milk by providing the chilling facilities at the village level itself. They can be run on a profitable basis by the producers themselves through their cooperative organisations.

These are considered as the mini milk plants and these are established by the main milk plant management in order to have a regular supply of milk to the milk plant. These centres are generally located covering a group of villages or the cooperatives, which are very far off from the main plant approximately at distance of about 100 km or above. The milk collected by this chilling centre is cooled immediately to the minimum temperature of about 5 °C and then transported to the main plant through the milk tankers. The main objectives of these centres are:

1. To preserve the quality of the raw milk supplies and
2. To provide the supply of milk in bulk quantities to the processing dairy.

Location

1. It is guided by the availability of milk sufficient to manage the centre with optimum capacity.
2. There should be adequate potable water supply.
3. Availability of electricity and
4. Sewage disposal facilities.

Major Items of Equipment

1. Milk weighing tank with weighing scale.
2. Drop or Dump tank with cover.
3. Can washer.
4. Milk pump-Sanitary type.
5. Surface/plate cooler.
6. Refrigeration unit of suitable capacity.
7. Cold room of suitable capacity.
8. Milk storage tanker either vertical or horizontal of suitable capacity.
9. Milk testing unit or laboratory etc.

Operational Procedure

All the dairy procedures or practices that are carried out at the milk processing plant are being followed at this centre also. When the milk is received after its collection from various collecting centres, the milk is graded for acceptance/rejection by the organoleptic tests, weighed, sampled for testing, cooled and stored at a temperature of about 5 °C until it is despatched to the main milk processing unit.

Cooling of milk

Importance

Freshly drawn milk from the udder contains initially some microorganisms and their numbers might increase during subsequent handling. The common milk microorganisms grow best between 20-40 °C. The bacterial growth is invariably accompanied by deterioration in market quality due to the development of off flavours, acidity etc. and one method of preserving the milk to increase its shelf life is by prompt cooling to a low temperature around 5 °C below.

Effect of temperature

It has been reported in the literature that 10 °C is the critical temperature for the bacteria to grow further in the milk. An attempt has been made to show the bacterial growth factor in milk at different storage temperature in the table 7.3.

Table 7.3 Effect of storage temperature on bacterial growth in milk

Milk held for 18 hours at temperature °C	Bacterial growth factor
0	1.00
5	1.05
10	1.80
15	10.00
20	200.00
25	1,20,000.00

Source : Outlines of Dairy Technology-S.K. De

Multiply initial count with this factor to get the final count

In view of these facts, the freshly drawn milk, should therefore be promptly cooled to 5 °C or below and should be held at that temperature until it is processed further.

Principle

It is based on the heat exchange i.e. heating and cooling.

Methods

A In can or Can-immersion method

From carrying pails, the milk is poured directly into cans through a strainer, when the can is full, it is gently lowered into a tank/trough of cooling water. The most important factor to be considered is that the water level in the tank or the trough should be lower than the neck level of the milk in the cans in order to prevent the water being entered into the cans and thus contaminating the milk.

Advantages

1. In this not only the milk is cooled, but it also stays cool and
2. It requires a much smaller mechanical refrigeration unit.

Disadvantages

1. It cools the milk very slowly and
2. There is a danger of milk contamination in case the tank water enters milk in the can.

B Surface Cooler

This may be plain conical, spiral or horizontal tubular in shape although the last mentioned is now commonly used. The milk is distributed over the outer surface of the cooling tubes from the top by means of a distributor pipe or trough and flows down in a continuous thin stream. The cooling medium mostly chilled water, is circulated in the opposite direction through the inside of the tubes. The cooled milk is received below in a receiving trough, from which it is discharged by gravity or a pump.

Advantages

1. Transfers heat rapidly and efficiently.
2. It is relatively inexpensive and
3. It also aerates the milk, thus improving its flavour

Disadvantages

1. It requires constant attention for rate of flow, which must be neither too slow nor too fast, but it must be at a constant rate.
2. There is a greater chance of air borne contamination.
3. The cleaning and sanitization is not very efficient and
4. There is slight evaporation losses.

C In tank or bulk tank cooler

This method is extensively used in the developed dairying countries like U.S.A. Properly designed bulk milk tank coolers, which are normally run by a mechanical refrigeration system, will cool the milk rapidly to a low temperature 5 °C or below and automatically maintain this temperature during the storage period. Milk can be poured directly from the milking pail into the tank subsequently, the milk can be drawn into the cans or pumped into a tanker for despatch to the city dairy.

Advantage

It permits the collection of the milk from producers on alternate days.

Disadvantage

Relatively expensive in the initial equipment.

7.4 Transportation

Under the conditions prevailing in our country milk has to be regularly collected and transported twice a day is morning and evening from the milk producers.

7.4.1 Methods of Transport

These depend upon the carrying load, the distance of the collection centres and the prevailing local conditions and these particulars have been given in the table 7.4.

Table 7.4 Modes of milk transportation in India

Mode	Optimum load (kg)	Optimum distance (km)	Remarks
Head-load	15-25	3-8	Generally employed for small loads and distances. Important in hilly areas
Shoulder - sling	up to 40	3-6	Meant for heavier loads but for shorter distances than head-load.
Pack-animal	up to 80	6-10	Ponies, horses and donkeys usually employed.
Bullock-cart	300 - 400	10-12	Rather slow
Tonga	250 - 300	12 or more	Larger quantities transported, faster than head-load, shoulder-sling and pack-animal.
Bicycle	40 or more	15 or more	Quick and handy; easily accessible to milk producer's home.
Cycle rickshaw	150 - 200	10 or more	More carrying capacity than bicycle.
Boat	40 - 200	2 - 8	Only means of transport when rivers, etc. have to be crossed
Auto-rickshaw	250 - 500	15 or more	Greater carrying capacity and faster than cycle-rickshaw
Motor	1/2 ton to 3 ton	15 or more	Increasingly in use with more road building and improvement programmes

Table 7.4 *contd...*

Mode	Optimum load (kg)	Optimum distance (km)	Remarks
Railway Wagon	11 tonnes or more	80 or more	Great scope in future
Tankers (road or rail)	5 tonnes or more	80 or more	Great scope in future

Source : Dairy for all.com

Road Vs. Rail Transport

The individual advantages of each method have been presented in the table 7.5.

Table 7.5 Advantages of Road and Rail Transport

Type of Transport	Advantages
Road	(i) Loading and unloading possible directly at godown of seller and buyer
	(ii) Cheaper than rail over short distances
	(iii) Less time-consuming
Rail	(i) Cheaper than road over long distances,
	(ii) Larger quantity of milk can be handled at a time

Source : Dairy for all.com

Can Vs Tanker Transport

The individual advantages of each method have been given in the table 7.6 .

Table 7.6 Advantages of Can and Tanker transport

Type of Transport	Advantages
Can	(i) Handling small quantities of milk possible
Tanker	(i) Quicker mode of transport
	(ii) Lower cost
	(iii) Better temperature control
	(iv) Less risk of contamination
	(v) More time and labour saving
	(vi) Lower investment in cans
	(vii) Overall saving in detergents etc.

Source : Dairy for all.com

7.4.2 Types of Containers Used

There are several types of containers being used in the transportation of milk in different parts of the country are :

1. Baked earth.
2. Wood or bamboo.
3. Metal generally brass vessels.
4. Galvanised Iron (G-I) buckets.
5. Second hand tins - mainly vegetable oil or ghee tins.
6. Tinned iron or aluminium alloy used by organised dairies.
7. Plastic material made from HDPC.

Problems

The problems in relation to collection and transportation of milk are:

1. Milk is liquid, perishable and bulky.
2. Small and scattered milk production being remote in nature.
3. Tropical climatic condition.
4. Lack of proper transport facilities like road connectivity etc.
5. Lack of country wide organisations for milk collection and its transportation.
6. Vested interests among local milk merchants including the sweet manufacturers. Influence of middle man in the rural areas.

7.5 Milk Processing

Milk treatment is the preparation of raw milk including heat treatment as a precondition for milk processing. Processing of milk is the quality oriented activity of manufacturing and packing of dairy and dairy based products on the basis of treated milk. Milk processing includes all process activities and steps as well as the required machines, instruments and lines as well as their technology which are dealt in detail.

It can be observed that homogenisation process is not always a part of the treatment and it is used only when it is advantageous for further processing.

The milk which has been procured from the milk producers in the rural areas after chilling or in raw condition is transported to the main dairy for further processing. The various dairy processes which are to be carried after the milk is received at the receiving dock have been represented in the form a basic flow diagram or flow chart 7.1 for better understanding by the student of the dairy science.

Flow Chart 7.1 Basic flow chart of milk treatment/Processing

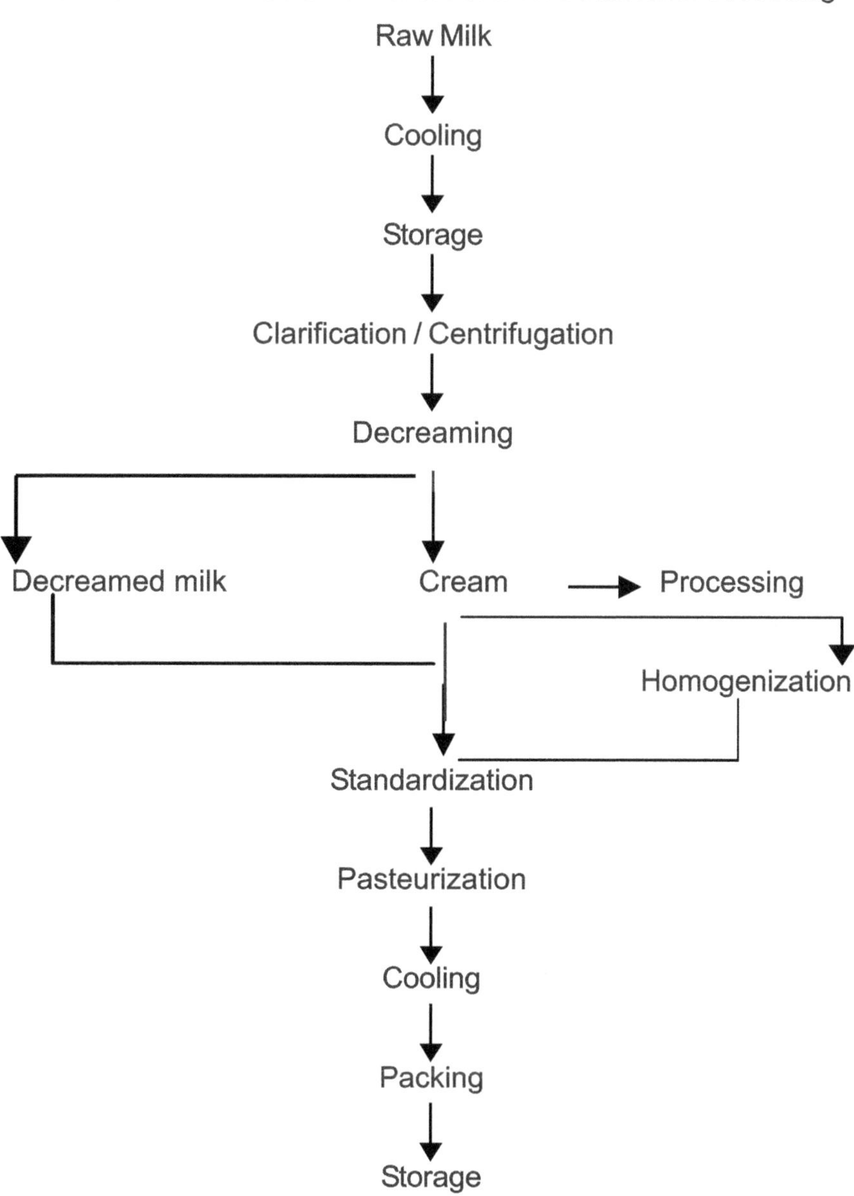

7.6 Receiving of Milk

Milk may be delivered to the main milk plant/dairy either in cans or road or rail tankers and the milk in these containers has to be graded, emptied, measured by weight or volume, sampled and bulked to provide continuity of supply to the pasteurizing equipment. Milk reception should be so planned and the equipment so chosen that the intake operations are expedited. This is especially important where large volumes of milk are received. Any delay will permit the deterioration of milk awaiting dumping, increase labour costs and may increase the operating cost of the can washer. The deliveries of milk should follow a planned schedule. If the milk is received continuously during the scheduled period, the operations in the plant will not be interrupted and the employees in the various sections will be fully occupied. The aim should be to complete milk reception within 3-4 hours especially in the tropical countries.

In the absence of any mechanical aids, the milk cans are off loaded manually to the tipping point, where the lids/covers are removed and the milk is inspected. They are then tipped manually and both the cans and lids are passed on to a can washer via a drip saver or drain rack, where a higher throughout is required, this procedure is mechanized and the cans are un loaded directly from the truck on to the conveyor which may be either power driven or by gravity roller and the tipping, sampling and weight recording are completely automatic.

The milk is generally measured by weight, it is tipped into the weight tank/pan, which is suspended from a weighing machine, the indicating dial of which is calibrated in weight and so placed that it can be easily seen by the recorder/checker. Two weigh tanks can be used for quick reception. The discharge valve has a large diameter so as to permit the rapid and instant emptying and should be easily controlled by the tipper/checker. The milk is discharged from the weigh tank into a dump tank placed immediately below it and from this the milk may be pumped continuously to a raw-milk storage tank normally situated at a higher level to give gravity flow to the pasteurising unit.

The reception of milk from large rail or road tankers is primarily a matter of providing a covered area under which emptying and subsequent cleaning can take place. The tanker outlet must be connected to sanitary piping. The milk may be removed by a milk pump, situated at a lower level than the tanker or a compressed air line may be connected to the top of the tanker and the milk is forced out by air pressure. Washing and sanitization of the tanker should follow immediately after the milk is completely emptied. The measurement of the milk delivered by the tankers can be done either by using weigh bridge or a flow meter.

Milk which has been received from the milk chilling centres, has already been graded, weighed, sampled and cooled. It may be weighed and sampled again or the centre's report may be used. The latter procedure applies especially to the deliveries made through the tanker.

Market milk requires milk of a higher quality from the stand point of good health, flavour, sanitary quality, keeping quality and aesthetic quality, than is required by any other branch of the dairy industry. The quality of the incoming milk greatly influences the quality of the processed milks or manufactured dairy products. It is well known, that the sanitary quality of milk on the receiving platform/dock depends on its background on the farm or village conditions namely healthy animals, clean milk production, clean utensils, freedom from colostrum, prompt cooling and refrigerated transport. However, there is a need for systematic and thorough inspection of all milk supplies everyday by conscientious and experienced milk graders.

The milk which is received at the milk plant/dairy, it should be at 5 °C or below. The milk should be clean, sweet of pleasant flavour, free from off flavours and reasonably free from extraneous material, contamination with anti biotics, pesticides and other chemicals or metals is highly undesirable. No abnormal milk should be accepted. Acid development is objectionable, because not only does it indicate an excessive bacterial count but also reduces the heat stability of milk.

7.7 Milk Reception Operations

The milk after it was received at the receiving dock, the following operations are to be carried out for further processing which are sub devided into

(a) unloading
(b) grading
(c) sampling
(d) weighing and
(e) testing.

(a) Unloading: The motor trucks carrying the filled milk cans are backed up (or kept aside) to the receiving dock or unloading platform. The milk cans are then unloaded manually and if the level of the truck surface is in line with the platform, the unloading requires the least effort as there is no lifting of the cans up and down, but only pulling or dragging on to a level surface. The milk cans are then arranged/assembled for grading in a definite order, according to each supplier namely villages, contractor or patron etc. In the case of a milk tanker it should be positioned properly so that necessary connections can be made easily and conveniently.

(b) Grading: This refers to the classification of the milk on the basis of the quality, for price-fixing purposes because it is well known that the quality of the finished product depends on the raw material used. The milk grader is the key man for the proper selection of milk. The principle of grading is based on the organoleptic (sensory) tests such as those for smell (odour), taste, appearance and touch, acidity sediment etc., which are generally termed as the "platform tests".

Important

The term "Platform Tests" includes all those tests, which are performed to check the quality of the incoming milk on the receiving platform so as to make a quick decision regarding its acceptance or rejection. They are performed on each can or tanker of the milk with the object of detecting milk of inferior or doubtful quality, so as to prevent it being mixed with high grade milk. Sometimes the term "Rapid platform Test" is used to refer mainly to the organoleptic or sensory tests which take very little time to perform. The technique of grading of milk may be described as under:

(a) **Milk Tanker (Road/Rail):** Generally, the grading should have been carried out at the milk collection centres or the milk chilling centres, because the milk has been cooled to 5 °C or below, it was not possible to detect the off flavours. Only the appearance can be noted, as testing of raw milk is usually avoided and after thoroughly mixing it for 5-10 minutes a sample is taken for laboratory testing.

(b) **Milk Can:**The main tests applied to each can of milk consists of smell, appearance and temperature (touch) and other tests such as taste seldom carried out with raw milk and sediment might be used to substantiate the initial findings. Tests involving time, laboratory facilities and special techniques are best done by the quality control technician, for which a sufficiently large sample is properly taken as the milk is being received. Even though, the cans of milk have been despatched from the chilling centre, it is advisible to inspect each and every can separately. The various platform tests are described as under:

1. **Smell (Odour):** This furnishes an excellent indication of the organoleptic quality of milk. It can be ascertained very quickly. In carrying out the test, the cover of each can is removed, inverted, raised and brought to the nose. The odour or smell will be representative of that in the can. The top of the milk in the can may also be noted simultaneously for smell. By replacing the lid and mixing the milk in the can vigorously with the help of a plunger the test may be repeated. An experienced milk grader with a 'trained nose' usually relies to a great extent in the acceptance or rejection of the intake milk on the odour test alone and the milk should be free from any of flavours.

2. **Appearence:** Regular observation of the milk in each can after the odour test has been made, any floating extraneous matter, off colour or partially churned milk may be noted. The milk should be normal in colour, free from churned fat globules and reasonably free from any floating extraneous material.

3. **Temperature:** The temperature at which the milk is delivered often acts/ gives an indication of its quality. A daily check on the temperature of milk is helpful in grading the milk on the receiving platform. With experience, the grader can tell with a high degree of accuracy, whether or not the milk is sufficiently cold by touching the side of the can. A temperature of 5 °C or below is satisfactory.
4. **Sediment:** The sediment test is an indication of the visible foreign matter contained in the milk. It need not be conducted every day, but should be made frequently in order to ensure a clean milk supply. For this purpose a reliable sediment tester, by which the work may be expedited should be selected. Any method by which maximum sediment will be revealed should be considered satisfactory. A low sediment is desirable.

Important

The sediment test is used as a check on the milk production and handling methods on the farm. A low sediment need not necessarily reflect or mean a low bacterial count in milk although the reverse is often true. A clean sediment disc does not indicate clean methods, however it may be a pointer, rather, to the milk having been filtered/strained on the farms.

5. **Acidity:** Reference has already been made and pointed out that 'natural' or 'apparent' acidity of milk does not make the milk taste sour nor does it affect the normal properties of milk or jeopardize its quality or its behaviour towards processing heat. However, developed or 'real' acidity does adversely affect the quality of the milk.

Important

Determination of the titrable acidity of milk, which is equal to the natural acidity plus developed acidity in order to decide the acceptance or rejection of milk on the basis of an arbitrarily set limit cannot serve much useful purpose in our country today, because of the fact that the milk suppliers are freely adding neutralizers to the milk with the sole objective of reducing its acidity. Nevertheless, the 'acidity test' does have its proper place on the milk reception dock, and with the daily incoming milk, it is always well to have a certain acidity above which milk should not be accepted. A modified form is known as the 'rapid acidity test'.

6. **Lactometer Reading:** The addition of water to milk results in the lowering of its lactometer reading. Hence this test is applied for detection of adulteration of milk with water. It does not take much time, it is often used as a platform test in the milk collection or chilling centres in our country. However, this test has its own drawbacks.

(c) **Sampling of Milk:** The importance of securing an accurate and representative sample of milk for subsequent chemical and bacteriological analysis cannot be over emphasized. Strict precautions regarding sterility of the stirrer, sampler, and container etc., are required for obtaining a bacteriological sample, dryness and cleanliness of the above equipment should suffice for a chemical sample. The first and foremost prerequisite of sampling is thorough mixing of the milk. This can be done with a plunger or stirrer (agitator) operated either manually or mechanically in the milk in cans or tankers as the case may be. A representative sample may also be taken after the milk cans are quickly dump into the weighing tank, whereby it gets mixed thoroughly so that, a representative sample may be taken without further mixing.

Samples may be individual, composite mixture of two or more individual lots of milk or drip representing the entire days supply etc. Samplers may be dipper tube or proportionate, automatic vaccum drip etc., and their characteristics have been presented in the table 7.7.

Table 7.7 Characteristics of milk samplers

Type	Principle	Advantage	Disadvantage	Remarks
Dipper	Secures 10-15 ml milk	(i) Fairly fast and easy to work with (ii) Quite accurate when milk is mixed adequately before sampling	Inaccurate when wide variations exist in milk lots, both in quality and quantity	Most commonly used. Most useful for cream
Propor-tionate	Secures aliquot portion of milk	Most accurate	(i) Cumbersome to use (ii) Larger sample bottle needed	Not to commonly used (not so useful for cream)
Auto-matic vaccum	Secures aliquot portion by vaccum automatically	(i) Very fast in operation (ii) Very accurate	Expensive	Increasingly used in large market milk plants and product factories
Drip	Milk collections in drops in the sample bottle (which is kept under refrigeration)	Helpful in fat and SNF accounting of the total intake	Not useful for individual sampling	Useful in large product factories

Source : Outlines of Dairy Technology S.K. De

The milk must be kept sweet, if the composite sample is to be successful, while the sample is being assembled, which is accomplished by the usage of preservative. It is a good plan to place the preservative in the empty bottle before the milk is added. A wide mouthed glass/plastic bottle-must be made of HDPC and transparent with a stopper has been found to be the most reliable and practical container for keeping the composite samples of milk or cream. The common preservatives used are :

1. **Mercuric chloride or Corrosive sublimate:** This is very poisonous. It must be added in the form of tablets which are usually coloured red in order to prevent the milk being mistaken for food.
2. **Formalin:** This is a 40 percent solution of formaldehyde being liquid in nature; it is very convenient to handle. However, it interferes with the fat test.
3. **Potassium Dichromate:** This is not as effective as the above two, but it is easy to handle in dairy plants because it is available in tablet form.

Important: The composite samples should be stored in a cool place away from direct sunlight. Each bottle should be labelled properly with proper precautions.

(d) **Weighing:**This is an important and essential step in accounting for the receipt of milk, its disposal and making the payments etc. The milk in cans is dumped into the weigh tank, either manually or mechanically. The tank is mounted on scales and the scale dial is set at zero, when the tank is empty, thus enabling the operator to make a direct reading of the weight of the milk. Automatic printing of the weight is also now being very common. The weighing is facilitated by the use of the dial reading or some other indicating scale, rather than a beam scale. The milk in tankers either road or rail may be measured by volume by passing through a flow meter and its measurement is converted into weight by multiplying the volume with an agreed density. Another common alternative method is to use a weigh bridge, in the case of road milk tankers. The tanker is weighed once when it is full and again after it has been emptied. However, any mud present on the tanker should be washed off before weighing and adjustments to fuel or water should not be made between the weighings, and the driver and his assistants should be either on or off the vehicle on both occassions. The characteristics of measurement by weight and by volume have been presented in the table 7.8 for a better understanding.

Table 7.8 Measurement of milk by weight Vs Volume

Method	Characteristics
By weight	(i) Gives accurate reading, regardless of foam or temperature
	(ii) Involves considerable initial expense for both apparatus and its installation.
	(iii) Involves problems with maintenance.
By volume	(i) Not so accurate, as affected by foam and temperature, both influencing density.
	(ii) Lower initial expense
	(iii) Presents maintenance problems
	(iv) Definitely a factor to be considered in the overall picture of sanitation.

Source : Outlines of Dairy Technology S.K. De

(e) Testing: There is always some scope of doubtful quality, apart from the initially accepted or rejected lots of milk. All the accepted lots of milk have already been properly sampled. These, together with the samples of the remaining two categories have to be tested in the quality control laboratory for the final verdict of acceptance or rejection. Moreover, a record of the chemical and bacteriological quality of all the accepted milk has to be maintained for making the payments etc. The students and the teachers are advised to consult the Bereau of Indian Standards (BIS) which is previously known as the Indian Standards for "Methods of Test " ISI : 1479 (Part-I) 1960, (Part-II) 1961 and (Part-III) 1962 for further clarifications if any.

The common quality control tests that are generally performed in the dairy plant have been summerised and presented in the table 7.9.

Table 7.9 Quality control tests for milk and their significance

Name of Test	Purpose	Remark
Acidity	To determine final acceptance/ rejection of milk (on the basis of a pre-determined level)	Applied as such or in a modified form, as a platform test
Ethanol (Alcohol)	To determine the heat-stability of milk	Applied as a platform test
Alcohol-Alizarin	To determine both heat-stability and pH of milk	Applied as a platform test

Table 7.9 *Contd...*

Name of Test	Purpose	Remark
COB (Clot-on-boiling)	To determine the heat-stability of milk	Applied as a flatform test
Dye-reduction test (MBR or Resazurin)	To determine the extent of bacterial contamination and growth in milk	2-mts. Resazurin applied as a platform test
DMC (Direct microscopic Count)	To identify the types of micro-organisms present in milk	Applied as a laboratory test
SPC (Standard plate count)	To determine the extent of bacterial contamination and growth in milk	Applied as a laboratory test
Lactometer	To detect adulteration of milk with water	Applied as a platform test
Freezing point	- do -	Applied as a laboratory test
Fat and/or SNF	To make payment for milk received	Applied as a laboratory test

Source : Outlines of Dairy Technology S.K. De

7.8 Pre Heating

This term refers to heating before the operation, which follows immediately. The milk is preheated for efficient filtration/clarification. Preheating becomes essential if the incoming milk is cold, because the flow of milk is hampered. As the temperature of the milk increases, the viscosity decreases and more efficient filtration/ clarification results occur. The usual temperature of pre-heating is 35-40 °C and the equipment used may be a plate or tubular heater.

7.8.1 Filtration/Clarification

(a) **Object:** To improve the aesthetic quality of milk by removing the visible foreign matter. Which is unsightly and may therefore cause consumer complaints.

(b) **Principle:** Filtration removes all the suspended, foreign particles, by the straining process, whereas the clarification removes the same by centrifugal sedimentation.

(c) **Types of filters and Clarifiers:** There are two types of filters/clarifiers namely those which operate with cold milk and those operating with warm milk. Generally, the filter/clarifier which is operated with warm milk is in usage throughout the world, whereas the one which operates with cold milk has the following advantages and disadvantages.

Advantages

1. There is no need for pre-heating of the milk and
2. There is less possibility of the 'soluble' dirt going into the solution and
3. The only disadvantage in this method is the slow flow rate of milk.

7.8.2 General Construction and Operation of Filters and Clarifiers

(a) Filters

The important features of the filters are:

1. A filter cloth or pad or the desired pore size which can retain the smallest particles.
2. A frame or support to compress and hold the margins of the cloth or pad so that the milk can pass through the pores.
3. A perforated metal or any such other support for the cloth or pad, which will not tear or break under the pressure/force of the milk.
4. An enclosure to confine both the unfiltered and filtered milks in a closed system fitted suitably with an inlet and outlet connections for sanitary piping.
5. A means of distributing the incoming stream of milk so that it does not damage or tear any part of the cloth or pad by vigorous washing and
6. A design so planned that filter cloths or pads can be changed quickly, and all parts are easily accessible for washing. Where continuous operation is essential, or where large volumes/quantities of milk are processed, two or more filters are used so that the operations need not be inturrupted when it becomes necessary to change the filter cloth. The frequency with which the cloth is changed will depend upon the temperature of the milk, the amount of foreign matter present in the milk etc. It is always better to use the filter cloths only once because a washed cloth is not that effective resulting in efficient filtration, besides being a source of contamination.

(b) Clarifiers

Generally the appearance and the construction of the clarifiers are very much quite similar to the centrifugal cream separators. However, the main differences between these two are:

1. In clarifiers there is only one outlet, whereas in the cream separator, there are two outlets one for the cream and the other for the skim milk.
2. The discs of the clarifier bowl are smaller in diameter, so as to provide a large surface/space for the accumulation of the slime than the cream separator and

3. The milk distribution holes are at the outer edge of the discs in the clarifier whereas the holes are near the centre in the cream separator.

7.8.3 Location of Filters/Clarifier in the Processing Line

It is the general practice to locate the filter in the raw milk line before the milk enters the pasturising unit in some cases, while in the others the filter is located at a convenient place/part in the regeneration section where the temperature of the milk is around 50-60 °C. The clarifier may be located in any one of the following places in the processing line as shown in table 7.10

Table 7.10 Location of the clarifier/filters

Location	Type of clarification
Between	
1. Reception and storage tank	Cold
2. Storage tank and pasteurizer	Cold
3. Pre-heater and pasteurizer	Warm
4. Regeneration and Heating section of HTST	Warm
5. Heating section and Holding tube of HTST.	Warm

Source : Outlines of Dairy Technology - S.K. De

7.8.4 Relative Efficiency

The clarification removes sediment much more efficiently than the filtration, and the clarifier remove still finer particles that escapes the filters.

7.8.5 Composition of Clarifier Slime

The material which was retained on the filter cloth includes the suspended foreign particles, milk fat, protein and some leucocytes including bacteria. Whereas the slime that accumulates in the clarifier bowl consists of foreign matter, milk proteins, leucocytes, fragments of secreting cells from the udder, fat, calcium phosphate, and other ash, bacteria and occassionally red blood corpuscles. The amount of clarifier slime is influenced by the amount of foreign matter, the condition of the udder, the stage of lactation, the bacterial count, and the acidity of milk, the clarifying temperatures, the velocity of the bowls and the amount of milk that runs through the bowl or the length of time the bowl is run.

The composition of the clarifier slime varies considerably due to the factors stated above. The separator slime is usually considered to be identical with that of the clarifier slime. The typical composition of the clarifier/separator slime reported in the literature has been given in the Table 7.11.

Table 7.11 Composition of clarifier/separator slime

Constituent	Moisture (%)
Water	67.3
Total solids	32.7
Fat	1.1
Protein	25.9
Ash	3.6
Lactose	2.1

Source : Outlines of Dairy Technology S.K. De

Important

The removal of clarifier slime does not affect the composition of milk to a significant extent, the loss in solids resulting from clarification is usually 0.01 percent or even less.

7.8.6 General Remarks

1. Both filtration and clarification tend to decrease the depth of the cream layer that will form on the milk and this effect becomes more pronounced as the processing temperature increases.
2. The keeping quality of the milk is not improved/increased either by filteration or by clarification.
3. Milk should not be filtered or clarified after pasteurization because the process may contaminate the pasteurized milk.

Important

Bactofugation is the process by which 99 percent of the bacteria in milk are removed by centrifugal force. It is claimed that this method triples the shelf life of the market milk. In general, the bacteria from the milk are removed with two centrifugal clarifiers in a series, the first one operating at a high velocity of 20,000 rpm. The process is considered as a supplementary to the pasteurization, however, it is still necessary to destroy the bacteria which have not been removed even by this process.

7.9 Cooling and Storage of Raw Milk-Cooling (In the Dairy Plant)

Introduction

The importance of cooling and the effect of temperature has already been discussed earlier. The milk is chilled to 5 °C or below, immediately after it is received at the dairy plant and stored at this temperature, to prevent the deterioration in its bacteriological quality during the interim period.

7.9.1 Methods

(a) Surface Cooler

These coolers are either individual units or cabinet type in which two or more individual units are assembled in a compact manner. Generally these coolers are larger in size than those used either on the farm or chilling centres due to the fact that large volumes of milk is being handled at the milk plant.

(b) Plate coolers

These are generally used in the bigger dairy plants/industry where large quantity of milk is being handled on a large scale for continuous cooling. The plate coolers consists of a number of thin, flat, grooved, stainless steel plates, sealed at the edges with a gasket and clamped tightly within a press. The spaces between the plates are occupied alternately by the milk and the cooling medium-chilled water or brine. Thus one side of each plate is exposed to milk and the other side to the cooling medium. Additional plates may be added to provide increased capacity for handling the milk at a nominal cost.

Advantages

1. Cooling (heat exchange) is very quick and efficient.
2. The milk is not exposed to air borne contamination.
3. There are no evaporation losses.
4. Cleaning and sanitization is easy.

(c) Internal Tubular Cooler

It consists of a stainless steel tube of about 2.5 to 5.0 cm in diameter, surrounded by a similar tube forming a concentric cylinder for the continuous cooling of milk. Several such tubes are connected in a series to obtain sufficient cooling. The cooling medium flows in the opposite direction to the milk flow.

Advantages

1. Cooling is quite efficient
2. There is no exposure to air borne contamination.
3. There is no evaporation loss

Disadvantages

1. The efficiency of cooling is lower as compared to the plate cooler.
2. It needs a larger floor space as compared to the plate cooler.

(d) Jacketed Vat/Tank

This method of cooling the milk can be adopted for batch cooling especially when the milk is being handled in small quantities. It consists of tank within a tank, with the space in between the two being used for circulation of the cooling medium, by either pump or main pressure. An agitator is provided to move the milk, which is in the upper tank for rapid cooling.

Disadvantages

1. Cooling efficiency is very low.
2. Too much agitation of milk is required in order to keep the milk to stay cool, which causes churning and impairs the creaming property of the milk.

7.9.2 Storage

Introduction

The modern milk plants have the capacity to hold both raw and pasteurized milks for much longer period than before. Generally the storage capacity of the milk is equal to a day's intake, and thus allows a more nearly uniform work-day for processing and bottling operations with less dependance on the time for receiving the raw milk. Storage tanks are used in milk plants for the storage of raw, pasteurised or processed products, often in very large volumes. The term 'processed' is generally used in a wider sense in dairying, which includes, pasteurization, homogenisation evaporation, drying etc. Therefore, it is desirable to use the specific term, when only one process is meant. In view of the longer periods of holding, the storage tanks occupies an important position among the most important items of equipment. They must be designed for an easy sanitation, preferably, by the circulation cleaning method. In addition, the tanks should be insulated or refrigerated, along with a mechanical agitator to provide an adequate and homogenous mixing in a gentle manner thus preventing the churning as well as incorporation of air, by maintaining the required temperature throughout the holding period.

Objectives

1. To maintain milk at a low temperature so as to prevent any deterioration in quality prior to processing and/or product manufacture.
2. To facilitate bulking of the raw milk supply, which will ensure uniform composition.
3. To allow for an uninterrupted operation during processing and bottling and
4. To facilitate standardization of the milk.

7.9.3 Types of Storage Tanks

1. **Insulated or Refrigerated:** In the former there are 5 to 7.5 cm of insulating material between the inner and outer linings, whereas in the latter, the space between the two linings is used for circulation of the cooling medium. Another variation of the refrigerated type is the cold wall tank.
2. **Horizontal or Vertical:** Horizontal storage tanks requires more floor space and less head space, whereas the latter requires less floor space and more head space. Modern circulation cleaning methods have made very large vertical storage tanks very practical.

3. **Rectangular/Cylindrical or Oval:** Of these, the first i.e. the rectangular suffers from the disadvantage of having dead corners during agitation whereas the other two do not have this kind of disadvantage as there are no corners.
4. **Built for gravity flow, air pressure or vaccum operation:** The first kind of storage tank is more common, however, the air pressure is used same times in order to evacuate the product. This requires spacial construction of the storage tank for greater strength than is necessary for normal operations under gravity flow i.e. atmospheric pressure.

7.9.4 Location of the Storage Tank

The storage tanks are located on the upper floor or on a specially constructed platform at a higher position. In this system the milk is pumped from the receiving room to the floor above. Then the milk flows by gravity to the pre-heater–filter or clarifier, pasteurizer, cooler and the bottling or milk filling machine. In the another system, the milk is pumped from the storage tanks through a preheater and filter into pasteurizer. Then it may flow by gravity to the cooler or it may be pumped to the cooler while it is hot.

Fig. 7.1 Milk Storage Tank

7.9.5 Parts of the Storage Tank

The storage tank generally consists of the following parts.

1. Sight glass
2. Light glass and lamp
3. Ladder
4. Manhole
5. Agitator
6. Outlet valve
7. Inlet with proper lid/cap
8. Airvent
9. Safty valve
10. Legs
11. Indicating thermometer
12. Volume meter

7.10 Standardisation

Definition

Standardisation of milk is defined as the adjustment of the fat and/or solids not fat which refers to either raising/increasing/or lowering their percentages to a desired value, so as to confirm to the legal standards or other requirements that are prescribed.

Procedure

Standardisation of milk can be carried out either by the addition of milk or cream with a higher or lower fat percentage than that of the material to be standardised and at times even the addition of skim milk will suffice the purpose. To solve the problem, it is necessary to find the relative amounts of the original material and the standardizing material to be mixed together to give a product with the desired fat content. It is easy to calculate the exact amount of each which must be mixed together to give a certain weight of the finished product or the exact amount of standardising material needed to use up a given weight of milk or cream when the relative amounts/proportions have been determined. A simple method, called the "Pearson's square" can be used to calculate the relative quantities of the materials involved in a standardisation problem. It should be remembered that all the measurements are based on these calculations are by weight and not by volume.

The Pearson's Square Method is as follows

Draw a square and place in the centre of it the percentage of fat required for standardization. Place the material with high fat percentage than the one required for standardisation at the left hand upper corner. Similarly the other or the second material with a low fat percentage than that is needed for standardisation at the lower left hand corner. After placing both the values, now substract the number in the centre from the larger number which is at the left hand upper corner of the square and place the remainder at the diagonally opposite right hand lower corner. In the same manner substract the smaller number on the left hand lower corner from the number in the centre and place the remainder at the diagonally opposite right hand upper corner. The numbers on the right hand side now represent the number of parts of each or the original materials that must be blended to make a product with a fat test given by the number in the middle of the square. The number at the upper right corner refers to the parts of material whose fat test was placed at the upper left corner, and the number at the lower right corner refers to the part of the material whose fat test was placed at the lower left corner. When the numbers on the right side of the square are added and thus the obtained value will represent the parts of the finished product with the fat test given by the number obtained in the middle of the square. The most important factor to be borne in mind is that while selecting the products that are to be standardised must have different fat percentages in that one with a higher and the other should be lower, further this method is applicable with two ingredients only. The following illustration will give the details of the calculation using the 'pearson's square' method of standardisation of milk.

Example 1

500 kg standardised milk testing 4.5% fat is to be prepared by mixing whole milk with a fat test of 6 percent and skim milk testing 0.2 percent fat. Calculate the whole milk and the skim milk to be required for the standardisation.

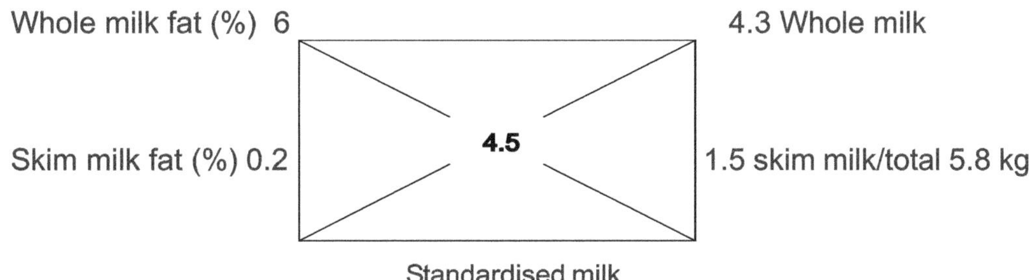

For 5.8 kg standardised milk whole milk required is 4.3 kg.
Therefore for 1 kg of standardised milk it will be 4.3/5.8
and 500 kg the whole milk required will be = 4.3/5.8 x 500 = 370.686 kg
similarly the skim milk required will be = 500 – 370.686 = 129.314 kg
The requirement of skim milk can also be calculated in the following manner.
For 5.8 kg standardised milk, the skim milk required will be = 1.5 kg
For 500 kg of standardised milk 1.5/5.8 x 500 = 129.31 kg
the requirement of skim milk will be = 129.31 kg

Example 2

Find out the quantity of whole milk and skim milk to make 1000 litre of toned milk of 3 percent, with the wholemilk having 4 percent fat and 0.3 percent skim milk.

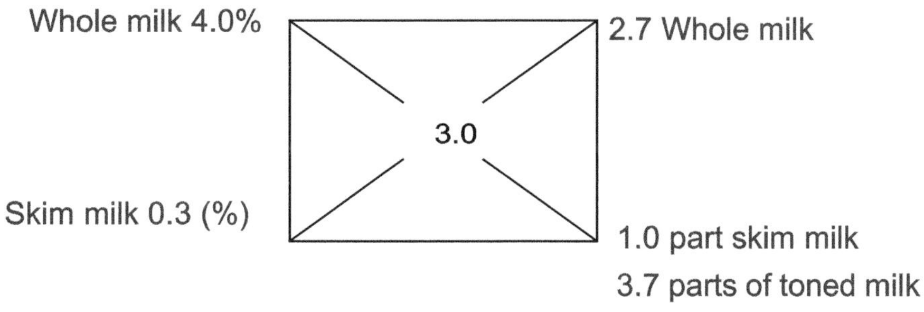

3.7 parts of toned milk contains 2.7 parts of whole milk

Therefore 1000 litres of toned milk will contain = 2.7 × 1000/3.7

= 729.73 litres

3.7 parts of toned milk contain 1 part of skim milk.

Therefore 1000 litres of toned = 1 × 1000 / 3.7 = 270.27 ltres skim milk

Thus for 1000 litres of toned milk - 729.73 ltres wholemilk

- 270.27 ltres skim milk.

The same quantities of whole milk and skim milk can also be calculated arithmetically for the standardisation of the milk as under using the fat percentages as stated below.

Example 3

100 kg of milk testing 5% fat and 8.5% S.N.F is to be standardised to 3.0% fat and 8.5% SNF with the addition of skim milk testing 0.2% fat and 8.5% SNF. Find out the quantity of skim milk needed for the purpose.

Calculation

Let us assume 'x' kg of skim milk will be needed for the purpose of standardisation. Standardised milk = initial whole milk + skim milk = (100 + x) kg.

100 kg of standardised milk contains fat = 3 kg

Therefore, (100 + x) standardised milk will contain fat = 3/100 × (100+x)

100 kg of skim milk contains fat = 0.2 kg

Therefore 'x' kg of skim milk contains fat = 0.2 / 100 × x

= 0.002 x

Now to find out the value of 'X' in terms of fat content.

5 + 0.002 x = 3/100 × (100 + x)

= 3 + 0.03 x

or 5 – 3 = 0.03 × –0.002 × = 0.028

Therefore 2/0.028 = 71.43 kg of skim milk

Example 4

Calculate the quantity of whole milk testing 6.0% fat and skim milk testing 0.2% fat to prepare 500 kg of standardised milk with 4.5%

Calculation

Let us assume the whole milk to be taken = x kg

Therefore, the skim milk required will be = (500 – x) kg

Fat content of 100 kg whole milk = 6 kg

Therefore 'x' kg of whole milk will have fat = 6/100 x = 0.06 x

The fat content of 100 kg skim milk will be = 0.2 kg

Hence (500 – x) kg of skim milk will have fat = (500 – x) × 0.2 / 100

= (500 × 0.2 /100) – (0.2x / 100) = 1.0 – 0.002 x

Fat content present in 500 kg standardised milk will be = 500 × 4.5 kg

Total fat = 0.06 x + 1.0 – 0.002 x = 22.5 kg

= 0.06 x – 0.002 x = 22.5 – 1.0

= 0.058 x = 21.5

x = 21.5 / 0.058 = 370.69 kg

Therefore whole milk required = 370.69

and skim milk 500 – 370.69 = 129.31

Example 5

10,000 kg of whole milk with a fat percentage of 5 is to be converted into 25000 kg of toned milk having 3 percent. The mathematical derivations adopted will be as follows:

Total fat required = 25000 × 0.03 = 750.0 kg

Fat in the whole milk = 10,000 × 0.05 = 500.0 kg

Therefore the amount of additional fat required = 750 – 500 = 250.0 kg

i.e. Butter oil.

Amount of skim milk powder of 9% SNF required = 15000 × 0.09 = 1350 kg

The fat content of butter oil to be added = 12.5 percent

Requirement of butter oil to be added = 250 / 0.125 = 2000 kg.

Therefore the requirement of reconstituted skim milk = 15000 – 2000

= 13000 kg

Finally we find out that to prepare the 3 percent toned milk from 10,000 kg of whole milk with 5 percent fat to 25000 kg we require 2000 kg of butter oil with a fat percentage of 12.5 and 1350 kg of skim milk powder of fat free but 9% SNF mixed with 13000 kg of water.

Example 6

How many kg of 28% cream and 3% milk will be required to make 500 kg of a mixture testing 4% fat.

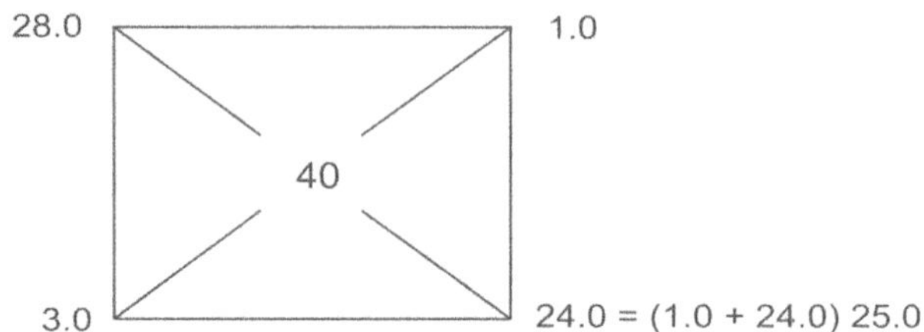

1:25.0 = C : 500 where C is the amount of cream required in kg

or 1 /25 = C / 500 or C = 500 / 25 = 20 kg.

Milk (3%) = 500 – 20 = 480 kg

Proof/Check

500 kg of 4% milk contain = 500 × 4 ÷ 100 = 20.0 kg

20 kg of 28% cream contain 20 × 28 ÷ 100 = 5.6 kg fat

480 kg of 3% milk contain 480 × 3 ÷ 100 = 4.4 kg fat = 14.4 + 5.6 = 20 kg

7.11 Homogenization

History

It is derived from the greek word homos which means = same and genos = kind this meaning same kind which is a process of converting two immiscible liquids into an emulsion. An emulsion is a type of colloidal substance which is microscopically dispersed throughout another substance continuously and when both the substances are liquids then the colloid is termed as emulsion. Homogenisation is a process used to make a mixture of two mutually non-soluble liquids the same throughout. This is achieved by turning one of the liquids into a state consisting of extremely small particles distributed uniformly throughout the other liquid. Sometimes two types of homogenisation have been distinguished namely primary homogenization – when the emulsion is created directly from separate liquids and secondary homogenization – when the emulsion is created by the reduction in the size of the substance/particles to droplets in an existing emulsion.

One of the oldest application of homogenization is in milk processing which is achieved by mixing massive quantities/amounts of milk procured from different animals and milk collection centres on dairies in order to produce a more consistant raw milk prior to processing by forcing the same at high pressure through a small aperture/hole thus creating a constant. The main objective of this is to reduce to prevent and delay the natural separation of cream from the rest of the emulsion. The fat in the milk normally separates from the water and collects at the top. The homogenization process breaks the fat globules into smaller size of 2 μ or less thus preventing its separation and allowing the sale of non-separating milk at any fat specification. Yet another method of homogenization uses extruders, hammer mills or colloid molls to grind the solids. Milk homogenisation is an important and essential tool of the milk food industry to prevent creating various levels of flavor and fat concentration. Another application of homogenization is in soft drinks like 'cola' products. The reactant mixture is rendered to intense homogenization to as much as 35,000 psi so that various constituents do not separate out during storage or distribution.

7.11.1 Process of Homogenization

The homogenisation refers to the process by which the fat globules ranging from 4.0 to 15 μ in the milk are broken down into a smaller globules of uniform size of 2 μ or even less by forcing the milk through a homogenizer and thus resulting in the loss of its creaming property. This process increases the viscosity with the increase in the number of the fat globules of much homogeneous size. According to the United States Public Health Service the homogenised milk is defined "as the milk which has been treated in such a manner as to ensure break up of the fat globules to such an extent that after 48 hours of quiescent storage, no visible cream separation occurs in the milk and the fat percentage of the milk in the top/upper portion of 100 ml of milk kept in a quart, bottle or of any proportionate volumes in containers of other sizes does not differ by more than 10 percent of itself from the fat percentage of the remain of milk as determined after thorough mixing. In the efficiently homogenized milk, the fat globules are subdevided or broken down to 2 microns or even less in their diameter.

Merits

1. There is no formation of cream layer or plug.
2. The fat present in the milk after homogenisation does not churn due to rough handling or excessive agitation.
3. It is better suited or adopted for bulk dispensing and mixing is not necessary.
4. It is more palatable perhaps due to the brighter appearence, heavier body and richer flavour.
5. It produces a soft curd which is better digested, hence recommended for infant feeding.
6. It is less susceptible to oxidised flavour development.

Demerits

1. There is an increase in the cost of production.

2. There is difficulty in salvaging/utilising the retained homogenised milk, due to the problem in the recovery of fat.
3. Appearance of sediment is of much greater degree.
4. The milk curdles on cooking or heating.
5. The milk is more susceptible to the production of activated or sunshine flavour defect.
6. There is an increased/greater tendency for the seepage of milk through the bottle caps even for pouches or sachets when there is a small hole in the sealing.

7.11.2 Homogenizer

This is a machine which causes the subdivision or breakage of the large fat globules. It consists of a high pressure piston pump which forces the milk at a high pressure and/or velocity through a narrow opening between the homogenizing valve and its seat. The fat globules in the milk are thereby sub-divided into smaller particles of more uniform size. The homogenising valve is held down by a heavy pressure spring against the seat of the valve. The valve and its seat are made of extremely hard material like stellite and the contact faces are grounded carefully so that the valve sits accurately on its seat. The Homogenisers are either single stage or double stage and there are, three distinct types and these are:

1. **High pressume type:** The piston forces the milk through an orifice/opening past the homogenising valve. This valve is loaded hydraulically or by a powerful spring with a hand wheel control. The pressure exerted by a loading system is balanced by the pressure in the milk which is being forced through a very small orifice. This causes the larger fat globules to break down and reform as smaller globules. The process of homogenisation is carried out at 60 °-70 °C and at a pressure ranging from 500-5000 pounds per square inch (PSI).
2. **Low pressure - Rotary type Homogeniser:** This type of homogeniser is so constructed that milk fat is subjected to grinding and shearing action under lower pressure, lower than that in high pressure 500 pounds per square inch. The action on fat globules is effective, but it is doubtful whether shattered fat globules are as small as those obtained by the high pressure machines or not.
3. **Sonic vibrator:** Milk is exposed to high frequency vibrations, processed in a device called as sonic vibrator or oscillator. This machine is not used now-a-days in the dairy industry for homogenisation.

Important

1. A viscolizer is a machine similar to a homogeniser but usually operates at a lower pressure and has similar openings.

2. A clarifixator is a machine developed in Sweeden, which not only clarifies but also homogenises the milk.

7.11.3 Factors Influencing Homogenization

(a) Temperature

At the time of homogenisation, the milk should be at a temperature above the melting of fat namely above 33 °C or 91°F because the fat should be in the liquid state for proper subdivision or breakage of the fat globules. The enzyme lipase should be in-activated, preferably prior to homogenisation or immediately afterwards. This can be achieved by heating the milk to a temperature of 55 °C or 131°F or above. In routine practice the milk is heated to 65-70 °C or 149-150 °F for homogenisation. The danger zone for lipase activity i.e. temperatures of 38-49°C or 100-120 °F should be avoided during homogenisation.

(b) Pressure of Homogenization

In a sigle stage up to 6 percent fat milk usually 2000 -2500 PSI pressure is sufficient. Higher pressures may increase the tendency for the milk to curdle when cooked due to the increased destablising effect on milk proteins. For liquid products with more than 6 percent fat a two stage homogenisation is necessary in order to prevent fat clumping with a pressure of 2000 PSi at the first stage and 500 PSI at the second stage.

(c) Sequence of events in Homogenization

The possible various sequences are as under:

1. Clarification, pre-heating, homogenization, pasteurization and cooling.
2. Clarification, pre-heating, pasteurization, homogenization and cooling.
3. Pre-heating, homogenization, clarification, pasteurization, cooling, which is recommended under Indian conditions.
4. Pre-heating, clarification, homogenization, pesteurization, cooling and
5. Pre-heating, clarification, pasteurization, homogenization and cooling.

(d) Important

The main considerations are:

1. All homogenised milk must be pasteurized either prior to or immediately after homogenisation.
2. The homogenised milk should preferably be clarified after homogenisation.
3. In order to avoid any post pasteurization contamination it is desirable to follow the sequence stated under item three (3) above.

7.11.4 Method of Manufacture

The Procedure for the manufacture of homogenised milk has been given in the flow chart 7.2

Flow Chart 7.2

Receiving milk

↓

Cooling to 5 °C and bulk storage

↓

Pre-heating (35 - 40 °C)

↓

Filtration/Clarification

↓

Cooling to (5 °C)

↓

Standardizing and storage (5 °C)

↓

Pre-heating (60 °C)

↓

Homogenisation (2500 PSI) (60 °C)

↓

Pasteurization (Holder or HTST)

↓

Cooling (5 °C)

↓

Bottling/Tetra pack sachets

↓

Storage (5 °C)

Pasteurization

7.12 History

The process of heating wine for the purpose of preservation has been known in China since 1117 and was documented in Japan Written by a series of monks between 1478 and 1618. Much later in 1768 an Italian priest and scientists research proved that a product could be "Sterile" after thermal processing. A less aggressive method was developed by the French chemist Louis Pasteur during 1864 in order to find out a remedy to the frequent development of acidity in the local aged wines. He found that experimentally that it is sufficient to heat a young wine to only about 50-60 °C (122-140 °F) for a short period of time in order to kill the microbes and that the wine could be aged subsequently without sacrificing the final quality. In honour of pasteur the process became known as Pasteurization and was originally used as a way of preventing wine and beer from souring and it would be many years before milk was pasteurized.

Milk is an excellent medium for the growth of the microbial population and when stored at ambient temperature bacteria and other pathogenic organisms may soon proliferate. The US centres for disease control says improperly handled raw milk is responsible for nearly three times more hospitalization than any other food borne disease source thus making it one of the world's most dangerous food products pasteurization of milk was suggested by Frariz Von Soxhlet in 1886. In the early 20th century Milton Joseph Rosenau (1912) established the standards i.e., low temperature slow heating at 60 °C (140 °F) for the Pasteurization of milk. Different states in USA began enacting mandatory dairy pasteurization laws with the first in 1947 and in 1973 the US federal government required pasteurization of milk used in any interstate commerce. The shelf life of refrigerated pasteurized milk is greater than that of the raw milk.

Pasteurization derives its name after the eminent french scientist Louis Pasteur, who during his research studies (1860 - 1864) observed that heating certain liquids, especially wines to a high temperature between 122 - 140 °F or 50 - 60 °C improved their keeping quality by killing the spoilage organisms. The first pasteurization test was completed by Pasteur and Claude Bernard in 1862. The application of this process gave rise to the new term "Pasteurization" which soon became current in the technical language. Even though Louis Pasteur poineered the studies on heat treatment for preservation, pasteurization of milk has been carried out and was first attributed to Dr. Soxhlet of Germany in 1886.

Pasteurization of milk ensures the destruction of the pathogenic microorganisms present in the milk. In this process the milk is heated to a particular temperature for a definite period of time as it is related to the thermal death point of the most of the microorganisms that are pathogenic. The destruction of these pathogens is essential due to the following reasons namely:

(a) to safe guard the public health,

(b) to extend the keeping quality of fluid milk by destroying most of the milk souring bacteria, and

(c) to ensure good manufactured milk products.

Fig. 7.2 Milk Pasteurization Unit

Fig. 7.3 Refrigeration

7.12.1 Definition

Pasteurisation has been defined by the U.S. Public Health Service, "The term pasteurization shall be taken to refer to the process of heating every part of milk or milk products to at least 143 ° to 145 °F (63 °C) and holding at such temperature for at least 30 minutes or to at least 160 °-161 °F (72 °C) and holding at that temperature for at least 15 seconds in approved and properly operated equipment". After pasteurisation, the milk is immediately cooled to 5 °C (41 °F) or below.

It has been pointed out earlier that the milk production in our country is scattered in the rural areas, where it is difficult to excersize strict supervision over all the milk supplies received either at the chilling centres or at the milk plant. Therefore it becomes a necessary event to pasteurise the milk in order to make it safe for human consumption by destruction of all the pathogenic microorganisms and to improve the keeping quality. However, any improvement of nutritive value is of the slightest extent.

Many objections have been raised from different quarters with regard to the pasteurization of milk and some of these objections have been summerised below.

1. In our country, pasteurization is not necessary, because of the fact that milk is invariably boiled on receipt by the consumer housewife even though the milk is pasteurized.
2. Pasteurisation reduces significantly the nutritive value of the milk.
3. It reduces the cream line or cream volume, which is not acceptable to the majority of the consumers.
4. The quality of milk especially which is of low quality is masked by the pasteurization.
5. Pasteurised milk will not clot/coagulate with rennet, which is a disadvantageous to the consumers for the preparation of paneer in their homes.
6. Pasteurisation encourages slackening of the efforts for the sanitary/clean milk production by the producers.
7. Pasteurisation when carried out in a careless manner without following all the safety measures it gives a false sense of security to the consumers.
8. Pasteurisation destroys the pathogenic bacteria only but not the bacterial toxins that have been produced and released into the milk.

In order to decide the quality of the pasteurisation which has been carried out at the milk plants certain standards have been fixed which are to be satisfied. The following considerations were involved in the formulation of the standards of pasteurisation.

1. **Bacterial Destruction:** All the pathogenic microorganisms must be destroyed. Any heat treatment i.e. the temperature time combination, which kills the *tuberculosis* organisms is expected to destroy all other pathogenic

organisms that have gained their entrance into the milk. *Mycobacterium, tuberculosis is* considered as the most heat, resistant among the pathogens, which has been chosen as the index organism for pasteurisation.

2. **Cream line Destruction:** The consumer judges the quality of the milk on the basis of the cream line. The cream line or cream volume is reduced progressively with the increase in the temperature time combination of heating/ pasteurization.
3. **Phosphatase Test:**The enzyme must be destroyed completely by the pasteurisation. The phosphatase test is used to detect the adequecy of the pasteurisation. Presence of any traces of the enzyme phosphatase in the milk is considered as inadequate in the pasteurisation process.

Therefore the standards of pasteurization were such as to ensure:

1. Total destriction of the pathogenic microorganisms.
2. Negative phosphatase test and
3. Least damage to the cream line.

The *tuberculosis* bacteria are destroyed by the heat treatment which is slightly lower than the temperature required for the inactivation of the phosphatase enzyme. Thus the pasteurization when carried out at a heat temperature above that of the temperature required for the phosphatase inactivation and yet it is below the temperature required for the reduction of the cream line has been presented in the table 7.12.

Table 7.12 Pasteurisation requirements

Particulars	30 minutes	15 seconds
To kill T.B. Organisms	138 °F / 58.9 °C	158 °F / 70 °C
Inactivation of Phosphatase	142 °F / 61.1 °C	160 °F / 71.1 °C
Requirements of Phosphatase	143 °F / 61.7 °C	161 °F / 71.7 °C
	145 °F / 62.8 °C	(since 1956)
Reduction of cream line	144 °F / 62.2 °C	162 °F / 72.2 °C

Source : Outlines of Dairy Technology - S.K. De

Important

1. Even though pasteurisation is considered as a health measure it has become actually a commercial expedient.
2. Pasteurisation cannot be considered or taken for grant as a fool proof measure because it is not a cure for all.
3. Post pasteurisation should be avoided.

7.12.2 Pasteurization Process

Pasteurization process is an application of mild heat treatment of liquid foods both packaged and unpackaged where products are typically heated to below 100 °C. The heat treatment and cooling process are designed to inhibit a phase change of the product. The acidity of the food determines the parameters like time and temperature of the heat treatment as well as the duration of shelf life. The parameters also take into account nutritional and sensory qualities that are sensitive to heat. In low acid foods (pH 74.6) such as milk the heat treatments are designed to destroy pathogens and spoilage organisms (yeast and molds). It is important to note that all spoilage organisms are destroyed under pasteurization parameters thus requiring refrigeration.

Definition: Pasteurization of milk is defined as the process of heating milk to a predetermined temperature for a specified period without recontamination during the entire process. The predetermined temperature usually depends on the heat resistance of spoilage micro-organisms that the pasteurization programme is targeting to destroy.

Objectives:

1. ***Public health aspect:*** The chief objective of milk pasteurization is to destroy the pathogenic bacteria that could have a public health concern and with the destruction of these micro-organisms the milk and milk products become safe for human and public consumption.
2. ***Keeping quality aspect:*** Pasteurization eliminates destructive bacteria and enzymes that could cause spoilage of the product leading to a prolonged shelf life of the milk.

Pasteurized milk is commercially sterile which means that they are not entirely rid of bacterial and compounding their reservation with another method usually the refrigeration. The amount of heat temperature and length of time used in pasteurization depends on the "Thermal death time of the targeted micro-organisms. The minimum combination should target the most resistant pathogen such as Coxiellae burnettii. While deriving the thermal death time of any micro-organism the temperature must remain constant and holding time varied to kill the specified number of bacterial cells. All the micro-organisms are not destroyed through this process and still exists in the pasteurized products in very low concentration e.g., E.Coli Lysteria and Salmonella. Refrigeration retard the further growth of the bacteria.

(a) **Thermal Death time: (D value):** It is the duration or time taken to kill certain bacteria at a given temperature.

The decimal reduction time – 'D value is the amount of time under specified conditions to reduce microbial population by one decimal. This time varies and is dependant on the temperature and target micro-organism one decimal reduction (1D) is equivalent to a population drop by one log cycle or 90% reduction.

(b) **Z-value:** It is the measure of change in the rate of death due to change in temperature. It is the change in temperature required to change the decimal death time by one log cycle or one decimal (1D). It is the measure of bacterial sensitivity to heat treatment. It is the change in temperature that will reduce the 'D' value by the factor 10 and the Z-value is obtained by plotting the two 'D' values against temperature.

(c) **'F'-Value:** It is the duration/time that will take to kill known bacterial population. It is usually set at 12 log cycles (12D) to kill the most resistant mesophile spores in a food sample. Different micro-organisms have different 'D' values. However these 'D' values follow a negative slope when plotted on a graph.

7.12.3 Methods of Pasteurization

Broadly speaking pasteurization can be categorized as either low or high temperature method, of pasteurization and both these can be either batch or continuous processes. Low temperature pasteurization method is generally concerned with the safety of food and aims at the killing of all the pathogenic micro-organisms thus resulting in the reduction of all the spoilage types in the milk. Low temperature method can assume various temperature-time combinations such as 63 °C/30 minutes or 72 °C/15 seconds, thus preserving most of the physicochemical properties of the milk. On the contrary the high temperature pasteurization method aims at the killing of the vegetative pathogenic and spoilage bacteria as well as denaturizing as much serum protein as possible. Because of the denaturing of the proteins they will not be able to separate and does not allow 'syneris' to occur high temperature pasteurization method involves intense heating and may involve temperature time combination regimes such as 140 °C/2 seconds 85 °C/30 minutes or 90 °C/20 minutes which aims at the destruction/denaturing of serum proteins with an intention to avoid "Synerisis".

The choice or type of pasteurization method depends on several factors which may not be limited to:

1. Intended purpose of the Pasteurized milk
2. Access to the sophisticated equipment
3. Volume of the milk to be Pasteurized and
4. Target of the micro-organism etc.

Whatever may be the purpose/objective a person can choose to carryout either normal or ultra pasteurization. Normal pasteurization will preserve the milk for a period of about 2 to 3 weeks under proper storage conditions whereas ultra pasteurization will preserve the milk even upto 1 year period. Under normal proper storage conditions.

There are several methods of Pasteurization that are being in use/followed at various dairy plants as mentioned below:

1. **Thermization:** In this process the milk is heated to between 57 °C to 68 °C and holding for a period of 15 minutes thermization targets pathogenic bacteria while learning the good bacteria in the product. The low temperature does not alter the structure and taste of the milk.
2. **In the bottle pasteurization**
3. **Batch/Holding pasteurization:** It is also known as low temperature – Long time (L.T.L.T) method. In this process the milk is heated to 63 °C and holding the same for 30 minutes. The extended holding time may cause some alterations in the milk protein structure and taste. This method consists of:
   (a) Water-Jacketed vat
   (b) Water spray type and
   (c) Coil vat type
4. **High temperature short time (HTST) also called as Flash Pasteurization:** In this method the milk is heated between 72 °C to 74 °C for a period of 15-20 seconds of time and targets the resistant pathogenic bacterial spores (Clostridian botulinum spores).
5. **Ultra high temperature (UHT) pasteurization:** In this process the milk is heated between 135 °C to 140 °C for a period of 2 to 4 seconds. This extreme high heat targets the "Coxiella burnettii" which causes 'Q' fever in human beings. This act kills all the vegetative forms of bacteria and the milk can be stored for a period of about 9 months to 1 year under normal storage conditions.
6. **Canned Sterilization:** This is a wet treatment of canned milk products in an autoclave/specialized treatment chambers and are heated between 115 °C to 121 °C for a period of 10-20 minutes.
7. **Electric pasteurization**
8. **Vaccum pasteurization – vacreation**
9. **Stassanization and**
10. **Uperisation or ultrapasteurization**

The detailed procedure of individual methods of pasteurisation are briefly given as under.

7.12.3.1 In the bottle pasteurisation

Bottles filled with raw milk and tightly sealed with special caps are held at 63-66 °C or 145-150 °F for 30 minutes, then the bottles pass through water sprays of decreasing temperatures which cool both the product as well as the bottle also. The only advantage of this method is that it prevents the possibility of post pasteurization contamination.

Disadvantages

1. Very slow rate of heat transfer.
2. Bottle breakage risk is greater.
3. In order to allow the expansion of milk during the heating, there is a need to use the over sized bottles to provide an extra space.
4. Special types of water tight caps have to be used. Presently this method has become obselete, and has become an academic importance only. However, the in-the bottle sterilization of milk is still widely prevalent in some parts of the world.

7.12.3.2 Batch/Holding pasteurisation

This is also called the low temperature - long time (LTLT) method. The milk is heated to 63 °C/145 °F for 30 minutes and promptly cooled to 5 °C or below. In this method the heating is done indirectly where the heat moves through a metal wall into the product for heating and out of the product for cooling. The pasteurizer may be of three types namely.

(a) **Water Jacketed Vat:** This is a double walled container on all the sides and bottom and the space in between the two is utilised for circulating the hot water or steam under partial pressure for heating and cold water for cooling. The outer lining wall is generally insulated to reduce the heat loss. This vat in all respects resembles that of a thermos flask except for the difference that vaccum is created by removing the air between the two walls of the flask and is sealed. The heat exchange takes place through the wall of the inner lining. The difference between the temperature of the heating water and the milk is kept to a minimum. The milk in the vat is agitated by slowly moving or revolving paddles or propellors. At the time of heating the vat cover is left open for the escape of the off flavours and when holding the cover is closed. During the holding period an air space/foam heater generally steam prevents surface, cooling of milk. This is also known as a multipurpose or multi process vat because of its flexibility in the usage.

(b) **Water spray type:** In this method water is sprayed in a thin film through the perforated pipe over the surface of the tank holding the product. The product in the tank is agitated in the similar manner as explained above. Rapid heat transfer is provided by the quickly moving continuous thin film of water.

(c) **Coil Vat type:** The heating/cooling medium is pumped through a coil placed either in a horizontal or vertical position and the coil is turned through the product which agitates the product also, however, additional agitation may be necessary. The main disadvantage in this system is the difficulty in the cleaning of the coils which accounts for the decline in their usage and the efficiency in heating and cooling the product is the milk.

Important

This method of pasteurisation i.e. the vat heating/cooling is very slow and involves too much agitation, resulting in churning, and impairment of creaming properties and this operation can be efficiently carried out by plate or tubular heat exchangers. A continuous flow of milk can be obtained with multiple vat installations in the vat holding system, however, this may encourage the growth of thermophilic organisms over a long period of time and also it requires a large floor space for these multiple vats to be installed in the dairy plant.

7.12.3.3 High Temperature Short Time (HTST) Pasteurisation or Flash Pasteurisation

This method was first developed by A.P.V. Co. in the year 1922 in United Kingdom. This is the modern method of pasteurising the milk and invariably being used by all the major dairy plants, where large volumes of milk are handled. The HTST pasteurizer gives a continuous flow of milk which is heated to 72 °C [161 °F] for 15 seconds and immediately promptly cooled to 5 °C or below. The advantages of this system are as under.

1. Maintaining the rigid quality control measures on both the raw and finished product, it has the capacity to heat treat the milk quickly and adequately.
2. It requires less floor space for its installation.
3. Lower initial cost.
4. Packaging of milk can begin immediately as soon as the pasteurisation process begins and thus permitting the more efficient use of labour for packaging and distribution.
5. The unit is cleaned easily and sanitized and is well suited or adopted itself well to the C.I.P. system of cleaning.

6. Lower operational costs due to the better utility of the use of regeneration at its fullest capacity.
7. The capacity of pasteurisation can be increased at a nominal cost.
8. The milk losses are at the bearest minimum.
9. The development of the thermophilic organisms is not a problem.
10. In between the process can be interrupted and restarted quickly depending on the necessity.
11. Automatic precision controls ensure positive pasteurisation.

The system also has got some disadvantages which have been enumerated below:

1. This system is not suitable or well adapted to handle small quantities of several liquid milk products.
2. Constant attention or supervision is required for the possible damage of the gaskets, and lack of sanitation measures.
3. Complete drainage is not possible without losses, which exceeds those when compared to the holder system.
4. Automatic, precision control instruments are required for its operation due to the fact that the margins of safety in product, sanitary control are very narrow.
5. The pasteurisation efficiency of high thermoduric count of raw milk is not as great as it is when the Holder system is used.
6. There is a possibility for the greater accumulation of milk stone in the heating section due to the high temperature of heating.

7.12.3.3.1 HTST-Milk Flow

Raw cold milk which is stored in a constant level storage tank at 4 °C is drawn into the regenerator section of the pasteurizer, where it is warmed to approximately 57 °C to 68 °C by the heat given up by hot pasteurized milk flowing in a counter current direction on the opposite side of the thin stainless steel plates. The raw milk which is still under suction passes through a positive displacement timing pump which delivers it under positive pressure through the rest of the HTST system.

The raw milk is forced through the heater section where hot water on opposite sides of the plates heat milk to a temperature of 72 °C. The milk at this pasteurization temperature and under pressure flows through the holding tube, where it is held for at least 16 seconds. The maximum velocity is governed by the speed of the timing pump, diameter and length of the holding tube and surface friction. After passing temperature sensors of an indicating thermometer and a recorder controller at the end of the holding tube, the milk passes into the Flow Diversion Device or Valve (FDD or FDV). The FDV assumes a forward flow position if the milk passes the recorder controller at the preset cut in temperature (72 °C). The FDV remains in normal position which is in diverted flow, if the milk has not achieved the preset cut in temperature. The improperly heated milk flows through the diverted flow line of the FDV back to the raw milk constant level tank. Properly heated milk flows through the forward flow part of the FDV to the pasteurized milk regeneration section, where it gives up its heat to the raw product and in turn cooled to approximately 32 °C - 9 °C. The warm milk passes through the cooling section where it is cooled to 4 °C or below by coolant on the opposite sides of the thin stainless steel plates. The cold pasteurized milk passes through the vaccum breaker which is at least 12 inches above the highest raw milk in the HTST system and then into the storage tank for further packaging and distribution.

The efficacy of the heat exchange process become important to the food, beverage and the dairy industries in the light of 1. The rising cost of energy and 2. The environmental concerns associated with the production of energy. The heat exchanger is therefore designed to optimise the energy consumed by pre-heating the incoming raw product using the hot pasteurized product. The heat regeneration is determined from the ratio of the heat load in the regeneration zone to that of the total system and it is typically expressed as a percentage.

A plate and frame heat exchange unit consists of a series of plates mounted in a frame and clamped together. The space between the adjacent plates create flow channels and the unit is arranged to allow hot and cold fluids to enter at the four corners. This design provides a highly efficient counter current flow yielding heat transfer coefficients significantly higher than other designs. A corrugated Chevron or herringbone pattern pressed into each plate gives strength and rigidity as well as increasing turbulance in the flow channels. The net result of this design feature is to enhance further the rate of heat transfer. The plate and frame exchanger may be designed to produce wide channels to avoid fouling and blocking with the more viscous fluid products. However the relatively narrow channels in the plate and frame units may be a disadvantage of the products contain particulates in case crystallization occurs during the process.

7.12.3.3.2 Flow of milk through the plate heat Exchangers (PHE) for effective pasteurization

1. Chilled milk from the bulk milk tank at 4 °C moves into the balance tank from where it is pumped into the regeneration section of the PHE.
2. At the regenerative heating section of the PHE. Chilled milk receives heat from the already pasteurized milk leaving the section. Heat exchange occurs

across the PHE plates in a counter current motion of the two fluids of different temperatures. The regenerative heating rises the temperature of milk to about 40 °C to facilitate easy standarisation. Heating then continues to 60 °C to facilitate easy homogenization of the fat globules.

3. After clarification, standarisation and homogenization milk passes into the heating section where milk exchanges heat with steam across the PHE plates. The steam heats the milk to over 72 °C with is the perfect H.T.S.T pasteurization condition.
4. Once the milk has attained the pasteurization temperature, it moves into the holding tubes. The lengths of these tubes have been calibrated with the flow rate to ensure that the milk stays at the temperature for atleast 16 seconds. This time is sufficient to destroy the target pathogen according to the 'D' values.
5. If the milk fails to attain the required temperature, the flow diversion valve (FDV) diverts its flow back to the heating section to ensure that the temperatures are sufficient to kill the target pathogens and their spores.
6. The milk when fully pasteurized it moves back to the regenerative heating section to rise the temperature of the incoming chilled milk. In this process the temperature of the outbound pasteurized milk drops to about 32°C.
7. The pasteurized milk then moves to the cooling section of the pasteurizer where chilled water/or the coolant further lowers the temperature of 4 °C.

7.12.3.3.3 Advantages of regenerative heating

1. Utilization of the incoming chilled milk to cool the outgoing hot pasteurized milk thus increasing the efficiency of the PHE.
2. A smaller amount of heat energy is required to heat the milk to pasteurization temperature because the heating does not start from 4 °C of the chilled milk.
3. Reduces the amount of time required for the pasteurization of milk.

Note: When starting the process of pasteurization in the PHE the milk is circulated in the heating section until it attains the required temperatures before the beginning of the regenerative heating.

7.12.3.3.4 Holding time

When fluids move through a pipe, either of two distinct types of flow can be observed. The first is known as turbulent flow which occurs at high velocity. The second type is streamline or laminar flow which occurs at low velocities. The "Reynolds number" is used to predict whether laminar or turbulent flow will exist in a pipe.

Re < 2100 laminar

Re > 4000 fully developed turbulent flow.

There is an impact of these flow patterns on holding time calculations and the assessment of proper holding tube lengths.

The holding time is determined by timing the interval for an added trace substance (salt) to pass through the holder. The time interval of the fastest particle of milk is desired. Thus the results found with water are converted to the milk flow time by formulation since a pump may not deliver the same amount of milk as it does water.

7.12.3.3.5 Milk supply or Flow

The following steps or stages are involved, as the milk passes through the HTST pasteurisation unit or system:

(a) Balance tank

(b) Pump

(c) Regenerative heating

(d) Holding

(e) Regenerative cooling and

(f) Cooling by chilled water or brine solution.

Provision is also made in the circuit for the incorporation of the filter/clarifier, homogeniser etc as and when required/desired. However, there is some variation in the usage or order/sequence of these steps in different milk processing plants.

7.12.3.3.6 Important parts-and their Functions

(a) **Float Controlled Balance Tank (FCBT):** The balance or constant level tank facilitate a constant supply of milk and is equipped with a float valve assembly, which controls the level of the milk nearly constant, ensuring uniform head pressure on the product leaving the tank. The over flow level must always be below the level of lowest milk passage in regenerator. Thus it helps to maintain a higher pressure on the pasteurisation side of the heat exchanger. The balance tank also prevents the entry of air into the pasteurizer by placing the top of the outlet pipe lower than the lowest point in the tank and creating downward slopes of at least 2%. The balance tank provides a means for re-circulation of diverted or pasteurized milk. Maintains a constant supply of milk for feeding the raw milk pump and also receives any sub-temperature milk that is diverted by the Flow Diversion Valve (FDV).

(b) **Pump:** The timer pump draws the milk through the raw regenerator and pushes it under pressure through the pasteurization regenerator. It governs the rate of flow through the holding tube. It must be interwined so that it operates only when the FDD or FDV is fully forward or fully diverted and must be fall rate. A centrifugal pump with magnetic flow meter and controller can also be used.

It may be either a rotary positive pump located between the regeneration and heater or a centrifugal pump with a flow control device to ensure a constant output after FCBT is being used.

(c) **Plates:** The plate heat exchanger which is also called as paraflow is commonly used in the HTST system of pasteurisation, especially for heating to the temperatures which are below the boiling point of milk. The plate heat exchanger is a compact, simple cleaned and inspected unit with very ease. The plates may be used for heating, cooling regeneration and holding, which are supported in a press between a terminal block in each heating and cooling section. The heat moves from a warm to cold medium through these stainless plates. A space of approximately 3 mm is maintained between the plates by a non-absorbant rubber gasket or seal which can be vulcanised to them. These plates are numbered and must be assembled properly. They are tightened into place and are designed so as to provide a uniform but not excessively turbulant flow of products with rapid heat transfer. Corrugation for raised sections on the plates in the form of knobs, diamonds and channels help to provide the turbulant action required. Greater capacity is secured by adding more plates. Ports are provided at appropriate places, both at the top and bottom of the plates to permit both the product and the heating or cooling medium to flow without mixing with each other.

(d) **Regeneration Heating:** Heating and cooling energy can be saved by using a regenerator which utilizes the heat content of the pasteurized milk to warn the incoming cold milk.

The raw cold incoming milk is partially and indirectly heated by the hot out going milk i.e. milk to milk regeneration, which adds to the economy of the HTST process, because the incoming milk requires less heating by the hot water in order to raise its temperature for holding.

Filter: Different shapes of filters units are placed after the pre-heater or regenerative heating section in order to connect directly to the HTST system. These units, using a 40-90 mesh cloth, are usually cylindrical in shape and generally two filters are attached, but they are used one at a time. This permits a continuous operation, the flow being switched or diverted on to the other while replacing a filter.

(e) **Holding:** It must slope upwards 1/4" per feet in direction of flow to eliminate air entrapment so that nothing flows faster at air pocket restrictions. The holding tube or plate ensures that the milk is held for a specified time, which is not less than 15 seconds, at the pasteurisation temperature of 72 °C (161 °F) or more.

Flow Diversion Valve (FDV): This valve routes or diverts the milk after heat treatment acting in two ways namely:

1. When the milk has been pasteurised properly it allows the milk to flow in the forward direction through the unit.
2. On the other hand if the milk has not been pasteurised properly, i.e. in which the temperature does not reach the prescribed limit then it is automatically diverted back to the FCBT for re-processing. It is usually operated by air pressure working against a strong spring. In case if there is a fall in the temperature the air pressure is released and the valve snaps and closes immediately and once the temperature is regained, the air pressure builds up and the valve opens allowing the forward flow of milk. The system is arranged in such a manner that any failure of air or electricity move the valve in the diverted position.

Important

The flow of the unpasteurised milk can also be stopped with a 'pumpstop', which stops automatically the milk pump motion, when the temperature of the product drops or falls below the desired level. The pump stop restarts the operation and allows the flow of milk to continue when the proper temperature is reached.

(f) Vacuum Breaker

The vacuum breaker is located at the place where the pasteurized product is discharged which breaks to atmospheric pressure. It must be located greater than 12 inches above the highest point of raw product in system. It ensures that nothing downstream is creating suction on the pasteurized size.

(g) Booster Pump

It is a centrifugal "stuffing" pump which supplies raw milk to the raw regenerator for the balance tank. It must be used in conjunction with pressure differential controlling device and shall operate only when timing pump is operating, proper pressures are achieved in regenerator and system is in forward flow.

(h) Regeneration Cooling: The pasteurised hot out going milk is partially and indirectly cooled by the incoming cold milk similar to the process, which has been stated under regeneration heating i.e., milk to milk regeneration. In the similar manner this again adds to the economy of the HTST process. In fact when the pre-cooled raw milk is received, the high degree of regeneration which is about 72-85% allows the water cooling to be dispensed entirely.

(i) Control Panel: It contains instruments, controls, FDV mechanism and holding system all centralised in one moisture proof panel. The lower half of the panel forms an air insulated, chamber which carries the holding tube.

Hot water set: Hot water is circulated through the heating section of the machine to maintain the correct milk temperature reasonably within the very fine limits.

Automatic Control devices : These include:

(a) **Steam Pressure Controller:** It maintains a constant hot water temperature for heating milk accurately to the required pasteurization temperature. It acts as a reducing valve in the steam supply line so as to give a constant steam pressure.

(b) **Water temperature controller:** It regulates the amount of steam entering the hot water circulatory system.

(c) **Milk temperature recorder:** This records the temperature of the milk in the form of a graph, leaving the holding tube/plate. This is an electric contact instrument i.e. a thermocouple which operates either the FDV or the milk pump, automatically preventing milk from leaving the holding section at sub level temperatures. Both the frequency and duration of the flow diversion and the temperature of the milk leaving the holder are recorded on the thermograph or otherwise called as recording chart by means of two separate pens. The "check thermometer" is placed near the milk temperature recorder.

7.12.3.3.7 Pressure in the System

It is important to maintain a higher pressure on the pasteurization side of the heat exchanger for the purpose of the continuous pasteurization. In the event that a pinhole leakage develops in the thin stainless steel plates, maintenance higher pressure of the pasteurized milk by at least one (1) PSi than the raw milk in the ragenerator, it prevents the contamination of the pasteurised milk with that of the raw milk. This pressure differential is maintained using a timer pump in simple systems and differential pressure controllers and back pressure flow regulators at the chilled pasteurization outlet in more complex systems. The position of the timer pump is crucial so that there is suction on the raw regenerator side and pushes milk pressure through pasteurised regenerator. There are several other factors involved in maintaining the pressure differential.

1. The balance tank overflow level must be less than the level of lowest milk passage in the regenerator.
2. Properly installed booster pump is all that is permitted between balance tank and raw milk regenerator.
3. No pump after pasteurised milk outlet to vaccum breaker.
4. There must be greater than 12 inch vertical rise to the vaccum breaker.
5. The raw regenerator drains freely to balance tank at shut down.

The normal pressures maintained in the HTST system are as under :

Pasteurised milk	-	15 PSi
Raw Milk	-	14 PSi
Heating/Cooling medium	-	12-13 PSi

7.12.3.3.8 Holding Time Test

The holding time of a HTST Pasteuriser is the flow time of the fastest particle of milk at a prescribed temperature through the holding section. The holding time is calculated between the points at which the heated milk leaves the heating section and reaches the FDV. The efficiency of pasteurisation in the HTST system depends both on the correct maintenance of temperature as well as on the holding time. Therefore the latter should be checked periodically. Several methods are used for the determination of the holding time namely:

(a) the electrical conductivity method of a salt solution,

(b) the dye injection method, and

(c) the electronic timer method etc.

7.12.3.4 Electric Pasteurisation

The modern electric pasteuriser consists of a regenerative pre-heater. The milk is brought here at about 120 °F (48.89 °C) in the pasteurizer proper. The pasteuriser consists of a rectangular, vertical chamber of two feet height and about two centimeters thickness in cross section. The two sides are made up of carbon electrodes kept at about 7-8 centimeters apart from each other. The milk is allowed to flow between these two electrodes where the temperature is raised to 161°F to 163 °F (71.7 °C - 72.8 °C) by the resistance offered to the passage of a 110 volt. A.C. and the milk exposed at this temperature for 15 to 20 seconds, after which it is cooled immediately to the cooling temperature.

7.12.3.5 Vaccum Pasteurization (Vacreation)

This refers to pasteurisation of milk or cream under reduced pressure by direct steam. The process was first developed in Newzealand by M/s Murray Deodorizers Ltd., who called the equipment a 'Vacreator' and the process "Vacreation". It was designed to remove the feed and other volatile flavours from cream and to pasteurize it for butter making.

The Vacreator consists of three stainless steel chambers, connected to one another for steam heating and vacuum treatment with continuous flow of the product. The product in the form of fine droplets enters the first chamber of the vacreator, where pasteurisation occurs the chamber is operated under a vaccum of 5 inches mercury (Hg) which maintains a temperature of 90-95 °C, where the steam fed from the top, falls by gravity to the bottom of the chamber. Then the product and some free steam are moved from the bottom of the first chamber to the top of the second chamber, where the temperature is maintained at 71-82 °C under a vaccum of 15-20 inches Hg. A portion of the steam which was previously added is removed and the product moves down through the chamber. Some of the tainting substances and other 'off' flavours are removed by heat and vaccum treatment. The product then moves on to the third chamber at 43 °C by maintaining a vaccum pressure of 26-28 inches Hg and here more water and off flavours are removed. A multistage centrifugal pump removes the product from the third chamber. Altogether, it takes about 10 seconds for the product to move continuously through the unit. Steam of uniform pressure, normally ensured by a uniform pressure reducing and regulating valve should be provided. It must not contain any boiler compounds, otherwise, they may contaminate the product.

7.12.3.6 Stassanisation

This method of pasteurisation is carried out in a tubular heat exchanger consisting of three concentric tubes. It was invented in France by Henry Stassano and is used to a considerable extent in France, Danish, Italian and other milk plants. The principle of its operation is the heating of milk to the desired temperature by passing it between two water heated pipes through the narrow space of 0.6 to 0.8 mm. The milk is heated to about 74 °C (165 °F) for 7 seconds and then promptly cooled as usual.

7.12.3.7 Uperization

This is a shortened form of 'Ultra Pasteurisation' which has been developed in Switzerland. In this process the milk is heated with direct steam up to 150 °C (302 °F) for a "fraction of a second" or 'no hold' time which is a continuous process and the method of manufacture of uperised milk is as follows.

7.12.3.8 U.H.T Process

History: Heat treatment is the most commonly applied technique to provide a safe and shelf stable milk. The first system involving indirect heating with continuous flow (125 °C for 6 min) was developed and manufactured in 1893. A continuous flow direct heating method of mixing steam with milk at temperatures of 130 °C to 140 °C was patented in 1912. However, due to lack of availability of commercially viable aseptic packaging systems to pack and store the product this technology could not make any breakthrough till 1950. In 1953 the A.P.V pioneered a steam injection technology involving direct injection of steam through a specially designed nozzle which rises the temperature of the product instantly under the brand name "Uperiser" and the process is called uperisation and the milk was packed in sterile cans. In the year 1960 the A.P.V launched the first commercial steam in fusion system under the brand name "Palarisator".

In Sweeden Tetra pack launched the tetrahedral paper bound cartoons in 1952 and made a commercial breakthrough in the year 1960. With the advancement of technology in combining cartoon assembling and aseptic packaging technology followed by the international expansion. In the aseptic processing the product and the package are sterilised separately and then combined and sealed in a sterile atmosphere whereas in contrast to canning where the product and the package are first combined and then sterilized.

The U.H.T is a food processing technology that sterilized the liquid food chiefly, the milk by heating it above 135 °C (275 °F) at which temperature the microbial spores that are present in the milk are killed when the milk is exposed for about 1 to 2 seconds of time. The U.H.T milk was first developed in 1960 and because generally available for consumption in 1970. The heat used during the U.H.T process can cause "Maillard browning" and can bring a change in the taste and smell of the milk and milk products. U.H.T milk packaged in a sterile container if not opened has a typical unrefrigerated shelf life when kept under normal room temperature for a period of about Six (6) to Nine (9) months.

7.12.3.8.1 U.H.T Process

There are two ways in which the U.H.T process can be carried out 1. In container sterilization and 2. U.H.T processing and these two methods produces a commercially sterile product. In the container sterilized milk has a typical cooked flavor with a slight brownish colour. On the other hand the U.H.T processed milk has much less cooled flavour and very little brownish colouration if any. The U.H.T processing is performed in a complex production plants and there are three types of exchangers that are in use namely 1. Plate exchangers 2. Tubular exchangers and 3. Scraped surface exchangers. For higher efficiency pasteurized water or steam is used as the medium for heating the exchangers themselves accompanied with a regeneration until which allows the reuse of the medium and thus saving the energy.

The major steps in the U.H.T process are as follows:

1. Preheating of milk with or without holding time
2. Homogenisation. For indirect system
3. Heating to sterilization temperature
4. Holding at sterilization temperature
5. Initial cooling
6. Homogenisation – Alternative position for direct or indirect systems
7. Final cooling and
8. Aseptic packaging

The preheating step of milk takes the temperature from 5 °C to 90 °C using the hot milk of post sterilization as the heating source in tubular or plate heat exchangers. This heat regeneration step is very important for the efficient usage of energy in the U.H.T processing plant which varies with the type and nature of the equipment plant machinery. In some plants the milk is held for some time in the holding tube after pre-heating. The major reason for this being the reduction of the amount of fouling or deposit formation in the subsequent heat exchangers. Further it can also have a major effect on the quality of the final product by inactivating the natural milk enzymes contained in it.

The final heating step to the required sterilization temperature is achieved by either direct of indirect systems. In the direct system milk is heated by direct contact with culinary super heated steam whereas in the indirect system heat exchangers are employed in which the super heated steam heat the milk indirectly through the stainless steel barrier in the form of either a tube or plates. The direct system can be either an injection type in which the stream is infected into the milk or an infusion type, where the milk is infused into a chamber of super heated steam. The major difference between these two method, is the rate at which the milk is being heated. The direct system takes less than one second of time to heat the milk from the pre-heating temperature to the sterilization temperature, whereas the indirect system the time taken to achieve the same objective varies from several seconds to minutes depending on the type of equipment/machinary that is being used in the plant. The major consequence of this difference is that for the same bacterial effect the direct system produces much less chemical change in the milk constituents as compared to with the indirect system.

The milk when attains/reaches the sterilization temperature enters a holding tube where the temperature and the time taken to pass through the holding tube are the nominal conditions which are generally cited/fixed for a U.H.T. Process for e.g., 140 °C for 5 seconds. This is a convenient convention and it does not give any true picture of the heat process to which the product is subjected. Many chemical and microbiological changes occur in the heating step immediately before as well as in the cooling step immediately after the sterilization step. Therefore these sections of the plant should be taken utmost care and account in addition to the sterilization holding tube while considering the extent of these changes.

The initial cooling of the product in the direct system is achieved very rapidly into a vaccum chamber where it suddenly loses the temperature and vapourizes. This process is referred to as "flash cooling". Which reduces the risk of thermal damage removes some or all of the excess water obtained through the contact with steam and removes some of the volatile compounds which negatively affect the quality of the product. The cooling rate and the quantity of water removed is determined by the level of vaccum which must be calibrated carefully and while doing so returns the temperature of the product close to the temperature from which it was heating usually around 75 °C. In the final cooling step in the direct system and in both cooling steps in indirect system. The heat from the hot milk is transferred to the cold milk in the preheating/heat regeneration steps.

If the fat is present in the product such as in the case of whole milk, a homogenization step is included. This is carried out at 60-70 °C either before or after the sterilization step. However it has been found that milk processed by direct heating process has to be homogenized down stream to break up aggregates of protein which form during heating and thus causes an astringent taste to the milk.

The aseptic packaging is a very crucial step. The Product must be transferred after cooling to the final package and the package is sealed without introducing even one bacterial cell. In majority of the commercial plants the product is held in an aseptic tank before it is sent to the aseptic packer. Various packaging types are available but most common are paper board and multilayer plastic. The packages are sterilized before being filled usually with hot hydrogen peroxide followed by hot air in order to remove the residual peroxide.

7.12.3.8.2 Changes During Storage

Keeping milk in good condition at room temperature for longer periods upto 12 months is a major challenge because of the myriad changes which can take place. The flavour changes through the progress of the 'Maillard' reaction and through oxidation by dissolved oxygen in the milk. The major flavour compounds produced are methyl ketones and aliphatic aldehydes but a large number of flavour compounds are generated. Other flavours which may develop during the storage are due to the action of heat resistant bacterial enzymes, which may be present in the raw milk and survive that U.H.T treatment. These include lipases which breakdown the fat and form free fatty acids. Some of which have strong flavours and proteases which breakdown protein to produce peptides some of which are bitter.

Another change which can be brought by proteases is what is known as age gelatin where the milk thickens during the period of long storage and eventually turns into a gel similar to 'Yogurt'. This undesirable defect can be caused by the heat resistant bacterial enzymes but it can also be caused by 'Plasmin' a naturally occurring protease in the milk which is quite heat stable and can remain active in the U.H.T milk. Recently it has been found to be inactivated by same U.H.T pre-heating conditions, which is an excellent reason for the inclusion of the holding in the pre-heating section of the U.H.T plants.

7.12.3.8.3 Advantages of U.H.T Process

The high quality raw milk which is received at the milk plant is clarified immediately chilled and stored in the storage tanks. In the first stage/part of the uperisation process the milk is pre-warmed to 50 °C and deaerated to remove most of the dissolved oxygen and volatile off flavours by vaccum treatment. In the second phase, the milk is preheated to about 80-90 °C and then heated in the uperisation chamber with high pressure steam to around 150 °C for 1/2 to 3/4 second. After this heating, the product moves into an expansion chamber at near the atmosphere pressure, thereby forcing some evaporation of moisture. The product is then moved into a cooler and then into the storage tanks. The advantages claimed by the uperized milk are as under:

1. Long keeping quality period.
2. Removal of feed and other volatile 'off' flavours.
3. Appreciable homogenisation effect.
4. Reduction in acidity.
5. Efficient destruction of microorganisms.
6. Effect of uperisation on nutritive value and flavour no greater than that of pasteurisation.

7.12.3.8.4 Disadvantages of U.H.T

1. There is a possibility of alternation of the milk proteins affecting the properties of milk when used to make other food products.
2. High temperatures inactivate the enzymes that protect the product increasing the risk of spoilage.
3. Elevated temperatures cause "Maillard" reaction which discolours the product making it undesirable to the consumers.
4. High temperatures alters the protein structure and in parts a cooked flavour to the milk.

7.12.3.8.5 Alternatives to Pasteurisation

Various types of treatment have been proposed from time to time as alternatives to pasteurisation and these include

(a) the Hofius process.
(b) electronic heating.
(c) Ultra-violet irradiation and
(d) Ultra-sonic vibration etc.

7.13 Bottling/Bottle Filling

The usage of bottles for the filling of the pasteurised milk and its sale to the consumers has become obselete and almost all the major milk processing plants have dispensed with this system. Therefore, this has become only of an academic interest since the milk now-a-days is sold in the form of sachets using poly propylene or polythene sheet. The filling of milk into the bottles is done by two ways. They are (1) Gravity fillers and (2) Vaccum fillers.

7.13.1 Gravity Fillers

These consists of six different parts namely

(a)The drive, (b) Bowl, (c) Filler valves, (d) Carrier, (e) Capper and (f) Star wheels

The circular bowl receives the milk to be bottled and the level is kept constant by a float valve on the inlet pipe. Filling valves are attached radially to the bottom of the bowl. The bottle area fed either by hand or directly from the battle washer by a conveyor. They are mechanically centred into lifters which are located directly under the filling valves and revolve with them. These lifters raise automatically as the filler revolves and the mouth of the bottle is forced against a tightly fitting rubber valve. The raising bottle pushes up the valve and the milk flows down into the bottle. As the filling valve almost completes its revolution, the lifter on which the bottle is carried is lowered automatically thus closing the valve, which remains in closed condition until the next bottle opens it. The filled bottle is then transferred to a capper, where it is sealed automatically and the lifter which at present is in the lowered position is ready to receive another empty bottle to for filling, when the bottle is being filled the air which is displaced by the incoming milk escapes through a vent tube, which extends from the bottom of the valve sleeve to the point above the milk level in the bowl. The height of the bowl tank is adjusted by the operator to suit the size of each bottle that is to be filled. Bottles are automatically discharged from the capper on to a conveyor, which delivers them to an accumulating table from where they are loaded into the crates either manually or mechanically. This system has advantages and disadvantages also which are as under:

Advantages

(a) It is relatively simple to operate.

(b) Maintenance is not complicated and

(c) Easily and swifty cleaned.

The Disadvantages

1. Slow in filling hence suited for a limited or small milk plants which handles as small volume of milk.
2. Leakage losses are very high may be due to badly sealed bottles, bottles with chipped mouth, faulty valve etc.

7.13.2 Vacuum Fillers

These consists of two types namely, the vaccum assisted or single bowl and straight forward vacuum or double bowl types. In the former type, the typical gravity bowl, which has open vent tubes and conventional gravity valves is closed with an air tight cover. In the double bowl or straight forward vaccum system, there is a rotary bowl and a float bowl. The float bowl is slightly below the level of the tops of the filling head. When the bottle is raised against the rubber ring on the filling head a seal is formed and the air inside the bottle is immediately drawn-out through the vertical vacuum pipes, and the milk is drawn from the float bowl through the milk pipes to the filling head and is then released into the bottle. Foam is drawn off through the vacuum pipes into the vacuum tanks. Excess milk collected in the vacuum tank automatically goes into the float bowl, towards the end of the revolution of the filler, the lifters on which the bottles are carried, are lowered and the seal is broken. Any milk that remains in the milk pipe is syphoned back into the float bowl and that remains in the vacuum line is drawn back into the vacuum tank. The main advantages of this system are:

1. Rapid filling
2. It will not fill the bottle which is having a chipped mouth or bad seal thus saving the milk and
3. There is no leakage of milk through the faulty valves,
   whereas the disadvantages of the system are
   (a) The maintenance of the equipment is much complicated.
   (b) It is not easy to operate because of the complicated nature of the equipment and
   (c) The cleaning of the equipment is also cumbersome and more time consuming.

7.13.3 Caps and Capping

Generally the capping machine is often incorporated into the filler and in any case, the working of the copper must synchronise with the filler. The milk bottle cap or closure has three main functions to perform which are:

1. To retain the milk in the bottle
2. To protect the pouring lid from contamination and
3. To seal the bottle against tampering which refers to the removal or replacement of the milk from the bottle without being giving any evidence from the appearance of either the bottle or the milk within it.

The caps may be:

1. The cord board discs, which are impregnated with a moisture proof layer of either paraffin wax or polythene.
2. Aluminium foil caps and
3. Crown corks.

The cardboard discs with separate hoods were the first to be introduced, but now they have become obselete, Aluminium cap is most commonly used either preformed or formed in place, both types having their own advantages and disadvantages. The foil may be 0.05 to 0.15 mm in thickness and of 50 mm in width as per the ISI or BIS standards IS : 1705-1960. The crown corks are generally used for sterilized milk which are made up of lacquered tin plate, the inner surface of which is lined with water proof paper/polythene and are more expensive. All types of caps or closures can be printed or embossed with names or trade marks and can be coloured for the purpose of coding.

7.13.4 Inspection of Filled Bottles

The filled milk bottles should be inspected for dirt etc., by rotating them as they are removed from the machine before being placed in the creates.

7.13.5 Decrating or Recrating of the Bottles

Removal of the used, returned dirty bottles from the crates is decrating and refilling them with the bottles filled with the pasteurised milk is recrating are the most back-breaking and labour consuming operations in the dairy. Both decrating and recrating machines looks the same. The decrator lifts the empty bottles by vacuum operated rubber grippers and rejects those which are damaged. The recrator also lifts the bottles by compressed air operated rubber grippers. In decrating, the crates of empty bottles must be correctly positioned at the first instance before lifting whereas in recrating, a special marshalling machanism allows the bottles from the filler to assemble in correctly positioned groups ready to be picked up for transfer to the crates.

7.13.6 Crate Stacker

The crates containing the filled bottles are stacked in this stacker thereby relieving the labour of another back breaking job.

7.14 Packaging

Glass bottle is the traditionally accepted packaging medium for the retail distribution of milk, however, this has become obsolete due to the increased operational costs, as well as due to the introduction of the simple service paper/plastic containers which have been increasingly adopted to serve the same purpose. The invention of the plastic has revolutionised the packaging system almost in every commodity either solids or liquids. As far as milk packaging is concerned, paper is a carrier for a waterproof layer of either polyethylene or wax. The provision for 5 -10 percent of synthetic fibres is said not to influence the printing or the folding capacity of paper but rather to provide better wet strength. Polyolefines such as polypropylene and polyethylene have dominated in the milk packaging, used either singly or blended to offer a wide range of materials, plastics have superseded paper in the recent years. The polyolefines of interest are coating grades for cartons, film grades for sachets and extrusion or blowing grades for bottles.

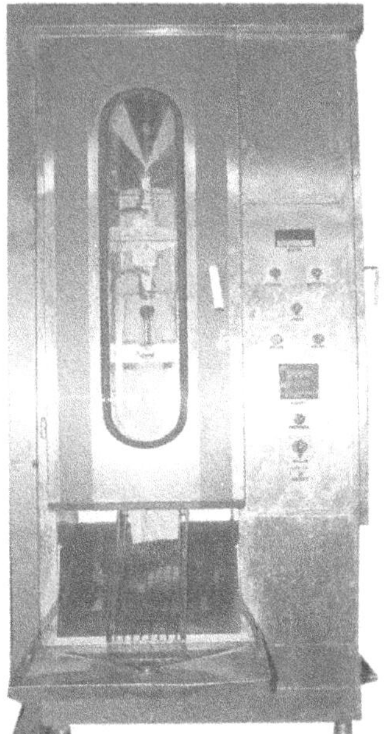

Fig. 7.4 Pouch Filling Machine

Cartons are commonly used for both bulk milk and retail sale. They are made up of combinations of food grade paper and wax or plastics. The advantage of cartons in the distribution are:

1. Maximum utilisation of space in the vehicles and storage.
2. Ability to carry attractive printing and
3. Convenience for the purpose of stacking milk on super market or distributor shelves.

The carton systems which are in common use are :

Perga - U.K.

Pure Pak - USA

Zupack, and BlockPack - W. Germany and

Tetra Pack - Sweden

When filling the cartons either pre-formed or precut blanks may be used. Pre-formed cartons are supplied in a fully erect form ready for filling, whereas in the pre-cut system the blanks are supplied in a knocked down shape and the final carton is setup formed, filled and sealed on one machine Sachets are flexible water proof bags and the systems which are in common use are

Polypack and Bertopack- Switzerland

Finnpack - Finland

Milk pack Rotapack - W. Germany and

Prepac - France

The sachet filling of milk usually flows a form/fill/seal system. The sachets are formed from a reeled film over a shoulder and tube forming sealer. Their size is changed simply by pushing a button without stopping the machine. They are filled by a time regulated valve accurate with + 3 ml per litre, Ultra violet light is used to sterilize the inside of the film.

7.14.1 Glass Bottle Vs Paper/Film Package

The relative advantages of paper/film packaging of milk over bottling are given as under:

Advantages

1. Light in weight, easy to handle and there is no danger of breakage.
2. The distribution costs are lower, because it occupies less space, lower pay load, due to lightweight and there is no collection of empties.
3. Bottle washing costs are eliminated, thus there is saving in equipment, detergents, steam and recurring charges on the replacement of bottles in lieu of the broken ones.
4. Tamper proof.
5. Effective sales messages can be printed.
6. Less noisy during filling, sealing and transportation.
7. No exposure to sunlight.
8. Filling machinery is compact and occupies much less floor space.
9. No container deposit is needed either from the distributor, retailer or the consumer.

Disadvantages

1. Costs are higher per unit of milk distributed.
2. There is no visibility of the product.
3. Difficult to remove cream.
4. Inspection of milk for sediment is not possible.
5. Regular supply of special paper or film is essential.
6. Not so easy to open and
7. In some cases leakage may occur.

7.15 Storage

In any milk plant, it is necessary to provide, refrigerated cold storage rooms, where the milk can be stored until delivery. The temperature of the cold storage room should be 5 °C or below so as to check the bacterial growth and to avoid the spoilage of the milk.

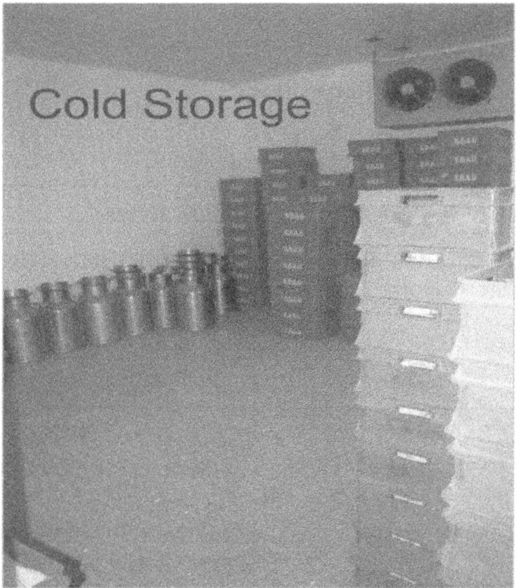

Fig. 7.5 Cold Storage

In order to have a better understanding pertaining to the better shelf life of the milk and different milk products during the period of storage by the producers, processors, marketing personnel, retail sellers as well as the consumers an attempt has been made to present the recommended storage time and the temperature in the Table 7.13.

Table 7.13 Recommended storage times and temperatures for milk and other milk products

		Safe Storage Time	
Dairy Food	**How to Store**	**On Refrigerator Shelves 35-40 °F (2-4 °C)**	**In Freezer 0 °F (– 18°C) or below, properly packaged**
Pasteurized Fresh Whole or Skimmed Milk, Sweet Cream, Flavored Milk Drinks	Refrigerate immediately in original container. Keep container closed	Unopened cartons for 30 days. Opened for 1 week. Check code date.	Do not freeze. (Change of texture, body appearance. Separation of fat).
Sweetened & Condensed Milk (opened)	Keep covered	1 week	Do not freeze
Whipped Topping	Keep covered	3 months in aerosol can. 3 days prepared from mix. 2 weeks bought frozen (once thawed)	Do not freeze
Yogurt	Keep covered	7-10 days	Do not freeze
Sour Cream, Butter, Milk, Cultured Milk	Refrigerate immediately in original container. Keep container closed.	2 weeks. Up to 2 months for salted butter (2 weeks for unsalted butter)	6 to 9 months for salted butter; up to 5 months for unsalted butter. Do not freeze others.
Soft Custards, Milk Puddings, Cream and Custard Fillings for Cakes and Pies	Cool cooked dishes quickly and refrigerate within 2 hours. Refrigerate cold dishes immediately after preparation.	5-6 days	Do not freeze

Table 7.13 *Contd...*

		Safe Storage Time	
Dairy Food	**How to Store**	**On Refrigerator Shelves 35-40 °F (2-4 °C)**	**In Freezer 0 °F (– 18°C) or below, properly packaged**
Ice Cream	Store in original container in freezer or transfer to a moisture-vapor-proof container	Do not store here	2-3 weeks
Natural Hard Cheese and Semi-Hard Cheese and Processed (Cheddar, Swiss, Parmesan, Brick, Bleu, etc).	Refrigerate in original package and over wrap tightly in aluminum foil, plastic wrap or plastic bag tightly closed to ab=void drying	1 month, If mold forms, cut off.	Freezing affects texture (makes it crumbly). Still suitable for cooking. Thaw in refrigerator. Do not freeze soft cheeses.
Soft Cheese (cream, cottage, limburger, camembert)	Refrigerate tightly covered	1 week	Do not freeze (Can freeze cream cheese. Texture may change).
Cheese Spreads	Refrigerate tightly covered	1 month	Do not freeze
Evaporated Milk (opened)	Refrigerate tightly covered	1 week	Do not freeze
Homogenized, Reconstituted Dry Nonfat and Skimmed	Keep containers tightly closed. Do not return unused milk to original containers	1 week	Do not freeze

Source: Newer knowledge of Dairy products. National Food Safety database. (Internet)

7.16 Distribution

The last and final stage of the market milk industry is the distribution of milk and the others are the preparatory procedures to place the product into the hands of the consumer. The quality of the product alone will not assure its wide distribution which should be planned and executed intelligently. The distribution facilities consists of:

1. The physical equipment and personnel required for transporting the products from the milk storage rooms to the consumer/retailer.
2. Sales promotion personnel and
3. Advertising

The successful distribution programme requires

1. A product of high quality.
2. attractive package
3. Courteous salesmen with neat and clean habits.
4. delivery equipment of pleasing appearance.
5. efficient use of men and equipment.
6. effective advertising.

Bottled/packaged milk distribution system has been adopted invariably by all the developed countries. Unpacked or loose milk distribution is quite prevalent in developing countries as in India, whereas, the dis-organised sector of the Indian dairy industry follows the loose milk distribution and the organised sector persues, the pattern of developed countries. A notable exceptions are the Mother Dairies established in the metropolitan cities, which have adopted the bulk milk lending systems developed by N.D.D.B. in the year 1974. These dairies have set up milk vending booths in various parts of the city. In each both coin or token operated milk vending machines, popularly known as the push button mini-dairy have been installed. The milk holding capacity of these machines varies from 1000 to 1300 litres each. The consumer is expected to bring his/her own container large enough to hold the milk and exchanges money for metal tokens. The container is to be placed under the tap. On inserting the token into the slot the button lights up and on pressing the button, the first half litre of milk flows out and the process is repeated to get the second instalment of milk by inserting another token and so on. The dairies claim that this method enables it to pay more to the milk producers on account of the savings effected in the cost of packaged milk distribution, yet operates the dairies on a no-profit and no-loss basis.

7.16.1 Organisation of Routes

It varies with the size and the type of business. In a small plant the same trucks can deliver both the wholesale and the retail goods, whereas in large scale operations of the major milk plant, wholesale and retail distribution are usually handled by separate personnel who have been entrusted with the responsibility. The wholesale routes handle larger volumes and have fewer stops than the retail outlets/routes and for an economical operation, the trucks must the utilized for a maximum number of hours per day for the distribution of the milk.

Three different methods are in vogue for the payment of route salesman which are:

1. Flat salary gives no incentive to sell more products, secure new customers etc.
2. Salary plus commission appears to be most satisfactory and
3. Straight commission

It attracts the retailers and wholesale dealers to increase the sales. It also encourages the drivers who own the vehicles and the route.

Different systems may be used for loading the trucks in checking out the routes. The trucks must be loaded rapidly so that the drivers are not delayed at loading stations.

The Principal Systems are

1. Loading directly from the storage rooms through one or more doors.
2. Usage of a long loading platform with conveyors from the milk storage rooms similar to that of the receiving dock or platform.
3. Loading platform trolleys in the storage room with orders for the different routes. The platform trollys are then wheeled on to the loading platform and finally loaded into the delivery trucks. The empty bottles/sachets which are unsold and brought back to the dairy along with the empty bottles have to be checked and verified with the sales slips or the delivery coupons which are kept on the platform conveyor or platform trolly.

The various sales outlets for both the wholesale and retail sales include:

(a) Home delivery
(b) milk booths or distribution depots
(c) departmental stores
(d) Soda fountains
(e) Coin vending machines
(f) Automatic dispensors
(g) Factories, Hospitals, jails, restaurants, hotels, hostels and schools etc. and
(h) Lunch counters.

There are various systems for the collection of the receipt of payments of milk which has been sold to the consumers, agents etc., which include credit, cash or advance of payment either coupons or monthly card and their relative merits and demerits have been presented in the table 7.14.

Table 7.14 Comparative Study of different collection systems for payment of milk

System	Merits	Demerits
Credit	(1) It is attractive to the customers as there is no immediate payment and looking up of money. (2) Rapid deliveries (3) Accounting process is easy.	(1) Losses through bad or default accounts. (2) Needs a special check on the amount given as credit and the payment made by the consumer.
Cash	(1) No losses through the bad/default account holders because the money is collected on the spot or cash down purchase.	(1) Slows down the deliveries because some of the customers may not bring correct change (2) Less attractive to the customers. (3) Counterfiet notes/currency may pose problem. (4) There is a risk in the carrying of large or huge amount of cash by the dairy staff.
Coupons	(1) There is no loss of any kind through the bad or default accounts. (2) It curbs the corruption especially on the part of the sales man and other dairy staff.	(1) Slows down the deliveries (2) It may result in the increase of sale of price of milk. (3) It is less attractive to the customers because their money is locked up as advance. (4) Accounting is cumbersome. (5) Special check is needed in order to prevent the misuse of the coupons lost or stolen from the rightful owner. (6) There are chances of fraudulent printing of the fake coupons.

Source : Outlines of Dairy Technology S.K. De

7.17 Judging and Grading of Milk

Judging of milk refers to the act of evaluating its palatability or eating quality on the basis of various attributes and grading refers to the classification of milk into various categories. The eating quality or palatability of a dairy product is determined by the organoleptic/sensory tests which includes all the five senses of sight, smell, taste, touch and sound, of these the taste and smell are the most important in judging and grading. The acceptability of a dairy product by the consumer is determined primarily by its palatability i.e., by the sensation of smell, taste, feel etc., which is experienced by the consumer when the product is tasted or consumed. It is a well known fact that the dairy products cannot be of a higher/good quality than the raw material from which they are made. Therefore the milk producers should have definite knowledge as to what constitutes the desirable and undesirable flavours in milk as well as the factors that are responsible in causing them, only then they will be in a position to produce the milk that would make high scoring of the finished products. The manufacturer of the dairy products or the milk processor should have the ability to discriminate against the objectionable flavours and manufacturing defects and also to recognize desirable flavours and to make up the characteristics because, these will enable him to make a product of good quality and consumer acceptability. The consumer should have adequate knowledge of desirable and undesirable flavours in the milk and milk products as these will enable him to purchase the same wisely to get his money's worth.

The score card, which has been developed and approved by the American Dairy Science Association, U.S.A. for the milk and cream for different variables has been given in the tabular form 7.15.

Table 7.15 Score card for milk and Cream (ADSA)

Items	Perfect score
Flavour (smell and taste)	45
Bacteria	35
Sediment	10
Temperature	5
Container and closure	5
Total	**100**

Source : Outlines of Dairy Technology - S.K. De

In the process of judging and grading of the milk the first and foremost objective is to obtain/secure a good representative sample aseptically by the standard procedure for the determination of the bacterial count. The sediment discs that are being used for the determination of the sediment should be compared with standard charts for scoring. However, sediment scoring of packaged milk is no longer done in standard contests. The cap or the closure should be carefully observed and the score should be given after having the score for the sediment. Then the glass bottles or the containers are examined for fullness, cleanliness and freedom from cracks and chips particularly about the pouring lip. The paper container is examined for cleanliness, freedom from leakages, smoothness and the adherence of a coating to its surface. Correct fillage can be determined only by actual measurement of the volume of milk in a graduated cylinder. The score for the flavour is, smell and taste and is carried out only when the above stated items have been considered and given the scores. The temperature of the milk should be around 16-21 °C (60-70 °F). The milk should be well mixed before it is sampled in a small clean beaker or a paper cup. As soon as the milk sample has been taken, sip the milk but do not swallow a sufficient amount of about 5-10 ml to yeild a normal taste reaction and yet one can smell sufficiently to permit its easy manipulation in the mouth. Roll it in the mouth and observe the flavour and sensation and then expectorate it. Note the after taste as well and by placing the nose directly over the milk when it has been shaken any off smell may be detected.

7.17.1 High Grade Market Milk - its Requirements

In order to produce milk of high grade market quality the following points must be given due consideration.

1. The container should be neat and clean and should contain the full volume of milk represented. The milk in the bottle should be protected from contamination by a well made, well seated, water proof cap, which protects the pouring lip fully. The bottle itself should be bright, shining free from dirt, dust etc. It should not be cracked, or chipped, particularly on the pouring lip. The paper or plastic containers should be clean, and fresh with no leakage, pronounced bulging etc.
2. The milk must be delivered at 10 °C or below.
3. It should have the least amount of sediment.
4. It should have a low bacterial count. The high quality milk is generally low in bacteria, but the low count bacterial milk need not always have the best flavour.
5. The flavour of the milk should be pleasantly sweet and should have neither a foretaste nor an aftertaste other than what is imparted by its natural richness.

7.17.2 Manufacture of Different Grades of Milk

It is always preferable that there should be a single minimum standard of quality for all (types) milk, whether it is to be used directly as fluid milk or for the manufacture of various products, however, this is not so in practice. Therefore the following classifications may be used in order to grade the milk depending upon the quality of the milk.

1. **Grade - I:** Milk with a clear, pleasant flavour MBR time over 5 1/2 hours and practically having no sediment on the sediment discs having tolerance for slightly feed, flat or salty off flavours.
2. **Grade - II:** Milk having off flavours such as definitely feed, flat or salty, slightly barny, bitter, foreign, malty, metallic, musty, oxidized or rancid, or very slightly weedy. MBR time between 2 1/2 and 5 1/2 hours or milk which shows a medium amount of sediment.
3. **Grade - III:** Milk having off flavours such as definitely barny bitter, foreign, malty, metallic, musty, oxidised, or rancid, or slightly high acid. M.B.R. time between 20 minutes and 2 1/2 hours or definitely high in sediment.
4. **Reject or No grade:** Milk with markedly high acid - rancid, weedy or foreign flavours. MBR time of less than 20 minutes and containing an extremely high amount of sediment or any abnoxious foreign matter.

Important

The use to which each grade of milk is put will depend upon public health regulations concerning conditions of production and upon the specific character of the product. In general the following categorization may be made.

Grade - I: Market milk, sterilized milk, evaporated milk, sweetened condensed milk, milk powder (whole or skim). Infant food, cheese etc.

Grade - II: Butter, Ice Cream, Flavoured milks, Fermented Milks, Khoa, Chhana/Paneer, Butter Oil, Ghee etc.

Grade - III: Ghee, Casein etc.

7.17.3 Flavour Defects-Causes and Prevention

Generally the normal good quality milk has a pleasant flavour but sometimes it develops some defects in its flavour known as "Off flavour" when it has an undesirable smell, foretaste or aftertaste and in case the mouth does not feel clean and pleasant after it has been tasted. The flavour defects may arise due to faulty methods of production, processing and storage.

The common flavour defects in milk, their causes and prevention have been summarised and given in the table 7.16.

Table 7.16 Flavour defects, their causes and prevention

Flavour defects	Causes	Prevention
Barny	(i) Improper ventilation of milking byre/barn; (ii) Milk not properly covered during production.	(i) Proper ventilation of milking byre/barn; (ii) Keep milk properly covered during production.
Bitter	(i) Intake of bitter weeds by milch animals ; (ii) Using late lactation milk	(i) Eradicate offending weeds (ii) Use normal lactation milk
Cooked	Overheating of milk	Avoid overheating of milk
Feed	Feeding of milk-tainting feeds (such as silage) within 3 hours before milking.	Feed milk-tainting feeds (such as silage) soon after milking
Foreign	Addition or absorption of foreign smelling substances in milk.	Avoiding contact of milk with foreign smelling substances.
High-acid/Sour	Excessive lactic acid development (due to considerable growth of lactic acid producing microorganisms).	Store milk at 5 °C (40 °F) or below (to check bacterial growth and acid development).
Malty	Growth of *stor. lactis var, mattigenes microorganisms* in milk.	Store milk at 5 °C (40 °F) or below to check bacterial growth.
Rancid	Fat hydrolysis due to lipase action.	Inactivate lipase by proper pasteurization of milk.
Oxidized, Oily, Metallic, Tallowy	Fat oxidation due to : direct contact of milk with copper or iron, exposure of milk to light etc.	(i) Tin milk-holding vessels properly : or use aluminium alloy/stainless steel as milk-contact surfaces ; (ii) De-aerate/vacuumize pasteurized milk.
Salty	Milk of animals suffering from mastitis, or far advanced in lactation.	Avoid mastitis or late lactation milk.
Weedy	Intake of milk-tainting weeds (within 3 hours of milking).	(i) Eradicate milk-tainting weeds; (ii) Vacuum pasteurization of milk.

Source : Outlines of Dairy Technology-S.K. De

Important

In souring of milk, lactic acid is the main fermentation product, but the sour smell is not caused by lactic acid, which is non-volatile. Volatile substances such as acetic acid, formic acid, propionic acid acetaldehyde, acetone, diacetyl and methyl acetyl carbinol are responsible for the characteristic sour smell of the milk.

Chapter 8
Dried Milk or Milk Powder

Drying is usually applied to make a durable product that is easy to handle and after re-constitution with water is very similar to properties to the original material. The drying process ensures that the milk solids are obtainable as a dehydrated powder which can be stored for long periods and readily re-constituted into normal milk. Removal of water is expensive especially with respect to energy. The powdered milks differ chiefly with regard to the percentage of fat contained. They also differ in respect of the physical state in which the fat is present. Some dried milks contain added sugar while alkali is added to others in order to improve the solubility of the powder. Considerable variations do exist between individual powdered milks, as to the ease with and the extent to which they can be reconstituted by the addition of water into a fluid resembling milk. Very often it happens that the physical and chemical properties of the milk solids are so changed during the processing that subsequent mixing with water will no longer produce a fluid which resembles milk even in a remote degree. Thus the casein may be coagulated and will not assume a colloidal condition on reconstitution. Roller dried powder is rather darker in colour than the spray dried variety while the brownish tints indicate that the drying temperature has been excessive and resulted in caramelisation of the milk sugar. Spray dried powder consists of spherical grains whereas the grains of roller dried variety are irregular in shape. The fat of spray dried variety consists of small globules and when reconstituted the fat is restored to much the same condition as it is found in the original milk. On the contrary the fat of the roller dried milk powder is present in the form of large drops which when reconstituted with the water tend to separate out and float upon the surface in an oily layer. The factors which exert the most effect upon the properties of dried milk are:

1. The precondensing process
2. The drying temperature

Dry milk provides a means of handling the excess milk supply in the dairy plant/ factory during the flush season whereas in the lean season, that meant for production of dry milk can be diverted to market milk.

Thus the dry milk may be called as the balance wheel of the dairy industry. The objectives or purposes of drying milk and milk products are as follows:

1. To remove the moisture/water so as to reduce bulk, thereby effecting a saving in storage space and packaging costs.
2. To reduce the cost of transportation.
3. To improve the storage life of the product
4. To provide a product which can be utilized for many food manufacturing operations.
5. To retain the natural properties of the original raw material as far as possible.

Dried milk or milk powder is the product obtained by the removal of water from milk by heat or other suitable means to produce a solid containing 5 percent or even less moisture. The dried product obtained from whole milk is called dried whole milk or Whole Milk Powder (WMP) and that from skim milk is known as (SMP) or Non-Fat Dry Milk (NFDM). Specific names have been given to the various dried milk products.

8.1 Whole Milk Powder-PFA or Legal Standard

According to PFA Rules (1976), whole milk powder is the product obtained from the cow or buffalo milk or a combination thereof or from standardized milk by the removal of water. It may contain calcium chloride, citric acid and sodium citrate, sodium salts of ortho phosphoric acid and poly phosphoric acid not exceeding 0-3 percent by weight of the finished product and 0-01 percent of butylated hydroxy anisole by weight of the finished product. Such addition need not be declared on the label. Milk powder should contain not more than 5.0 percent moisture and not less than 26.0 percent fat. The total acidity expressed as lactic should not be more than 1.2 percent. The standard plate count may not exceed 50,000/g and the coli count may not exceed 90/g. The maximum solubility index should be 15.0 for a roller dried and 2.0 for the spray dried product.

Important

The process of drying must be mentioned on the label. The spray dried product must be packed in hermitically sealed containers when the net quantity exceeds 510 g.

8.2 Skim Milk Powder (SMP)

According to PFA Rules (1976) skim milk powder is the product obtained from the skim milk of cow or buffalo milk or a combination thereof by the removal of water. It may contain added calcium chloride, citric acid, and sodium citrate, sodium salts of orthophosphoric acid and poly phosphoric acid not exceeding 0.3 percent by weight of the finished product. Such addition need not be declared on the label. Skim milk powder may not contain more than 1.5 percent milk fat and moisture may not exceed 5.0 percent. The total acidity expressed as lactic acid should not exceed 1.5 percent. The standard plate count should not exceed 50,000 / g and the coli count must not exceed 90/g. The maximum solubility index should be 15.0 for roller dried and 2.0 for a spray dried product.

Important

The process of drying must be mentioned on the label. The spray dried product must be packed in hermetically sealed containers when the net quantity exceeds 510 g.

The specifications for the dried milks according to the B.I.S. (Institution - IS 1165) (1967) and the FSSAI Standards for milk powder and cream powder, skim milk powder along with partly skim milk powder have been presented in table 8.1 and 8.2 respectively for the better understanding of the students as well as the dairy enterpreneures.

Table 8.1 BIS Specifications for dried milks

Characteristics	**Requirement for**	
	Whole milk powder	**Skimmed Milk Powder**
Flavour and odour	Good	Good
Moisture (% wt) (Max)	4.0	5.0
Total milk solid (% wt)	96.0	95.0
Solubility :		
(a) Solubility index (Max) ml.		
(If roller-dried)	15.0	15.0
(If spray-dried)	2.0	2.0
(b) Solubility (% wt) (Min)		
(If roller-dried)	85.0 ml	85.0 ml
(If spray-dried)	98.5 ml	98.5 ml
Total ash (on dry basis) (% wt) (Max)	7.3	9.3
Fat (% Wt)	Not less than 26.0	Not more than 1.5
Titratable acidity (% lactic)	1.2	1.5
Bacterial count per g (Max)	50,000	50,000
Coliform count per g. (Max)	90	90

Table 8.2 FSSAI (2017) standards for milk powders and cream powder

Sl.No	Parameter	W.M.P	Partly S.M.P	S.M.P	Cream Powder
1.	Moisture[⊗] Maximum % (m/m)	5.0	5.0	5.0	5.0
2.	Milk fat % (m/m)	Minimum 26.0 and less than 42.0	More than1.5 and less than 26.0	1.5 (Maximum)	42.0 (Minimum)
3.	Milk protein [⊗⊗] in milk S.N.F minimum % (m/m)	34.0	34.0	34.0	34.0
4.	Titrable acidity maximum (me 0.1 NaOH for 10 g S.N.F)	18.0	18.0	18.0	-
5.	Insolubility Index maximum	2.0	2.0	2.0	-
6.	Total ash Max. % (m/m) on moisture and fat free basis	9.3	9.3	9.3	-
7.	Scorched particles Maximum	Disc. B	Disc. B	Disc. B	Disc. B

Note: [⊗]The moisture content does not include water of crystallization of the lactose, the milk S.N.F content includes water of crystallization of the lactose.

[⊗⊗] Protein content is 6.38 multiplied by the total nitrogen determined.

Similarly the American Dairy Milk Institute (ADMI) standards also have been presented in the table 8.3.

Table 8.3 ADMI Standards

Particulars	Instant	Spray		Roller	
	US extra	**US extra**	**US Stand**	**US extra**	**US Stand**
		Skim milk powder			
Moisture (%) NMT	4.25	4.0	5.0	4.0	5.0
milk fat (%) NMT	1.25	1.25	1.50	1.25	1.50
Titratable acidity (%) NMT	0.15	0.15	0.17	0.15	0.17
Scroched particles (mg/g) NMT	15.0	15.0	22.5	22.5	32.5
Solubility index (ml) NMT	1.0	1.2	2.0	15.0	15.0
Standard plate count (per g) NMT	30,000	50,000	100000	50000	100000
Coliform count (per g) NMT	90	--	--	--	--
		Whole milk powder			
Moisture (%) NMT	2.25	2.25	3.0	3.0	4.0
Milk fat (%) NLT	26.00	26.00	26.00	26.00	26.0
Solubility index (ml) NMT	0.5	0.5	1.0	15.0	15.0
Scorched particles (mg/g) NMT	15.0	15.0	22.5	22.5	32.5
Titratable acidity (%) NMT	0.15	0.15	0.17	0.15	0.17
Standard plate count (per g) NMT	35,000	50,000	100000	50000	100000
Coliform count (per g) NMT	90	--	--	--	--
Copper (ppm) NMT	1.5	1.5	--	1.5	--
Iron (ppm) NMT	10	10	--	10	--
Oxygen (% vol) NMT (If gas packed)	2	3	--	3	--

NMT - Not more than NLT - Not less than

Source : Dairy Science & Technology Pieter Walstra et.al

Composition

The composition of dried milk depends upon two variable factors namely.

1. The composition of the raw milk used in manufacture.
2. The effect of the processing upon the removal of the moisture. The average percentage composition of the whole milk powder and skim milk powder is given in the table 8.4.

Table 8.4 Composition of Dried milks (Percentage)

Constituents	Type of dried Milk	
	Whole milk powder	**Skimmed Milk powder**
Moisture	2.0	3.0
Fat	27.5	0.8
Protein	26.4	35.9
Lactose	38.2	52.3
Ash	5.9	8.0

Source : Text Book of Dairy Chemistry - Edgar. R Ling

Milk Drying Systems/Processes

The systems and processes of milk drying can be classified and presented in the flow chart 8.1.

Flow Chart 8.1 Milk Drying Process

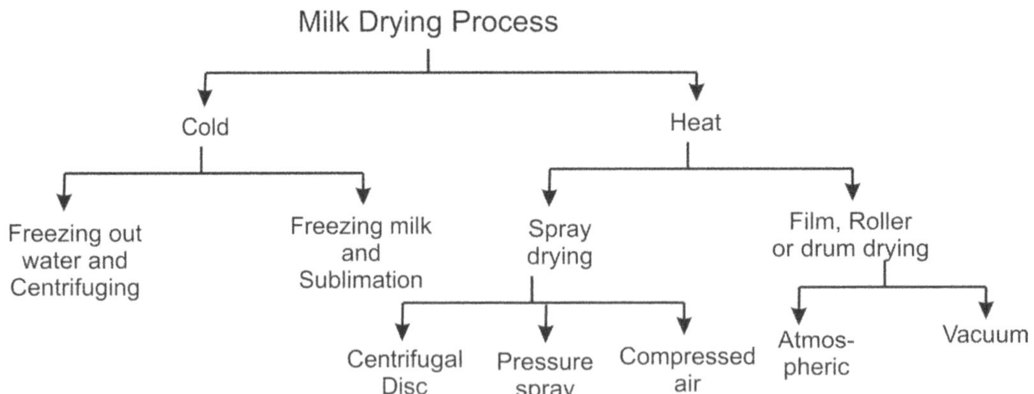

8.3 Drying of Milk Freezing

A food is generally cooled or frozen to retard spoilage in many parts/regions of the world. The fresh milk is routinely cooled to about 5 °C in order to extend the shelf life by days or weeks. Cooling of milk is also applied for specific purposes, such as inducing fat crystallization or enhancing the creaming tendency. Freezing is mostly carried out to make a specific product such as ice cream and occasionally as a concentration process. Freezing is sometimes used to substantially retard deterioration of fresh milk or liquid milk products.

8.4 Cooling of Milk causes Several Changes, the most Important Ones being

1. The growth of most of the microorganisms is much slower, including the changes induced in milk by their metabolism.
2. There is retardation of almost all the chemical and enzymatic reactions.
3. Autoxidation of lipids is enhanced presumably because of the activity of the enzyme superoxide dismutase is decreased.
4. Changes in solubility and association of salts occur and the pH increases.
5. The viscosity is increased with an enhanced susceptibility to attack by plamin.
6. There is an alteration in the fat globule membrane and its structure, which are irreversible.
7. Cold agglutination of fat globules occurs which enhances the creaming rate.
8. Partial crystallization of the tri-glycerides in the fat globules will occur.

8.5 Freezing

Milk when cooled starts to freeze at about 0.54 °C if no under cooling occurs. Concentrated milk naturally has a lower freezing point. Pure ice is formed, and the remaining milk thus becomes concentrated, thereby, decreasing the freezing point further. The lower the temperature the higher the proportion of the water that freezes. The change in composition of the remaining solution is considerable. The ionic strength may be almost tenfold higher, the calcium ion activity is slightly increased and the quantity of calcium phosphate in the casein micelles is also increased despite the decrease in pH.

The remaining liquid can attain a glassy state, upon freeze concentration lactose exhibits rather slow nucleation and crystal growth and upon rapid cooling its crystallization will fail to occur. The viscosity of the liquid will increase markedly thereby resulting in the crystallization of lactose less likely changing the liquid into a glassy state, preventing the further crystallization of ice. This special glass transition at maximum freeze concentration is characterised by a special glass transition temperature and unfrozen water content.

Lactose, when does not crystallise, it acts as a cryoprotectant. Aggregation of casein micelles does not occur if lactose crystallization has been prevented by freezing rapidly. Freezing and thawing of whole milk and especially of cream generally causes partial coalescence or clumping of fat globules because of the growing ice crystals pressing the globules together. It can be prevented by prior homogenisation and rapid freezing. In practice rapid freezing is achieved in a scraped surface heat exchanger. The liquid flows through a tube that is deeply cooled at the out side causing the ice to form at the inner surface. The ice is removed immediately by the rotating stirrer, provided with knives that scrape over the surfae. In this way a mixture of concentrated milk product and ice crystals results. By removing the crystals concentrated milk can be obtained. This process of freeze concentration is however, much more expensive and more limited in possibilities than removing water by evaporation. It is occassionally used to concentrate a liquid containing aroma compounds because these may become lost during concentration by evaporation.

8.6 Drying of Milk-Heat Application

The methods employed for drying of milk by the application of heat are:

1. Film drying
2. Spray drying

In the film drying process three methods are in vogue each one possessing certain advantages. These methods are:

1. **Atmospheric-Roller Drying:** In this process the drying rollers are exposed to the air and drying takes place at atmospheric pressure. A thin film on the order of 0.1 mm of milk, skim milk etc., is dried on a large rotary metal cylinder or drum, which is steam heated internally. Very often two drums are set up side by side at a very small distance apart. The water evoporates within a few seconds, which is possible due to the high drying temperature over 100 °C. The dried film is scraped off from the drum by means a steel knife, collected and ground. Considerable change to the product due to heating occurs, mainly because of scraping off which is always imperfect and accordingly, a part of the milk is repeatedly being wetted and dried. The quality of the powder can be improved by using a vaccum roller drier in which the milk is dried at a lower temperature, but this method is expensive.

8.6.1 Vacuum Roller Drying

In this process the whole unit is enclosed in an air tight chamber in which high vacuum is maintained. The result of this arrangement is to increase the rate of evaporation, while in addition the drying operations are carried out at much lower temperatures than are required when the open roller process is employed.

The moisture content in the milk evaporated depends upon the degree of vacuum maintained in the chamber. The results obtained from the use of the vacuum roller drier are:

1. The oxidation of the milk by air is reduced and there is a sterile and inert atmosphere in the drying chamber.
2. Dry heating at low temperature, the food value is maintained.
3. The product is easily soluble in water.
4. The plant is compact and there is little loss of powder.
5. The conditions of drying and processing can be closely monitored and controlled.

8.6.2 Band Film Drying

This process also is an improvement of roller drying with a comparatively new idea and is likely to become more popular. Two methods are used as follows.

(a) The milk spread in thin layers on trays drying being carried out by hot air in an enclosed chamber.

(b) The milk is spread on an endless belt instead of on rollers and is passed through a hot air chamber in which vacuum is usually maintained. This method is most commonly employed for whey drying.

8.6.3 Foam Drying

Air or nitrogen is injected into the concentrate, under pressure and the mixture obtained is heated in vacuum. Many gas cells are formed in the concentrate which soon turns into a spongy mass that can subsequently be dried fairly quickly. The process can be carried out batchwise where the concentrate is in shallow trays or continuously on a conveyer belt. The dried cake is ground to a voluminous, easily soluble powder. The powder quality is excellent due to the low drying temperature applied. The process is expensive and is only applied for special products such as some infant formulas. An advantage of this method is that it can be applied to un-homogenous products.

8.6.4 Freeze Drying

In this process a thin layer of the liquid is frozen, where upon the ice is sublimated under a high vaccum. A voluminous cake is left because the space of the ice crystals is now occupied by holes and is subsequently ground. A batch processing or a continous operation in a high vacuum belt drier can be applied. The method is expensive. Damage due to heating does not occur but that is also true of spray drying if performed skilfully. A drawback is that the fat globules are subject to partial coalescence, and this causes freeze dried whole milk powder to show segregation after its reconstitution. Freeze drying is suitable for processing in small quantities and is applied in the drying of lactic starters etc.

8.6.5 Spray Drying

The basic principle of spray drying consists of atomizing the milk preferably. Pre-heated and concentrated to form a spray of very minuted droplets (fog like mist) which are directed into a large suitably designed drying chamber, where they mix immediately with a current of hot air. Owing to their large surface area, the milk particles surrender their moisture practically and instantaneously and dry to a fine powder, which is removed continuously.

Spray drying is the common process/method. There are several variants, however the following are the essential process steps that are always involved.

1. *Air heating*: The air heated to about 200 °C and leaves the drier at 100 °C.
2. *Atomizing the concentrate in the air*: This produces such small droplets that they will dry very quickly with either a spinning disk or a pressure nozzle.
3. *Mixing hot air and atomized liquid*: Drying occurs correspondingly. Air and liquid usually enter the drying chamber, co-currently and are mixed so intensely that the air cools very rapidly. Consequently, the larger part of the drying process occurs at temperatures not much over the outlet temperature.
4. *Separating powder and consumed drying air*: Cyclones are commonly used. In the drying of a liquid several stages can be distinguished. In milk products, a solid like material is obtained at a water content near 8 percent, whereas a powder with say 3 percent is desired. Traditionally one process step included in both drying stages, though in freezing the temperature must be raised during the final drying to complete it within a reasonable time. In spray drying advantage is often taken of separating the final drying which is generally achieved in a fluid bed apart from the main process.

The flow diagram of the spray drying process has been given in the flow chart 8.2.

The method of manufacture by drum process of whole milk powder (WMP) and skim milk powder (SMP) combined is shown in flow diagram 8.3 for WMP and SMP.

Flow Diagram 8.2 Spray Drying System

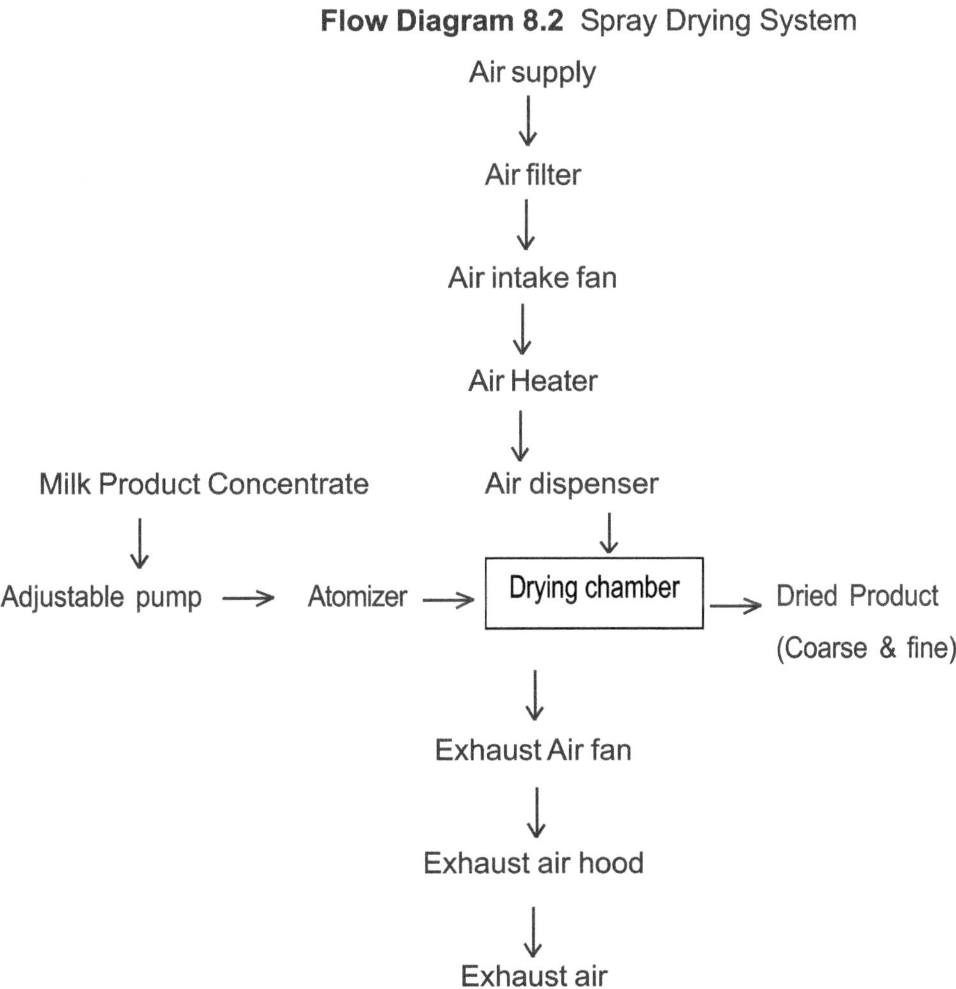

Flow Diagram 8.3 Method of manufacture of WMP and SMP

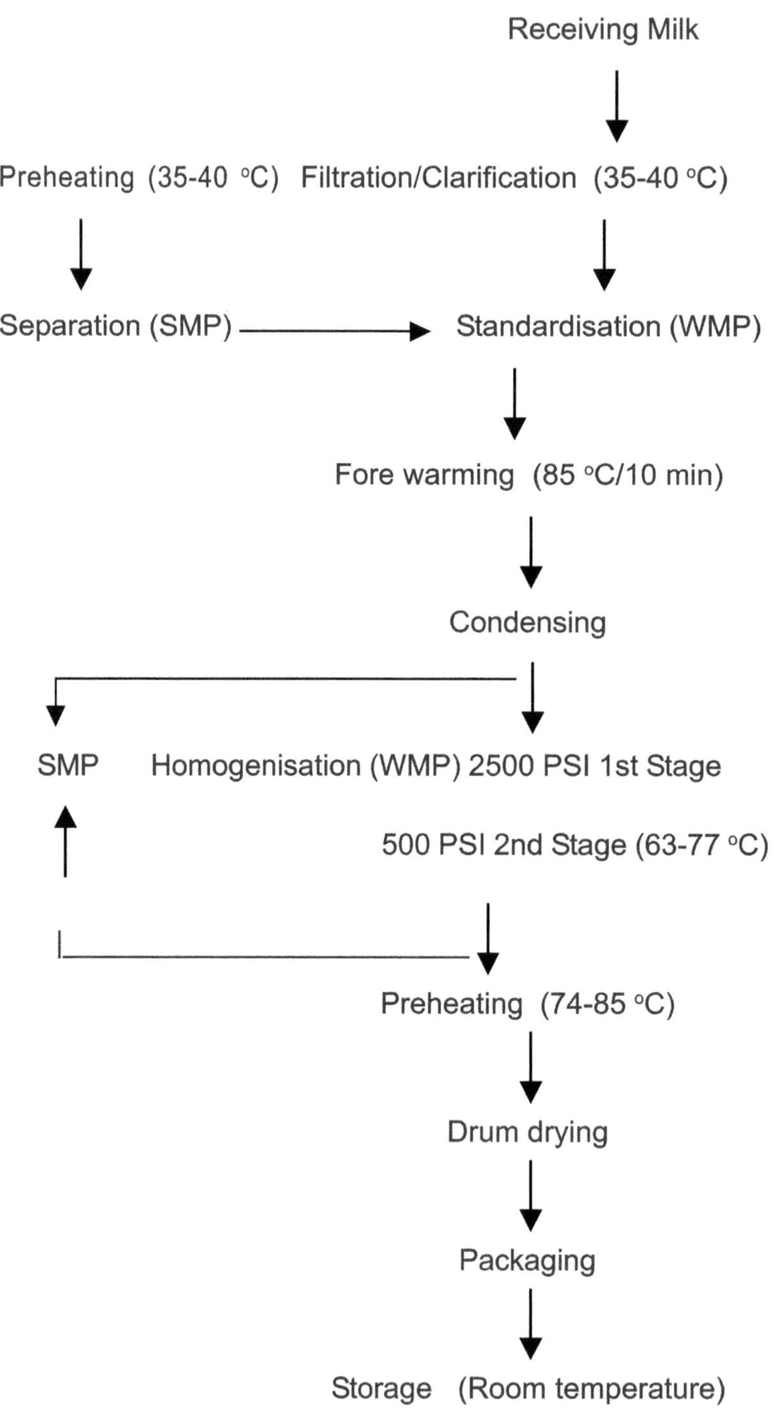

Method of manufacture by spray drying process of whole milk powder (WMP) and skim milk powder (SMP) combined: SMP is shown in Flow diagram 8.4

Flow Diagram 8.4 Manufacture of spray dried WMP and SMP

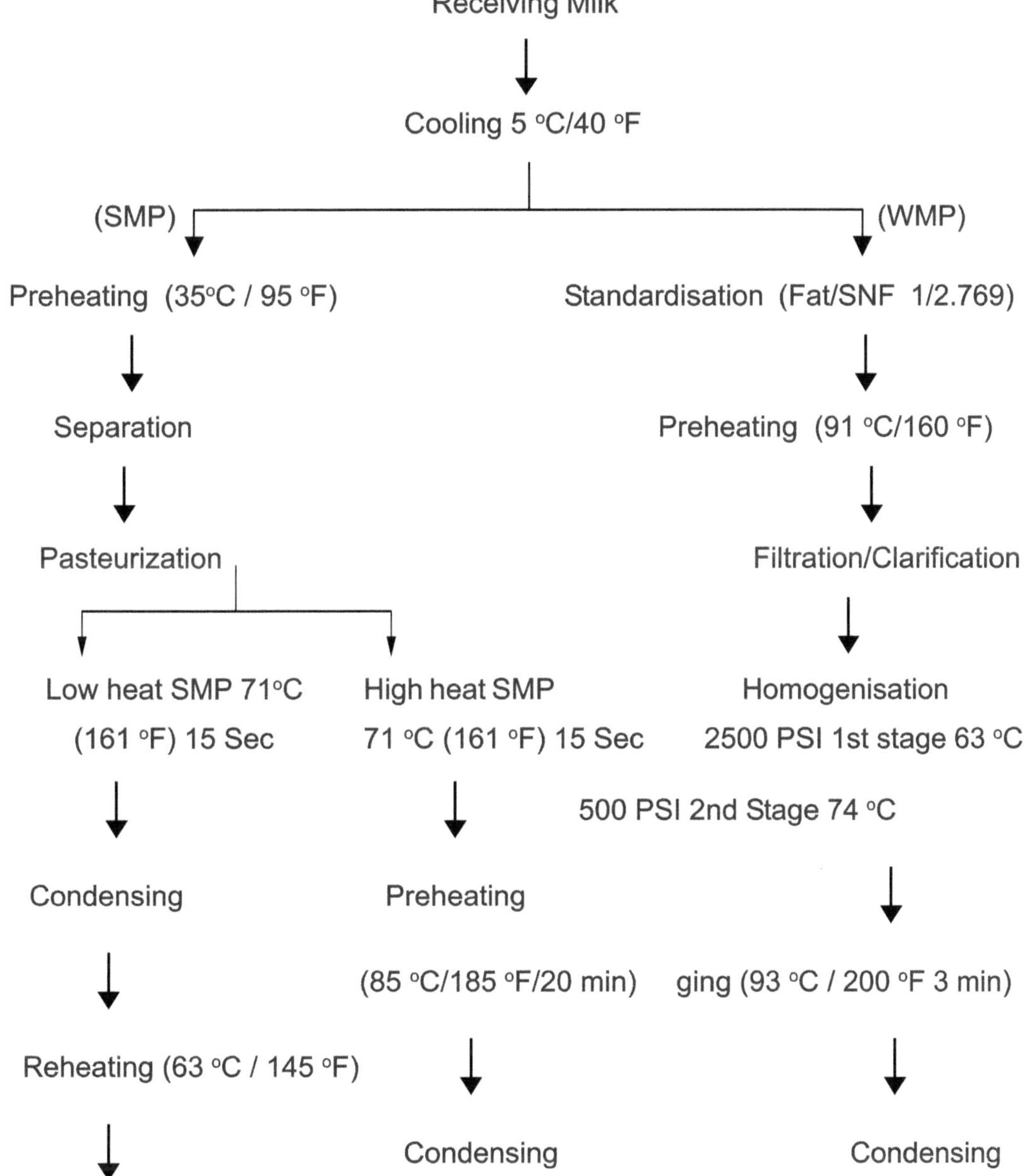

Flow Diagram 8.4 ***Contd.....***

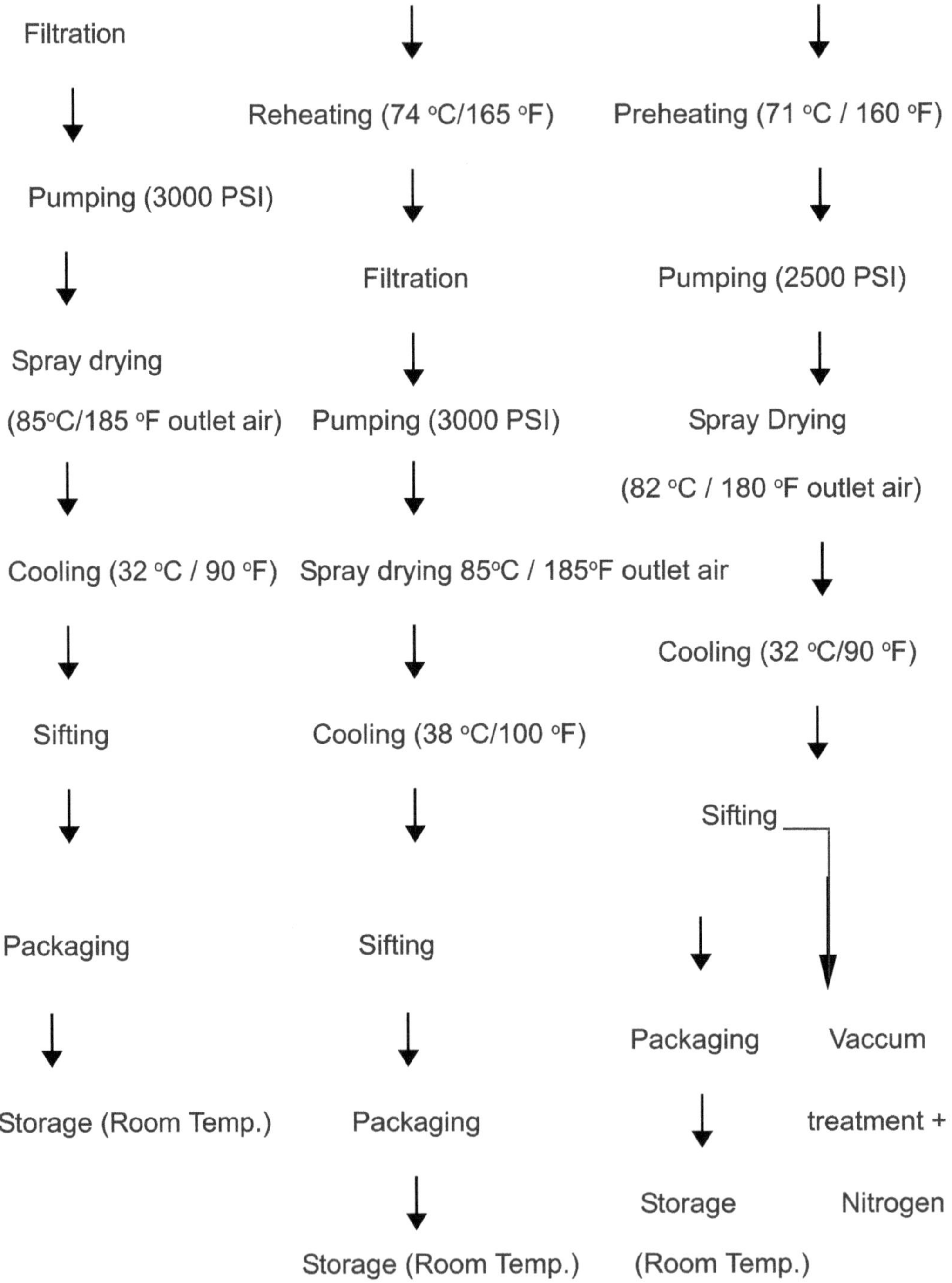

A simplified diagramatic presentation of spray drying process of milk has been presented in Fig. 8.1.

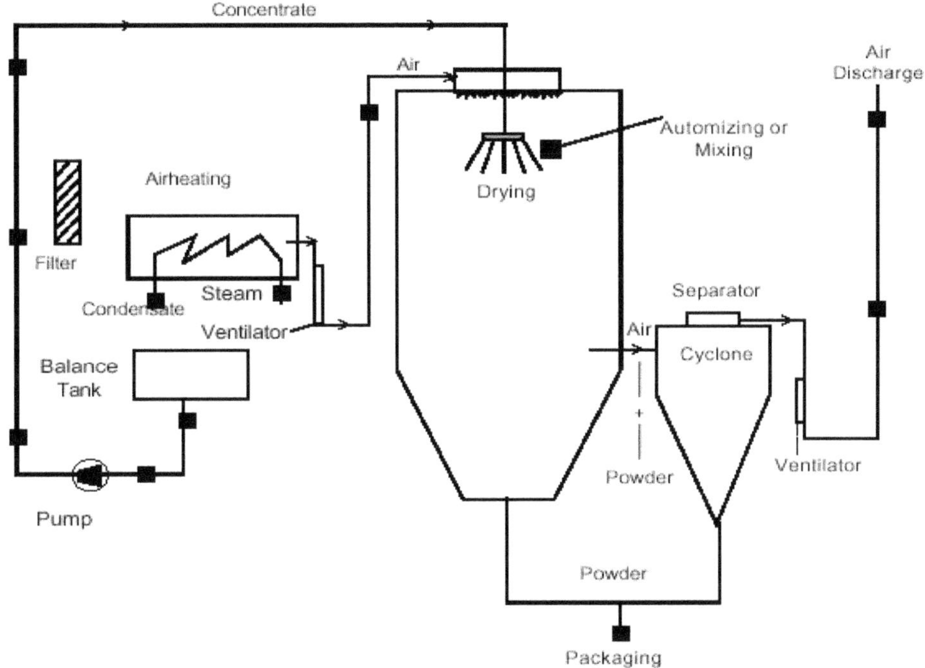

Fig. 8.1 An example of the spray drying process

8.7 Configuration of the Drier

Generally a wide range of configurations are applied according to type of raw material used, product specifications and local possibilities. Naturally, minimization of operating expenses is desired, and this is not simple when the drier is to be used for a range of different products. The main variables are the following.

8.7.1 Heating of the Air

Generally heat exchangers are used. Steam is the classical medium. Hot gas also is being used very often obtained by burning natural gas predominantly methane. Another possibility is hot oil. Direct electric heating of the air can also be applied.

8.7.2 Automization and Air Inlet

A spinning is used for the atomization of the liquid. The drops leave the disc in a radial and horizontal direction. The drying chamber must be wide to prevent the droplets reaching and thereby fouling the wall of the chamber. Currently, spraying nozzles are more commonly applied. Except for very small driers, several nozzles arranged in one or more clusters are installed. The drops leave the nozzles, in a roughly downward direction. The air inlet is generally in the centre of a cluster also directed downward. The drying chamber can have a smaller diameter and often has a greater height as compared to driers with disk atomization. The shape of the chamber is designed in such a way that a mixing of air and droplets and the coarse of drying of the droplets are optimal.

8.7.2.1 Powder Air Separation

This is primarily achieved with cyclones arranged in different ways. In a drier configuration, which rotates strongly the lower part of the drying chamber also acts as a cyclone and most of the powder is discharged below. The powder often is cooled before packaging. The air has to be cleaned before it is returned to the atmosphere. The outlet air is still hot and part of the heat can be transferred to the inlet air in a heat exchanger.

8.7.2.2 Aggregation of Powder Particles

This can be achieved by nozzle spraying when the nozzles are arranged in such a way that the sprays overlap. Another method is to return the fines in a region where the drops are still fairly liquid. A third method is re-wetting in a fluid bed. In all these situations, sticky powder particles are caused to collide with each other.

8.7.2.3 Atomization

It is aimed at forming droplets fine enough to dry quickly, but not so fine as to escape with the outlet air after having been dried. Moreover, a very fine powder has undesirable properties because it is hard to dissolve. The advantages of disk atomization are the following:

1. The droplets formed are relatively small.
2. The disk does not readily become clogged.
3. Disk atomization is still practicable at high viscosity, highly evaporated milk can thus be processed.

A drawback is that many vacuoles are formed in the particles, further more, the droplets are flung away perpendicularly to the axis of the disk and accordingly, the chamber has to be wide enough to prevent the droplets from reaching the wall. With pressure nozzles, generally the liquid is forced through a small opening at a high pressure, after it has been given a rotary motion. Advantages of the nozzles are its simple construction, the possibility to adjust the angle of the cone shaped spray of the atomized liquid and a low vacuole content in the powder particles. When drying milk, the fat globules are disrupted into much smaller ones, about as they would be in a homogeniser, after all the pressures applied are comparable. Nozzles cannot be used for liquids that contain solid particles because they readily become clogged. The maximum liquid viscosity tolerated is somewhat lower than for a spinning disk. A given nozzle can only be used at a narrow range in capacity.

8.7.2.4 Two Stage Drying

Spray drying is relatively expensive. In principle, a better efficiency can be achieved by increasing the concentration factor of the milk before atomization and by applying a higher inlet temperature, but these measures can lead to heat damage of the product. Alternatively the powder can be separated from the air before it is completely dry, while additional drying occurs outside the drying chamber. The powder may be discharged from the drying chamber after it has become so dry as to have lost its stickiness. The final drying is often achieved in a fluid bed drier. A layer of powder deposited on a perforated plate can in principle be fluidized by blowing air through the layer from below. The powder layer is expanded containing a high volume fraction of air, in such a fluidized bed, allowing the mixture to flow, almost like a liquid, if the perforated plate is tilted slightly. The conditions for fluidization are:

1. that the particles are larger than about 20 μm and that their size distribution is not very wide and
2. that the air flow is evenly distributed over the bed and has a suitable velocity.

Another method of two stage drying is carried out in the fuller mat drier. The first stage is conventional spray drying. Atomization is by nozzles and the flow of air and drop spray are co current. The partly dried drops fall on a moving perforated belt. The spent air is removed through the powder bed formed on the belt and is sent to cyclones. The fines removed by the latter are added to the second section of the machine, where it agglomerates with the powder on the belt. In the third section, the hot air is blown through the bed for final drying. The drying material then reaches the fourth section where it is cooled. The powder particles are strongly aggregated and form a porus cake which falls from the belt in large lumps. These are gingerly ground and the resulting powder is packed.

8.8 Instantization

It refers to the process by which the dried milk and milk products are made soluble instantaneously when mixed with water. The chief purpose of this process is to improve the rate and completeness of the reconstitutability of dry milk and milk products. This process affects the wettability, sinkability, dispersability as well as the solubility of the particles, however the total solubility is not improved. This is mainly applied to skim milk powder/non-fat dried milk. However, whole milk powder and other milk fat containing dry dairy products are also being instantized.

8.9 Keeping Quality

It refers to the shelf life of dried milk or milk product to the time i.e. in days or months, it retains its edible qualities after production. Several factors affect the keeping quality of the dried milk in an adverse manner causing decreased solubility, rancidity, tallowiness, stale flavour or oftensive odours and in some cases, the development of a brown colour. The rate of spoilage is influenced by the following factors such as

(a) Composition
(b) Quality of milk used in production
(c) Care during production
(d) Care during handling and storage
(e) Manufacturing conditions especially the heat treatment
(f) Moisture content
(g) metalic contamination
(h) Conditions of packaging including the methods of materials and
(i) Storage conditions.

8.9.1 Changes during Storage

The changes that occur during the storage of the dried milk and milk products are mainly chemical in nature because the microorganisms do not generally proliferate in the dried milk. There are two major types of storage deterioration which are

1. Those affecting the fat and
2. Those affecting the lactose

Changes in fat or fat decomposition may be due to oxidation and rancidity (Hydrolytic). Several factors are responsible and influence the rate of milk fat oxidation which include:

1. Storage temperature
2. Presence of copper and iron
3. Acidity
4. Sunlight

5. Pre-heating
6. Addition of anti oxidants
7. Gas packing

Rancidity may be due to the hydrolysis of fat through lipase enzymes, leading to the production of free fatty acids such as butyric. However, this defect is of minor importance in fat rich dried products.

8.9.2 Lactose Protein Changes

The dry milk or milk products may develop an off flavour, off colour and solubility, all due to lactose protein changes during storage.

Defects

The defects may occur either in the whole milk powder or skim milk powder which may arise due to low quality milk and faulty methods in production, storage of these products. The defects are due to

(a) Flavour : These include
1. Stale/old
2. Oxidised/tallowy
3. Rancid
4. Scorched/Burnt

(b) Body and texture which includes the fineness and homogenecity
1. Lumpy
2. Caked

(c) Colour and appearance
1. Browning/Darkening of colour
2. Scorched/Burnt appearance
3. Lack of uniformity/homogenecity in appearance.

8.10 Dehydration-Concentrated Milk

Milk, skim milk and whey etc can be concentrated i.e. part of the water can be removed. The main object or purpose is to diminish the volume and to enhance the keeping quality of the product by this process. Water can be removed from milk by evaporation, in addition to water, volatile substances especially the dissolved gases are also removed by this process. Generally evaporation is carried out under reduced pressure, therefore, decreased temperature, in order to prevent damage caused by heating. Removal of water to a low level, such that product becomes solid like and is called drying. Drying is achieved by vaporization of water generally from the concentrated milk. The different methods of concentrating the milk by dehydration has been presented in the Flowchart 8.5.

Flow Diagram 8.5 Different methods of Dehydration of Milk

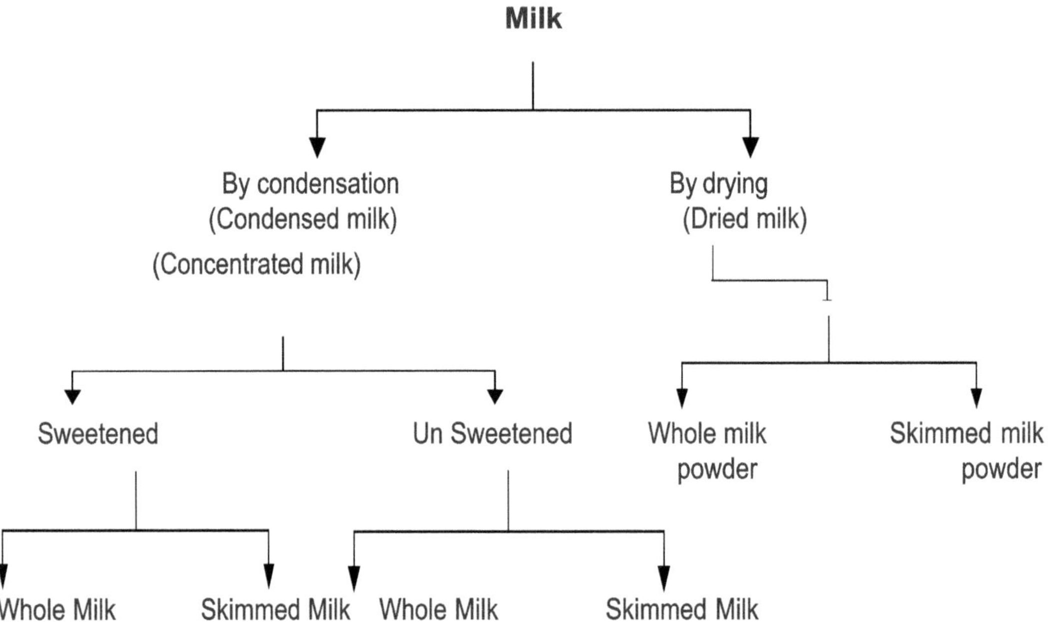

Unsweetened condensed milk is generally termed as evaporated milk. The removal of water from milk causes numerous changes in the properties of milk apart from the increase of most of the solute concentrations. The changes also depend on other conditions such as heat treatment and homogenisation before concentrating. Some important changes in properties are as follows :

1. Decrease in the water activity.
2. Increase in the hygroscopicity.
3. Change in the salt equilibrium.
4. Changes in the confirmation of proteins.
5. Changes in the physico-chemical properties.

6. The Rheological properties are affected, and
7. Decrease in the diffusion coefficients.

Condensed milks are the products obtained by evaporating a part of the water of whole milk, or fully or partly skimmed milk with or without the addition of sugar. The term "Condensed" milk is commonly used and refers to the "full cream sweetened condensed milk", whereas the term "evaporated milk" is generally used to refer the "full cream unsweetened condensed milk". Similarly skimmed milk products are called as "Sweetened condensed skim milk" and "Unsweetened condensed skim milk" respectively.

Various condensed milks have been specified in accordance with the P.F.A. Rules (1976) as under :

8.11 Sweetened Condensed Milk

It is the product obtained from cow or buffalo milk or a combination thereof or from standardised milk by the partial removal of water. It may contain added calcium chloride, citric acid, and sodium citrate, sodium salts of orthophosphoric acid and polyphosphoric acids not exceeding 0.3 percent by weight of the finished product. Such additions need not be declared on the label. Sweetened condensed milk should contain not less than 9.0 percent milk fat and not less than 31.0 percent milk solids and not less than 40 percent cane sugar.

8.12 Un-sweetened Condensed Skim Milk

This product is also similar in its constituents as that of sweetened condensed milk except that the difference being skimmed milk is used instead of whole or standardised milk in its preparation. The sweetened condensed skimmed milk should contain not less than 26.0 percent of total milk solids and not less than 40.0 percent cane sugar. The fat content should not exceed 0.5 percent by weight in accordance with the PFA Rules (1976).

8.13 Composition

The composition of sweetened condensed milk varies in accordance with the type of milk used during the manufacture. The FSSAI (2017) standards for different sweetened condensed milks has been presented in the table 8.5.

Table 8.5 FSSAI (2017) standards for sweetened condensed milks

Sl.No	Parameter	Sweetened condensed milk	Sweetened condensed partly skimmed milk	Sweetened condensed skimmed milk	Sweetened condensed high fat milk
1.	Milk Fat % (m/m)	8.0 (Minimum)	more than 1.0 and less than 8.0	1.0 (Maximum)	16.0 (Minimum)
2.	Milk solids Minimum % (m/m)	28.0	24.0	24.0	-
3.	Milk solids Net Fat Minimum % (m/m)	-	20.0	-	14.0
4.	Milk protein[⊗]in milk S.N.F. Minimum % (m/m)	34.0	34.0	34.0	34.0

[⊗]protein content is multiplied 6.38 with the total nitrogen determined

The BIS specifications for the sweetened condensed whole or skim milk have been given in the table 8.6 below.

Table 8.6 BIS specifications for Condensed milks (IS : 1166 - 1973)

Characteristics	Requirements Condensed whole milk	 Sweetened Skim milk
Total milk solids (% Wt min.)	31.0	26.0
Fat (% Wt)	Not less than 9.0	Not more than 0.5
Sucrose (% wt) min.	40	40
Acidity (% lactic) Max	0.35	0.35
Bacterial Count (Per g)	500	500
Coliform count (Per g)	-ve	-ve
Yeast and Mould count (Per g) Max	10	10

8.13.1 Physico-chemical Properties

These include

1. Specific Gravity/Density
2. Freezing Point
3. Colour and Flavour

8.13.2 Colour and Flavour

The condensed milks are tends to darken their colour and develop a cooked flavour due to the exposure to heat during manufacture and storages. The darkening of colour is generally referred to as browning discolouration, which results from the interaction of the amino compounds with sugar which is knownas the "Maillard-type browning reaction". The brown pigment is called 'melan ordin' and the cooked flavour is attributed to the sulfhydril (-SH) compounds formed during heat treatment. The development of brown colour in milk and milk products is objectionable and the evaporated milk is particularly susceptible to this effect. The important factors affecting the browning discolouration and cooked flavour production in fluid milk system are as under:

1. Heat treatment
2. concentration of total solids
3. pH
4. Storage temperature and time
5. Oxygen content
6. Various added compounds

8.13.3 Viscosity : which is again influenced by

1. Concentration
2. Composition
3. State of protein and fat
4. pH and salt balance
5. Temperature of fore warming/Preheating.

8.14 Method of Manufacture-(Sweetened Condensed Milk)

It is the milk that is concentrated by evaporation of which sucrose is added to form an almost saturated sugar solution after which it is canned. The high sugar content is primarily responsible for the keeping quality of the product as well as for its fairly long shelf life even after the can has been opened. The milk is highly concentrated and the product is highly viscous due to high sugar content. The product is some what glassy in its appearence because the fat globules show little light scattering as the refractive index of the continuous phase is almost equal to that of fat. The turbidity of the product is also largely due to lactose crystals.

The typical manufacturing process for sweetened condensed milk has been given in a flow diagram.8.6.

Flow Diagram 8.6 Manufacture of sweetened condensed milk

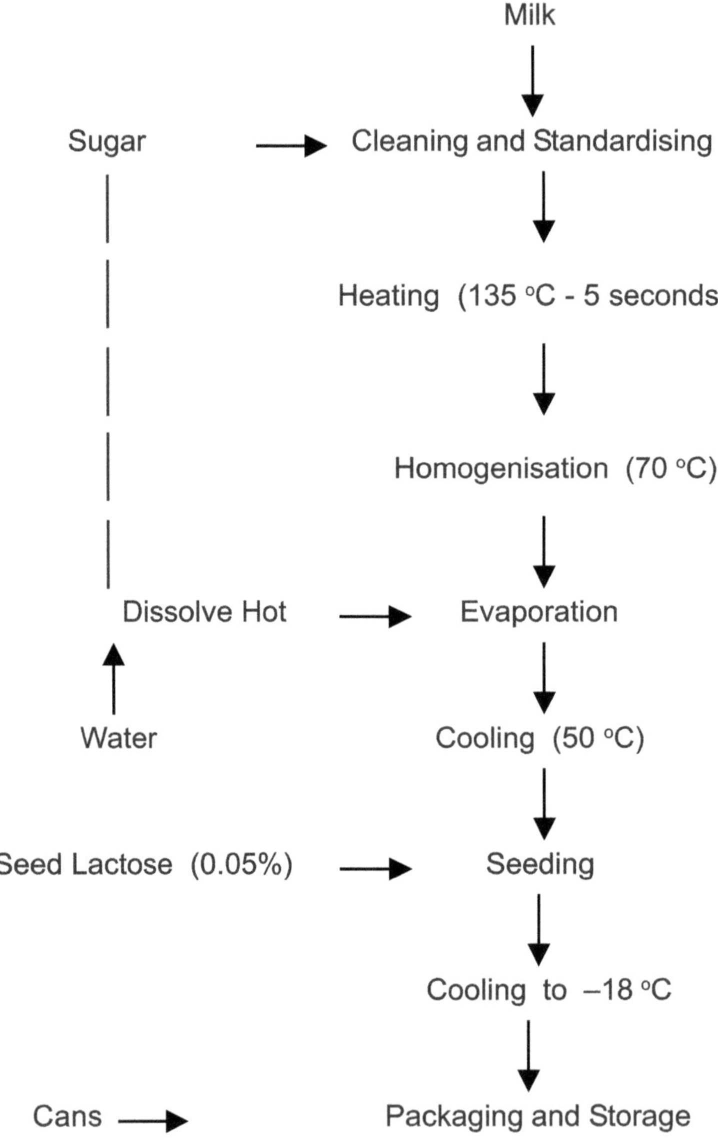

8.14.1 Heating

Pathogens and the potential spoilage organisms present in the milk must be killed. Among the enzymes, the lipase present in the milk should be inactivated primarily however, bacterial lipases are not inactivated, if present, they can cause severe rancidity. Deterioration due to proteinases has not been reported. The heating intensity considerably affects the viscosity and also age thickening and gelation of the product. Therefore the actual heat treatment must be adjusted to these properties. Generally the UHT heating at about 130-140 °C is applied.

8.14.2 Homogenisation

Homogenisation is not carried out or done very often because, creaming is not a major problem. However, at present the sweetened condensed milk is made less viscous, which exhibits less thickening than previously. The density difference between fat globules and continuous phase is large. The homogenisation is done at a low pressure.

8.14.3 Sugar

It can be added to the original milk. The amount of sugar to be added is adjusted readily, and is pasteurized along with the milk. However, this process/procedure, causes a fairly extensive "Maillard" reactions during heating and evaporation with a faster age thickening. As an alternative concentrated sugar solution, which is sufficiently heat treated to kill any osmophilic yeasts, is added at the end of the evaporation step. The sugar should be refined and be devoid of invert sugar to prevent excessive Maillard reactions.

8.14.4 Concentration

Usually this is done by evaporation. A falling film evaporator is generally used to remove the bulk of water and a circulation evaporator to remove the remainder. Relatively high temperatures up to 80 °C are often applied, which implies a lower viscosity in the evaporator, but a higher initial viscosity of the final cooled product. The low water content of the sweetened condensed milk implies a high viscosity and boiling point. Evaporation in continously operating equipment with many effects is therefore not easy resulting in the ready occurence of the fouling. It is difficult to adjust the desired water content accurately which is mostly monitored by means of refractive index.

8.14.5 Cooling and Seeding

Formation of large lactose crystals must be avoided in these steps. The condensed milk must be cooled to a temperature at which the lactose is super saturated so that the sead lactose does not dissolve before it is added. However, the temperature must not be so low that spontaneous nucleation can occur before the seed crystals are mixed in. Cooling should be continued to crystalise the lactose after seeding.

8.14.6 Packaging

Packaging in cans is very common and these are covered with a lid and the seams are sealed. The cans and lids are sterilised by flaming. The packaging room must be supplied with air purified through the bacterial filters.

8.15 Keeping Quality

It mainly depends upon the following factors namely :

1. **Microbial Spoilage:** Sweetened condensed milk is not sterile and contains microbes and spores. The low water activity or the high sugar contents prohibits the growth of most but not all microorganisms. Generally the deterioration occurs due to osmophilic yeasts, which often cause gas formation i.e. bulging of cans a fruity flavour and coagulation of protein resulting in the production of ethanol thus making the product un-acceptable. Some micro cocci may grow in the sweetened condensed milk, although slowly provided the water activity and temperature are high. If they keep growing several off flavours may develop. As long as oxygen is present some moulds may grow.

2. **Chemical changes:** The main change in the sweetened condensed milk during the storage process is presumably age thickening and finally gelation. It is far more concentrated than evaporated milk. A difference with evaporated milk is that an initial decrease in viscosity before age thickening is not observed. The viscosity increases almost linearly with the time. The following are the main factors that may affect the age thickening.

    1. Type of milk
    2. Preheating of milk
    3. Stage at which the sugar is added
    4. Concentration factor
    5. Stabilizing salts
    6. Storage temperature

Brown discolouration is stronger as the storage temperature is higher. Auto oxidation of fat can occur because the packaged product contains a little oxygen and may not have been heated sufficiently to prevent the autoxidation. Obviously, contamination with copper must be avoided by taking rigorous precautions.

3. **Lactose Crystals:** The sweetened condensed milk contains around 38-45 g lactose per 100 g of water. Due to high viscosity, nucleation will be slow and only a few nuclei would be formed per unit volume of milk leading to large crystals. Satisfactory results can be reached by using seed lactose. The final size of the crystals in the product should not exceed 8 μm.

Defects

The defects in sweetened condensed milk may be due to

(a) Microbial

- (i) Mould buttons
- (ii) Bacterial thickening
- (iii) Bloats-microbial

(b) Non microbial

- (i) Sandiness due to lactose
- (ii) Sandiness due to sucrose
- (iii) Sugar separation (lactose)
- (iv) Age thickening (Physico-chemical)
- (v) Bloats (Physico-chemical)

Very rarely fat separation may occur in the condensed milk.

8.16 Evaporated Milk or Un-sweetened Condensed Milk

Un-sweetened condensed milk is generally called as evaporated milk and is obtained practically in the same manner as that of sweetened variety. The process although differs only in certain minor details, it differs from the sweetened condensed milk or condensed milk in that no sugar is added as a preservative. It is merely fresh milk of cow, buffalo or a combination thereof from which approximately 65 percent of the water content has been removed. The product is sterilized by heat after concentration in order to effect preservation and is sold in the hermitically sealed cans. Evaporated milk, although classified as condensed milk by legislation, it is obvious, that it is not the same product, from the treatment which it receives in its production/manufacture.

The term "evaporated Milk is generally used and refers to either "full cream un-sweetened condensed milk" or "Un-sweetened condensed skim milk".

Concentrated milks are liquid milk preserves with a considerably reduced water content. Water is removed by evaporation, preservation is achieved either by sterilization, leading to a product called evaporated milk or by creating conditions that do not allow the growth of microorganisms. The latter is generally realized by the addition of a large quantity of sucrose and exclusion of oxygen. The resulting product is called sweetened condensed milk. These products were initially meant for use in tropical regions in general where milk was hardly or not available.

Evaporated milk is sterilized, concentrated homogenised milk. The product can be kept without refrigeration and has a long shelf life and is completely safe for the user/consumer. The major problem with sterilization is the heat stability, the higher the concentration of milk, the lower the stability. In view of this, the concentrating cannot be done or increased by more than 2.6 times, which corresponds to a level of about 22% solids not fat in the evaporated milk. The flavour and nutritive value of the product are not greatly different from that of fresh milk after dilution.

According to PFA Rules (1976) un-sweetened condensed milk (evaporated milk) is defined as the product obtained from cow or buffalo milk or a combination there -of or from the standardised milk by partial removal of water. It may contain added calcium chloride, citric acid, and sodium citrate, sodium salts of orthophosphoric acid and poly posphoric acid not exceeding 0.3 percent by weight of the finished product. Such additions need not be declared on the label. Un-sweetened condensed or evaporated milk should contain not less than 8.0 percent milk fat and not less than 26.0 percent milk solids.

Similarly the un-sweetened condensed skim milk or evaporated skimmed milk is the product obtained from cow, buffalo skim milk or combination thereof by partial removal of water. It may contain added calcium chloride, citric acid, and sodium citrate, sodium salts of orthophosphoric acid and poly phosphoric acid not exceeding 0.3 percent by weight of the finished product. Such additions need not be declared on the label. Un-sweetened condensed skimmed milk should contain not less than 20.0 percent total milk solids. The fat content should not exceed 0.5 percent by weight.

8.16.1 Composition

The composition of the finished product varies according to the nature of the raw or basic material used in its preparation i.e. whole or skimmed milk. The FSSAI (2017) standards for each type of evaporated milk has been given the table 8.7:

Table 8.7 FSSAI (2017) standards for evaporated or concentrated milks

Sl.No	Parameter	Evaporated milk	Evaporated partly skimmed milk	Evaporated skimmed milk	Evaporated high fat milk
1.	Milk fat % (m/m)	7.5 (minimum)	More than 1 and less than 7.5	1.0 (maximum)	15.0 (minimum)
2.	Milk solids minimum % (m/m)	25.0	20.0	20.0	26.5
3.	Milk protein® in milk S.N.F minimum % (m/m)	34.0	34.0	34.0	34.0

Note: Protein content is 6.38 multiplied by the total nitrogen determined.

8.17 Food Value

Evaporated milk will keep indefinitely due to the sterilization process. The lactose is often caramelised or burnt during the process of sterilization giving the product a slightly brownish tinge besides imparting the peculiar flavour. The fat globules in the evaporated milk are broken up into minute particles by the homogenizing process. It is likely that only vitamins A and C are affected in a way by the process, because these substances being susceptible to oxidation. Evaporated milk is a valuable food for infants even though it is not complete in itself. Apart from human milk, the cow's milk is the best food for the infants without any doubt, however even this requires to be modified and supplemented with other foods in order to have a properly balanced diet for the child. Homogenisation of evaporated milk produces the minute fat globules which are characteristic of human milk. The curds of evaporated milk are small and flocculent and are readily and easily digested, because of the heating during its preparation. In addition to infant feeding evaporated milk is employed for a variety of purposes, because of its indefinite keeping qualities. Commercially the evaporated milk is widely used by the bakers, confectioners, sweet makers, and in restaurants etc.

8.18 Method of Manufacture

The manufacture of evaporated milk is a lengthy process which is carried out in the following stages.

It may be pointed out at this stage here, that the skimmed evaporated milk is also processed in the similar manner with that of the full cream or whole milk product. The various steps involved in the manufacture process of evaporated milk are:

1. Preparation of the milk
2. Fore warming
3. Condensing
4. Super heating
5. Homogenising
6. Cooling
7. Filling and sealing
8. Sterilization

The manufacturing process has also been outlined in the Flow Diagram 8.7.

Flow Diagram 8.7 Manufacture of Evaporated Milk

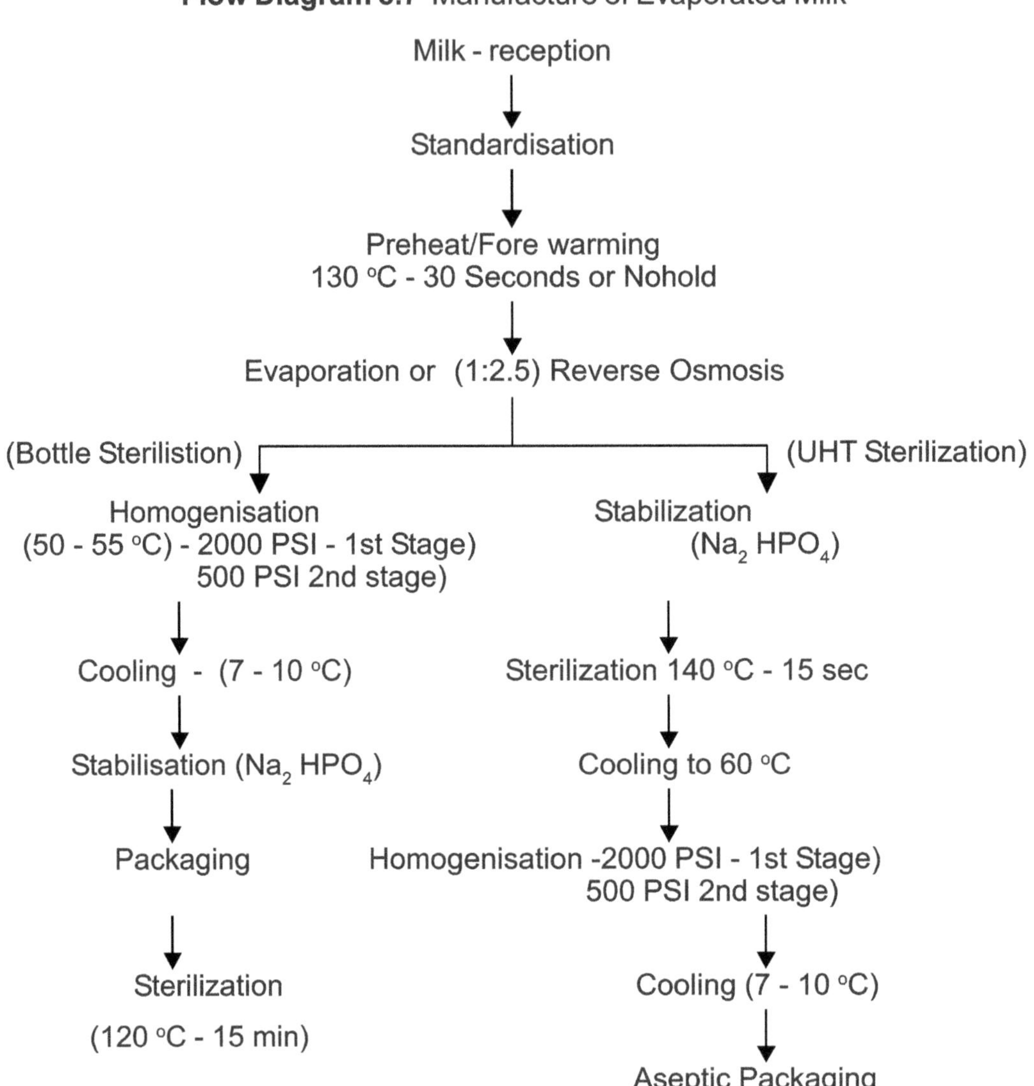

8.19 Preheating/Fore Warming

It serves to enhance the heat stability of the evaporated milk, inactivate enzymes and kill the microorganisms including a significant proportion of the bacterial spores that many present. The heating time and temperature is generally selected on the basis of heat stability. At present the UHT treatment preferred in general, because it reduces the number of spores that may present in the milk considerably hence a less intensive sterilization will suffice.

8.19.1 Concentrating

The milk is usually concentrated by evaporation, standardisation to a desired dry matter content is of much concern, because a higher concentration causes a lower yield and a poorer heat stability. The supply of either raw milk or steam is adjusted based on the continuous standardisation, which is usually applied through the determination of the mass density. As an alternative, the standardisation can be based on the determination of the refractive index. The milk can also concentrated by reverse osmosis which is carried out very rarely. After concentration the manufacturing process for in bottle sterilization and UHT sterilization evaporated milk may differ. The in bottle sterilization process is described first.

Homogenisation serves to prevent creaming and coalescence. It should not be too intensive because the heat stability becomes too low. Stabilisation is carried out to ensure that the evaporated homogenized milk does not coagulate during sterilization and at the same time does acquire a desirable viscosity. A series of sterilisation tests are often conducted on small quantities of the evaporated milk to which varying quality/amounts of stabilising salt (Na_2HPO_4) is added. These tests are essential because variation occurs among the batches of milk. The addition of salt essentially means the adjustment of pH. Further, processing must be postponed/stopped until the test results are available, because this necessitates cooling of the evaporated milk after its homogenisation and storing it for a while. However, long term storage should be avoided in order to prevent bacterial growth, moreover cold storage of the milk increases the tendency of age thickening. The stabilizing salt is added as an aqueous solution which dilutes the evaporated milk slightly. Therefore the milk is often concentrated somewhat too far and is re-standardised to the correct dry matter content during the stabilization process.

8.19.2 Packaging

It is very common practice to package the evaporated milk in cans. The tin plate of the cans is coated/provided with a protective layer of a suitable polymer to prevent iron and tin from dissolving in the product. After filling, the cans may be closed by soldering but mechanical sealing is preferred currently. Evaporated milk intended for use in coffee is usually packaged in bottles that are closed with a crown cork or screw cap.

Sterilization in can or in bottle can be applied batchwise or continuously. Machines that have rotary air locks in order to maintain the pressure may be applied for cans and hydrostatic sterilizers for the bottles. The sterilisation is primarily aimed at killing all the bacterial spores and inactivating plasmin lipases and proteinases from the psychotrophs should be absent from the raw milk because these enzymes would be in sufficiently inactivated.

8.20 UHT Process of Sterilization

It kills the bacterial spores more effectively than in bottle method of sterilization. The combination of preheating and UHT treatment of the concentrate suffices to inactivate plasmin. Preheating is also required to prevent excessive heat coagulation in and fouling of the UHT sterilizer. Some heat coagulation always occurs and the subsequent homogenisation also serves to reduce the size of the protein aggregates formed. Aseptic homogenisation must be applied. Indirect UHT sterilization in a tubular heat exchanger allows the pump of the homogeniser to be fitted before the heater and the homogenising valve behind it thus reducing the risk of re-contamination with the bacteria. The addition of stabilizing salt can often be omitted if the UHT sterilization processes applied/followed or the amount to be added is not so critical that sterilization tests must be carried out. It implies that the whole process from preheating up to and including asceptic packaging can proceed without any interruption.

8.20.1 Recombination

Manufacture of evaporated milk by means of recombination has been outlined briefly in the flow diagram 8.8. The skim milk powder used has to comply with strict requirements. The powder must have been made from skim milk that is heated so intensely that the recombined concentrated milk after its homogenisation is sufficiently heat stable. The count of bacterial spores should be so low that a moderate sterilization of the evaporated milk suffices. Sometimes up to 10 (%) percent of the skim milk powder is replaced by sweet cream butter milk powder to improve the flavour of the product. The copper and peroxide contents of the anhydrous milk fat should be low to avoid flavour deterioration. A high level of calcium in the water used can cause problems with heat stability, "Filled evaporated milk" is also made, where a fat different from milk fat generally of vegetable origin is used.

8.20.2 Properties of the Product

'Maillard' reactions are of paramount importance for the flavour and colours of the evaporated milk. Obviously the temperature and duration of the heat treatment during manufacture determine the initial concentration of the reaction products but on going Maillard reaction occur during storage especially at high temperature developing a stale flavour eventually off flavours due to auto oxidation need not occur.

The viscosity of the evaporated milk is often considered as important quality mark. Bacterial lipases and proteinases due to the growth of psycotrophs when present in the original raw milk, may remain active in the evaporated milk and leads to strong deterioration resulting in a soapy rancid and bitter flavours. Evaporated skim milk may even become more or less transperant due to proteinase activity. The nutritional value of evaporated milk can be decreased significantly as compared to the original plain milk. However these changes are comparatively far smaller when UHT heating is applied.

Flow Diagram 8.8 Manufacture of Recombined Evaporated Milk

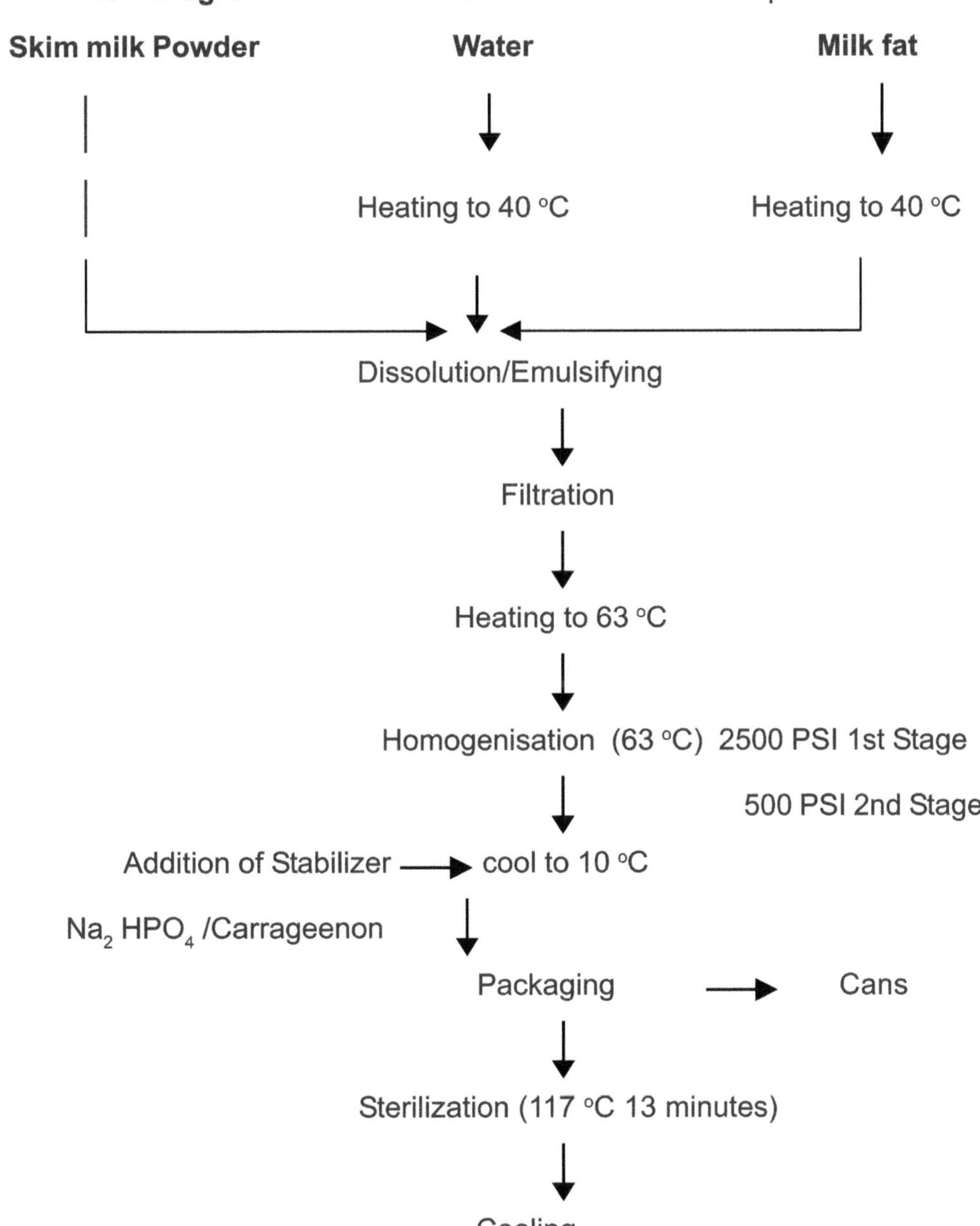

8.20.3 Heat Stability

Concentrated milk is far less stable during sterilization than non-evaporated milk and the fairly intensive homogenisation applied decreases the heat stability further. The heat stability of the milk is of paramount importance not only in the successful processing of milk but also in the manufacture of majority of the milk products. The application of heat is essential in the destruction of spore forming microorganisms with the simultaneous preservation of most of the desirable product characteristics. The milk should not coagulate during heating by the de-stabilisation of the milk proteins.

The heat stability of the milk may be defined in terms of the time required to induce coagulation at a given temperature such as 115 °C/239 °F. The term heat stability of the evaporated milk refers to the relative resistance of the milk to coagulation in the sterilizer. The coagulation of milk revolves around the coagulation of milk proteins. As already stated casein exists in milk as complex colloidal particles or miscelles in addition to casein protein. This system is generally referred to as the calcium, caseinate phosphate complex. As soon as the casein particles in milk join together to form large aggregates, coagulation takes place.

Factors influencing heat stability of milk

(i) Acidity and pH of milk

(ii) Concentration of total solids

(iii) Concentration of salts and ions

(iv) Preheating/Fore warming of milk

(v) Homogenisation

The Factors responsible for the control of heat stability are :

(i) Fore warming or preheating of whole milk

(ii) Heat treatment after condensing evaporation.

(iii) Addition of stabilizers-Chemical

(iv) Mineral Ion exchange treatment.

8.20.4 Defects in Evaporated Milk

The defects may be due to either to the presence of abnormal microorganisms or to changes of chemical or physical origin. The principal defects are:

1. Cooked flavour
2. Coagulation/curdling
3. Browning/discolouration
4. Age thickening
5. Fat separation
6. Feathering in coffee
7. Age thinning
8. Mineral deposit and
9. Bloat or gassy fermentation.

CHAPTER 9

Dairy Products

9.1 Cream

Cream represents the milk in which the normal proportion of the fat has been increased mainly at the expense of water which is removed during the process of separation or skimming. It is one of the most important products of milk because of its ready sale and also as the prime or main constituent of the butter and certain varieties of cheese. It furnishes the body with the heat and energy by virtue of its nourishment value. Cream may be defined as that portion of milk which is rich in butter fat and which has been separated by skimming or by other means usually a mechanical separator.

According to the PFA Rules (1976), cream excluding sterilized cream is the product of cow or buffaloe milk or a combination thereof which contains milk fat not less than 25 percent.

The official definition of cream in the United States of America is as under :

"Cream is that portion of milk, rich in milk fat, which rises to the surface of milk on standing, or is separated from it by centrifugal force. It contains not less than 18 percent of milk fat and not more than 0.2 percent of acid, reacting substances, calculated in terms of lactic acid".

Cream is not a definite substance and contains all the milk constituents in varying proportions. The quantity of protein and sugar present in the cream is approximately half that contained in average cow's milk even though the fat percentage may vary from 18 to 85 percent depending upon the method of separation. However, the solids not fat constituents occur in lower proportion than in milk. Broadly the cream may be classified into:

(a) market cream, which is used for direct consumption and

(b) Manufacturing Cream, which is used for the manufacture of various dairy products.

The different types of cream are:

(a) Table cream

(b) Light cream

(c) Coffee cream - contains milk fat ranging from 20-25 percent on an average

(d) Whipping cream

(e) Heavy cream contains milk fat ranging from 30-40 percent on an average, and

(f) Plastic cream - which an average contains 65-85 percent of milk fat.

Fig. 9.1 Ageing Vat

It may be noted that higher the fat percentage in the cream, the lower the solids not fat percentage and vice-versa. The formula for calculating the percentage of the solids not fat content in the cream is

$$\% \text{ SNF in cream} = \frac{100\% \text{ fat in cream}}{100\% \text{ fat in milk}}$$

Average chemical composition of different types of cream have been presented in the table 9.1 for the better understanding of the students of dairy science.

Table 9.1 Average analyses of various types of cream (% hof)

	Type of cream	Water	Fat	Protein	Lactose Solids	Ash	Total	SNF
1.	Coffee cream	76.60	15.20	3.10	4.50	0.60	23.40	8.20
2.	Fruit cream	67.93	24.44	4.04	2.96	0.63	32.07	7.63
3.	Medium cream	55.00	36.20	6.00	2.50	0.30	45.00	8.80
4.	Thick cream	37.62	58.77	1.83	1.46	0.32	62.38	3.61
5.	Clotted cream	26.10	67.50	4.90	1.00	0.50	73.90	6.40

Source : Outlines of Dairy Technology - S.K. De

9.1.1 Separation Methods

The butter fat in the milk is present in the form of small globules varying in their size, which resemble the minute droplets of oil and confirm to the physical law which states or determines that when a mixture of oil and water is allowed to stand without any disturbance in a quiescent state or manner, the oil will rise to the surface quickly. In the same manner, the milk when allowed to stand for a definite period of time, the butter fat, which is the lightest portion will rise to the surface, which is then skimmed or poured off, leaving behind the water serum called as skim milk containing the protein and the mineral salts. Thus the cream from the milk can be obtained either by gravity or centrifugal methods.

The basic principle of cream separation is based on the fact that the milk fat is lighter than the skim milk portion. The average density of milk fat is 0.93 and that of skim milk is 1.036 at 16 °C or 60 °F. Therefore, when the milk is considered to be a mixture of fat and water i.e. cream and skim milk, and are subjected to either gravity or centrifugal force, the two components stratify or separate from one another, by virtue of their difference in the densities.

9.1.1.1 Gravity Methods

There are two methods of separating cream from milk, which rely upon hand skimming and is being carried out in many farmhouse dairies. These are:

1. Shallow pan setting and
2. Deep setting methods.

The velocity or rate at which the fat globules rise has been expressed by the equation given below which is known as the stoke's law.

$$V = \frac{2G(ds - df)r^2}{9n}$$

Fig. 9.2 Cream Separator

where

V = velocity or rate at which a single fat globule rise

G = acceleration due to gravity

ds = density of skim milk

df = density fat globule

r = radius of the fat globule

n = viscosity of the skim milk

Thus from the stoke's law it will be observed that theoretically the velocity or rate at which the fat globules rises can be increased by:

1. Increase in the radius of fat globule
2. Increase in the difference between the densities of skim milk and fat and
3. Decrease in the viscosity of skim milk.

However, in general practice the important factors, which affect the rate of rise of cream in the gravity methods are as under:

1. ***Size of fat globule***: With the increase in the size of the fat globule, the rate at which the cream rises also increases. For example, in buffalo milk, the gravity creaming occurs at a faster rate due to the fact that the fat globules are larger in size as compared to that of cows milk.

2. ***Temperature***: Due to an increase in the temperature, the viscosity decreases and thereby increasing the velocity or the rise of the fat globules in creaming.
3. ***Clumping***: A clump or cluster of the fat globules acts like a single fat globule so far as the movement through skim milk is concerned, thereby increasing the effective radius of the globule size, which in turn increases the velocity or rate of rise of the fat globule in the creaming.
4. ***Addition of Adhesives***: Ultimately helps in increasing the rate at which the fat globule rise i.e. the velocity.

9.1.1.2 Shallow Pan System or Method

In this method the milk which is received is strained immediately into the shallow setting pans of 12 to 24 inches in diameter and having a depth of 2-4 inches and each pan approximately holding about 5 litres/kgs of milk. After filling, the pans are allowed to stand or settle in a cool place for 24-36 hours at a temperature of about 55 °F. The cream was removed from the surface of the milk by means of a shallow skimmer. Obviously, this process has several disadvantages such as:

1. The removal of cream is never complete,
2. The skim milk may retain fat ranging from 0.5 to 1.0 percent,
3. The production of cream is dependant upon the time for which the milk is allowed to stand,
4. The temperature of the milk,
5. The nature of milk fat,
6. The depth of skimming and
7. lastly the major disadvantage is the fact that the casein which is present in the milk, frequently curdles, thus hampering the cream rising and complete creaming of the milk.

9.1.1.3 Deep Setting Method

This method has been devised in order to overcome the difficulties or the disadvantages of the shallow pan system or method. In this method the milk is placed in cans having 8-10 inches in diameter and approximately 20 inches in depth. In order to prevent the souring or curdling of the milk the cans are cooled either by water or iced water. The cream may be removed by hand with a shallow skimmer, after the milk has been kept standing for the requisite period. Alternatively, the cans may be arranged in such a manner, that the skim milk can be withdrawn from the bottom with help of the taps provided without disturbing the cream. This method yeilds better quality cream and if carefully operated, not more than 0.2 percent of butter fat will be left behind in the skimmed milk.

When the cream is desired in reasonably large quantities, it is necessary to replace the methods previously described by the use of a mechanical or centrifugal separator either hand or power driven. In mechanical or centrifugal separation, fresh milk can be separated and there is no necessity for the milk to stand prior to separation.

The mechanical separation possesses many advantages over the old method of gravity setting of milk, the important being :

1. A greater percentage of butter fat is removed from the milk, and to be more practical, the entire amount being obtained.
2. The richness of the cream can be regulated and different varieties of cream can be produced.
3. Considerable saving of labour and space in the dairy can be effected.
4. The cream thus produced is perfectly fresh thereby resulting in a low bacterial count.
5. The separated cream possesses a low lactic acid content.

9.1.1.4 Mechanical or Centrifugal Method

In this process, the milk when enters the rapidly revolving bowl of a cream separator, it is immediately subjected to a tremendous centrifugal force which is 3000 to 6000 times greater than the gravitational force. Both the fat and skim milk are subjected to the centrifugal force, the difference in density affects the heavier portion more intensely than the lighter portion. Thus the skim milk is forced to the periphery, whereas the fat portion moves towards the centre. The skim milk and cream both form vertical walls within the bowl and are separated by being led through the separate outlets. The cream outlet is at a higher level than the skim milk outlet both being near the axis of rotation. The centrifugal separator bowl is expressed by the following equation.

$$F = K \times W \times R \times N^2$$

where

F = Centrifugal force

W = mass (weight) of the revolving body

R = radius of the circle in which body revolves

N = r.p.m. of revolving body

K = Constant

There are three standard types of mechanical or centrigual separators that are being used in the cream separation namely:.

1. **Disc Plate type:** This is the type of most general use. The machine is built up with a series of cones or discs which fit into a revolving bowl. The milk is fed in at the upper end of the machine, the cream being discharged from one opening while the skimmed milk leaves by another opening.
2. **Curved Plate Type:** This machine is similar in all aspects to that described above, but it differs in that instead of discs it possesses a series of curved plates directing the course of the milk in its passage through the milk bowl.
3. **Tubular Bowl Type:** This apparatus possesses a long tubular bowl and operates extremely at high speed. The milk passes/enters into the separator at the base and as it rises up the bowl, the cream is separated by means of centrifugal action. The completeness of the separation together with the quantity of cream and skimmed milk obtained, depend upon the speed of the bowl, the uniformity of the speed, and the temperature of the milk and the rate of inflow. The physical and chemical condition of the milk itself also exerts an influence upon the completeness of the separation.

Different types of separators are generally provided with means of adjustment, the passage of the milk through the apparatus being regulated in such a way that the operator retains a perfect control of the separation. The important parts of the centrifugal cream separator open bowl type are as under:

(a) Supply can
(b) Faucet
(c) regulating chamber (Float)
(d) Cream or milk screw
(e) Bowl shell
(f) Milk distributor
(g) Cream spout
(h) Skim milk spout
(i) Top disc
(j) Discs
(k) Bowl nut
(l) Rubber ring
(m) Spindle
(n) Set of gears and
(o) Crank handle etc.

The various types of cream separators are:

1. Warm milk separator
2. Cold milk separator
3. Hand driven separator (farm)
4. Power driven separator (Factory or Creamery)
5. Open bowl or gravity fed separator
6. Semi enclosed separator
7. Hermetically sealed or Air tight or pressure fed or foam less separator.

The merits and demerits of the different types of cream separators have been presented in the table 9.2.

Table 9.2 Merits and Demerits of different types of Cream separators

Type of Cream separator	Merits	Demerits
Warm milk	1. Close skimming 2. Higher capacity of operation	1. Lower viscocity of cream 2. More foam formation
Cold Milk	1. Less foam formation 2. Higher viscocity of cream 3. Higher quality of cream and skim milk.	1. Lower capacity of operation 2. Partial churning
Factory	Higher capacity	Expensive
Farm	Cheap	Lower capacity
Open bowl	Useful in small diaries	1. Lower capacity 2. Foam is produced
Semi-enclosed	1. It has some of the advantages of air tight separators	1. It has some of the disadvantages open bowl separation.
Air tight or Foam less	1. No foam is produced 2. Higher capacity 3. Fat percentage in cream can be regulated by external control without stopping the separator. 4. Cream and skim milk can be delivered to respective storage tanks without the aid of any additional pumps. 5. Keeps the products away from the contact with air. 6. The viscosity of the cream is increased greatly.	1. Expensive 2. Maintenance is complicated.

Source : Outlines of Dairy Technology - S.K. De

Stoke's law can also be applied with little modification to the centrifugal separation of cream as in the case of gravitational methods which is as follows:

$$v = r^2 \frac{(ds - df)}{n} N^2 R.K$$

where

v = velocity of movement of single fat globule

r = radius of fat globule.

ds = density of skim milk

df = density of fat

N = speed of bowl (r.p.m.)

R = distance of fat globule from the axis of rotation

K = constant

n = viscosity of the skim milk

Thus it will be observed from the above equation of the stoke's law that the speed or rate of cream separation can be increased by:

1. Greater radius of fat globule.
2. Greater difference in the density between skim milk and fat.
3. Greater speed of the bowl (r.p.m.) i.e. the centrifugal force.
4. Greater size of the bowl and
5. Lower viscosity of skim milk.

An attempt has been made to compare the different characteristics between the gravity and centrifugal methods of cream separation for the benefit of the students of dairy science. They have been summerised and presented in the table 9.3.

Table 9.3 Characteristics of gravity and centrifugal creaming methods

Particulars	Gravity method	Centrifugal Method
Nature of force causing separation	Gravitational	Centrifugal
Speed of separation	Extremely slow	Practically instantaneous
Direction of movement of fat and skim milk particles	Vertical	Horizontal
Bacteriological quality of cream and skim milk	Low	Higher
Fat percentage of cream	10 - 25	Varies from 18 - 85 and can be controlled
Fat percentage in skim milk	0.2 or above	0.1 or below
Scale of operation	Small	Large
Recovery of fat percentage in cream	Not more than 90	99 - 99.5

Source : Outlines of Dairy Technology - S.K. De

9.1.2 Physico-chemical Properties of Cream

The Physico-chemical properties of the cream are as under:

1. **Viscosity:** This is defined as the resistance offered by a liquid to flow, which is an important property from the commercial point of view because the consumer judges the richness of the cream from its viscosity.

However, the viscosity of the cream may be affected by the following factors such as:

(i) *Fat percentage*: Higher the fat percentage greater the viscosity and vice-versa.

(ii) *Temperature*: The higher the temperature the lower the viscosity and vice-versa.

(iii) Conditions of separation such as temperature.

(iv) *Homogenisation*: Single stage homogenisation increases the viscosity in direct relation to the pressure used however, the double stage homogenisation reduces viscosity.

(a) **Size of fat globules:** The viscosity of the cream depends partly upon the tendency of the fat globules which it contains to coalesce. This agglutination in turn depends upon:

   (i) Size of the globules and

   (ii) Agitation

(b) **Butter fat Content:** This naturally affects the viscosity, as the greater the fat content, the more fat globules will be present in the cream to form clumps.

(c) **Pasteurization:** This process tends to decrease viscosity, the higher the temperature the lower the viscosity of the cream.

(d) **Age and temperature:** Ageing of the cream at low temperatures results in the increase of the viscosity, even before the development of acidity sufficient to coagulate the casein.

(e) **Homogenisation:** Homogenisation increases the viscosity of the cream, the higher the homogenisation pressure the greater being the increase in viscosity. The homogenised cream has a tendency to feather in coffee and because of this reason high pressures of homogenisation are seldom employed.

(f) **Cooling:** The rate of cooling of cream exerts a decided effect upon viscosity. Slow cooling of either raw or pasteurized cream considerably increases the viscosity over that of the same cream when cooled rapidly.

2. **Whipping Quality:** It refers to the beating of cream to produce froth or foam (emulsion of gas/air in a liquid). Whipped cream is a special foam possessing remarkable stability which is used in cakes, ice creams, etc for decorative perposes. The main factors which affect the whipping quality of the cream are as under:

   (a) Fat percentage

   (b) Whipping temperature

(c) Separation temperature
(d) Ageing
(e) Homogenisation
(f) Acidity and
(g) Addition of stabilizers

3. **Specific gravity**
4. **Acidity:** The titrable acidity percent of all fresh cream should be consistant with the fat percentage of the cream. There is an inverse relationship between the percentage of fat and the percent of titrable acidity. The titrable acidity percent of freshly separated cream, is always lower than that of the milk from which it was separated.

$$\text{\% Titrable acidity of cream} = \frac{\text{\% serum in cream}}{\text{\% serum in milk}} \times \text{percent titrable acidity of milk}$$

$$= \frac{\text{100\% fat in cream}}{\text{100\% fat in milk}} \times \text{titrable acidity of milk}$$

In the process of centrifugal separation or production of cream, there are certain or important factors that may affect the percentage of fat. Some of these factors are:

9.1.3 Factors Affecting the Fat Loss in Skim Milk during Separation

A. 1. **Position of Cream or Skim Milk Screw:** The cream screw/outlet consists of a small threaded, hollow screw pierced by a circular orifice through which the cream emerges. This screw can be driven or pushed inside thus bringing it nearer to the centre of rotation or can be distanced or brought away from the centre of rotation by bringing the screw to the outer surface or out. The skim milk screw also works in the similar manner for the removal of the skim milk. Once the skim milk or cream screws are adjusted, under normal conditions, the cream separator delivers a definite ratio of skim milk and cream which is generally 90 : 10 or 85 : 15 by volume. When the position of the cream screw or skim milk screw is altered or changed, the ratio of skim milk to cream is also changed or altered resulting in either higher or lower fat test or percentage of the cream.

2. **Fat Percentage in Milk:** Practically all the fat present in the milk is contained in the cream, thus the separation of cream from high fat milk has a higher fat as compared to that from the low fat milk.
3. **Speed of the Bowl:** The fat percentage of the cream is dependant on the speed of the bowl. The higher the speed, the greater will be the centrifugal

force, hence the skim milk will also leaves the bowl more quickly or rapidly thus increasing the capacity of the skim milk discharge. This indicates that less cream is discharged with the same fat content consequently resulting in a higher fat percentage in cream will obtain.

4. **Rate of Inflow of Milk:** The discharge of the cream increases from the cream outlet. When the rate of inflow of milk increases and the skim milk discharge remains constant with constant centrifugal force.
5. **Temperature of Milk:** The lower the temperature of milk, the higher the fat percentage of the cream. The viscosity of both cream and skim milk are increased with low temperature however, the viscosity of cream increases proportionately as compared to that of skim milk. Thus the quantity of cream discharged is reduced due to the clogging of the bowl, thereby resulting in a higher fat test.
6. **Amount of water or skim milk added to flush the bowl:** The addition of more water or skim milk will result in an increase of the amount of cream produced, with the same quantity of fat showing a lower fat test. Thus all the factors stated above may affect or influence the fat percentage in the cream, and in the similar manner there appears to be some other factors which may affect or influence the fat loss in skim milk during the process of cream separation.

B. 1. **Temperature of Milk:** In order to obtain an efficient separation of the fat in the cream, the temperature of the milk should be above that of the melting point of the fat such that all the fat globules of the milk fat are entirely in liquid phase; for this reason, satisfactory temperature is around 40 °C or 104 °F for efficient separation. There is no marked increase in efficiency after 43-49 °C or 110-120 °F. The milk is invariably heated before separation to 35-40 °C in plate or tubular heaters for efficient separation which is known as "preheating or fore warming" of milk.

2. **Speed of the Bowl:** There will be more fat loss in skim milk due to a low rated speed, because of insufficient centrifugal force for efficient cream separation. However, the skimming efficiency does not increase in a considerable manner at higher or above rated speeds.
3. **Rate of Inflow of milk:** The milk which passes through the bowl too rapidly than the designated capacity of the separator in order to allow complete separation, thereby resulting in a higher fat loss in skim milk. On the contrary under feeding or slow rate of inflow of milk to the separator bowl does not increase the efficiency of separation significantly.
4. **Position of the cream screw:** The fat test of the cream can be varied without influencing the efficiency of the skimming, because a good separator is designed to give an efficient skimming within a fairly wide range of positions

of the cream screw. The position of the cream screw has little effect on the fat test of skim milk unless and until the fat test of the cream is more than 45-50 percent with majority of the cream separators. Separation of very thick cream at low temperatures may lead to higher fat losses through clogging of the bowl with viscous cream.

5. **Mechanical Condition of the Separator:** Mechanical condition of the separator which is not satisfactory may result in greater fat loss through skim milk, which include:

   (a) **Vibration of the Separator:** Vibration is caused by installing the separator on an insufficiently firm foundation, the bowl being out of balance, the bearings being worn out, the axis of rotation is not in exactly vertical position and so on reduces the efficiency of separation.

   (b) **Condition of the discs:** Discs which are in an unsatisfactory condition such as out of shape, dirty, scrached or having rough surfaces suffers a reduction in the skimming efficiency due to the uneven flow of the counter current streams of cream and skim milk between them.

   (c) **Amount of separator slime in the bowl:** Fat loss in skim milk increases due to the accumulation of too much slime which is caused not only by the disturbance in the even flow of the counter-currents of cream and skim milk but also due to the reduction in the centrifugal force due to the decrease in the effective diameter of the bowl.

Separator Slime

It is generally considered as identical with that of the clarifier slime, and consists inside the bowl shell of the cream separator. It is made up of foreign matter, milk proteins, leucocytes, fragments of the secreting cells of the udder fat, calcium phosphate and other minerals.

The average composition of separator slime has been given in the table 9.4:

Table 9.4 Composition of Separator Slime

Constituent	Moist Slime (%)	Dry slime (%)
Water	68.2	--
Fat	1.4	4.4
Protein	25.3	79.6
Lactose	1.8	5.6
Minerals	3.3	10.4

Source : Outlines of Dairy Technology - S.K. De

6. **Size of Fat Globules:** Generally the fat globules which are less than 2 microns in their size enters the skim milk, because they are not subjected to sufficient centrifugal force to be recovered in the cream. Thus greater the number of less than 2 micron size enters the skim milk, because they are not subjected to sufficient centrifugal fore to be recovered in the cream. Therefore greater the number of less than 2 micron size fat globules the greater the fat loss in skim milk.
7. **Temperature:** The higher the degree and temperature of agitation results in the disintegration of the normal fat globules into smaller ones, which escape the effect of centrifugal force thus resulting in more fat loss in skim milk.
8. **Presence of air in milk:** When the milk which is entrapped with air bubbles is delivered to the separator, the separation of air as a consequence of centrifugal force disturbs the counter-current streams of cream and skim milk between the discs, thus lowers the efficiency of separation. The effect of air in the milk is greater with hermetic than with non-hermetic type of cream separators.
9. **Acidity of Milk:** The higher or increased acidity of the milk, lowers the stability of the casein particles which in turn gets precipitated resulting in the clogging of the bowls, thus lowering the efficiency of separation.

Separation of efficiency is generally the main concern in the manufacture of skim milk powder, in skimming cheese, whey etc. Factors affecting the proportion of the globular fat escaping separation follow partly from the stokes equation or formula for the velocity (V) of a fat globule relative to the surrounding liquid.

The “Skimming Efficiency” (SE) of a cream separator refers to the recovery of the total fat from milk expressed as a percentage. The higher the fat percentage in milk and/or the greater the fat loss in skim milk the lower the skimming efficiency and vice-versa. In order to get the best index for the SE, the skim milk should be checked for its fat test periodically at regular intervals.

9.1.4 Yield of Cream

This can be calculated by the formula.

$$C = M \times \frac{f_m - f_s}{f_c - f_s}$$

Where

C = Weight of cream (kg)

M = Weight of milk (kg)

f_m = fat percentage of milk

f_s = fat percentage of skim milk

f_c = fat percentage of cream

9.1.5 Quality of Cream

The quality of the cream depends upon the quality of milk from which it is prepared. Thus cream is no better than the milk from which it is made and it is not possible to obtain good quality cream from the low grade milk. Therefore, in order to produce a high quality cream, the following steps are to be followed meticulously. They are:

1. Clean milk production.
2. Cooling of milk soon after production.
3. Separating milk under hygienic conditions.
4. Prompt cooling of cream and its storage at low temperature.
5. Transport of cream to the dairy under low temperature.

9.1.6 Defects in Cream

Cream in common with other food stuffs, may be subjected to certain defects or faults due to unsatisfactory processing or handling. The consumer requires a clean flavoured product which possesses a good keeping quality with natural colour, and sufficient viscosity giving reasonably a rich appearance. The principal defects that are generally encountered are:

1. **Poor Keeping quality:** The keeping quality of the cream is extremely important particularly, when the cream is to be stored under un-refrigerated conditions in the house for any length of time. Extended ageing of cream will result in the damages to its keeping quality. Whereas other factors which may affect the keeping quality are high bacterial count of the milk from which the cream is produced. Use of improperly sterilized or cleaned utensils, apparatus and other containers.
2. **Poor flavour:** Flavour is mainly dependent upon the quality of the initial milk supply, thus a wholesome flavour is possible only when fresh milk of good quality is separated. Bitter or rancid flavours may develop due to non-spore forming bacteria of soil and water origin resulting in spoilage at low temperatures. Tallowry flavours are due to the oxidation of fat rancid or bitter taints may arise due to the action of lipase the fat splitting enzyme.
3. **Creamplug formation:** This is caused by excessive agitation of the cream in partially filled cans or by high speed pumps. Cream must be cooled immediately and quickly avoiding the excessive heating always.
4. **Poor Colour:** It depends upon the solids not fat content and the size of the fat globules. Homogenisation, increases the colour and partially churned cream does not possess the colour of the cream with the natural emulsion.

5. **Thin body or Texture:** The various factors that are responsible causing this defect are:
   (a) Too much handling
   (b) Pumping or agitation
   (c) Separation at high temperature
   (d) Separating warm milk and
   (e) Ultra rapid cooling
6. **Serum Separation:** The important factors that are responsible for this defect are excessive pumping resulting in the partial churning of the cream, high temperature of separation and low fat content. This defect is evidenced by fat rising and serum separation because of the fact that the viscosity is not sufficient to retain the fat globules in position throughout the serum.
7. **"Oiling Off":** The causative factors are the freezing of cream during cooling, separating excessively hot or cold milk, pasteurization of very rich cream, very high temperature and excessive agitation of warm cream, high speed pumping by the centrifugal pumps and slow cooling.
8. **Feathery Cream:** This type of cream is highly acidic, possesses calcium or magnessium salts are in excess and cream homogenised at excessive pressures will feather when added to coffee.
9. **Poor whipping Quality:** This defect is sometimes encountered in whipping cream and is due to temperature and the type of whipping apparatus employed. Excessive ageing will cause this defect as with partial freezing and homogenisation.

9.1.7 Uses of Cream

The various purposes for which the cream may be utilized are :

1. For direct consumption as table/coffee/whipping cream.
2. In the preparation of special dishes.
3. In the production of plastic, frozen, and cultured (sour) creams.
4. In the manufacture of butter cheese, ice cream, butter oil and ghee and
5. For creaming the cottage cheese.

9.2 Ice Cream

Generally, Ice-cream is being considered as a luxury food item in this country, however it has came to be an universally accepted staple diet of the western countries especially the United States. The production of ice-cream in increasing quantities

Fig. 9.3 Ice Cream Plant

forms a useful outlet for the surplus milk produced by the dairy industry apart from its food value. The in creasing popularity of ice-cream undoubtedly provides an excellent and profitable means of utilising the large quantitites of unsaleable surplus liquid milk. Very scant information is available pertaining to the manufacture or production of ice-cream, however the available information indicates that it was originated in Europe. In 1860, Carlo, Gatti an Italian, was the first person to introduce the product ice-cream to the public in the United Kingdom by manufacturing it by means of hand freezer, in the manner still practised today by many small vendors. The trade was carried out by the small vendors in this manner until the World War-I i.e. (1914 - 1918). The modern ice-cream manufacturing machinery was imported from America, inorder to produce the ice-cream on a large scale basis in the United Kingdom, thus bringing or making a beginning for the remarkable evolution and improvement in the ice-cream plant in its manufacture due to the ever increasing demand from the public.

Some of the factors, that have contributed for the development of the ice-cream industry in the developed dairying countries are as:

1. The perfection of mechanical refrigeration and its application to the food industry.
2. Improved manufacturing methods and equipment such as homogenisers, over run testers, continuous freezers, packaging machines etc.

3. Availability of more and better quality ingredients and the growth or development of knowledge concerning ice cream manufacture resulting in a better product.
4. Extensive advertising of the product.
5. Changing economic conditions, better wages, more purchasing power and a high standard of living amongst the consumers and
6. Improved storage facilities for ice cream both at home, as well as at manufacturers, distributors and retailers premises.

It is extremely difficult to define the term "Ice-cream" in a satisfactory manner, more particularly because of the variation in kind and quantity of the materials used in its manufacture. There are many definitions given by different organisations/ Associations which have been given here under such as the definition given in a standard dictionary is "Ice-cream is cream milk or custard which has been sweetened and flavoured and which has been frozen by means of a freezing mixture".

Ice-cream may be defined as a frozen dairy product made by suitable blending and processing of cream and other milk products together with sugar and flavour with or without stabilizer or colour and with the incorporation of air during the freezing process".

According to PFA Rules (1976) , "Ice cream is the frozen product obtained from cow or buffalo milk or a combination thereof or from cream and/or other milk products with or without the addition of cane sugar, eggs, fruits, fruit juices, preserved fruits, nuts, chocolate, edible flavours and permitted food colours. It must contain permitted stabilizers and emulsifier not exceeding 0.5 percent by weight. The mixture must be suitably heated before freezing. The product should contain not less than 10 percent milk fat, 3.5 percent protein, and 36 percent total solids. However, when any of the aforesaid proportions contain fruits, or nuts or both, the content of milk fat may be proportionately reduced but may not be less than 8 percent by weight. Starch may be added to a maximum extent of 5 percent with a declaration to that effect on the lable or the composition of the mix". According to the Ice-cream Association of Great Britain and Ireland.

"A frozen product containing not less than 8 percent of milk fat and not less than 10 percent of milk solids not fat".

The county Medical Officer of Health of Fife once has stated that.

"Ice cream was a kind of trade-mark for a frozen mixture of unknown composition, the contents of which were only known to the makers and which might vary from day to day. Ice Cream has no recognised composition and probably the great majority of ice creams had no cream as the word was originally understood".

The United States Department of Agriculture has defined the composition of the finished ice-cream which is based upon and is as follows :

"Ice-cream is a frozen product made from cream and sugar with or without a natural flavouring and containing not less than 14 percent of milk fat".

"Fruit ice-cream is a frozen product made from cream, sugar, fresh fruit and sound, non-rancid nuts and contains not less than 12 percent milk fat".

The definition given by the New York Sanitary code (1941) is:

"The pure, clean and wholesome frozen product made from milk products and with or without the use of sugar, water, eggs, harmless flavouring or colouring or added stabilizer composed of wholesome edible material and in the manufacture of which freezing has been accompanied by agitation of the ingredients". Under the code the ice cream must possess the following minimum composition.

1. Milk fat not less than 10 percent by weight.
2. Total milk solids not less than 18 percent by weight.

Fancy Ice-creams must not contain less than

1. 8 percent by weight of milk fat.
2. 14 percent of weight of total solids.

A close study or observation of all the above stated definitions will reveal that no mention has been made to the bacterial content of the Ice-cream. This is a serious ommission. If the ice-cream is to be given the definition officially, it is most essential that this definition should indicate not only the chemical composition of the product but should also fix a practical bacteriological standard.

Ice-cream was once considered as a seasonal delicate food is now universal and available irrespective of the season, due to the fact that it has become almost an every day dish possessing a real food value. The nutritive value of the ice-cream mainly depends not only on the composition but also on the products from which it is prepared. Generally it may be stated that the food is really nutritious when it contains the following compounds namely

1. Proteins or nitrogenous compounds which are essential for the body building.
2. Fats to supply heat and energy.
3. Sugar constituents which are essential for heat as energy producers.
4. Mineral matter or salts such as calcium and phosphorus that are essential for body growing tissues of the body and
5. Vitamins - necessary for normal growth.

Ice-cream contains all the above essentials as in the case of milk, but it is enriched with the added cream and sugar as compared to milk therefore ice cream may be defined as an excellent protective food, if constituted suitably and it does not mean that one's diet should consist solely of ice-cream only. It is only indicative or suggestive that the ice-cream contains all the elements upon which the life may be sustained. Ice-cream is particularly regarded as the tissue building and as a source of energy generation, so far as its actual nutritive value is concerned and is more useful as compared to a large variety of foods which are supposed to be the foods of the highest value. Ice cream is of particular value in the diets of sick and under nourished persons because it is highly palatable and is easily digested because of the fact that the fat has been finely dispensed by homogenization. Ice-cream contains two to three times as much fat and slightly more protein than does milk. In addition, it may contain other food products such as fruits, nuts, eggs, and sugar, which enhance its nutritive value as food. The protein content of Ice-cream also rates high both in quantity and quality, which are largely derived from the milk and a small amount from the stabilizer gelatin and from eggs when they are used in the mix. It is an excellent source of food energy, having twice to three times the fat content of milk and more than half of its total solids being sugar, the energy value of ice cream is very high.

There is no standard classification of ice cream, that has been adopted by the industry even in the developed countries. However, some kind of classification of the frozen deserts can be made according to the type of ingredients used in this preparation/manufacture.

Ice Cream
- → Plain
- → Chocolate
- → Fruit
- → Nut
- → Milk ices or milk lollies
- → Ices
- → Sherbet
- → Fancy moulded
- → Novelties
- → Soft Ice Cream or Softy

9.2.1 Composition

The Ice cream composition is generally expressed as a percentage of its constituents such as percentage of milk fat, milk solids not fat or serum solids, sugar, stabilizer, total solids etc. The composition varies according to the locality and market and it is very difficult to establish and standard composition for a manufacturer. After giving due consideration to the legal standards or requirements, the quality of the product desired, raw materials available, plant procedures, trade demands, cost of production and competetion, there appears to be a choice of the product of minimum, average or high fat/solids composition. Some manufacturers may choose to produce only one of these products, whereas others may like to produce any two and still others would like to produce all the three categories such as economy brand, a good average product as a trade brand or deluxe high quality product. The percentage of milk fat varies more than any other constituent in the preparation of the ice cream. An increase in the fat content results in the reduction or decreae of the milk solids not fat content in order to avoid the "Sandiness" i.e. the crystallization of the milk sugar or lactose in the finished product. The average chemical composition of a standard ice-cream mix has been given in table 9.5.

Table 9.5 Approximate composition (Percentage by weight) of some types of ice-cream

Consultant	Dairy Ice Cream	Ice Milk	Sherbet	Ice lolly
Milk fat	10	4	2	0
Non fat milk solids	11	12	4	0
Added Sugar	14	13	22	22
Addetives	0.4	0.6	0.4	0.2
Edible energy KJ / 100 ml	390	300	340	370
*% over run	100	85	50	0

* means the relative increase in volume by air incorporation or beaten in

Source : Pieter Walstra et. al

The average chemical composition of some commercial varieties of ice-cream that is being marketed by the manufacturers has been given in table 9.6, whereas the ice cream mix according to BIS specification has been given in table 9.7. Further an attempt has been made to present the FSSAI (2017) standards for different varieties of ice cream, frozen dessert and frozen confection have been presented in table 9.8 for the better understanding of the students of dairy science and the dairy enterprenuers.

Table 9.6 Approximate composition of commercial ice-cream (as a percentage)

Type of Ice cream	Milk fat	Milk solids not fat/ serum solids	Constituents		Total Solids
			Sugar	Stabilizer & emulsifier	
Economy Ice Cream	10	10 - 11	13 - 15	0.30 - 0.50	35 - 37
Good average Ice cream	12	11	15	0.30	37.5 to 39.0

Source : Outlines of Dairy Technology - S.K. De

Table 9.7 BIS Specifications of Ice Cream (IS : 2002 1964)

Characteristics	Requirements
Weight (g /litre) (min)	525
Total solids (% wt) (min)	36.0
Milk fat (% wt) (min)	10.0
Acidity (% lactic acid) (max)	0.25
Sucrose (% wt) (max)	15.0
Stabilizer/emulsifier (% wt) (max)	0.5
Standard Plate count (Per g)	Not more than 2,50,000
Coliform Count (Per g)	Not more than 90
Phosphatase Test	Negative

Role of the Various Components

Milk fat is of special importance for the flavour and for a solid structure to be formed during freezing and hence for consistency, appearance and melting resistance. A high fat content leads to a dry, almost a grainy texture, whereas a low fat content gives a smooth homogenous, somewhat a slimy texture. Fat gives stability to the ice cream but impairs the whipping ability.

9.2.2 Milk Solids Not Fat (MSNF)

Milk solids not fat contribute to the flavour and are responsible for part of the freezing point depression and for an increased viscosity. The protein partly serves to stabilize the foam lamella during the process of air incorporation. It is essential for the formation of fat globule membranes during homogenisation. Lactose can crystallize at low temperature and the crystals formed should be small in size in order to prevent the "Sandiness". Finally the cooling should be quick during freezing and the temperature fluctuations should be avoided afterwards.

Sugar, very often sucrose is essential for the taste and for the freezing point depression. Too little sugar may result in the formation of too much ice and also makes the ice cream overly or excessively sweet. In order to overcome this part of the sucrose may be replaced by a substitute such as glucose syrup or corn syrup by about 25 percent, which are less sweet and leads to a greater freezing point depression per kg sugar. The sugar also causes a higher viscosity, especially when most of the water has been frozen. However, the most important role of the sugar is that it causes far less water to freeze than otherwise would be the case with the result the consistency of the ice cream is softer and the mouth feels less cold.

Table 9.8 FSSAI (2017) Standards for ice cream, kulfi, chocolate, ice cream, softy ice cream frozen dessert and frozen cenfection

Sl. No.	Parameter	Ice cream, kulfi, or chocolate ice cream or softy ice cream	Medium fat ice cream or kulfi or chocolate ice cream or softy ice cream	Low fat ice cream or kulfi or chocolate ice cream or softy ice cream	Frozen desert or frozen confection	Medium fat frozen dessert or frozen confection	Low fat frozen dessert for frozen confection
1.	Total solids (max) % (m/m)	36.0	30.0	26.0	36.0	30.0	26.0
2.	Weight minimum (g/L)	525.0	475.0	475.0	525.0	475.0	475.0
3.	Milk fat % (m/m)	10.0 (minimum)	more than 2.5 and less than 10.0	2.5 (maximum)	10.0 (minimum)	more than 2.5 and less than 10.0	2.5 Maximum
4.	Milk protein[⊗] Minimum % (m/m)	3.5	3.6	3.0	3.5	3.5	3.0

[⊗]Protein content is 6.38 multiplied by the total nitrogen determined

Note: 1. In case where coating, base or layer of non-dairy ingredients forms separate part of the product only the frozen dessert or frozen confection part shall conform to the respective composition.

2. When any part of frozen dessert or frozen confection is offered for sale in contravention of the requirements the standards prescribed for these types of frozen desserts or frozen confection as per this item shall apply.

The table 9.9 shows the FSSAI (2017) standards for milk ice or milk lolly.

Table 9.9 FSSAI (2017) standards for milk ice or milk lolly

Sl.No	Parameter	Milk ice or milk lolly
1.	Total solids minimum % (m/m)	20.0
2.	Milk fat maximum % (m/m)	2.0
3.	Milk protein®maximum % (m/m)	3.5

®Protein content is 6.38 multiplied by the total nitrogen determined

Note: In case where base or layer of non-dairy ingredients forms a separate part of the product only the milk ice or milk lolly portion shall conform to the above composition.

The role of the stabilizer or more popularly and properly known or called as the thickening agent is not quite clear yet. Among those used are gelation, alginate carrageenan, pectin, locust bean gum, guar gum, Xanthan, carboxymethyl cellulose and mixture. These substances, of course affect the consistency and consequently, also the heat transfer during the freezing. However, these agents may cause the consistency of the product to become somewhat slimy in the mouth.

Emulsifier generally is not needed when more than sufficient protein is present during homogenisation, as it does not play any significant role in foam formation. It serves to stimulate the fat globules to clump and to become attached to the air bubbles. The emulsifiers used include, egg yolk, monoglycerides. Polyoxyethylene sorbitan esters and esters from citric acid and monoglycerides.

Flavouring agents are self explanatory and increases the acceptability of the ice-cream. Naturally the ice crystals are essential for the consistency and for the coolness in the mouth. Moreover, the low temperature causes the sweetness to be less intense. The freezing should be fast and there should not be any fluctuations in the storage temperature such that the crystals should not be too large.

Incorporation of air has important roles to play such as:

1. To make the ice cream light otherwise it would be too rich.
2. To soften its consistency thus making it deformable in the mouth and
3. To moderate the coldness by lowering the rate of heat transfer, otherwise the ice cream would be far too cold in the mouth.

Ice cream is manufactured according different types or varying methods depending upon the nature and type of the premises selected. The methods of manufacture may be classified as under :

1. Home production/Home made ice cream.
2. Small vendors/small scale production.
3. Large scale manufacture or factory production.

9.2.3 Home Production

Ice cream manufactured in the home is usually frozen custard or milk and corn flour frozen by any one type of home freezers that are now available in the market, since it is generally made in small quantities for immediate and private or personal consumption only, the control of home made ice cream is relatively unimportant. The various steps that are involved in the manufacture of ice-cream on a small scale domestically has been detailed in a flow diagram 9.1 for better understanding.

Small Vendors

The mix generally being used by these vendors consists of:

1. Milk and cornflour with Vanilla flavouring.
2. Milk with ice cream mix or powder and flavouring
3. Curstard slightly flavoured but manufactured in larger quantities than is used for home production.

Some of the small manufacturers may also use eggs, dried eggs or egg substitutes as a filler. Some of the common ice-cream powders contain gelatine, finely ground sugar and quite often cereal fillers.

The starchy ingredients are usually mixed into a smooth paste, with the addition of milk or other liquid during the preparation and the whole mix being warmed. The flavouring substances and sugar are added during this warming process and the mix is constantly stirred in order to prevent the formation of lumps and scorching till the boiling point is reached. After simmering the mix for 10 minutes at this stage, the mix is strained through a cheese cloth and then allowed to stand over night, which is a general practice among these small vendors who do not possess the necessary apparatus for the rapid cooling of the mix. When eggs are to be incorporated as a constituent of the mix, the yolks are separated from the white, and are then washed and stirred in after the simmering has been completed, while the white is added on the following day before freezing the mix.

The mix thus prepared is frozen on the next or following day in the morning prior to the delivery. A hand operated freezer, which is usually rotated at a speed of 100-150 revolutions per minute (rpm) is used for this purpose. The speed of rotation, however, depends upon the size and type of freezer employed. The freezing agent is usually a mixture of ice and salt in the ratio of 10 : 1 : 1 (10 Ice - 1 Salt - 1 Water)

Flow Diagram 9.1 Manufacture of Ice cream

Selection of Ingredients

Composition of the mix / Figuring the mix

↓

Making the mix

↓

Pasteurizing the mix (65 °C or 155 °F for 30 min)

↓

Homogenisation of the mix
(1st stage - 2500 - PSI, 2nd stage - 500 PSI)

↓

Cooling of the mix (0 - 4 °C / 32 - 40 °F)

↓

Ripening / ageing of the mix (0 - 4 °C / 32 - 40 °F) (4 - 24 hrs)

↓

Freezing the mix (– 4 to – 5 °C) or (23 - 25 °F)

↓

Packaging of ice cream

↓

Hardening of Ice cream (– 23 to – 29 °C) or (– 10 to – 20 °F)

↓

Storage of Ice cream (– 23 to – 29 °C) or (–10 to – 20 °F)

and the resultant brine temperature is in the neighborhood of 21°F. Once the freezing operation is commenced it should not be ceased until the ice-cream has attained the required consistency, very rapid freezing is not desirable with such type of freezers and when rotation of the freezer becomes difficult, the operation must be stopped otherwise, there will be a loss of over run because of the loss of air contained being worked out.

9.2.4 Large Scale or Factory Production of Ice-cream

The flow chart or the diagram given indicates that there are several steps or stages in the manufacture of the ice cream on a large scale by the ice-cream factories which will be described very briefly.

1. **Selection of Ingredients:** The first and fore most important aspect in the manufacture of ice cream is the selection of the raw materials that are needed and depends upon:

1. Availability of milk products
2. Perishability of the products
3. Convenience in handling the products
4. Effect on flavour, body and texture of the ice-cream
5. Cost and
6. Equipment available.

The most important factor in the successful manufacture of the ice-cream undoubtedly is the selection of good quality ingredients. A clean, fresh creamy flavour in ice-cream can be secured only by the use of products which have been carefully produced and handled. The ice-cream ingredients may be grouped into two categories namely.

1. **Dairy products-which include:**
   - **(a) Fat the source of which may be**
     1. Sweet cream-which is most desirable
     2. Frozen cream
     3. Plastic cream
     4. Unsalted butter and
     5. Butter Oil
   - **(b) Milk Solids Not Fat (MSNF) : Which include**
     1. Skim milk
     2. Skim milk powder which is most frequently used either in spray dried or flaked form
     3. Condensed Skim milk either plain or sweetened and
     4. Sweet cream butter milk.

In addition to those mentioned above the following ingredients which supplies both the fat and the milk solids not fat are also being used in the manufacture of the ice cream depending upon their availability and the economy in the cost of production of ice cream. They are:

1. Whole milk
2. Whole milk Powder
3. Condensed whole milk either plain or sweetened and
4. Evaporated milk

2. The Non-Dairy Products: That are generally incorporated in the ice-cream mix during the manufacture or production of ice cream on large scale or by the ice cream factories are as under:

(a) Sweetening agents such as:

1. Cane or beet sugar the most common being sucrose
2. Corn Sugar (Dextrose)
3. Corn Syrup Solids (Dextrose maltose)
4. Corn Syrup
5. Invert Sugar (glucose + fructose) and
6. Sachccharin

(b) Stabilizers which include:

1. Gelatin
2. Sodium alginate, a modified form of vegetable origin which is being used very widely.
3. Guargum-which is of Indian origin and
4. Others

(c) Emulsifiers - Mono or di-glycerides of fat forming fatty acids

(d) Flavours : These include:

1. Vanilla-the most popular flavour
2. Chocolate
3. Strawberry
4. Pine apple
5. Lemon
6. Banana
7. Mango and
8. Orange etc.

(e) Colours

1. Yellow
2. Green
3. Pink etc.

(f) Egg solids-to improve the whipping quality and or ability

(g) Fruits and nuts such as

1. Apple
2. Banana
3. Mango

4. Pine apple
5. Grape
6. Almond
7. Pistachio
8. Cashewnut
9. Walnut and
10. Ground nut etc.

9.2.5 Figuring the Mix/Composition of the Mix

In the manufacture of ice cream confirming to the legal standards, a knowledge pertaining to the calculation of the ingredients of ice cream mix is helpful in balancing the mix properly in establishing and maintaining the uniform quality. The ice-cream mix may be devided into two groups of categories namely simple which require the least calculation and are made of ingredients each of which supplies one constituent, whereas the complex mixes are more difficult to calculate. Before calculating the mix it is necessary to:

1. Decide upon the composition of the mix to be made.
2. Decide on the amount of mix to be made in the batch at one time.
3. Choose from the available ingredients those that will give the desired quality characteristics and composition at the lowest cost
4. Be familiar with the composition of the ingredients to be used.

Cream, plastic cream, or butter which are high in fat is generally incorporated as one of the ingredients because the ice-cream contains a higher percentage of fat than the fluid milk. Similarly the cream also contains more solids not fat than milk, substances which are high in these solids such as condensed skim milk, condensed whole milk or skim milk powder are required. The fluid milk is usually used as the basic ingredient in the ice-cream mix. Calculation of the mix involves finding the required weights of these ingredients along with the necessary sugar and stabilizer.

A typical calculation method which is applicable to all types of mixes along with the solution has been given hereunder.

Problem

Prepare an ice cream mix containing fat 10 percent, serum solids 12 percent, sugar 14.5 percent and stabilizer 0.3 percent with the following ingredients. Whole milk testing 7.2 percent fat and 9.0 percent serum solids cream testing, 35 percent fat and 5.6 percent serum solids, skim milk powder testing 0.5 percent fat and 98 percent serum solids sugar and stabilizer.

Solution : Let

Whole milk required = X kg

Cream required = Y kg

and skim milk powder required = Z kg

Hence,

$$X = \left[\frac{7.2}{100}\right. + Y\left[\frac{35}{10}\right. + Z\left[\frac{0.5}{100}\right. = 10 \text{ - I (Fat Equation)}$$

$$X = \left[\frac{9}{100}\right. + Y\left[\frac{5.6}{100}\right. + Z\left[\frac{98}{100}\right. = 12 \text{ - II (Serum Solids Equation)}$$

X + Y + Z + 14.5 + 0.3 = 100 – III (Weight Equation)

Solving the above equations

X = Y = Z =

9.2.6 Properties of the Mix

The properties of the ice-cream mix are of utmost practical importance which include:

1. **Viscosity:** It is considered as an important property of the ice-cream mix and seems essential for proper whipping and the retention of the air. There are two types of viscosity that exists in the ice-cream mix namely, apparent viscosity which is a thickened condition that disappears with agitation and basic viscosity which remains after the apparent viscosity disappears. The viscosity of the ice-cream mix is influenced by:
   1. Composition
   2. Kind and quality of the ingredients
   3. Processing and handling of the mix which include
      (a) Pasteurization
      (b) Homogenisation and
      (c) Ageing
   4. Concentration of the total solids and
   5. Temperature
2. **Acidity (Lactic) and pH:** The normal acidity of the Ice-cream mix is dependent upon the serum solids content which is calculated by the formula

$$\text{\% acidity of mix} = \text{\% acidity of milk} \times \frac{\text{\% Serum Solids in mix}}{\text{\% Serum Solids in milk}}$$

The normal pH of the mix is about 6.3 and as a thumb rule method, the acidity of the mix can be determined by multiplying the serum solids with 2 and then deviding it by 100. Acidity and pH are related to the composition of the mix and an increase in the milk solids not fat results in an increase of the acidity lowering the pH.

3. **Mix Stability:** It refers to the resistance to separation or stability by the milk proteins in an ice-cream mix. Instability results in the separation of milk proteins as coagulated or precipitated material in the mix, thus giving the ice-cream a curdled appearance on melting. This defect may be the result of various factors that affect the colloidal stability of the milk proteins such as:
   (a) High mix acidity
   (b) Low citrate and phosphate content
   (c) High calcium and magnesium content
   (d) High homogenisation pressure
   (e) High heat treatment (Pasteurizing temperature)
   (f) Low ageing time/period.
   (g) Destabilizing effect of freezing
   (h) Particle size, charge and hydration
4. **Specific Gravity:** The specific gravity or density of an ice-cream mix varies with its composition and may range from 1.05 to 1.12, which can be calculated by using the following equation.

   Specific gravity at 16 °C or 60 °F is

$$\text{Sp gr} = \frac{100}{\dfrac{\%\,\text{Fat}}{(0.93)} + \dfrac{\%\,\text{Sugar} + \%\,\text{MSNF} + \%\,\text{Stabilizer}}{(1.56)} + \dfrac{\%\,\text{Water}}{(1.0)}}$$

5. **Surface Tension:** It is related to the attraction of the molecules of a liquid with each other at its surface. The greater the attraction between the molecules the higher the surface tension fluid vice versa. The normal surface tension values of ice-cream mix may vary from 48-53 dynes sq. cum at 20 °C or 69°F.
6. **Freezing Point:** It is dependent on the soluble constituents and varies with its comparison. The constituents which affects the freezing point directly of the mix are (a) Sugar (b) milk sugar (c) milk salts (d) any other substance that are in true solution. The normal freezing point of the ice cream mix of normal composition is about 27.5°F.

7. **Whipping Rate:** The whipping ability according to the present hypothesis is based on the tensile strength and the strength of the lamella i.e. the walls around the air cells. High whipping rate means the ability to whip rapidly giving a high over run and now it is definitely known that the differences in the whipping ability cannot be explained on the basis of viscosity. The whipping ability is improved by a high processing temperature, proper homogenisation and ageing the mix for 2-4 hours. The rate of whipping is measured by calculating the "Over Run" at every minute interval, while the mix is being frozen in a batch freezer. Under normal conditions within 3-5 minutes after the freezing process commences, the mix is frozen and within 7 minutes an over run of 90 percent is obtained. In some of the mixes, which have a more rapid whipping rate, 90 percent of over run is obtained in about 5 minutes or even less, whereas mixes requiring 8 minutes or more time to reach the 90 percent over run are considered to have a slow whipping rate.

9.2.7 Preparation of the Mix

The milk products and other ingredients selected must be of good quality and combined so as to produce the desired body and delicately blended flavour in order to make or prepare a good ice-cream. It is essential that the mixing process may be carried out in a speedy manner, such that a good quality product with the properly balanced mix otherwise a poor quality product may be produced. All the ingredients should be accurately weighed or measured. The mixing process may range in its scope from the small batch operation to the large automatic continuous operation, where many of the ingredients are metered into the batch. Modern computers can be used to control the cost and quality of ice cream mixes by the mathematical technique known as linear programming or least cost analysis.

In the preparation of the ice-cream, all the liquid ingredients are placed in the jacketed vat provided with a mechanical power stirrer commencing the agitation and heating simultaneously at once. The dry ingredients including skim milk powder sugar and stabilizer are added agitating the liquid material and before the temperature reaches 49 °C (120 °F). In order to avoid the lumpiness and proper suspension of the dry ingredients, they may be mixed thoroughly with part of the sugar before adding to the liquid or by slowly adding these substances to the liquid. When gelatin is used as the stabilizer, it is always best to add it after thouroughly mixing with an equal amount of sugar and before the temperature of the liquid material reaches 49 °C/120 °F. Similarly when sodium alginate (Dariloid) is used as the stabilizer, it should be added to the liquid mix only when the temperature has reached to at least 66 °C or 150 °F. Butter, plastic cream, frozen cream and other frozen products are intended to use in the preparation of the ice-cream, they should be cut into very small pieces in the first instance and should be added only after giving sufficient time allowing them to melt completely before the pasteurizing temperature is reached.

9.2.8 Pasteurizing the Mix

Pasteurization of the ice-cream mix is considered as highly desirable, essential and compulsory because it destroys all the pathogenic or disease producing organisms thus safe guarding the health of the consumer.

The merits of the Pasteurization are:

1. It renders the Ice-cream mix completely free from all the pathogenic micro-organisms.
2. It helps to give a uniform blend by dissolving the ingredients of the mix.
3. It gives an improvement to the flavour of the mix.
4. It improves the keeping quality and
5. It gives a more uniform texture and body to the product.

The BIS specifications (IS 2002 : 1964) for the pasteurization temperature time combination of ice-cream mix are:

Batch method - 68.5 °C (155 °F) for not less than 30 minutes.

HTST method - 80 °C (175 °F) for not less than 25 seconds.

9.2.8.1 Homogenisation of the Mix

The main or sole object of homogenisation is to make/reduce the fat globule size to a very small diameter preferably not more than 2μ. In order to make a permanent and uniform suspension of the fat in the ice-cream mix which is essential. The advantages are:

1. Homogenisation prevents the separation of fat during ageing process.
2. It produces a more uniform ice-cream with a smoother texture.
3. It improves the whipping ability.
4. It shorterns/reduces the ageing period/time.
5. It decreases the risk of charring occurring in the freezing process or freezer and
6. It leads to the use of a limited amount or less quantity of stabilizer.

Generally the mix is homogenised at temperatures ranging from 63-77 °C (145-170 °F). Homogenisation increases the formation of clumps of fat globules. Viscosity and the freezing time in the batch freezers at low temperatures. The homogenisation pressure generally depends upon several factors such as: (a) viscosity (b) composition of the mix (c) stability of the mix (d) temperature used and (e) the type and make of the homogeniser.

A homogenisation pressure of 2500-3000 PSI with one value or 2500-3000 PSI at the first stage and 500 PSI in the second stage generally gives good results for an average mix with 3 to 12 percent fat.

9.2.8.2 Cooling and ageing the Mix

It is very essential to cool the mix immediately after homogenisation to the temperature ranging from 0-5 °C or 32- 40 °F, by holding the mix in the ageing tanks until used.

Ageing refers to holding the mix at a low temperature for a definite period of time before freezing which is generally carried out by using either surface or cabinet type coolers. Under normal commercial conditions the ageing process may be 3-4 hours which has the following advantages.

1. It improves the body and texture of ice-cream.
2. It improves the whipping ability/capacity of the mix.
3. It increases and gives a maximum over run.
4. It increases the melting resistance.

9.2.8.3 Freezing the Mix

In the manufacture or preparation of the ice-cream, freezing is one of the most important operation, which follows immediately after the ageing process because the quality, palatability and yield of the finished product are depend. The freezer or freezing process has three functions namely:

1. To cool the mix to a low temperature so that the product may be hardened.
2. To incorporate the pre-determined amount of air into the mix to provide proper swell or over run and
3. To mix flavours, fruit etc into the mix as it is being frozen.

There are several types of freezers which are classified as:

1. Batch freezer - Horizontal, Direct expansion using either Freon or Ammonia.
2. Continuous freezer - Same as above.
3. Soft serve freezer - Batch and Automatic continous freezers of the direct expansion type.

Rapid freezing is essential in order to obtain a smooth product with the formation of small ice crystals. A continuous freezer accomplishes this task in few seconds whereas, the batch freezer takes 6-10 minutes. It is desirable to freeze the ice-cream as stiffly as possible and yet have it liquid enough to be able to draw it out from the freezer. There are several factors which influence or affect the freezing time and are mainly due to:

I. Mechanical which inturn depends upon

(a) Type and make of the freezer
(b) Condition of the freezer wall and blades
(c) Speed of dasher
(d) Temperature of the refrigerant
(e) Velocity of refrigerant passing around the freezing chamber
(f) Over run desired
(g) Temperature at which the ice-cream is drawn.
(h) The rate at which the freezer is unloaded

II. Character of the Mix

(a) Composition of the mix
(b) Freezing point of the mix
(c) Acidity content of the ingredients
(d) Nature of the ingredients particularly those representing the fat
(e) Processing method of the mix
(f) Nature and amount of flavouring materials incorporated or added to the mix.

9.2.9 Freezing Process-changes taking Place

The main important function of the freezing process is to freeze a portion of the water and to incorporate the air into the mix which involves lowering of the temperature from ageing temperature to that of freezing point and cooling the ice-cream from the temperature at which it is drawn from the freezer to the hardening room temperature. The ice cream mix when put into the freezer, its temperature drops very rapidly taking less than a minute or two, by removing the heat before the formation of the ice crystals. In the mean time the rapid agitation reduces the viscosity by partially destroying the gel structure and also by breaking up of the fat globule clusters with the incorporation of the air into the mix.

The liquid water present in the mix changes to ice-crystals, appears in the mix, when the freezing point is reached. These crystals of ice are practically pure water in the solid state/form and thus resulting in the concentration of sugar and other solutes. The increase in the concentration of those solutes results in the slight decrease of the freezing point of the liquid portion such that the temperature is further lowered before the formation of any more ice crystals, until the concentration is so great that further freezing will not occur. The influence of type of freezer on the freezing time and temperature as well as the percentage of water frozen in ice-cream mix at different drawing temperatures has been presented in the tables 9.10 and 9.11.

Table 9.10 Effect of freezing time and drawing temperature for different freezers

Type of freezer	Approximate freezing time (to 90 percent over run)	Approximate drawing Temperature	
		°C	°F
Batch	7 Min	– 4 to – 3	24 to 26
Continuous	24 Sec	– 6 to – 5	21 to 22

Source : Dairy Science & Technology, Pieter Walstra et.al

Table 9.11 Effect of different drawing temperature (freezing) on the percentage of water frozen in the ice-cream mix

Temperature °C	Temperature °F	Percentage of Water frozen	Temperature °C	Temperature °F	Percentage of Water frozen
– 4	25	33	– 7	20	59
– 4.5	24	41	– 7.5	19	62
– 5	23	47	– 8	18	64
– 5.5	22	52	– 8.5	17	67
– 6.0	21	56	– 26	– 15	90

Source : Dairy Science & Technology, Pieter Walstra et.al

During the first phase of the freezing process accounts for the freezing of 33 to 67 percent of water depending on the drawing temperature whereas in the second phase of the hardening process may account for freezing of another 23-57 percent which is again dependent upon the drawing temperature.

9.2.9.1 Over run

In the manufacture of ice-cream overrun is extremely important and all possible care must be taken to ensure that too high over run is avoided otherwise the ice cream will melt in the mouth before it is tested. The over run is defined as the volume of ice-cream obtained in excess of the volume of the mix which is generally expressed as the percentage and the increased volume is mainly composed of the air that has been incorporated during the process of freezing. The amount of air which is to be incorporated into the mix depends upon the composition of mix, the way it is processed, texture and palatability necessary to a good quality product. In order to secure uniform over run and yield the following points needs more attention.

1. Uniformity in refrigerant temperature and rate of flow of refrigerant.
2. The use of over run testers.
3. Uniform make etc of the freezer for operator.

4. Limited number of freezers per operator not too many freezers.
5. Hopper system for filling containers in case the batch freezers are used.

The control of over run is very important and should be nearly constant as much as possible from batch to batch and day to day. There are two basic or fundamental methods for calculating percentage of over run namely by volume and by weight.

1. $\% \text{ Over run} = \dfrac{(\text{Volume of Ice - cream}) - (\text{Volume of Mix})}{\text{Volume of Mix}} \times 100$

2. $\% \text{ Over run} = \dfrac{\text{Weight of Ice - cream} - \text{Weight of Mix}}{\text{Weight of Mix}} \times 100$

9.2.9.2 Hardening

This process may take place either before or after packaging, and it is of considerable importance because the fine texture of the ice-cream that is leaving the freezer may be destroyed by improper hardening. The frozen cream is stored at a temperature of –18°C or O°F to - 30°F and it should be allowed to stand for at least 12 hours before being sold or distributed for sale. Quick hardening is desirable because slow process of hardening favours the formation of large ice crystals and coarseness. The factors which may effect the hardening time are:

1. Size and shape of the ice cream packages
2. Speed of (cool) air circulation.
3. Temperature of cooling air
4. Section of the hardening room
5. Temperature of the ice-cream drawn from the freezer.
6. Composition of the mix and
7. Percentage over run in the ice-cream being hardened.

After the ice cream is hardened it may be marketed immediately or it may be stored for a week or two at the most. The operation of the storage rooms is similar to that of the hardening room. The temperature should be maintained uniformly within the range of – 23 °C to – 18 °C (– 10°F to 0°F) and the packages should be piled very close, to delay the changes in the ice-cream temperatures.

Hardening rooms should be insulated heavily with 6 to 9 inches of compressed cork in the form of slabs which are joined with hot bitumen which is odourless.

Forced air circulation should be provided because this will ensure rapid hardening and the texture of the final product will be superior.

9.2.9.3 Packing

As already indicated, the ice cream may be packed either before or after hardening. Generally the ice-cream when it is drawn from the freezers is collected in the containers that give the desired shape or size for convenient handling during hardening and marketing processes. The requirements for the package of ice-cream are:

1. Protection against contamination.
2. To give an attractive appearance in order to promote the sales.
3. It should be easy to open and for reclosure.
4. It should not pose any problem in its disposal.
5. It should give protection against loss of moisture and temperature fluctuations.

9.2.9.4 Judging and Grading

The Judging and grading procedure of the ice-creams includes:

1. Sampling which is taken at random at temperature ranging from – 15 °C to – 12 °C from the hardening room and allowing it or placing it in the dispensing room/cabinet several hours prior to judging. The sequence of observations to be made during the process of judging are:
   (a) Examination of the container for any package defects
   (b) Colour of ice-cream for its intensity and uniformity as well as its matching/ compatability with the flavour.
   (c) Sampling the ice-cream by observing the manner in which the product cuts, evenness of cutting, resistance offered during cutting, presence of ice particles, and its melting properties.
   (d) Judging the ice-cream for its taste, texture and flavour characteristics.
   (e) Observing the melting properties such as creamy, curdled, foam or watery.
   (f) Lastly determine the bacterial count.

An ideal ice-cream of a high grade should be packaged in an attractive container possess a typically pleasant and desirable flavour, having a close, smooth and uniform body and texture with desirable melting properties, natural uniform colour and a low bacterial count.

The score card as recommended by the American Dairy Science Association (USA) is given in the table 9.12.

Table 9.12 Score card for ice cream (ADSA)

Particulars/Items	Perfect Score
Flavour	45
Body and texture	30
Bacteria	15
Colour and package	5
Melting Quality	5
	Total 100

Source : Outlines of Dairy Technology - S.K. De

9.2.10 Defects in Ice-cream

It is very common practice that many defects are noticed in the finished products during the manufacture of the ice-cream. These defects may arise due to the usage of low quality ingredients, improper mixing or faulty methods of manufacture, package and storage. Some of the common defects have been summarized and are given below.

I. Defects due to flavour : these include:
   - (a) High
   - (b) Low
   - (c) Acid/sour
   - (d) Bitter
   - (e) Cooked
   - (f) Flat
   - (g) Un-natural
   - (h) Rancid
   - (i) Oxidised/Oily/metallic/Tallowy

II. Defects in the body of the ice-cream
   - (a) Crumbly
   - (b) Soggy
   - (c) Weak

III. Defects in the texture of the ice-cream
   - (a) Buttery
   - (b) Coarse/icy

(c) Fluffy

(d) Sandy

IV. Defects in Melting quality

(a) Curdy melt down

(b) Foam melt down

(c) Slow melting

V. Colour defects

Un-natural - due to wrong selection of the colour

VI. Miscellaneous : Shrinkage in the body of the ice-cream.

9.3 Butter

Butter is considered as the staple food and is generally made from cream by churning and working contains about 80 percent fat which is partly crystallized. The art of butter making has a long history which was generally confined to the farmhouse and is considered as an article of commerce and a sign of wealth. The butter, typically is a product originating from regions having a temperate climate.

Butter may be defined as a concentrate fat, obtained by churning cream, and gathering the fat as a compact mass and then working on it. According to the PFA (1976) rules the table or creamery butter is the product obtained either from cow, buffaloe or a combination thereof. It is a product prepared from cream or curd obtained from either cow, buffaloe or a combination thereof with or without the addition of salt and annatto or carotene as colouring matter. It should be free from other animal fats, wax and mineral oils, vegetable oils and fats. It must contain not less than 80 percent milk fat by weight, not more than 3 percent by weight of common salt and not more than 1.5 percent by weight of curd. Diacetyl may be added as a flavouring agent which should not exceed 4 ppm, calcium hydroxide, sodium bicarbonate, sodium carbonate, sodium poly phosphate may be added but must not exceed 0.2 percent by weight of butter as a whole.

There are many varieties or kinds of butter which are found in the market and differs with each other with the type of cream used as well as the variations in the manufacturing process. The different kinds of butter may or may not be salted unless and otherwise specifically mentioned. The various or different kinds of butter are as under.

1. **Pasteurized cream butter:** This is made from pasteurised sweet cream which has a mild flavour as compared to the butter which is made from the similar type of cream and is not pasteurized.

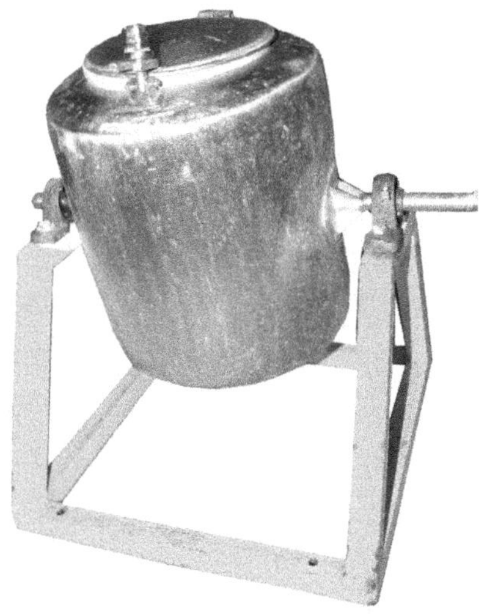

Fig. 9.4 Butter Churn

2. **Ripened cream butter:** It is made from the cream which has developed a pleasant delicate, aroma before churning by ripening and this delicate flavour is referred to as "Real butter flavour".
3. **Un ripened cream butter:** As the name suggests it is made from unripened cream and the flavour of such butter is usually mild.
4. **Salted butter:** Butter to which salt has been added.
5. **Un salted butter:** It contains no added salt.
6. **Sweet cream butter:** In this case, the acidity of the churned cream does not exceed 0.20 percent.
7. **Sour cream butter:** Made from cream which has more than 0.20 percent acidity.
8. **Fresh butter:** It refers to the butter which has not undergone or subjected to cold storage condition. Generally the fresh butter is not kept for more than three weeks.
9. **Cold Storage butter:** In this case the butter is stored at a temperature of about –18 °C (0 °F) for some time. Generally the cold storage butter is from one to six months old when offered for retail selling.
10. **Dairy butter (USA):** It is made on form conditions from unpasteurized sour cream which has not been standardised for acidity. This butter generally has a sour flavour due to high acid content of the cream.
11. **Creamery butter:** It is made in a creamery or a dairy plant, which is more uniform in its quality as compared to the dairy butter.

9.3.1 Composition

The table or creamery butter, according to PFA rules (1976) should contain not less than 80 percent fat, not more than 1.5 percent curd and 3.0 percent common salt. The composition of buttter varies slightly according to the principles and practice of manufacture, however, the average composition of the Indian butter and that of U.K. have been given for comparison in table 9.13.

Table 9.13 Composition of butter (%)

Constituents	Indian	U.K
Butter fat	80.2	82.5
Curd	1.0	-
Casein	--	1.5
Lactose	--	2.0
Salt	2.5	2.0
Water	16.3	12.0
	100.0	100.0

Source : Outlines of Dairy Technology - S.K. De

Further air attempt has been made by the authors to up date the knowledge pertaining to the standards of butter in accordance with the FSSAI (2017) act as published in the Gazette-govt. of India for the benefit a better understanding of the students of dairy science in the table 9.14.

Table 9.14 FSSAI (2017) standards for butter

Sl.No	Parameter	Table Butter	White Butter cooking butter
1.	Moisture maximum % (m/m)	16.0	-
2.	Milk fat minimum % (m/m)	80.0	76.0
3.	Milk solids-Not fat maximum % (m/m)	2.0	-
4.	Common salt minimum % (m/m)	3.0	-

Note: 1. where butter is sold or offered for sale without any indication as to whether it is table butter or white butter the standards of table butter shall apply

2. The extracted fat from butter shall meet the standards for R.M. value and butter refractometer reading as prescribed for ghee.

The method of manufacturing/preparing the butter from raw milk/cream has been summerised and presented in a combined form as a flow diagram 9.2.

Flow Diagram 9.2 Method of Manufacture of butter

Receiving milk
↓
Grading, Weighing, Sampling, Testing
↓
Skim Milk ← Skimming/Separation (Centrifugal)
→ Cream

Receiving of Cream
↓
Grading, Weighing, Sampling, Testing
↓
Neutralization
→ Cream

Cream
↓
Standardisation (35 - 40% Fat)
↓
Pasteurization (82 - 85 °C - No hold) or Vacreation

Pasteurization/Vacreation → Cooling (20 - 22 °C)
↓
Ripening (20 - 22 °C)
→ Ageing (5-10 °C)

Pasteurization/Vacreation ↓
(Cooling (5 - 10 °C)
↓
Ageing (5-10 °C)
↓
Churning
↓
Separating → Butter Milk
↓
Butter Grains
↓
(Water 12 °C) → Washing
↓
(Kneading)
↓
(Water salt) → Working Standardising → Butter Milk
↓
Working Standardising → Butter
↓
Packaging & Storage (–23 to 29 °C)

Cold Storage (7-10 days)
↑
Homogenising
↑
Packaging

The most important basic or specific requirements for the product and its manufacture are as under :

1. **Flavour:** Avoid the 'off' flavours of the fat especially those caused by the lypolysis, and due to volatile contaminants.
2. **Shelf life:** Spoilage by the microorganisms may result in several off flavours such as putrid, volatile acid, yeasty, cheesy, and rancid. Lipolysis causes a soapy rancid flavour and in cultured cream butter spoilage generally involves molds and yeasts. Auto oxidation of the fat can occur due to prolonged storage even at low temperatures (–20 °C) leading to a fatty or fishy flavour.
3. **Consistency:** Butter derives its firmness largely from the fat crystals that are aggregated into a close network in order to retain its shape, likewise oiling off should not occur. On the contrary the butter should be soft enough so as to be spreadable easily on the bread. The consistency sometimes may cause problems, because the firmness and the spreadability are strongly dependant upon the composition of the fat and temperature.
4. **Colour and Homogeneity:** Very rarely these pose problems.
5. **Yield:** Some amount of fat is lost in skim milk and also in butter milk. If the water content is below the legal standards or limit this also amounts to loss of yield.
6. **By-Products:** In general some quantity of butter milk is desirable but not always because of the lack of demand. Sour cream butter milk can be used as a beverage or as animal feed due to its poor keeping quality because of the development of an oxidized flavour. Sweet cream butter milk can be incorporated more readily in some products.

9.3.2 Cooling and Ageing of Cream

Cream is cooled by lowering the temperature and aged by holding at this low temperature for a few hours after pasteurization in order to make the churning possible. The cream will not churn until and unless the fat globules has at least partially solidified or crystallized. Under Indian conditions, the optimum temperature for the cooling and ageing of cream preferably should be 5 - 10 °C or 41 - 50 °F. High cooling and ageing temperatures of cream shortens the period of churning, resulting in the yield of large fat losses in butter milk, producing a butter a relatively soft bodied butter. Excessive cooling of cream abnormally to low temperature and also ageing it near about the same temperature makes the fat globules very firm/solid such that the coalescence with each other making difficult during the churning process. On the contrary, low cooling and ageing temperatures prolong the churning period, decrease fat losses producing a firm bodied butter which has a satisfactory standing up capacity. The ageing period of cream should be at least 2-4 hours preferably over night or 15-16 hr, during which the crystallization of fat in the fat globules should be completed satisfactorily. The optimum temperature for cooling and ageing of cream depends upon:

(a) Composition of fat
(b) Size of fat globules
(c) Fat percentage of cream
(d) Period or time of ageing
(e) Temperature of churning and
(f) acidity of the cream.

It is important to note that 'shock' or immediate cooling of warm cream with ice or cold and chilled water is not conducive to proper crystallization because the temperature rises subsequently due to latent heat. Owing to this, there is an exhaustive churning, resulting in an unsatisfactory, weak bodied butter.

9.3.3 Ripening of Cream

It refers to the fermentation of the cream with the help of the desirable starter cultives. Fermentation in the strict sense means the decomposition of the carbohydrates i.e milk sugar by the desired microorganisms only. The most important fermentation is the production of lactic acid or souring of milk, whereas the less important are those for the production of flavour etc. The main objective of ripening of cream is to produce butter with a pleasing pronounced, characteristic flavour and aroma uniformly from day to day and to obtain exhaustive churning with a low fat loss in the butter milk. The typical flavour of butter from ripened cream is mainly due to the effect of diacetyl and to a little extent due to acetic and propionic acids. There is no diacetyl production in the sweet cream. The normal diacetyl is produced from its mother substance acetyl methyl carbinol, and its normal content in the ripened cream butter on an average is about 2.5 ppm and rarely over 4.0 ppm.

9.3.4 Churning

Churning of cream consists of agitation at a suitable temperature until the fat globules adhere, forming larger and larger masses and until a relatively complete separation of fat and serum occurs. The object of churning cream is to produce butter. The fat exists in the form of an emulsion which is a continuous phase in milk/cream, which is fairly stable and there is no formation of butter as long as it remains intact. The factors that contribute towards the stability of the fat in skim milk emulsion are:

1. Force of surface tension.
2. Phenomenon of adsorption.
3. Electric charge on the fat globule and
4. Viscosity of the cream

Churning of cream : good churnability refers to and is dependant upon:

(a) *Ease of churning* refers to clear breaking stage and churning until the grains of butter are of the correct size.

(b) *Completeness of churning* means the exhaustiveness of churning, which also refers to fat losses in butter milk.

(c) *Satisfactory washing* - is possible only when the butter grains are sufficiently firm.

(d) *Optimum churning period* - which in turn affects the firmness of the body and exhaustiveness of churning.

There are several factors, which influences the churnability of cream and body of butter which are as under.

1. Chemical composition of the fat
2. Size of the fat globules
3. Viscosity of the cream
4. Temperature of cream at the time of churning.
5. Fat percentage of cream
6. Acidity of cream
7. Load of churn
8. Nature of agitation
9. Speed of churn
10. Pre-churning holding period.

9.3.5 Theories of Churning

There are three main theories on the churning process of cream into butter namely.

1. Fisher and Hooker's phase reversal theory.
2. Rahn's foam theory and
3. Kings modern theory

Fisher and Hooker's Phase Reversal Theory

Churning is a process of phase reversal i.e. changing an oil in-water type emulsion such as cream to a water in oil type emulsion similar to butter in accordance with this theory. Agitation of cream in the churning process results in the coalescence and clumping of the fat globules till the ratio of the surface area to the volume of the fat units becomes so small such that the butter milk no longer contain all of them in a stable form. The fat in water emulsion then suddenly breaks yielding butter grains and free butter milk. The main drawback in this theory is that the butter is not a true water in fat emulsion. Microscopic studies have revealed that a proportion of fat globules in butter are still intact in the worked butter.

Rahn's Foam Theory

The presence of foam or froth is essential for churning accordance with this theory. It postulates that there is a foam producing substance present in the cream or milk, which gradually solidifies when the cream or milk is agitated. Foam is created during the churning period/process. The fat globules, due to the surface tension effects are subjected to concentrate and clump on the foam bubbles. The foam producing substance assumes a solid nature/character resulting in the collapse of the foam and the fat globules then coalesce forming the butter. Foam formation is not desirable or required in the continuous butter making process.

King's Modern Theory

In accordance with this theory, the true explanation of what is happening or taking place during the process of churning appears to be a combination and midway between the two earlier theories. The modern concept of the machanism of the churning process may be summarised as under:

1. The fat is present as clusters or chumps of fat globules and within each globule it is present partly either in solid or liquid form in the cooled cream at the churning temperature.
2. Churning breaks up the clusters and results in foam/froth formation. The fat globules begin to concentrate to some extent in the film around the air bubbles in the foam thus bringing them into close contact with each other.
3. The movement of the fat globules over one another in the foam film and the direct concussion between them causes a gradual wearing away of the emulsion protecting surface layer of the phospholipid protein complex. Then the globules adhere together with each other forming larger and larger particles or clusters and eventually become visible as butter grains/granules. As these granules are formed, they enclose some of the air from the foam and the fat granules are still present mainly in the globular form.
4. The working of butter grains causes the globules to move over one another and some of them yield up a portion of the liquid fat under the effect of friction and pressure, whereas the others are broken up. There is enough free liquid fat present, to enclose all the water droplets and bubbles as well as the intact fat globules.

9.3.6 Churning Operation/Procedure

1. **Preparation of the Churn:** A new churn especially a wooden one requires careful pre-treatment before its use whereas an old churn either wooden or metal requires proper sanitization and cooling in order to make it clean. Scalding with boiling water and cooling immediately with cold water in order to prepare all dry utensils and must be kept wet until their usage. This treatment of the

wood prevents the cream and butter from sticking to it. The churn should be rinsed out with water at a temperature between 50 - 60 °F which is being used for hand churning; the average being 52 - 54 °F.

2. **Filling of cream into the Churn:** The cream should be strained in order to remove the lumps and the chance objects. The amount of cream that is being filled should be preferably slightly below the rated capacity of the churner. The cream for butter making should not be too thick and it should contain about 30% fat and should flow easily of a 'Scotch hand'. The consistency can be adjusted by the addition of cold water. The temperature of the cream should be the same as that of the Churn between 50-60 °F.

3. **Addition of butter colour:** In order to maintain the uniformity of yellow colour in butter throughout the year for consumer satisfaction some standard colours are being added and under Indian conditions it varies from 0-250 ml or more per 100 kg of butter fat. The butter colour should preferably be added to the cream in the churn. The butter colour should have the following properties.

   (a) It should be harmless i.e. non-toxic

   (b) Free from of flavours

   (c) It must be concentrated and permanent

   d) It must be oil soluble.

   There are two types of butter colours namely

   1. Vegetable origin - e.g. Annatto, Carotene
   2. Mineral origin - e.g. Yellow AB and Yellow OB.

9.3.6.1 Operation of the Churn

After filling the churn with the cream not more than one third full of its capacity and with correct consistency and temperature, the churn must be rotated initially for 5-10 minutes. The revolving of the churn should be done in an even manner. A small valve or vent is provided to the churn, which is pressed at intervals during the churning process and the purpose of this ventilator is to maintain an even pressure. The liberated gas is removed once or twice by opening the churn vent and the cream sample is drawn for the fat test. The pressure increases due to the liberation of gases during the churning process and if not released by the ventilator, the increased pressure may cause the plug of the churn to be forced out to resulting in the wastage of the cream. During the churning process, invariably there is an increase in the temperature from 1-3 °C. The indicator glass will show, when the cream thickness and churn feels heavier to handle hence should be turned more slowly which is accompanied by foaming. The "breaking stage" appears when the cream breaks away from the 'spy glass' which becomes more clear. At this stage, the fat in the skim milk emulsion breaks and very small butter granules of the size pin head make their appearance. Sometimes it becomes necessary, especially in the tropics to add ' break water' in order to reduce the temperature of the contents of the churn as well as to control the body of the butter. The amount of break water and temperature depends upon the reduction of the churning temperature required and the churning process is continued further till the butter grains are reached the required/ desired size.

9.3.7 Difficulties/Problems of Churning

In general certain problems are being encountered during the process of churning of cream in order to make the butter, they are:

1. Excessive hardness of fat
2. Small size of the fat globules
3. Excessively thin cream
4. Over loading of the churn
5. Excessively low churning temperature
6. Excessive foaming or abnormal cream.

It has already been stated that some factors are responsible for the fat losses that may occur during the process of cream separation. Similarly there are some factors, which are responsible for the fat loss in the butter milk also during the process of butter making or churning and these include:

1. Cream fat percentage at the time of churning
2. Size of the fat globules
3. Acidity of the cream at the time of churning
4. Physical properties of fat i.e. soft or hard fat
5. Conditions of cooling and ageing of cream
6. Conditions prevalent at the time of churning.

9.3.7.1 Washing of Butter

The rotation of the butter churner or the churning process is stopped when the butter granules about the size of the wheat grains are produced. If the granules are allowed to form into large lumps, the removal or elimination of the butter milk is difficult resulting in the reduced keeping quality of the butter. The butter milk is removed through the sieve of the plug hole of the churn and should be replaced by an equal amount of water to wash the butter. The purpose of washing the butter is:

1. To remove the loose butter milk adhering to the butter grains so as to reduce the curd content thus resulting in the improved keeping quality of the butter. Generally unwashed butter grains contains about 1.1 to 1.5 percent of curd, whereas the washed butter contains curd ranging from 0.6 to 1.0 percent.

2. To correct the defects in the firmness of the butter by proper adjustment of wash water temperatures and
3. To decrease the intensity of certain of flavours, which are the resultant of poor quality cream etc.

The butter milk, when drained completely chilled water is added to the butter grains in the churn. The amount of chilled water added should be equal to the amount of the butter milk drained and the temperature of the chilled water added must be or usually 1-2 °C lower than the churning temperature. After a few revolutions, the wash water is drained out and in general one washing is sufficient for good quality butter, however it is always better to give a second washing such that the drained wash water coming out of the churn is quite clear, indicating that the butter grains have been washed completely and free from butter milk. The quality of the wash water should be clean and bacteriologically and chemically safe. It is always better to use freshly pasteurised and cooled water for this purpose.

9.3.7.2 Salting and Working or Kneading

Salting refers to the addition of salt conforming to the BIS specifications IS - 1845 -1961 with the objective:

1. To improve the keeping quality of the butter.
2. To enhance the taste of the butter and
3. To increase the over run.

The amount of salt added to the butter is at the rate of 2-2.5 percent of the butter fat. The various methods of salting are:

1. Dry salting
2. Wet salting and
3. Brine salting

The brining of butter has two advantages namely

1. No streakiness occurs in the butter from uneven salting and
2. Salt dissolved in water reduces the temperatures of the water by three or four degrees, which is an asset in hot weather conditions.

The temperature of final washing water and that of the butter should be approximately 50 °F before it is taken to the working table. The butter worker should be rinsed down either with water or brine at the same temperature as that of the churn or butter. Any marked difference in the temperature between the wood of the butter worker and the butter will result or cause the butter to 'stick' to the wood during the working process. The main objectives of the working or kneading of the butter are:

1. To dissolve completely, uniformly distribute and incorporate properly the added salt.
2. To expel the remaining butter milk if any and to control the moisture content.
3. To fully incorporate the added or make-up water in the butter and
4. To bring the butter grains together into a compact mass for convenient handling and packaging. It is important to observe that the moisture in the butter is reduced to droplets of microscopic size, which are mostly sterile in nature.

9.3.8 Method of Working/Kneading

The butter, which has been removed from the churner should be placed on the centre of the butter worker, and the wooden rollers carefully pressed. The butter can be handled during working by means of scotch hands. The butter is worked or kneaded in order to expel the moisture and to distribute the water droplets evenly in the butterfat as well as to consolidate to have a compact body a closely knit grain a tough waxy texture and an even distribution of salt. There should no free moisture on the trier plug drawn from the butter. The indicator paper develops coloured spots when the free moisture is present in the butter. The temperature should be controlled throughout the working process. Both over working and under working are harmful and should be avoided, the former damages the body and texture of butter whereas the latter produces a leaky butter. Working increases the air content of butter and the normally worked butter has an air content ranging from 0.5 to 10 ml averaging being 4.0 ml per 100 gm of butter.

The air content of butter is most important because it affects (1) the density of butter (2) the microbial spoilage and (3) the oxidative spoilage. Butter for sale should not contain moisture more than 16 percent according legal standards. Butter produced under farmhouse conditions on an average contains 13-14 percent moisture, when the butter is kneaded for about 5-6 times on the butter worker. Further, the moisture should be mopped up with a clean muslin cloth during the process. In most of the countries the legal standards for the moisture content of butter is 16 percent, whereas in other countries including India it is 80 percent fat.

9.3.8.1 Packaging and Storage

The butter which has been manufactured and is intended for the sale on commercial basis must be removed from the butter worker either manually, or by gravity or by mechanical means and should be shaped attractively into butter bricks depending upon the demand from the consumers. The main objectives of the packaging of the butter are:

1. To offer protection against contamination and damage due to mutilation.
2. To protect the butter against loss in weight and degradation due to deterioration of flavour.

3. To provide ease and safety of transport.
4. To provide a convenient form of disposal and
5. To identify the contents and increase the sales appeal, messages, or appearance etc. The temperature of commercial cold storage of butter ranges from –23 °C to –29 °C or –10 °F to –20 °F. During commercial storage of butter invariably there is some deterioration in the flavour sometimes developing a fishy flavour in salted acid butter. Chemical degradation plays a major part as compared to the bacteriological degradation which is very negligible.

9.3.8.2 Judging and Grading of Butter

The product is usually exhibited in different sizes or blocks for its sale to the consumers. The person who is judging or grading must examine for general appearance effect of packaging in the first place. This is followed by colour, texture and moisture etc. The score card which has been developed and has been approved by the American Dairy Science Association (ADSA) U.S.A. and the score board used by the British Dairy Farmers Association (BDFA), U.K. has been given in the table 9.15

Table 9.15 Score Card of Table Butter

Particulars	ADSA Perfect Score		BDFA Points
Flavour	45	Flavour	50
Body and Texture	30	Texture colour	20
Colour	10	Freedom	10
Salt	10	From moisture	10
Package	5	Packaging	10
Total	100		100

Source : Outlines of Dairy Technology - S.K. De

9.3.9 Defects or Faults in Butter

Butter in common with other feed stuffs may exhibit a variety of defects or faults due to numerous causes, which are chiefly due to the errors in manufacture and to undiscoverable defects that are present in the original raw material. These defects may be classified into two different categories namely.

1. Faults in flavour and
2. Faults in appearance

1. **Faults in Flavour:** Butter may exhibit a variety of taints due to improper handling of the raw materials or of the finished products.

The common taints are:

(a) Sourness (b) Bitterness (c) Food taints
(d) Staleness (e) Rancidity (f) Insipid Taste

(g) Tallowy flavour (h) Fishiness (i) Saltiness
(j) Scalded Taste (k) Heated taste (l) Wooden taste
(m) Oiliness (n) Cheesiness

Most of the abnormal flavours or taints are accentuated by heating.

2. **Faults in appearance:** Butter may possess a poor appearance because of the following factors:

(a) Faulty colour (b) Streakiness (c) Mottled Appearance
(d) Leaking Butter (e) Poor Texture (f) Presence of Yeasts and Molds.

9.3.9.1 Keeping Quality

Definitely it can be stated that butter is one of the best foods with regard to keeping qualities known today. However, the term "Keeping" must be regarded as a relative term because butter will not maintain its original condition for an indefinite period. It was possible to manufacture butter, which would keep for sixty nine months even before the artificial refrigeration has been introduced as well as the important role played by the bacteria in the decomposition of the food are recognised. It was discovered that butter even when stored under cold storage refrigerated conditions is subjected to some kind of spoilage. This was attributed to the acid of sour cream, which acted upon the lecithin in the butter and produced the fishy flavour. This process was greatly accelerated by the presence of metal, traces of which can scarecely be avoided because of the fact that the milk and cream until churned are constantly in contact with such substances.

In spite of all these defects, butter keeps extraordinarily well, under the conditions it is produced and distributed are considered. The reason underlying the satisfactory keeping qualities of the butter is its composition. Over working is however harmful because the texture suffers, but when the product is worked to the optimum limits the bacterial decomposition may be reduced by about 40 percent.

9.4 Cheese

In the past, the production and manufacture of cheese did not develop in our country due to the fact that animal rennet, was not acceptable to the vast majority of the population consisting of orthodex vegetarians since they consider that the slaughter of the cow is a sinful act. However, the discovery of a plant seeds known as 'Withania Coagulans' which coagulates the milk has helped and resulted in the development and production of cheese or paneer cheese like butter acts as a balancing wheel of the dairy industry. The word "Cheese" is commonly used as a collective term for a widely variable products such as fully or partly ripened cheese made with rennet, acid curd cheese, fresh cheese, and even processed cheese. According to FAO/ WHO (1978) the definition of cheese:

"Cheese is the fresh or matured solid or semisolid product obtained by coagulating milk, skimmed milk, partly skimmed milk cream, whey cream, buttermilk or any combination of these materials through the action of rennet or other suitable coagulating agents and by partially draining the whey resulting from such coagulation".

Fig. 9.5 Hardening Tunnel

According to the PFA Rules (1976) cheese (hard) means the product obtained by draining after the coagulation of milk, with a harmless milk coagulating agent, under the influence of harmless bacterial cultures. It shall not contain any ingredients that are not found in milk, except the coagulating agent sodium chloride, calcium chloride not exceeding 0.02 percent by weight annatto or carotene colour and may contain emulsifiers and/or stabilizers namely citric acid, sodium citrate or sodium salts orthophosphoric acid and poly phosphoric acid not exceeding 0.2 percent by weight, wax used for covering the outer surface should not contain anything which is harmful to health. In case the wax is coloured only permitted food colours may be used. Hard Cheese shall contain, not more than 43.0% moisture and not less than 42.0 % milk fat of the dry matter. Hard Cheese may contain 0.1% of sorbic acid or its sodium, potassium or calcium salts or 0.1 % of nicin.

The drainage of whey is an essential aspect of the definition. Concentrated products obtained with the removal of water are considered as milk products. Numerous varieties of cheese have evolved over the ages. The development of a type or a variety of cheese is closely dependant on the local conditions including the amount and type of milk available, the climate, and other economic and geographic factors. The technological experience and the findings, that have accumulated by trail and error, and the ideal manufacturing conditions have led to a product being tasty, easy to handle and durable. Several types of cheese have changed their character greatly, over the ages, generally retaining their original names and very often the technological know how has been exported.

Irrespective of the country of origin and methods of manufacture, all cheeses contain/possessess certain characteristic features in common such as:

(a) They are made from the milk of mammals of certain specific species.
(b) The first stage in the manufacture of the cheese is souring/ripening of the milk.
(c) The second stage is the clotting/coagulation by rennet or by similar enzyme preparations.
(d) The third stage is the cutting or breaking up of the coagulum or the Junket in order to release the whey.
(e) The fourth stage is the consideration or matting of the curd and
(f) Lastly the fifth stage is the maturing/curing of green cheese in a special type of containers.

All these five stages which are stated above are common to all varieties/types of cheeses, however, the conditions may vary/differ considerably. Some of the main factors that are responsible for the differences in the preparation of final cheese depend upon :

1. The type of milk used for the preparation of cheese.
   E.g: Cow milk, Buffaloe milk, Goat milk etc.
2. The degree of souring and type of curd and nature of the souring organism used in the preparation.
3. The temperature of rennetting and subsequent “Cooking” of the curd in the whey.
4. The method and fineness of cutting or breaking up of the curd.
5. The treatment of the curd after its separation from the whey.
6. The milking and salting of the curd before placing it in the hoop or mould.
7. The pressure applied to the green cheese.
8. The time, temperature and the relative humidity of the curing.
9. Special treatments such as pricking and stabbing, bathing in brine, and surface treatment in order to produce a specific type of coat/cover to the cheese bricks.

Even though all these factors/variables mentioned above are under the control of the cheese producer, they exert an influence on the physical, chemical and microbiological changes that are taking place simultaneously and successively in the milk, coagulum and the cheese. The cheese making/preparation from the scientific view point, is essentially controlled syneresis of the rennet milk coagulum, and the expulsion of the moisture is being affected by:

1. The development of acid - with the pH falling from 6.6 to 5.0.

2. Increasing the temperature to 31 °C (88°F) for renneting with subsequent rise to about 38 °C for cooking of the curd followed by cutting the curd repeatedly. The main objective of all these combined factors is to give an effect to decrease the level of moisture from 81% in milk to about 40% in the green cheese.

It has been stated that more than five hundred (500) varieties of cheese are being manufactured in different countries, while new types are being evolved continually. These are prepared in different ways and ripened by variety of organisms. However only a small proportion/number of numerous varieties, that are most important commercially are considered to be distinctive types, possessing flavours and characteristics that are particularly peculiar to them only.

9.4.1 Types/Classification of Cheese

Already it has been mentioned that several varieties of cheese exist, however only the better known types will receive attention. Cheese can be classified by several systems, and one such method of classification of cheese is as under.

1. According to geographical conditions/considerations i.e, the origin or place from which it is first produced and marketed.
2. Based on the type of milk used in the manufacture of cheese e.g. Cow milk, Buffaloe milk, Goat milk etc.
3. Based on the method of manufacture and general appearance.
4. Based on the physical and Rheological properties.
5. Based on the chemical analysis and
6. Based on the microbiological properties.
   Another system of classification of cheese is as follows:

1. Hard cheeses: Which consists of

(a) Cover philly
(b) Cheddar
(c) Cheshire
(d) Derby
(e) Gruyere
(f) Gouda
(g) Crane
(h) Lancashire
(i) Leicester
(j) Swiss
(k) Edam
(l) Buttermilk

The moisture content of these types of cheeses relatively low, the curd is compressed very heavily to expel the air content and the mass ripens in a uniform manner. The growth and action of extraneous organisms are inhibited to a great extent with the result that such type of cheeses possess good keeping qualities.

2. Blue Veined Cheeses: Several varieties of cheese are brightened by blue veined moulds and are highly regarded and only three of the following varieties are name made. These cheeses possess a high moisture content and have a soft curd. They are not only ripened by the normal organisms, but also by the growth of moulds throughout their substance. The types of cheeses that are included in this category are

(a) Dosset (b) Gorgonzola
(c) Roque fort (d) Stilton (e) Wensleydale

3. **Soft Cheeses:** These cheeses are extremely palatable, and very profitable provided sufficiently good market facilities exists. Little equipment is required for its manufacture and owing to the high moisture content the yield of curd is also high. This type of cheese is usually small in size, is not pressed and quickly ripened. The whey is never drained fully from the curd which is allowed to drain naturally. It is usually prepared from the fresh and sweet milk and the acidity is developed after the addition of rennet and is not subjected to either heat or pressure and the whey is allowed to drain naturally therefore they contain a considerable quantity of moisture. The principle varieties of soft cheese are:

(a) Bondon (b) Brie
(c) Cambridge (d) Camenbert
(e) Colwick (f) Coulommier
(g) Gervias (h) Liniburger
(i) Neufchatel (j) Parmesan
(k) Port L Eveque

4. **Special Varieties:** There are two types of cheeses that are present and cannot be classified under the three previous headings and these include:

(a) Cream cheese
(b) Crust less cheese

9.4.2 Composition of Cheese

The composition of cheese varies considerably. No standard of composition has been fixed for it, however in U.S.A. and in Canada the following standards are in force.

U.S.A

1. Plain cheese-(cheddar type) - 50 %- Milk fat in dry substance. Not more than 39 % water. Fat content of many special cheeses limited.
2. Cream cheese - 65% milk fat.

Canada

1. Full cream cheese - 45 % milk fat in dry substance.
2. If fat content is below 45 % the article must be labelled as skimmed milk cheese.

Cheese contains fat, casein, mineral salts and water, the percentages of each of the constituent vary according to the type/variety of the cheese in question. The absence of any compulsory standard enables cheese to be sold, that may contain very little fat which has been manufactured either from whole or skim milk with any proper/appropriate indications being made given to the purchasers/consumer by the manufacturer.

The FSSAI (2017) standards as published in the gazette govt. of India for the cheese and cheese products has been presented in table 9.16 for the better understanding by the students and the dairy entrepreneurs.

Table 9.16 FSSAI (2017) standards for cheese and cheese products.

Product	Moisture, Maximum, % (m/m)	Milk fat, Minimum, % (dry basis)	Lactose, Maximum, % (m/m)
(i) Cheese			
(a) Hard-pressed cheese	39.0	48.0	-
(b) Semi-Hard-Cheese	45.0	40.0	-
(c) Semi-Soft Cheese	52.0	45.0	-
(d) Soft Cheese	80.0	20.0	-
(e) Extra Hard Cheese	36.0	32.0	-
(f) Mozzarella Cheese	60.0	35.0	-
(g) Pizza Cheese	54.0	35.0	-
(ii) Extra Hard Grating Cheese	36.0	32.0	-
(iii) Named variety cheeses			
(a) Cheddar	39.0	48.0	-
(b) Danbo	39.0	45.0	-
(c) Edam	46.0	40.0	-
(d) Gouda	43.0	48.0	-
(e) Havarti			
- Havarti	48.0	45.0	-
- 30% Havarti	53.0	30.0	-
- 60% Havarti	60.0	60.0	-
(f) Tilsiter			
- Tilsiter	47.0	45.0	-
- 30% Tilsiter	53.0	30.0	-
- 60% Tilsiter	39.0	60.0	-
(g) Cottage Cheese and Creamed Cottage Cheese	80.0	*	-
(h) Cream Cheese	55.0	70.0	-
(i) Coulommiers	56.0	46.0	-
(j) Camembert			
- 30% Camembert	62.0	30.0	-
- 40% Camembert	59.0	40.0	-
- 45% Camembert	57.0	45.0	-
- 55% Camembert	52.0	55.0	-

Table 9.16 *Contd...*

Product	Moisture, Maximum, % (m/m)	Milk fat, Minimum, % (dry basis)	Lactose, Maximum, (m/m)
(k) Brie	56.0	40.0	-
(l) Saint Paulin	56.0	40.0	-
(m) Samsoe			
- Samsoe	44.0	45.0	-
- 30% Samsoe	50.0	30.0	-
(n) Emmental	40.0	45.0	-
(o) Provolone			
- Smoked	45.0	45.0	-
- Unsmoked	47.0	45.0	-
(iv) Cheese products			
(a) Processed Cheese	47.0 (50% for chiplets, packed sliced processed cheese), when sold in a package other than tin	40.0	5.0
(b) Processed Cheese Spread	60.0	40.0	5.0

*Milk fat, minimum 4% (m/m) for creamed cottage cheese.

9.4.3 Food and Nutritive Value

Cheese is prepared from its parent substance milk which contains fats, salts, proteins, and sugar together with the vitamins A and B_2 which are particularly rich. Although the cheese is much richer than the milk from which it is prepared, it contains less sugar because the bulk of which passess out in the whey and can be described as a concentrated form of milk. It is an economical food which is easily digested by the body being changed into body tissue or energy. Cheese can be used in innumerable ways and is extremely palatable and it does not need to be refrigerated in order to keep it, and can be used in its natural state irrespective of the weather conditions. Cheese is manufactured either from whole milk or skimmed milk, therefore it is obvious that the food value of skimmed milk cheese is not comparable with that of a whole milk cheese. Not withstanding its lower food value, there is nothing to distinguish the whole from skimmed milk, cheese when kept for sale in the market. Therefore there is an urgent need to fix a standard for the fat content for all types of cheeses and should be labelled properly for their primary constituent, such that the consumer knows the type of cheese that he is purchasing by paying a correct price.

Cheddar type cheeses are characterised by the mixing of salt with the curd before it is pressed into a coherent loaf, which considerably retards the growth of lactic acid bacteria. Due to this most of the lactose in the curd, should have been converted before the curd is salted, therefore the curd making requires a longer period of time. Earlier, when cheese was made from skim milk, it was usually left for creaming for a long time that it turned sour. Naturally the curd was also acidic and could thus be salted before pressing and now-a-days most of the cheeses of this type are made from un-soured milk. The salt is relatively homogenously dispersed, through the fresh cheese, and due to this, the cheese can be made in large loaves, which is desirable to prevent the water loss by vapourisation and to minimize the curing costs. These cheeses are typically hard with a long shelf life without a surface flora.

Flow Chart 9.3 Manufacture of Cheese (Cheddar Type)

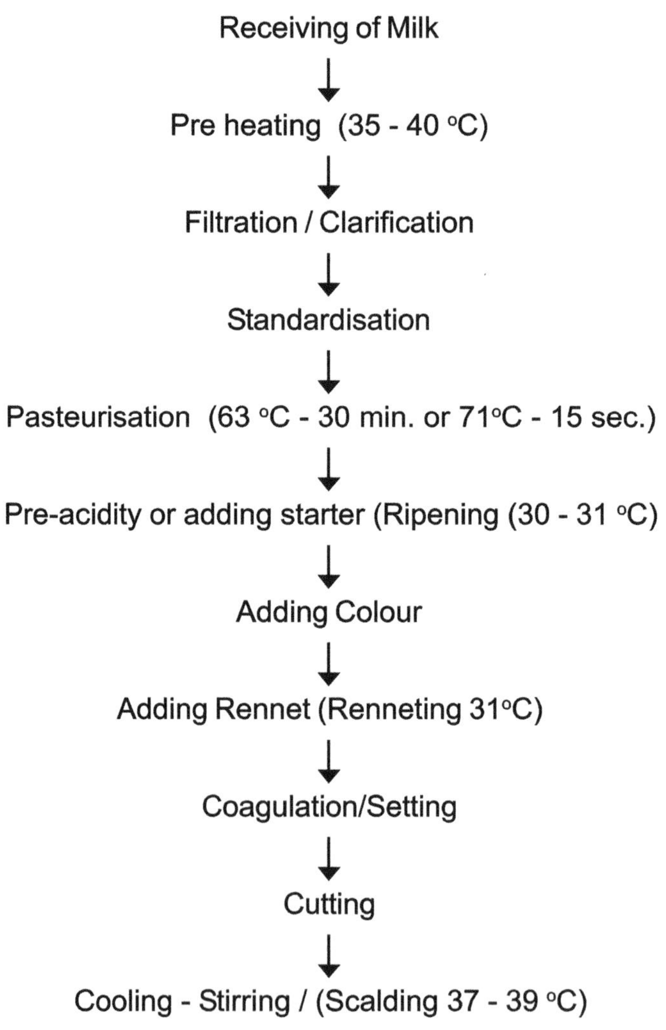

Flow Chart 9.3 ***Contd......***

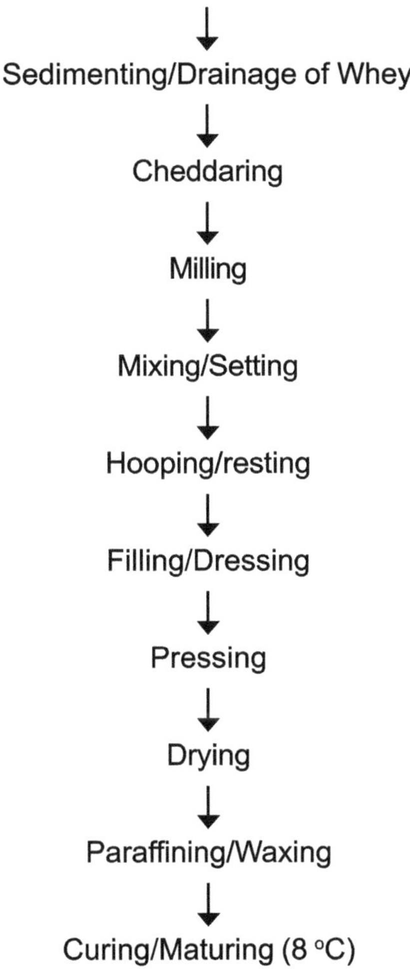

The flow diagram/manufacturing process given in flow Diagram 9.3 represents a somewhat traditional method, though the time from adding starter to the milling was originally often very longer. The end point of the curd treatment in the vat including 'Cheddaring' was assessed by determining the acidity of the whey. Now-a-days a fixed time schedule is usually maintained thus reducing the processing time.

Manufacture of cheese has also been machanised to a high degree. At present the cheese milk is generally pasteurized though a slightly milder heat treatment can be applied. Addition of high percentage of starter is being practiced very much in the present day manufacture of cheese, hence the pre-acidification is practiced very little now-a-days.

The fatty acid composition of the milk fat of different animal species varies, which affects the flavour of the cheese, cheese made from sheep, goat or buffaloe milk has a fairly white colour because these milks contain hardly any carotene.

Thermalizing serves to prevent psychrotrophs from forming lipase, pasteurisation is meant to kill the pathogens and spoilage microorganisms. Usually the milk is pasteurised shortly before rennetting, which re-establishes the rennetability, which is considerably reduced by prolonged cold storage. Moreover, pasteurisation also ensures that the fat is again in liquid state, thus preventing the partial coalescence of the fat globules. Bactofugation and addition of nitrate serve to prevent butyric acid fermentation. At present the nitrate often is not added to the milk, but to the mixture of curd and whey after a good deal/amount of the whey has been removed/drained. This is partly to obtain the largest possible amount of whey without any added nitrate. However, greater amount of nitrate should be added in case, bactofugation is omitted.

Excessive heat treatment results in the precipitation of a part of calcium salts present in the milk, which causes a slower rennetting with the addition of 0.01 to 0.03 percent calcium Chloride to the milk.

9.4.4 Ripening or Adding Starter to the Milk

Ripening of cheese includes all the chemical changes occurring in the cheese, some of which begin before the curd making is finished. The structure and the composition of the cheese change, hence there is also a change in its organo leptic properties. Manufacture and maturation of cheese seems to be impossible without lactic acid bacteria, hence in majority of the cases, they are added to the milk as a starter which is the heart in the manufacture of the cheese. Ripening or souring of milk also refers to the development of acidity in milk from the time it is received in the cheese vat until rennetting. The usual time to add the starters before all the milk has been received in the vat and the amount of starter added is to the extent of 0.5 to 1.0 percent of the milk and the temperature of addition is 30-31 °C (80 - 85 °F). Ripening helps in:

1. The desirable formation of the curd.
2. Establishing the desirable and favourable micro-flora.
3. Checking the growth of undesirable microorganisms and
4. The controlling of the moisture in the cheese.

Generally, when the colour is used it is added just before rennetting. The colour of cheese is usually an alkaline solution of annatto. Rennet and colour should not be mixed together before being added to the milk.

9.4.4.1 Rennetting

Addition of rennet to the milk in the preparation of cheese is commonly called as rennetting or setting. Rennet is a crude extract or preparation obtained from the abomasum or true stomach of the young calf commercially. It contains two principal enzymes viz. renin and pepsin. Renin is an extremely powerful clotting enzyme without proteolysis leading to bitterness in cheese. Rennet is available commercially as a liquid or powder or tablet forms. The essential properties of the commercial rennet are high activity stability of constant strength, and a reasonable bacteriological purity. It should be stored in a closed vessel in a dark room at a temperature below 10 °C. It is a sulphur containing protein, being an enzyme it is easily destroyed by heat, many chemical substances and also under some physical conditions. It is very sensitive to alkali and heating to 70 °C at a pH 6-8 to 7 - 0 will destroy the enzyme rennet in about 14 minutes. There are several factors which affect the action of renin and these include:

1. **Temperature:** Renin is inactive below 20 °C, active between 30-45 °C, the optimum being 41 °C and above 50 °C the activity falls off rapidly.
2. **Acidity:** Alkalinity retards the action and clotting of milk considerably by rennet.
3. **Clacium Ions:** Initially they have no or little effect but with the coagulation of milk they are sensitive to the changes in the calcium ion concentration.
4. **Inhibitory Substances:** Colloidal substances such as albumin serum peptone etc. interfere in the coagulation of milk by the rennin.
5. **Homogenisation:** Homogenisation has an accelerating effect on rennet clotting decreasing the curd tension.
6. **Heating of Milk:** Heat causes the removal or precipitation of the calcium ions and also destroys the rennin, thus resulting in the difficulties in the clotting of milk.
7. Rennet preparations other than calf rennet such as goat and lamb rennets; plant enzymes like 'Withania Coagulans' ficus, papain etc.
8. Difference in the behaviour between animal and vegetable rennet.
9. **Bacterial rennet:** The bacterial enzymes derived particularly from the aerobic sporeformers, '*Bacillus Subtilis*' and negative rods *'serratia marcescens'* are fairly recent new developments in cheese technology.
10. **Addition of rennet:** The rennet is added when the acidity has increased from the initial level by 0.02 percent and the ideal temperature under normal conditions is 30 °C (86 °F) for the raw milk and 31°C (88 °F) for the pasteurised milk. The amount of rennet which should be added depends on

    (a) strength of the rennet (b) temperature of milk

    (c) acidity of the milk and (d) composition of the milk.
11. **Coagulation:** It refers to the changing of milk from the liquid state to a semi solid junket. The curd is ready for cutting, when it gives an even split and pulls away from the sides of the container.

9.4.4.2 Cutting of the Curd

The curd is cut into small cubes with a special knife when it is compact and firm. The curd is tested to determine its readiness for cutting, by inserting the finger and lifting upwards through the curd. A clear break, with the liberation of clear whey, indicates that the curd is ready for cutting. The curd is cut in a clean, even fashion and the whey is allowed to drain immediately a thin film forming upon the surface of the cubes, which show contraction. The syneresis of the cheese curd refers to the expulsion of whey and contraction of the curd which depend upon:

(a) Cutting (b) Temperature (c) Acidity
(d) Agitation (e) time and (f) Salt

9.4.4.3 Scalding/Cooking

It refers to the heating of the curd cubes, which begins within 15 minutes of cutting. The temperature and time required depend on the variety of cheese. To begin with the heat is applied slowly. The rate of heating is such that the temperature rises to 32 °C in about 15 minutes and thereafter to the maximum cooking temperature (37-39 °C) at the rate of 1 °C for every 4 minutes. The heating of the cubes of curd reduces them to approximately one half of their original size.

9.4.4.4 Drainage of Whey

It refers to the removal of whey from the vat however, before the whey is drained off, its acidity is tested, while the cubes settle to the bottom of the vat. The whey is then drain off until level with the curd on the bottom of the vat and again tested for acidity. When the percentage of acidity is satisfactory, the remainder of the whey is allowed drain off and the curd is gathered in each side of the vat in order to facilitate easy draining which is most important. The curd is considered ripe when it contains the correct percentage of acid and moisture.

9.4.4.5 Cheddaring

It refers to the combined operations which include packing, turning, piling and repiling of the curd cubes. The curd cubes are closely together, after the bulk removal of the whey which is known as packing and results in the formation of long slab of curd, and can be cut with a cheese knife into small blocks or strips. The curd blocks are handled without any breakage and are rolled up side-down in the vat, which is called as turning and is carried out for every 15 minutes till the curd is ready for milling and setting. The packed blocks of curd are turned and kept or laid one over the other in a row of two or three which is called piling. When the position of the curd blocks is altered and is known as repiling.

The cheddaring operation generally lasts for two hours even more, and is very important not only from the moisture view point but also for improving the body and texture. This requires much labour because it had to be done in such a way as to prevent the curd from cooling unevenly and too fast, thereby ensuring uniform and fast acid production. The current mechanized cheddaring processes require less time and far less labour than the traditional method. The curd mass will only flow readily when its pH is lower than 5.8 and the temperature is not too low, and causes a 'fibrous' curd structure, which has been assumed to be essential in order to obtain a characteristic cheddar with a nutty and buttery aroma. The end of cheddaring process/ operation is indicated by:

1. Titrable acidity of 0.30-0.35 percent
2. *Hot iron test*: A piece of steel is heated to redness and allowed to cool until black. A small piece of curd is squeezed, dry and is then placed on the hot iron and pulled gently away. When strips of curd form between the curd and the iron. The strips should be firm and thread like.
3. *pH*: The pH should be around 5.4.

9.4.4.6 Milling

In this process the curd is cut into strips or small pieces about the size of a finger. Milling too finely leads to excessive loss of fat and of fine curd in the press whey. The main objectives of the milling are :

1. To promote the further removal of whey.
2. To enable quick distribution of salt in the curd.
3. To prepare the curd for pressing into final form.
4. Other benefits which include are deodorization, cooling of curd, through rinsing if needed, uniform composition of cheese.

   The salt is mixed with the curd and about 10 minutes is allowed for salt absorption or mellowing otherwise excess salt would be lost with the whey, which in normal circumstances contains about 50% of the added salt. Milling and pressing causes a considerable expulsion of whey. The salt should be distributed evenly, and this is hard to achieve. The salt imparts flavour to the cheese and acts as a preservative. The objectives of salting the cheese are :

1. Further removal of whey
2. Hardening and shrinkage of curd
3. Retarding further formation of lactic acid
4. Checking undesirable fermentation and
5. Providing desirable quality characteristics.

Under normal conditions the salt is added to the cheese:

1. About 15 minutes after milling
2. When the hot iron test shows threads of 1 to 2 cm long.
3. When the acidity is 0.4 to 0.5 percent and
4. When the pH is 5.4 to 5.0

9.4.4.7 Hooping

This process refers to the curd being placed in hoops or moulds in which the cheese curd is pressed into its final shape. Some cheeses are not pressed, being moulded and allowed to drain naturally, whereas the others are packed and pressed into moulds. It is usual to heat the curd to a temperature of about 30-32 °C, which is placed into cheese hoops lined with prepared cheese cloths. Wooden cheese hoops/ moulds have been almost and entirely superseded by those constructed with tinned steel strengthened with strong wrought iron hoops. Pressures applied for varying periods, which depends upon the nature and variety of the cheese to be manufactured. This operation refers to the forcing the particles of milled and salted curd in the hoops into the smallest possible space with the objective of giving a final shape to the cheese.

The curd is pressed in two stages viz

1. **Preliminary or first Processing:** In this stage the cheese was given a shape and reduces to almost its final volume for about 30-60 minutes and the cheese is dressed.
2. **Whereas the final pressing** after dressing lasts for about 6-20 hours with an average period of 15 hours. There are different types of cheese presses, which are

   (a) Screw type
   (b) Pneumatic type
   (c) Hydraulic type and
   (d) Spring type - In this the full pressure is maintained all the time.

9.4.4.8 Drying

The drying process involves removal/taking the cheese out of the hoop which is neat, clean, uniform in size and regular in shape and is done for the formation of rind. It is also carried out for the stamping of the cheese, which include the date of manufacture, batch number, variety, name, etc for its identification and record. The temperature of the drying room where the cheese is kept for drying is maintained at 12-16 °C (55 - 60 °F) with a relative humidity of 50 percent initially for few days. In order to dry the cheese on both sides and ends to form the desired rind, it is turned at 24 hour intervals.

9.4.4.9 Paraffining

This process or operation refers to the dipping of the cheese for few seconds in a melted paraffin bath, thus forming a thin coating on the surface of the cheese by the paraffin. The objectives of paraffining the cheese include,

1. To reduce the loss of moisture during curing.

2. To prevent the excessive mould growth even though it is not a perfect mould proof agent.
3. To protect it against the insects.

The different kinds of paraffin, that is being used in the paraffining of the cheese are:

1. Paraffin wax
2. Micro crystalline wax and
3. Flexible waxes etc.

There are several factors which determine the moisture control in the cheddar cheese and needs their consideration are:

1. Coagulation period
2. Size of the curd particles
3. Temperature of cooking
4. Dry stirring before cheddaring
5. Salting
6. Treatment of curd after salting prior to pressing.

9.4.4.10 Curing

The storage of cheese for at least 2-3 months at a given low temperature ranging from 0 -16 °C, during which, the physical, chemical and the bacteriological properties are changed profoundly, resulting in the development of a characteristic flavour, body and texture is termed or defined as curing/souring/ripening/maturing. Sometimes the term, 'green cheese' is applied generally to the hard pressed cheese in the early stages of ripening before the development of the characteristic flavour, body and texture of the ripened cheese. There are two systems of curing the cheddar cheese viz. cold curing and warm curing. In the cold method or system of curing the cheese the temperature is maintained ranging from 0 - 4 °C with an average of 1.5 °C. Humidity is about 75 percent with the duration ranging from 3 to 12 months. The quality of the cheese obtained will have a mild flavour and the defects of bacteriological nature are minimised. In the warm curing, the temperature is maintained ranging from 10-16 °C with an average of 12.5 °C and the humidity is about 85 percent. The duration of curing in this method varies from 15 days to 2 months with the development of sharp flavour while the bacterial defects are exaggerated. However, soon after curing the cheeses are to be stored at 0.5 °C in order to prevent over ripening.

The changes that occur during the curing of the cheddar cheese are :

(A) Physico-chemical which inturn consists of:

(a) Flavour (b) Body (c) Texture and (d) Chemical

(B) Microbiological: The precautions that are to be taken during the curing process of the cheese are:

(i) Flavour

(ii) Body and texture finish etc.

(iii) Temperature and humidity in the curing rooms.

(iv) The walls, floors, shelves, racks etc in the curing rooms for the sanitary conditions.

1. **Selection of the cheese i.e.,** cheese with good quality should be selected and there should be no damage to the rind.
2. **Supervision:** The cheese should be examined at periodical intervals during the process of curing which include the following:

During the process of curing, there is some loss in weight of cheese due to slight shrinkage which is natural, however, excessive shrinkage should be prevented. Shrinkage is caused mainly by "loss of moisture". The factors that cause the shrinkage or loss of moisture in cheese are:

1. Temperature of curing/shortage.
2. Relative humidity of curing/storage room.
3. Size and shape of the cheese
4. Moisture content of cheese and
5. Paraffining of the cheese

In the same manner, there are several factors, that affect the rate of cheese curing which include:

1. Time
2. Temperature
3. Moisture content of cheese
4. Size and shape of cheese
5. Salt content of the cheese and
6. Amount of rennet added for the coagulation of milk.

9.4.5 Cottage Cheese

It is defined as a soft unripened cheese, which is usually made/prepared from skim milk which has a mildly acid flavour. It consists of small particles or flakes of curd which has a meaty consistency when made from rennet curd. Creamed cottage cheese has cream mixed into it so that the finished product contains not less than 4 percent fat and both of these varieties are salted generally.

The method of manufacture of cottage cheese commercially has been given in flow diagram 9.4.

Method of Manufacture

Flow Chart 9.4 Method of Manufacture of the Cottage Cheese

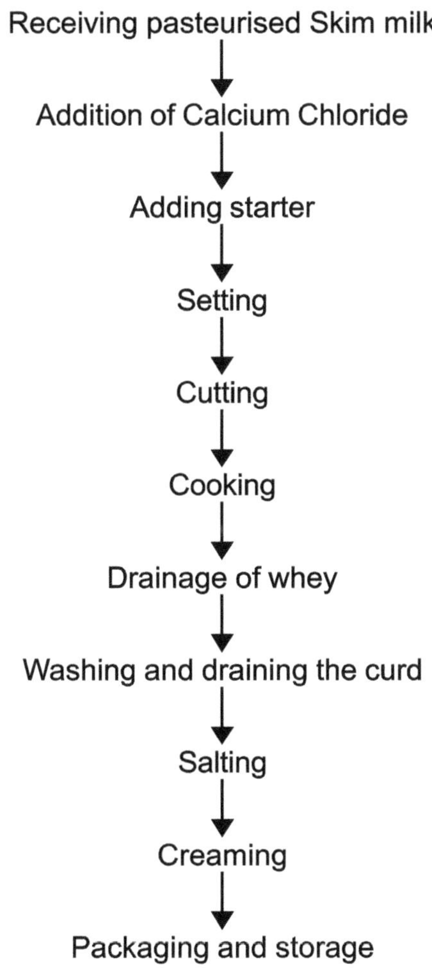

9.4.6 Processed Cheese

It may be defined as a modified form of natural cheese prepared with the aid of heat by blending one or more lots of cheese, except certain types such as cream, cottage cheese etc with water, salt, colour and emulsifier into a homogeneous plastic mass which is usually packed while hot.

According to the PFA rules (1976) processed cheese refers to the product obtained by heating cheese with permitted emulsifiers and/or stabilizers such as citric acid, sodium citrate, sodium salts of orthophosphoric acid and poly phosphoric acid with or without added condiments and acidifying agents namely venigar, lactic acid, acetic acid, citric acid and phosphoric acid. Processed cheese may contain not more than 4.0 percent of anhydrous permitted emulsifiers and/or stabilizers provided that the content of anhydroms in organic salts in no case exceeds 3.0 percent of the finished product. It should not contain more than 47.0 percent of moisture. The milk fat content should not be less than 40 percent of the dry matter. Processed cheese may contain 0-1 percent sorbic acid or its sodium, potassium or calcium salts or 0.1 percent of nicen either singly or in combination.

The details pertaining to the manufacture/preparation of processed cheese has been given in the flow diagram 9.5 for the better understanding of the procedure by the students of dairy science.

Flow Chart 9.5 Manufacture of processed cheese

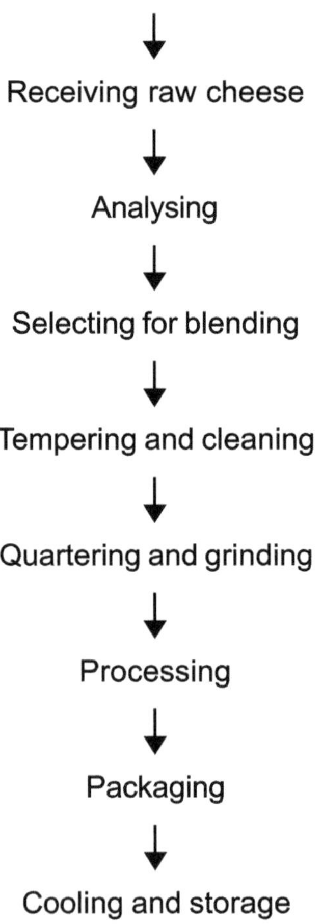

9.4.7 Packaging

It refers to the process where the cheese is being placed in a protective wrapper or container for transport or storage. Different types of packaging materials such as coated or lined paper, aluminium foil, polythene, cellophane, nylon, polyester, pliofilm, cryovac etc and cans, lined with plastics, bags, pouches etc. The materials that are generally used for the packaging of the cheese must satisfy the following requirements namely:

1. afford to give general protection
2. must prevent moisture loss
3. should improve the appearance
4. must protect the cheese against microorganisms and
5. must be able to prevent oxygen transmission.

9.4.8 Storage

The cheese under natural conditions must be stored at low temperatures, preferably 0-1°C in order to ensure good quality. High temperature leads to evaporation of moisture, growth of unwanted moulds and taint producing bacteria and other faults, processed cheese may be stored at 5-10 °C.

9.4.9 Defects/Abnormalities in Cheese

Cheese is particularly liable to abnormalities in its substance because such defects are originating from the following sources.

1. Raw milk of abnormal composition.
2. Faulty methods of manufacture.
3. Bacterial action.

In practice, it is found that these causes are closely allied, rendering it difficult to classify the defects properly into their appropriate sections. However, the common defects that are generally encountered in the cheddar cheese can be classified or grouped based on the following namely:

1. Colour defects
2. Finish and appearance
3. Body
4. Texture
5. Flavour and
6. Mouldy.

9.5 Butter Oil

Removal of all the water including the solids not fat content from the cream or butter and the fat concentrate thus obtained is referred or known as butter oil, which is a convenient method for the preservation of the fat in the absence of refrigerated storage facilities. The terms milk fat, anhydrous milk fat, dry butter fat and dehydrated butter fat are synonymously used with butter oil however the raw material used for the preparation of butter oil generally is cream. The FSSAI (2017) standards as published in the Gazette Govt. of India for the butter oil has been presented in the table 9.17 updating the present knowledge about the butter oil to the benefit of the dairy students.

Table 9.17 FSSAI (2017) standards for butter oil - ghee

Sl.No.	Parameter	Milk fat butter oil	An layers milk fat an layers butter oil	Ghee
1.	Moisture maximum % (m/m)	0.4	0.1	0.5
2.	Milk fat minimum % (m/m)	99.6	99.8	99.5
3.	Butyro-refractometer reading at 40°c	40.0 to 44.0	40.0 to 44.0	as per the table for ghee
4.	R.M. value minimum	28.0	28.0	-do-
5.	Polenske value	1.0 to 2.0	1.0 to 2.0	-
6.	FFA as oleic acid Max%	0.4	0.3	3.0
7.	Peroxide value (Milli equivalent of oxygen 1 kg fat) Maximum	0.6	0.3	-
8.	Baudouin Test	Negative	Negative	Negative

The butter oil is manufactured by using different methods and these are:

1. Direct evaporation, either at
   (a) Atmospheric pressure or (b) Under vacuum
2. Decantation
3. Centrifugation followed by vaccum drying
4. Direct from cream by de-emulsification and centrifugation. However, in the first three methods butter is used as the raw material for the preparation of butter oil.

9.5.1 Direct Evaporation Method

Generally this is a batch method/process which may be carried out either at atmospheric pressure which resembles the Indian method of a ghee preparation/ production or under vacuum.

The butter is placed in a large open vat/pan which is heated through a stream jacket or coil which is initially applied at a slow rate till the butter is melted, and is gradually raised to 108-110 °C stirring the melted butter till the evaporation of all the moisture. Then the residual butter fat is drained by filteration to remove the curd particles. The product thus obtained has a darker colour compared with other methods. Direct evaporation under vacuum does not have any additional advantage, either in the efficiency or in the quality of the finished products.

9.5.1.1 Decantation

It is also a batch process where the butter is melted and allowed to stand undisturbed for a sufficient period of time. Until it stratifies into three layers consisting of top scum, middle fat and bottom serum. The fat layer is drawn off or removed from the top, but a clean separation can not be obtained in this process resulting in a low fat recovery. Therefore this method is not economical when carried out on a large scale.

9.5.1.2 Centrifugal Separators

This method gives/yields a high quality product and more or less a continuous method. Different variants of this method have been used for large scale production of butter oil from butter in many countries. The method of production of butter oil and the various steps involved has been presented in a flow diagram 9.6.

Flow Chart 9.6 Method of manufacture of Butter oil

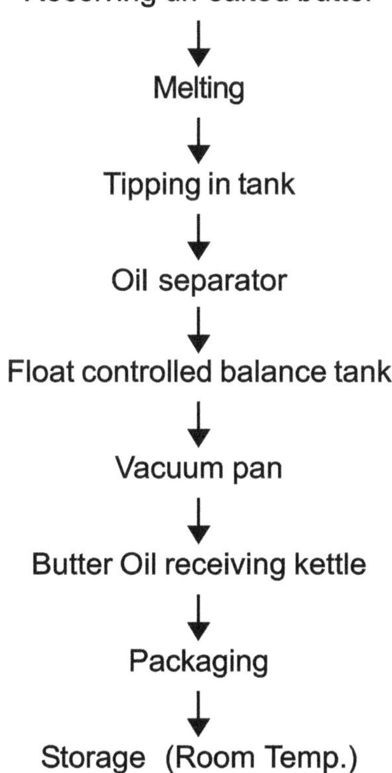

9.5.2 Directly from Cream by De-emulsification

As the name suggests, this process utilizes the principle of de-emulsification of cream used in various continuous butter making processes. The de-emulsified fat may be melted and clarified by centrifuging and vacuum heating. However, this method has not gained much significance in commercial application.

The cooling of butter oil is an important aspect and should be cooled and crystallized under careful control such that large number of fine crystals are formed. The desired results may be obtained by rapidly cooling (to 13-18 °C) and stirring the mass during forced crystallization.

9.5.3 Packaging

The butter oil should be packed by taking proper care to remove/exclude the oxygen, which is important to avoid the re-aeralion of the product, because of the fact that it comes out of the vacuum dehydrator in a partially or completely deaerated form. In tropical countries allowance must be made for expansion of the product, while deciding the size of the container for use. Tin cans are satisfactory for both bulk and retail packaging. The dehydrated butter fat develops a perceptible oxidised flavour, under atmospheric conditions in the course of a month. The shelf life is increased to about 3 months by the use of anti oxidant ethyl gallate in a concentration of 0.05 percent.

The keeping quality of butter oil mainly depends on the moisture content. It is more susceptible to oxidative deterioration than butter because some of the anti-oxidants present in the butter are lost in the butter serum. A high grade dry butter fat will retain its quality for at least 6-12 months, if properly packed and stored. The keeping quality of fat and its resistance to oxidation depends upon the following factors namely:

1. The copper and iron content.
2. Amount of air present.
3. The holding temperature.
4. Absence of light
5. Heat treatment during extraction and
6. Initial quality of butter fat.

The use of anti oxidants is not permitted under the Food and Health regulations of many countries. However, in India the permitted anti oxident is Butilated Hydroxy Anisole (BHA), and can be added at the rate of 0.02 percent to ghee and butter and at the rate of 0.01 percent to whole milk powder and partly skim milk powder by weight of the finished product. Anti-oxidants are substances which prevent or delay the process of development of an oxidized flavour in fat, when added. The natural anti oxidants of butter fat such as lecithin are retained by the direct evaporation process.

Usage of low quality milk, cream or butter with faulty methods of production, processing and storage of butter oil may lead to the development of certain defects.

The common defects that are generally encountered in the production of butter oil are listed below.

1. Brownish colour.
2. Coarse texture
3. Rancid flavour
4. Oxidised/Oily/metallic/tallowy flavour etc.

9.6 Ghee

The use of makhan (butter) and ghee was not new to the people of our country. According to the Hindu mythology, the ghee was being used extensively in the religious rituals such as "Yagnas" in which the fire god 'Agni' was usually appeased by offering the ghee. It has been mentioned in the epic "Sree Mad Bhagavatam" that lord Krishna used to steal Makhan along with his friends in his early boyhood days. The Rigveda, the oldest of the Vedas also contains some sanskrit Verses, mentioning the usage of makhan (butter) and ghee. The utilization of milk fat in the form ghee is best suited to our country and its unique position may be ascribed not only to its best form of preservation of milk fat, under tropical climate, but also as the only source of animal fat in the predominantly vegetarian diet. The large production of ghee is due to:

1. The concentration of milk production in rural areas, which are far away from the nearest urban consuming centres.
2. Lack of all weather and refrigerated storage and transport facilities.
3. Unfavourable climate conditions i.e. high temperature and humidity resulting in the rapid spoilage of milk.
4. The long keeping quality under tropical storage conditions and ordinary packing.
5. Market demand.

The Ghee may be defined as the clarified butter fat prepared chiefly from cow or buffaloe milk where clarify means "to make clear" a liquid or something liquified by removing the unwanted solid matter or impurities.

According to the PFA Rules (1976), ghee is the pure clarified fat derived solely from milk or from desi butter or from cream to which no colouring matter is added.

Fig. 9.6 Ghee Kettle

The authors made an attempt to present the revised standards according to the FSSAI act (2017) and published in the Gazette Govt. of India for the better understanding by the students of dairy science as well as the dairy entrepreneurs in table 9.18.

Table 9.18 FSSAI (2017) Standards for Ghee

Sl.No	Name of state or union Territory or location	Butyro refractometer recording 40°C	Minimum R.M. value	Poleuske value
1.	Andhra Pradesh, Telangana, Dadra and Nagar Haveli, Karnataka (areas other than Belgaum District)	40.0 to 43.0	24.0	-
2.	Andaman and Nicobar Islands	41.0 to 44.0	24.0	-
3.	Arunachal Pradesh, Assam, Goa Haryana (Cotton tract areas) Himachal Pradesh, Jammu and Kashmir, Kerala, Lakshadeep Maharashtra (other than cotton tract areas) manipur, meghalaya, mizoram, nagaland odisha, rajasthan (areas other than Jodhpur Distrct) Tripura, Uttarpradesh, Uttarakhand	40.0 to 43.0	26.0	-
4.	Bihar, Chandigarh, Delhi, Jharkhand, Haryana (other than cotton tract areas) punjab, West Bengal (areas other than Bishnupur subdivision) sikkim	40.0 to 43.0	28.0	
5.	Gujarat (cotton tract areas) Madhya pradesh (cotton tract areas) Maharashtra (cotton tract areas) Rajasthan (Jodhpur District) West Bengal (Bishmapur sub-division)	41.5 to 45.0	21.0	-
6.	Daman and Diu, Gujarat (areas other than cotton tract areas)	40.0 to 43.5	24.0	-
7.	Karnataka (Belgaum District) Madhya Pradesh (areas other than cotton tract areas)	40.0 to 44.0	26.0	-
8.	Chhattisgarh Puducherry	40.0 to 44.0	26.0	-
9.	Tamilnadu	41.0 to 44.0	24.0	-

Explanation:

1. By cotton tract is meant the areas in the states where cotton seed is extensively fed to the cattle and so notified by the concerned state governments. 2.Baudouin test shall be negative for all

The general chemical composition of the ghee is given in the table 9.19.

Table 9.19 Chemical Composition of Ghee

Sl. No.	Characteristics	Requirements	
		Cow	Buffaloe
1.	Milk fat	99	99.5 percent
2.	Moisture	Not more than 0.5 percent	Not more than 0.5 percent
3.	Un saponifiable matter (a) Carotene (ug/g) (b) Vitamin - A (IU/g) (c) Tocopherol (ug/g)	 3.2 - 7.4 19 - 34 26 - 48	 -- 17 - 38 18 - 37
4.	Free fatty acid (% Oleic)	Max 2.8 (Agmark)	
5.	Charred casein salts of copper and Iron etc.	Traces	

Source : Text Book of Dairy Chemistry - M.P. Mathur, D.D. Roy & P. Dinakar

9.6.1 Physico-chemical Constants of Ghee

The ghee is characterised by certain physico chemical properties and are found to be the basis for the fixation of the constants in order to define the chemical quality of the product as in the case of other fats and oil. These properties, show some variation which depend upon several factors and these have already been explained earlier, while discussing the properties of milk fat. Some of the important analytical constants or standards of mixed ghee produced under standard conditions have been given below.

1. Melting point - 28 °C to 44 °C.
2. Solidifying point - 15 °C to 28 °C
3. Specific gravity - 0.93 to 0.94
4. Refractive Index - B.R. Reading at 40 °C varies from 40 to 45.
5. Reichert Meissel (R.M) value. - It should not be less than 28 whereas in cotton seed tract the limit is 20.
6. Polenske (P) value - Normal value not more than 2
7. Cotton seed tract areas - 1.5.
8. Saponification value - Not less than 220
9. Iodine value - 26 - 38

9.6.2 Manufacturing Methods

The organised sector of the dairy industry employing the modern methods of technology, manufactures the ghee to the extent of about 10 percent only, whereas the desi or country method contributes about 90 percent of total ghee production in India. There are several different methods for the manufacture of ghee being practiced or in use which include:

1. Country or desi method.
2. Creamery butter method.
3. Pre-stratification method.
4. Direct cream method and
5. Continuous method.

9.6.2.1 Manufacture Flow Diagram

The method of manufacture of ghee by these various methods starting from milk has been given as a flow diagram 9.7.

Flow Diagram 9.7 Manufacture of Ghee

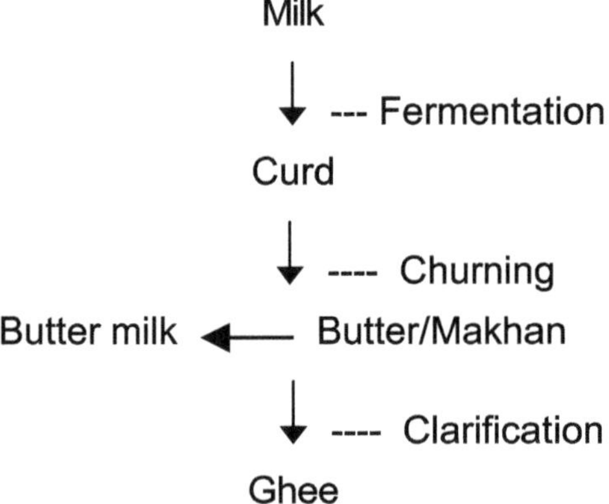

As already stated that ghee may be prepared from butter which is either desi or creamery or from cream. The pre-stratification and continuous methods also use butter. The method of manufacture of ghee by these various methods starting from milk has been given as a flow diagram 9.8.

Makkhan/Butter and ghee are essentially prepared under rural conditions on a cottage industry scale in every household of the milk producer. The lot of makkhan or butter either fresh or accumulated over a few days is generally taken in a suitable vessel and heated on low fire with continuous stirring in order to drive out the moisture. Further heating is stopped and the vessel is removed from the fire, when practically all the moisture is removed, a stage which is judged with experience by the house-wife. The vessel is kept undisturbed and the ghee is allowed to cool by settling the residue at the bottom, the clear liquid fat ghee is decanted into a suitable container for storage until usage or marketed.

Flow Chart 9.8

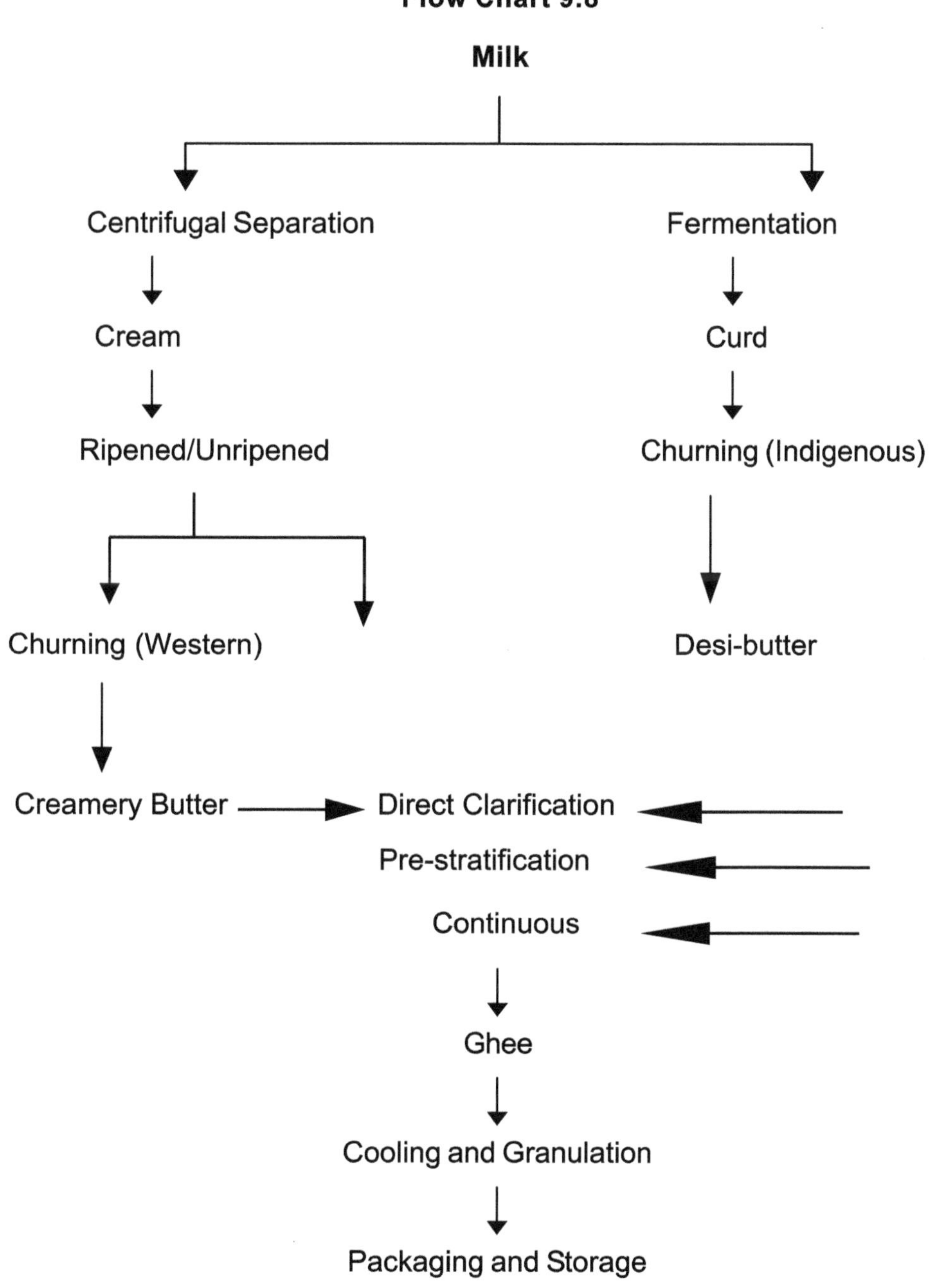

The yield of ghee from milk/cream/butter is influenced chiefly by:

1. Fat content of the raw material used for ghee making.
2. The percentage of fat recovered in ghee which in turn depends on the method of manufacture. The percentage of fat recovery in ghee by different methods has been given in the table 9.20 for comparison.

Table 9.20 Percentage of Fat recovery in ghee

Method of Ghee Making	Percentage of fat recovery in ghee
1. Country/Desi - traditional	80 - 85
2. Country/Desi - improved	88 - 90
3. Creamery butter	88 - 92
4. Direct Cream 40 - 50% fat	85 - 86
5. Direct Cream 65 - 85% fat	90 - 95

Source : Outlines of Dairy Technology - S.K. De

9.6.3 Keeping Quality

Ghee has a remarkable keeping quality unlike other indigenous milk products. Under optimum conditions of production, packaging and storage of ghee is expected to have a storage shelf life of about 6-12 months. The various factors which influence the keeping quality shelf life of ghee are as under:

1. Temperature of storage
2. Initial moisture content
3. Initial acidity content
4. Initial sediment content
5. Copper and Iron content
6. Method of packaging and
7. Exposure to light especially direct sunlight.

9.6.4 Grading

It is not possible for an average consumer/customer to judge the purity of a sample of market ghee by its appearance, taste or smell at the time of purchase. The quality as well as the purity of ghee can only be judged by the detailed physical and chemical analysis. Grading or classification of ghee according to quality, assures the customer hence it is obvious.

9.6.4.1 AGMARK Grading Scheme of Ghee

Agmark is an insignia or a symbol which was passed by the Indian legislature/ parliament in 1937 under the Agricultural Produce (Grading and Marking) Act with a view to develop an orderly marketing of the agricultural products on all India basis. This act provides a provision for the grading of the ghee on voluntary basis. The main objectives of the "Agmark" grading scheme for ghee are:

1. To assure the consumer a produce of pre-tested quality and purity.
2. To enable the manufacturers to produce a high grade product in order to obtain better returns and
3. To develop an orderly marketing of the commodities by eliminating the mal practices while transfering them from the producer to the consumer by the middle man.

9.6.4.2 Agmark Ghee Specifications

The original specifications which were introduced in 1938 have been modified from time to time. The ghee which was labelled as 'Agmark' can be freely marketed, in any part of the country even though the 'Agmark' specification may differ from those prescribed by the constituent states of the Indian Union. The present Agmark specifications for the ghee have been presented in the table 9.21.

Table 9.21 Agmark standards for Ghee

	Tests	All India	Regional	
			Winter	Summer
1.	Baudown		----------- negative ----------------	
2.	Phytosterol acetate		----------- negative ----------------	
3.	B.R. Reading (40 °C)	40 - 43.0	41.5 to 44.0	42.5 to 45.0
4.	R.M. Value not less than	28.0	23.0	21.0
5.	P Value	1.0 to 2.0	0.5 to 1.2	0.5 - 1.0
6.	Moisture percentage		---------- Not more than 0.3-----------	
7.	Free Fatty acid (% Oleic)			
	(a) Special grade - Red labelled Agmark		------ ----Not more than 1.4 -----------	
	(b) General grade - Agmark Green label		---------- Not more than 2.5 ---------	

Source : Text Book of Dairy Chemistry - M.P. Mathur, D.D. Roy & P. Dinakar

9.6.5 Defects

Certain defects are bound to arise in the ghee due to low quality raw material and some faulty methods/practices in the production and handling of the ghee. These defects are as under:

1. Flavour defects - consisting of
   (a) Smoky
   (b) Over cooked/burnt

Fig. 9.7 Continuous Ghee Kettle

(c) Under cooked
(d) Rancid
(e) Oxidised/oils/Metallic

2. Texture - Greasy
3. Colour - Burnt
4. Sediment - high burnt

9.7 Khoa

In the organised dairies it was the usual practice to convert the surplus milk into skim milk powder and ghee due to their convenience in handling, transportation and storage. However, the organised dairy sector is now concentrating and switching over to the value added traditional milk sweets/products. Under Indian tropical conditions there are certain specific benefits of these products because of their suitability, as stated hereunder :

(a) They do not require complicated and expensive equipment.

(b) These products can be produced easily under the village/rural conditions, thereby reducing the cost of transportation of the liquid milk.

(c) Helps in the generation of the rural employment and

(d) These products offer value addition.

Khoa is one of the important indigenous milk products, which is used as the base material for a variety of indigenous milk products. Conventionally, Khoa is prepared/manufactured by the continuous boiling of milk in an open Kettle/vessel, till the deserved concentration of 65-72% total solids and texture is acheived. The manufacture of Khoa is largely in the hands of private traders in the unorganised sector who are using the highly primitive techniques which are essentially based on

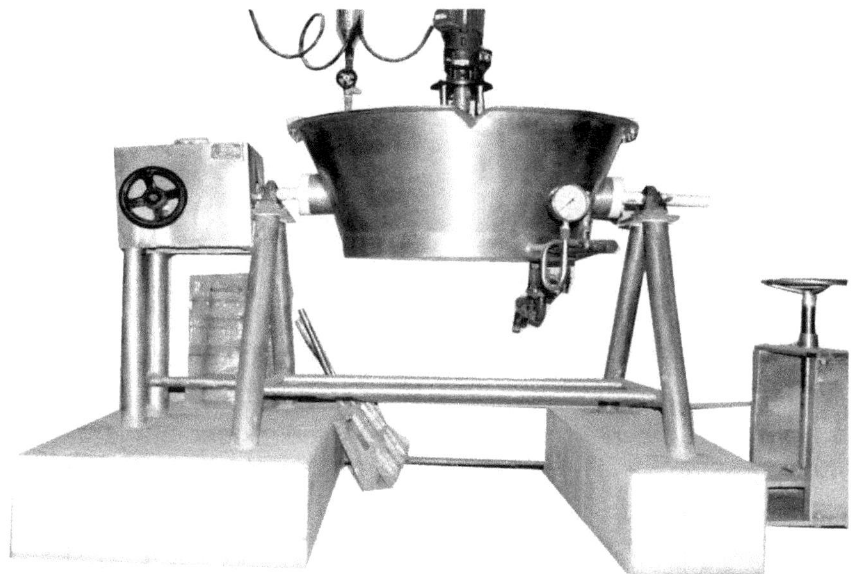

Fig. 9.8 Khoa Making Machine

their experience. The conventional methods of Khoa manufacture are though simple and cost effective, they suffer from several inherent limitations as given below:

(i) There is a wide variation in the chemical composition, microbial and sensory qualities from batch to batch.

(ii) The small scale batch production is unsuitable for commercial adoption.

(iii) Low heat transfer coefficients causing equipment to be bulky.

(iv) Un-hygenic operation due to atmospheric dessication.

(v) Excessive stress and fatigue on the operator.

(vi) Poor packaging

(vii) Limited shelf life of the product.

The technology of Khoa making has been standardised and continuous Khoa-making units have been developed, which are being used by many organised dairies as well as by some private large traders.

9.7.1 Standards of Khoa

According to PFA Act 2004 Khoa sold by whatever variety or name such as Pindi, Danedar, Dhap, Mawa or Kawa, means the product obtained from Cow, Buffalo, Goat or sheep milk, or milk solids or a combination thereof by rapid drying. Its milk fat content shall not be less than 30% on dry weight basis of the finished product. It may contain citric acid not more than 0.1% by weight. It shall be free from added starch, added sugar and added colouring matter. As per the PFA Act 2005, Khoa shall contain, not more than 50,000/gm total plate count, not more than 90 per gm coliform count and not more than 250/gm yeast and mold count, E. Coli, Aerobic, spore count and listeria monocytogenes shall be absent in 1 g of Khoa whereas Salmonella and Shigella shall be absent in 25 gm.

BIS Specifications : IS : 4883, 1980

Khoa is a heat coagulated milk product obtained by partial dehydration of milk of buffalo, cow, sheep and goat or a combination thereof. It shall not contain any foreign ingredient to milk except the addition of citric acid in Danedar Khoa, added to develop the desirable characteristic. The requirements of the three types of Khoa namely, Pindi, Danedar and Dhap as designated by the BIS have been presented in the table 9.22.

Further an attempt has been made to present the revised standards for khoa or mawa according to the FSSAI (2017) has been presented in table 9.23.

Table 9.22 BIS Requirements for different types of Khoa

S.No.	Characteristic	Requirements		
		Pindi	**Danedar**	**Dhap**
(i)	Total solids % by mass min	65	60	55
(ii)	Fat % by mass on dry basis min	37	37	37
(iii)	Total ash % by mass on dry basis max	6.0	6.0	6.0
(iv)	Titrable acidity as lactic acid % by mass max	0.8	0.9	0.6
(v)	Coli form count/g max	90	90	90
(vi)	Yeast and Mould count/g max.	50	50	50

The specific characteristics of the three varieties namely Pindi, Danedar and Dhap has been described hereunder separately.

1. **Pindi Khoa:** It is formed as a disc shaped pat and has a smooth and homogenous body and texture. The grains in it are very fine and of uniform size throughout the mass. The product possesses a characteristic heated/cooked flavour. It is used for the manufacture of different varieties of burfi and peda.
2. **Danedar Khoa:** It is characterised by its granular texture and irregular shaped body. The size of grains depends upon the amount of acidulant added and the acidity of milk used. Citric acid when added, should normally be less than 0.1% (preferably 0.02%). This type of Khoa is used as a base for the preparation of Kalakand, milk-cake etc., where granularity is desirable.
3. **Dhap Khoa:** This type is characterised by a loose and sticky body and a smooth texture. It is normally a pre-pindi stage and thus contains higher moisture. Dhap Khoa is purposely prepared for its use in preparations of Gulab Jamun and Pantua so that balls of smooth surface can be prepared.

Table 9.23 FSSAI (2017) standards for khoa or mawa

Sl. No	Parameter	Khoa
1.	Total solids minimum % (m/m)	55.0
2.	Milk fat minimum % (m/m) dry matter basis	30.0
3.	Total ash maximum % (m/m)	6.0
4.	Titrable acidity (as % lactic acid) Maximum %	0.9

Note: 1. It shall be free from added starch and added sugar

2. The extracted fat from khoa shall meet the standards for Reichert meissl value, Polenske value and Butyro-refracto meter reading as prescribed for Ghee.

9.7.1.1 Chemical Composition

The chemical composition of the Khoa which is being manufactured and available in the market shows great variation because it was produced largely by the private traders as there is no standard procedure for its adoption in the manufacture of Khoa. The chemical composition of Khoa produced under standard laboratory condition has been given in the table composition of khoa 9.24.

Table 9.24 Typical composition of Khoa (Percent)

Type of milk	T.S.	Fat	Protein	Lactose	Ash
*Buffaloe milk	78.40	30.50	17.70	23.90	5.90
*Cow Milk	80.70	25.20	15.80	33.50	4.10
*Mixed Milk	79.80	29.00	16.70	30.10	5.20
Market Sample	71.60	24.60	19.00	25.20	3.60

Prepared under laboratory conditions

Source: I.D.A. Publication

9.7.2 Physico-chemical Changes during Khoa Making

Preparation of Khoa involves three steps/actions namely heating of milk, concentration and vigorous stirring and scraping. During the course of these actions, the milk undergoes certain changes in its physico-chemical properties.

1. The removal of moisture from milk results in the concentration of milk solids and brings change in the status of milk constituents, particularly the proteins and minerals. These changes eventually alter the nature of milk from liquid to semi-solid or solid state. All the constituents, including lactic acid, in proportion to the degree of concentration.

2. As the temperature of the milk tends to rise above 77°C (90°C in the case of buffaloe total milk) the whey proteins particularly β lacto globulin and α lactalbumin undergoes varying degree of denaturation, aggregation and coagulation with the begining of their interaction with K casein.
3. The denaturation of the whey proteins results in the production of sulfhydroxil compounds, which gives rise to the characteristic cooked flavour in khoa and other heat desiccated products.
4. The combined effect of severe heat treatment and concentration and complexing of whey proteins with it, substantially increases the size of the casein micelles though the casein is quite stable at high temperatures, ultimately resulting in the loss of stability of casein micelle.
5. The coagulation of casein along with the aggregation of whey proteins results in the partial entrapment of the fat. The moisture also dispenses as fine droplets in the aggregates of protein.
6. The fat globular membrane is disrupted releasing some fat in free form due to vigorous agitation of the milk at higher temperature during desiccation. Formation of this free fat is very essential and important from the flavour and texture point of view.
7. Soluble calcium and magnesium salts are converted into insoluble form leading to their interaction with protein components when the milk is heated at high temperatures.
8. Heating of milk continuously at high temperature results in browning, which is attributed to the formation of Melanoidin Pigment. The interaction between aldose groups of lactose and the free amino group of casein which causes browning is called as Maillard reaction.
9. In Khoa the lactose is present as a super saturated solution, whereas in milk it is present in the dilute form.

9.7.3 Methods of Preparation

9.7.3.1 Traditional Method

(a) **Pindi Khoa:** Buffaloe milk is generally preferred and is taken in small quantities of five (5) litres in an open pan (Karahi) and boiled over a brisk, non-smokey fire. The milk is continuously and vigorously stirred in a circular motion by a ladle (Khunti) in order to avoid the burning of the milk solids. As the evaporation of takes place, the milk thickens, and the concentration, when reaches to about 2-8 times in cow milk and 2.5 times in the case of buffaloe milk, the thickened mass shows an abrupt change in its appearance and consistency indicating the commencement of the heat coagulation of the milk proteins. At this stage, vigorous stirring and dessication is continued till the viscous product reaches a pasty consistency and begins to dry up. The final product shows the signs of leaving the sides of the pan resulting in the formation of the bolus. This stage is called as pat formation and invariably achieved by working the contents up and down after the removal of the pan from the fire. Finally the finished product is transferred to the aluminium containers for moulding/ shaping. There are several limitations in the traditional method. Sometimes, certain undesirable flavours such as smoky and burnt may also develop and these can be eliminated with the adoptation of the improved/modern method of Khoa manufacture.

(b) **Danedar Khoa:** In this method, the milk of either buffalo or cow which has developed an acidity of about 0.17 to 0.18% as lactic acid. Sometimes citric acid of 0.01 to 0.02% is added to the milk at the boiling stage of the milk. This will result in the earlier coagulation of the milk proteins and development of large size grains which are needed for the manufacture of milk cake and Kalakand. The rest of the method is the same as for the Pindi Khoa.

(c) **Dhap Khoa:** The milk when reaches to a semi-solid state or starts leaving the sides or surface of the pan, the heating is stopped and the product is immediately taken out which at this stage appears to be rather weak and loose due to high moisture retention resulting in the production of Dhap Khoa.

9.7.3.2 Improved Traditional Method

(i) In this method, a stainless steel jacketed kettle, with steam supply is used for the boiling of milk instead of the open fire and the Karahi. There is a provision of controlling the heating with a regulatory steam valve, tilting mechanism and pressure valves etc.

(ii) Standardised buffaloe milk of 5.5 percent fat and 90% SNF is used in order to maintain the composition of the final product to meet the legal standards. The quality of the milk is also evaluated for its appearance, flavour, acidity, pH and for the presence of adulterants/neutralisers.

(iii) The conditions of desiccation in respect of temperature, speed of stirring and scraping and level of desiccation are controlled.

(iv) The Khoa is packaged immediately after its removal from the heating kettle and stored under sanitary conditions to prevent contamination, surface evaporation and spoilage.

9.7.3.3 Large Scale/Mechanised Methods

Several research workers have made attempts to upgradation of the technology in the separation/manufacture of Khoa aiming at the mechanisation process with limited success. A thin film scraped surface heat exchanger (TSSHE) system for the continuous manufacture of Khoa. The TSSHE units are arranged in a cascade fashion. In the first SSHE the milk is concentrated to about 40-45%. Total solids and finally to Khoa in the second SSHE. This unit is suitable for both the organised as well as the small dairies/enterpreneurs because normal milk is being used as the basic material.

The NDDB has developed an Inclined Scraped Surface Heat Exchanger (ISSHE) for the continuous manufacture of Khoa. In this machine concentrated milk of 42 to 45 percent total solids is used as the basic material. The inclination permits formation of a poor of boiling milk similar to traditional Karahi method and is critical to development of typical flavour and texture in Khoa. This unit has received a wide acceptance in the dairy industry. The capacity of both these units is about 50 kg Khoa/per hour. Many organised and private sector dairies have adopted these two continuous Khoa making plants. A cantherm convap scraped surface heat exchanger developed by Alfa level for heating and concentration of viscous sticky and lumpy products has been used for the manufacture of the Khoa by the Sugam Dairy, Vadodara. This unit handled about 500 kg concentrated milk per hour, however, the quality of Khoa produced by this process was inferior to the traditional product in terms of product flow, flavour, texture and consistency.

9.7.4 Factors affecting the Quality and Yield of Khoa

Generally fresh milk is preferred for the production of good quality Khoa, having the desirable organoleptic attributes and suitable for the preparation of different types of sweets. The average yield of Pindi Khoa produced from standard cow milk should be 18.5% whereas from buffaloe milk the yield is 23%. Similarly the yield of other two types, Danedar and Dhap made from buffalo milk is usually 24 and 25% respectively. The various factors that effect the quality and yield of Khoa have been presented briefly hereunder.

1. **Quality of Milk**
   - **(a) Species of animal:** Buffaloe milk is preferred to cow milk for the preparation of Khoa due to the following reasons.
     - (i) Organoleptic quality of buffaloe milk Khoa is much superior to that of cow milk Khoa, because of its pleasantly, sweet and creamy taste, total body and texture with uniform grains. On the contrary the Khoa prepared from cow milk has a salty taste due to the presence of higher amounts of chloride, and its body is hard and the texture is dry due to low fat content.
     - (ii) The free fat content, which provides some greasy appearance to the Khoa, which is considered as the desirable attribute for sweet making is also higher in buffaloe Khoa than the cow Khoa.

(iii) Due to higher total solid content, the yield of Khoa produced from buffaloe milk is also higher as compared to cow milk Khoa.

(iv) Buffaloe milk Khoa is more suitable than the Khoa prepared from cow milk for the manufacture of good quality sweets. The Khoa that is being prepared or produced by the organised dairy sector uses mixed milk and the yield of Khoa is between that of buffaloe and cow milk Khoa.

(b) Use of Colostrum: The chemical composition of colostrum is different from normal milk and has extremely poor heat stability. The Khoa has a deep yellow colour with big size clotted lumps hence unsuitable for sweet making.

(c) Use of Concentrated (high TS) milk: The use of concentrated with more than 31% Total solids requires less moisture removal as compared to whole milk for khoa making and is thus exposed to heat treatment for less time.

(d) Fat content of milk: In order to meet the legal standards according to PFA rules, it is essential that a minimum of 4% fat in cow milk and 5% fat in the case of buffaloe milk to prepare the Khoa. Many consumers prefer low fat products, low fat Khoa with the addition of fat replacers can be made. Normally, when protein concentrates are more frequently used to replace butter fat without much affecting the flavour and texture of the product.

(e) Titrable Acidity of Milk: Pindi Khoa with good flavour and texture is produced from fresh milk having acidity in the range of 0.14 to 0.16% as lactic acid; similarly largely the Danedar Khoa is produced with the milk having an acidity up to 0.18% developed by the addition of an acidulent is desirable.

(f) Adulteration of Milk: Due to the non-availability of good quality milk with an escalating prices of Khoa during the lean season and festivals prompts the Khoa manufacturers to use adulterated milk for its preparation. The milk after the removal of cream is substituted by the cheap vegetable fats and other ingredients such as starch and sugar in order to increase the yield of Khoa. The quality of Khoa prepared with such kind of milk is deteriorated very quickly and unsuitable for the preparation, the sweets of good quality.

2. **Speed of Stirring:** The speed of stirring should be optimum and depends on the type of machine/method. The optimum speed of stirring is about 100 rpm in the case of traditional method whereas it is 200 rpm in the continuous method. The optimum speed of stirring prevents the burning of milk solids, develops the desirable body and texture of Khoa.
3. **Temperature of Desiccation:** The milk should be maintained at the boiling temperature till it reaches a pasty consistency and then lowered to about 95 °C in order to obtain good quality Khoa. The heating is generally discontinued because continued heating at higher temperature at advanced processing stage results in undesirable flavour and texture in Khoa. Slow heating is not only more time consuming but also produces sandy texture and brown colour.
4. **Homogenization of Milk:** Generally milk is not homogenized for Khoa making. However, under certain conditions, such as use of recombined milk, or for increasing the fat by the addition of butter oil or butter, homogenisation becomes essential. Khoa prepared from homogenized milk is crumbly and lacks cohesiveness.The moisture retention in such Khoa is more, which results in softness and higher yield.

9.7.5 Packaging and Shelf Life of Khoa

The shelf life of the conventionally prepared Pindi Khoa is about 5 days whereas that of Danedar and Dhap varieties in merely 3 days. The Khoa which is prepared in the continuous form, the temperature is about 90 °C and is almost sterile except for some thermoduric spores. In the conventional method of preparation the temperature at the finishing stage is more than 85 °C. The post manufacturing contamination is responsible for the high microbial counts and poor shelf life of these products.

The three types of Khoa are prepared in for different end uses. Mainly the Pindi Khoa is marketed, whereas the producers themselves use the other two varieties, hence, the packaging of the two varieties are not much relevant. In the present existing trade practices, the Pindi Khoa, of one Kilo blocks, which is normally spherical or hemispherical in shape are packed in bamboo baskets covered with news papers or green leaves, covered with gunny bags for transportation. Different types of high barrier laminates, have been developed that are suitable for packaging the Khoa. For long distance refrigerated transportation and for bulk storage in cold storage of Khoa, polyester/eithylene vinyl alcohol/poly ethylene laminates are being used. Rigid plastic containers, Tin cans can also used advantageously for the bulk packaging of Khoa under vacuum and storage at room temperature.

The PFA Act also recommends, the addition of antioxidants and antimicrobial substances for improving the keeping quality and shelf life of the Khoa. Sterilization of polypack with gamma radiation before filling has been found to be beneficial. Vacuum packaging of Khoa could enhance the shelf life up to 120 days in refrigerated storage. Similarly good manufacturing practices (GMP) also play an important role in extending the shelf life of Khoa, producing a safe product. It is therefore essential to adopt GMP at all the stages of production, packaging and storage of Khoa. The Food and Drink Administration of United States (2003) has specified good manufacturing practices for manufacture, packaging and storage for the food products which can be adopted for Khoa very conveniently.

9.8 Paneer

Paneer is a highly popular traditional Indian dairy product obtained by acid and heat coagulation of milk involving the formation of large structural aggregates of milk proteins in which the milk fat and small fraction of whey along with soluble milk solids are entrained while the major portion of the whey separates from the coagulates. The solid mass which resulted from fusing together under pressure of the coagulated lumps is termed as paneer. Good quality paneer is characterized by a white colour sweetish, mildly acidic, nutty flavour, spongy body and a close knit texture which is highly nutritious, because it retains about 90% fat and protein, 50% minerals and 10% lactose of the original milk. The chemical composition of the paneer available in the market varies to a large extent, however, the product prepared by adopting the standard method, on an average approximately contains moisture 54%, proteins 18%, fat 25%, lactose 2%, and minerals 1.5%.

Paneer has many uses commencing with its composition in raw form to the preparation of several varieties of culinary dishes and snacks. The demand for the value added products with paneer as the base material is growing at a very fast rate particularly in the urban areas. Further there is need to tap the market potential for both domestic consumption as well as for export.

9.8.1 Principle

Paneer is obtained by heating milk to about 90 °C, acidifying the hot milk with the addition of citric acid solution followed by the removal of whey and pressing the curd before cooling the pressed mass in chilled water. Chemical and physical changes in casein and whey proteins, brought about by the combined influence of heat and acid treatment thus forming the basis of paneer making. The whey proteins have a definite configuration in their natural state, which is maintained by hydrogen bonding, hydration and secondary forces. The native configuration of whey proteins is disrupted, when the milk is heated to a temperature of 90 °C loosing their stability and forming a complex with casein present in the form of colloidal calcium caseinate phosphate micellae. The interaction of whey proteins with casein micellaes on heating milk at its natural pH may also increase the hydrophobicity of the llaes surface and reduce the hydration barrier against aggregation, thus allowing aggregation and geletion to occur. Increased or more rapid gelation or aggregation occurs more rapidly at a higher pH 5.4 to 5.7 in milk heated to 90 °C than in unheated milk. Paneer manufacture thus essentially involves the formation of co-precipitates due to complexing of heat, denatured whey proteins with the casein followed by their acid coagulation. The higher degree of co-precipitation the greater will be the total solids recovery and yield of paneer. The development of typical rheological characteristics of paneer could be due to the intensive heat induced protein-protein interactions, calcium linkages, fat entrapment in the fused mass forming as a result of pressing of the hot coagulates.

9.8.2 Conditions of Manufacture

There are several conditions which are to be fulfilled in order to obtain a good quality paneer in its manufacture. They have been described briefly hereunder.

1. **Type of Milk:** Best quality paneer is obtained from buffaloe milk whereas the cow milk product is criticized to be too compact, fragile and unsuitable for frying and cooking. Paneer with most desirable texture, attributes and maximum yield can be obtained from buffaloe milk wherein the fat and SNF are adjusted to 5.8 and 9.5 respectively.
2. **Heat Treatment:** High heat treatment of milk improves the solids recovery yield, flavour, body and texture characteristics of paneer. Coagulation of milk at 60 °C results in paneer with very weak and loose body unsuitable for frying in addition in low recovery of solids. Coagulation at 70 °C produces the desirable body and texture characteristics in paneer. Higher temperatures of coagulation induce greater solids recovery but lower yield due to increased moisture expulsion.
3. **Coagulation of milk:** The optimum pH of coagulation of buffalo milk at 70 °C is between 5.30 to 5.35 whereas that of cow milk is in the range of 5.20 to 5.25. The citric acid solution at 1% concentration results in the best product. The use of citric acid dissolved in naturally soured whey reduces the requirement of citric acid and increases the recovery of the solids without any loss in the quality of paneer.

4. **Incorporation of hydro colloids and other Milk solids:** In the conventional process of manufacture of paneer the coagulation of milk is carried out at 70 °C resulting in the increased solids recovery. However, at the same time, the yield is lower and the product is harder with lower moisture retention. The adverse effect on the body of paneer obtained on coagulation at higher temperature (90 °C) can be overcome still retaining the maximum total solids with the incorporation of certain hydro colloids. Incorporation of sweet cream butter milk solids to buffaloe milk also helps to improve the recovery of milk solids and retention of moisture thereby increasing the yield of paneer by about 1% without affecting the sensory quality.
5. **Drainage of whey and pressing of curd:** It is desirable to maintain temperature of curd at about 63 °C during drainage of whey, which can be achieved by accelerating the process. Curd is then filled into hoops lined with cloths and pressure is applied on top of individual hoops by placing a weight of about 45 kg (0.5 kg/cm) for 20-22 kg product. The hoop must have holes on all sides to facilitate expulsion of whey. Normally 15 minutes processing is enough to expel the whey and obtain a product with desired texture.
6. **Chilling of Paneer:** The paneer block is taken out of the press and cut into pieces of desired size, finally transferring them initially to an ordinary potable tank/vessel and finally to a chilled water tank maintained at a temperature less than 10 °C for about 2-3 hr. This will help not only in developing the desired texture, but also brings down the temperature of the paneer speedily. It is desirable to take utmost care to use water of good quality particularly from bacteriological point of view.
7. **Packaging:** Normally paneer blocks of desired size, are packaged in polythene pouches, heat sealed and stored under refrigerated conditions. Some of the manufacturers, have started vacuum packaging of paneer in laminated or co-extended films.
8. **Yield of Paneer:** The yield of paneer depends on the amount or percentage of total solids i.e. Fat and SNF in the milk and the retention of milk solids and moisture in it. Therefore, all the factors that affect the moisture and solids retention contribute to the yield of paneer.

9.8.3 Manufacture of Paneer

The most accepted method of manufacturing paneer has been presented in the flow diagram 9.9.

Flow Chart 9.9 Manufacture of Paneer

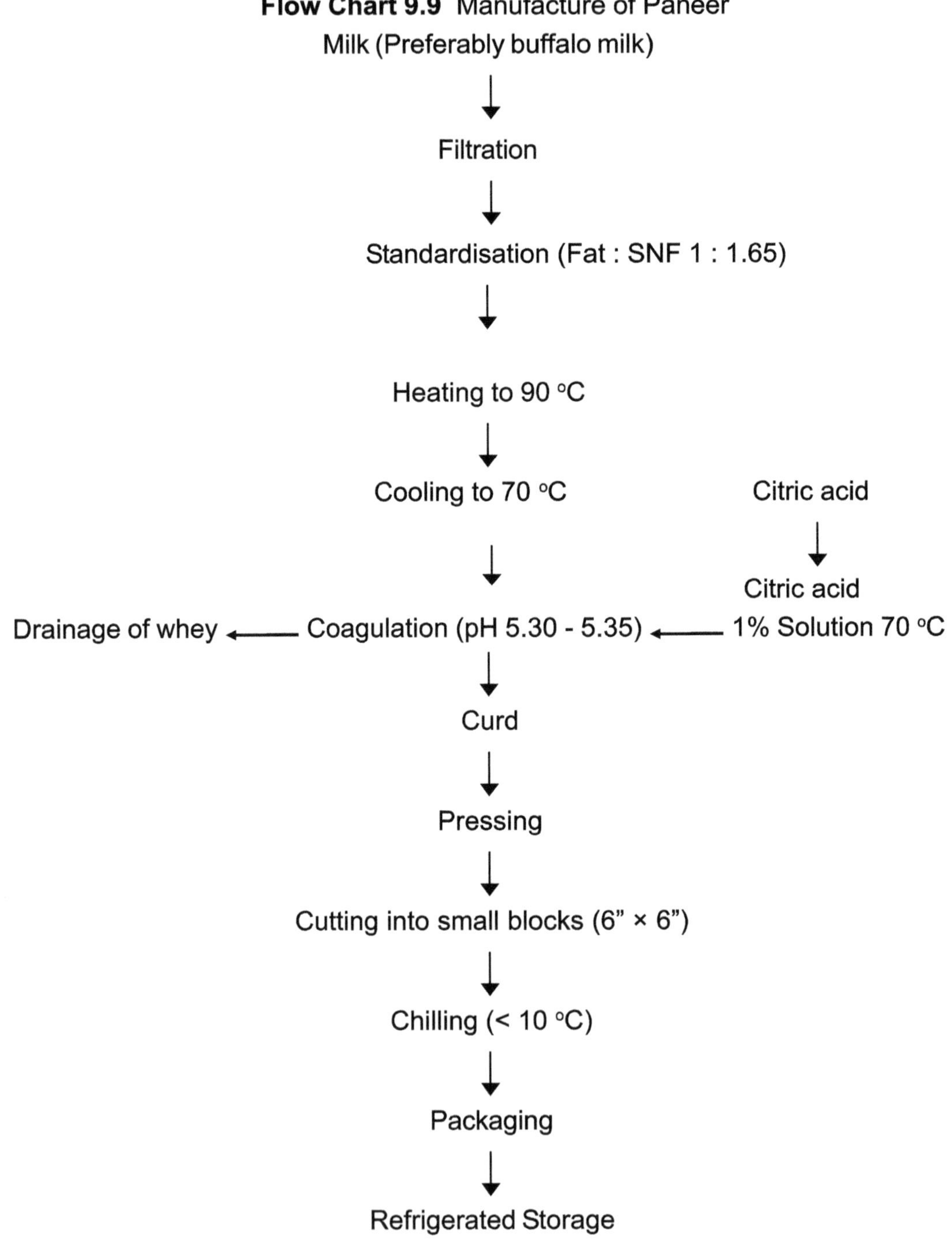

An attempt has been made to present the standards in accordance with the FSSAI (2017) for chhana or paneer for the benefit of the students of dairy science in table 9.25.

Table 9.25 FSSAI (2017) standards for chhana or paneer

Sl No.	Parameter	Chhana or paneer	Low fat chhana or paneer
1.	Moisture. Maximum % (m/m)	65.0 for chhana 60.0 for paneer	65.0 for chhana 60.0 for paneer
2.	Milk fat % (m/m) dry matter basis	50.0 (Minimum)	15.0 (Maximum)

9.8.4 Varieties of Paneer

Different variants of paneer have been available in the market which have been prepared/manufactured under certain circumstances such as health benefits and improvements in the nutritive value, modification of the normal paneer, and a product similar to paneer in sensory quality are sometimes desirable.

1. **Low Fat Paneer:** The paneer which is being prepared with conventional methods is quote rich in fat, and at the same time is also costly, sometimes making it unsuitable for those consumers who are conscious and desires to have a low fat paneer. It is now possible to prepare of a good quality from milk with fat content as low as 3.0%.
2. **Recombined and Reconstituted Milk Paneer:** Technologies have been developed to manufacture an acceptable quality of paneer from the re-constituted, whole milk powder and re-combined milk. However, the paneer made from re-constituted milk did not meet/confirm with the P.F.A rules/ requirements.
3. **Filled Milk Paneer:** Paneer of an acceptable quality has been prepared using buffalo skim milk and groundnut oil/vanaspathi, with a fat level of 5.5% in the filled milk and heating to 90 °C. The paneer thus obtained contained 16-18% protein, 22-33% vegetable fat and 55-56% moisture. Sensory quality of the product made from filled milk having 10% coconut milk was highly acceptable.
4. **Protein Enriched Filled Paneer:** A new process has been developed involving the supplementation of vegetable proteins in the form of calcium soy isolates or calcium groundnut solates to the mixture of skim milk and vegetable fat. This product appears to be superior nutritionally and economically as compared to the conventional paneer hence idealy suited to dietary management of consumers suffering from protein energy malnutrition.
5. **Use of Soy-Solids for Paneer Making:** A new technology has been developed for the preparation of paneer like product using the combination of

milk and soy-solids. According to this method the milk soy mixture is heated to an optimum temperature and coagulated using suitable coagulants. The coagulum is kept at optimum temperature for sometime till a clear and clean whey is visible. The curd is then molded and suitably pressed and can be used for the preparation of different culinary dishes. Thus paneer contained 21-24% total solids. 6.0-7.5% fat, 12-17% protein, 1.2-1.6% carbohydrates and 0.7 to 0.9% minerals.

9.8.5 In Package Paneer Process

Manufacture of paneer by membrane processing has a potential application. Ultra filtration (UF) when employed offers advantages like access to mechanisation, uniform quality, improved shelf life and increased yield with a nutritionally better product. The process involves standardisation and heating of milk followed by ultra filtration with the removal of majority of the lactose, water and some minerals along with the permeate. Ultra filtration of milk and the removal of permeate, is almost equal with the amount of whey obtained with the coagulation according to the conventional method. The concentrated mass obtained by UF Process contains about 40% total solids was cold acidified to get the desired pH and is dispensed into containers with automatic dispensing machines. The filled containers were then subjected to texturization by microwave heating in a microwave oven. This can also be achieved in a continuous process by using microwave tunnels, comprising of a series of magnetrons under which the product moves continuously on a conveyor belt.

9.8.5.1 Mechanisation Manufacture of Paneer

The important equipments needed for commercial manufacture of paneer using the conventional process include milk storage tank, cream separator, steam jacketed vat or cheese vat, curd press and heat sealing/packaging machine. The heating and coagulation of milk is carried out in a stainless steel double walled vat stirring the contents manually with a ladle. In the case of a large vat the stirring operation is done with the help of mechanical agitator/stirrer provided with a motor. A drain valve is provided at the bottom of the vat to drain the primary or the free whey after coagulation, whereas the secondary whey which is entrapped in the curd is removed during subsequent pressing separately. Steam supply line is equipped with a safety valve and a pressure guage and a water line is provided at a junction in the steam line to supply the hot water into the jacket of the vat.

9.8.5.2 Continuous Process

Batch production employing the traditional process is suitable only on a small scale, which often results in an inconsistant quality and poor hygienic quality. A continuous paneer making system has been developed recently by the scientists of NDRI, which is based on the twin apron, conveyor design. In this system, the unit operations involved in paneer making have been mechanized. The system consists of:

1. Supply of pre-conditioned milk at a desired flow
2. Acid supply and dosing
3. Transverse Jet Mixer Reactor (TJM-R) and
4. Continuous dewatering and matting system for the production of the paneer cake of 3 cm thick. It is designed to produce about 80 kg paneer/hour.

The technology that is available at present for the production of coagulum is based on the traditional manufacturing technique of coagulating the milk by acid in a steam jacketed vat. As traditional process is purely batch type, always there is a scope for the variation in the quality from one batch to another. On the other hand, the present continuous process developed at NDRI complete mixing occurs prior to the start of coagulation, thus leading to greater recovery of the product with better textural and uniform quality. The coagulant supply unit consists of stainless steel tank with agitator and steam is used for the indirect heating of the acidulent. Both the units are connected through a TJM-R, wherein effective mixing of milk and acidulant takes place finally the coagulum is fed into the de-watering unit.

The continuous de-watering and matting unit consists of twin apron conveyor to carry out primary and secondary filtration in the same machine. The lower conveyor is flanged type while the upper one is unflanged type. The flanged unit has perforations for drainage. Additional fine filter of appropriate size is placed over the conveyor to prevent the loss of the solids.

9.8.6 Preservation of Paneer

Generally the shelf life of paneer is about 7 days under refrigerated conditions even though the freshness is lost within 4 days. The heat treatment given to milk is more than enough to destroy all the pathogenic and spoilage microorganisms. Paneer blocks obtained after pressing are immersed in the water for cooling. It is during this period the microorganisms might establish themselves in the product and proliferate during storage. The spoilage in paneer primarily occurs due to the growth of the microorganisms. A greenish yellow slime formation on the surface and the discolouration is accompanied by an off colour. It is only the surface that gets spoiled early whereas the interior remains good for a longer time. The various methods of preserving the paneer have been described below.

1. **Chilling:** Rapid chilling of paneer is very much essential to arrest the growth of the microorganisms. Generally the manufacturers of paneer do not reduce the temperature to bring it down sufficiently (5-10 °C). Usually the temperature of the paneer remains in the range of 20-30 °C at the time of packaging. Such type paneer when transferred to a refrigerator or cold store it will take reasonably quite sometime to cool down to the desired/optimum temperature. The micro-organisms get fully established by that time resulting in the spoilage of the product. The bacteriological quality of the chilled water should be very high.

2. **Brining:** Paneer dipped in 5% brine solution has a longer shelf life of about 20 days as compared to the control which was kept well for 7 days at 6-8 °C. The sensory attributes are rated higher for salted samples. The paneer is generally salted and spiced before consumption, therefore salting at the time of dipping will be advantageously used in extending the shelf life.
3. **Chemical Preservatives:** The amended PFA act 2004 permits the usage of anti microbial such as sorbic acid and its salts, benzoic acid, propionic acid and its salts and Nisin as the preservatives in the storage of the paneer.
4. **Freezing:** Storage of paneer at sub-zero temperature viz –13 °C and –32 °C for about 120 days, there is no affect on the flavour and appearance, however, the body and texture have shown deterioration and the product becomes crumbly on thawing. Blast freezing has recently been used to enhance the shelf life of paneer. The product can be stored under frozen conditions per more than one year without any deterioration in its quality.
5. **Vacuum Packaging:** Vacuum packaging of paneer in laminated pouches results in the increase of shelf life to about 30 days without any deterioration in the quality of body and texture with better slicability.
6. **Heat Sterilization:** The shelf life of the paneer does not improve noticeably by the various treatments, even though, refrigeration improved the shelf life markedly. Heat sterilization of paneer is an effective treatment for improving the shelf life of paneer at room temperature. The perception of an oxidised flavour renders the product unacceptable afterwards. A slightly cooked flavour accompanied by maillard browning the intensity of which increases slightly during storage is noticeable in the product.
7. **Hurdle Technology:** Formulation gravy and paneer packed in retort pouches and subjected to heat treatments is an attractive convenience food. According to hurdle concept, preservation parameters like water activity, pH, redox potential, heat treatment etc., represses hurdles to microbial growth. A single hurdle can not preserve most of the foods alone without affecting their sensory and nutritional properties. For instance, paneer can not be effectively preserved by lone hurdle of heat treatment or dehydration or pH, because its body and texture and nutritive value get adversely affected. By using three or more hurdles together not only the damage to sensory properties is kept to minimum, but also their synergistic action is exploited.

9.9 Chhena

The indigenous dairy products have enjoyed an eminent position in indian food ethos, with their origin going back to the dawn of Indian civilization. They are intertwined into our way of life and form the part and parcel of social life, festivals and ceremonies. The concentrated and preserved milk solids in the form of milk sweets, provide sound nutrition and novelty of flavour and texture to consumers. According to a study it has been reported that about 53.5 percent of milk produced in our country is converted to various products in which, the share of indegenous products is about 50 percent.

The indigenous products of the country are grouped into:

1. Heat desiccated products such as Khoa and Khoa based products/sweets and burfi, Kalakand, Peda, Rabri, Kulfi and basundi etc.
2. Heat acid coagulated products such as paneer and Chhena and Chhena based sweets eg. rasogolla, paneer, sandesh, Chhena poda etc.
3. Fat-rich products like ghee, makkhan.
4. Fermented products like, dahi, Srikhand, Shrikhand wadi and
5. Milk based puddings or desserts like Kheer, Savian, Payasam, Pal Payasam etc, of these heat desiccated products are popular in the northern and western regions, heat-acid coagulated products in the eastern and northern regions, whereas the fermented and fat rich products are popular almost in all parts of the country.

Chhena is one of the heat acid coagulated products of milk and is extremely popular in the eastern parts of the country. The pattern milk consumption indicates that about 6 percent of milk produced in the country is converted into Chhena. The market volume of Chhena based sweets is about 1 million tonnes worth Rs.7000 crores. Chhena forms the base and filler material for a large variety of Indian sweet delicacies like rasogolla, rasamalai, sandesh, chhana kheer, cham-cham, chhena poda etc.,

9.9.1 Chemical Composition

Chhena (or paneer) is defined as the product obtained from cow or buffalo milk or a combination thereof by the addition of lactic or citric acid according to the PFA rules 2006. In accordance with the PFA and BIS, chhana should not contain more than 70% moisture, and milk fat should not be less than 50% of the dry matter. The average chemical composition of chhana obtained from various sources has been presented in the table 9.26 and also depends upon the type of milk, condition of coagulation, types of coagulant and its strength, thermal treatments given to the milk prior to acidification and method of straining.

Table 9.26 Average Chemical composition of Chhena obtained from various sources

Source	Constituents %					Reference
	Moisture	Fat	Protein	Lactose	Ash	
Cow milk	53.4	24.7	17.6	2.2	2.1	Anantha Krishnan et.al (1964)
Buffalo milk	51.6	29.6	14.5	2.4	1.9	- do -
Laboratory	58.8	22.0	17.3	--	--	Singh & Ray (1977)
Goat Milk	55.3	23.5	17.2	2.2	1.6	Jagtap et.al (1973)
Cow milk	53.1	24.8	17.8	2.2	2.1	Kumar et.al (1982)
Buffalo milk	51.7	29.7	14.4	2.3	1.9	- do -
Market milk	62.3	16.1	17.1	2.2	2.2	- do -
Cow milk	55.7	23.5	16.4	2.2	2.0	Sen (1986)
Buffalo milk	54.9	24.3	17.3	2.1	2.2	Singh (1994)

Source : I.D.A. Publication

Chhena may be called as a variety to fresh soft cheese, however it is somewhat different in its characteristics due to the differences in the temperature of coagulation and subsequent handling. Low coagulation temperature is employed for cheese, whereas, boiling of milk is usually required for chhana preparation. Unlike cheese freshly prepared chhana is free from acidic smell, because much lesser amount of acid is used at higher temperature of coagulation. The typical average chemical composition of Chhena, Paneer, and other cheeses has been given in the table 9.27 for comparison.

Table 9.27 Typical chemical composition of Chhena, Paneer and other cheeses

Details	pH	(Constituents %)				
		Moisture	Protein	Fat	Lactose	Ash
Chhana (Buffalo milk)	5.4	51.7	14.4	29.7	2.3	1.9
Paneer	5.8	51.3	17.3	27.0	2.3	1.9
Queso Blanco	5.2	55.0	23.0	15.0	2.5	1.9
Mozandla	5.2	54.0	22.1	16.1	2.0	0.7
Cheddar	5.6	37.0	25.0	32.2	--	1.2

Source : I.D.A. Publication

9.9.2 Traditional Chhena Making Process

Chhena is prepared in the cottage industry and in the process milk is heated to its boiling point in an open pan or in a steam jacketed vessel, which is allowed to cool to 75-80 °C. The citric acid at 1-2% saturation is added to the milk with slow and constant stirring with the help of a ladle. The process is continued until the milk precipitates into lumps, which settle down at the bottom involving the formation of large structural aggregates of milk protein, in which milk fat, other colloidal and soluble solids are entrained with whey. The clear whey floating on the top is filtered through a muslin cloth so as to obtain chhena. The traditional method of chhena manufacture has been given in flow chart 9.10.

Flow Chart 9.10 For Chhana Manufacture from Buffalo Milk.

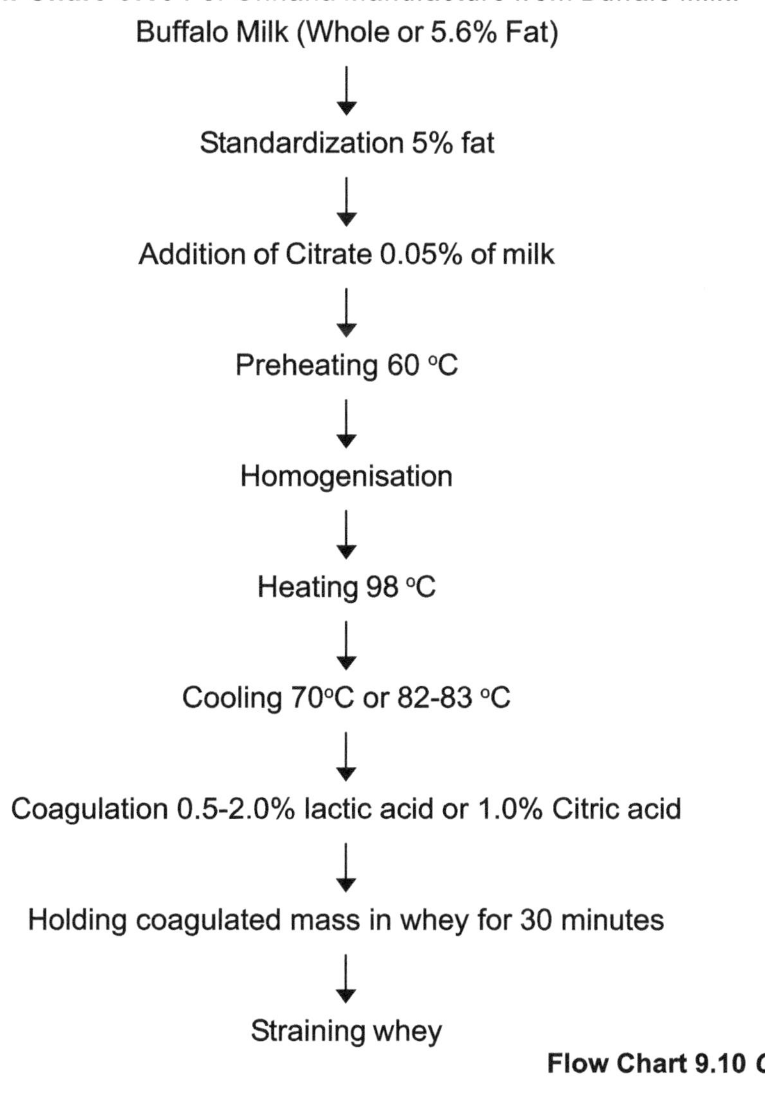

Flow Chart 9.10 ***Contd......***

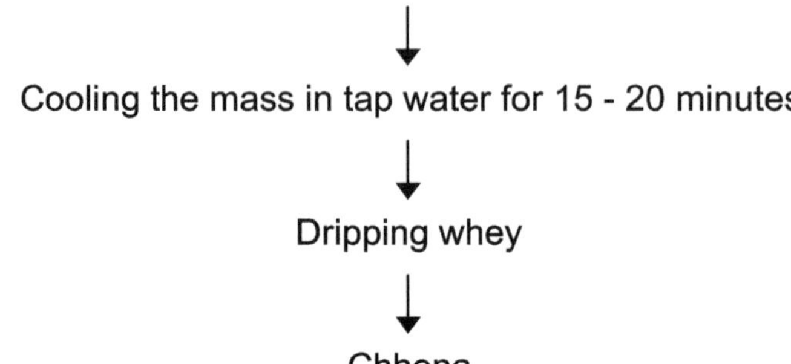

The traditional method of chhana making is cumbersome, time consuming, labour intensive and difficult to attain high quality standards. Some of the limitations of the existing practice have been enumerated briefly hereunder:

1. Heating and cooling rate of milk prior to acid coagulation is not carried out optimally.
2. Heat treatment of milk and its acid coagulation is carried out batchwise in an open vat.
3. The process consumes long time.
4. There is no uniformity in the process parameters resulting in the variations in the quality of chhana from batch to batch.
5. Heating, cooling and coagulation of milk are carried out as three independent unit operations.
6. Cooling of heated milk to 70 °C its acidification, involves wastage of time energy, and the coolant, and
7. Human intervention during the various unit operations leads to microbial contamination of chhana and reduces its shelf life.

Consequently, the chemical composition and organoleptic properties of chhana and chhana based sweets vary to a great extent. Absence of cold storage facilities and cold chain during the distribution leads to the quality deterioration of chhana due to its high moisture content. Therefore, the traditional method of chhana making is too rudimentary to meet the large scale demands, hence, there is a need to develop time-energy and/or cost efficient process suitable for industrial production and packaging of chhana based products.

Factors affecting the quality and yield of chhana

The body and texture, recovery of total milk solids and yield of chhana are influenced by usual factors such as type of milk, coagulant and its strength, heat treatment given to the milk before acidification as well as the method of straining the whey.

9.9.3 Mechanisation Process for Chhana Making

Inspite of large production and profit margine on chhana and chhana based sweets, knowledge base available on the technology of production, packaging and preservation of chhana is rather limited. In response to the changing life style and emerging consumer demand for food safety and extended shelf life of the product, mechanization of chhana and chhana based sweets is a necessity of the day. Mechanization would also help to improve produce value added by-products and reduce energy consumption.

Mechanization of manufacture of chhana and chhana based sweets requires:

1. Identification of industries and their involvement right from start,
2. Development of bench scale devices and optimization of process variables and
3. Developments of Pilot Plant equipment and their commercial production. The following parameters should be taken care of for successful mechanization.
   (a) Existing process flow and material balance sheet with optimum process parameters should be prepared.
   (b) The flow process should be modified to replace the manual operations by machines. The machines should be designed and fabricated as to have 100% cleanable conforming to the hygienic design standards.
   (c) Process automation should be employed. Wherever possible to ensure uniform quality product, and
   (d) Linkage between R and D organisation, equipment, manufacture and processors should be established.

NDRI Karnal has developed a proto type unit for continuous, dewatering and matting of acid coagulated milk curd to produce paneer in a continuous manner. The system consists of a plate heat exchanger, milk coagulation system and curd pressing system together with other accessories such as balance tank, pump, flow meter and temperature indicator.

Indian Institute of Technology, Kharagpur has demonstrated a continuous, chhana making unit of 60 l/hr of milk. In the unit, dosing of milk and acid is carried out by using a duplex plunger pump. Helical coil heat exchanger is used for heating the milk prior to acid coagulation.

After mixing the acid with the heated milk, the mixture is allowed to move through a vertical column which gives the required residence time for the separation of milk solids to chhana. Pulsating motion given to the coagulated milk acid mixture by the dosing pump facilitates the formation of lumps in chhana, which is one of the derived characteristics of chhana from textural point of view. The coagulated chhana whey mixture, after it is discharged from the top of the column is moved over an inclined strainer through which the whey is removed. Cooling of chhana can be carried out by using a water spray on the chhana mass, which moves over the straining unit of the device. The water spray can reduce the lactose content in chhana and extend its shelf life. The unit so developed can be combined with equipments for mechanized production of rasogolla, sandesh, chhana poda or other Indian milk sweets derived from chhana.

Shelf life of Chhana

The average shelf life of chhana is about 3 and 12 days at 24 °C and 7 °C respectively. Freshly made chhana wrapped in vegetable parchment paper could be preserved in good condition for 3-4 days at 21-27 °C and for about 10 days in a refrigerator.

9.9.4 Innovation Process in Chhana Making

Ultra Filtration Technique

This process has been successfully used for the concentration of milk by removing most of the soluble milk solids. Process automation and control are some of the advantages of this process. The process and flow of material for chhana making by ultra filtration method has been presented in flow diagram 9.11.

9.9.5 Dry Chhena Powder

As already stated chhana is an extremely perishable product and in order to increase/ extend the shelf life of chhana, technology has been developed for production of chhana powder by using tray roller and spray drying processes. The product could be reconstituted to chhana, suitable for sandesh making.

9.9.6 Active Packaging for Chhena

Development of suitable packaging is essential for extending the shelf life of chhana based products. Addition of compounds extending the shelf life to packaging films (active packaging) rather than adding them to food can be used to provide continued product stabilization.

9.9.7 Chhena based Sweet Products

The major strength of chhana and chhana based sweets, is the low capital investment and high profit margin as compared to the western dairy products. The rapidly growing demand and assured market for the chhena based sweets present a great opportunity for the organised dairy sector to produce and market these products. Attempts have been made by various institutions and R & D organisations to develop or modify the equipments for mechanized production of chhana and chhena based products. Emphasis has also been primarily laid upon characterisation of sensory, rheological and physico-chemical properties of the products and developments of technologies and use of cost effective packaging materials for extending their shelf life.

Flow Chart 9.11 Process of Chhana making by Ultra Filtration method

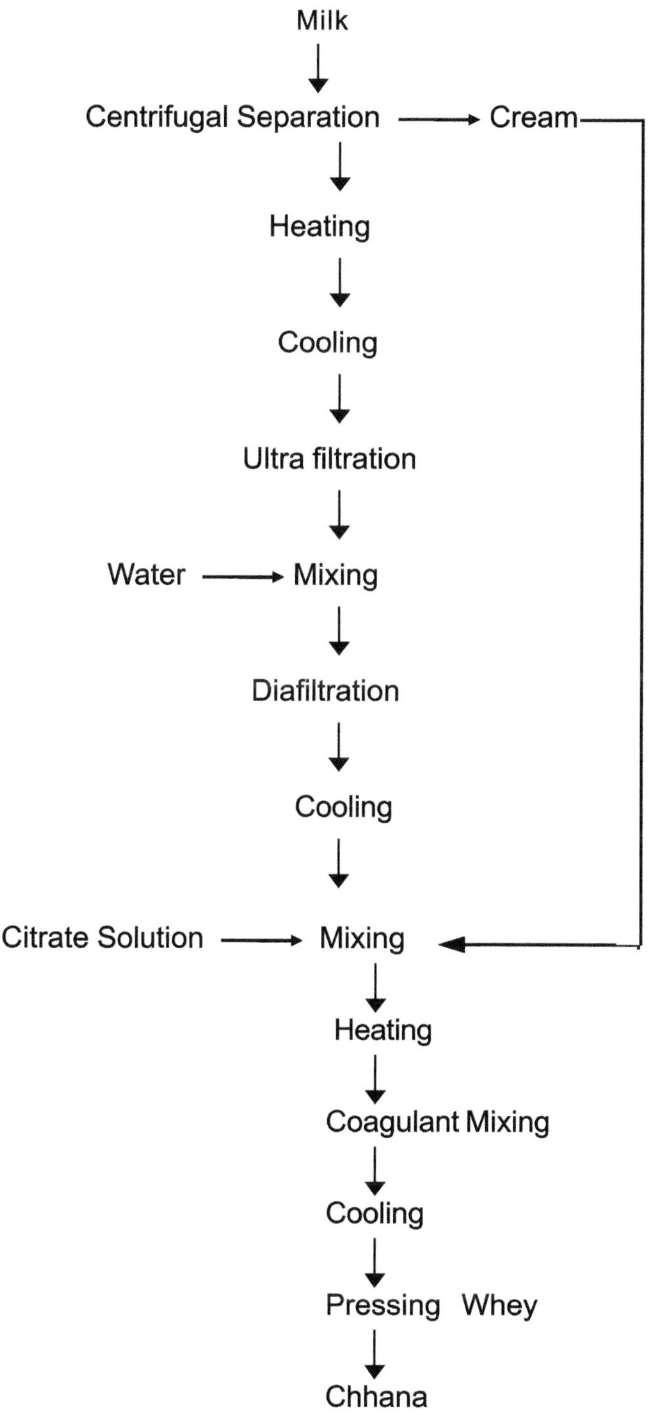

9.9.7.1 Rasogolla

The base material for the preparation of rasogolla is chhana and is prepared by the texturisation of milk proteins under controlled cooking of chhena balls in boiling sugar syrup to form an inter woven fiber like structure. The controlled cooking in sugar syrup imparts a chewy and spongy characteristics to rasogolla balls by improving their syrup holding capacity. The average chemical composition of rasogolla has been presented in the Table 9.28.

Table 9.28 Average chemical composition of Rasogolla (%)

Constituent	Rasogolla		BIS Standard
	Cow Milk	Buffalow Milk	
Moisture	50.0	52.8	55-0 (max)
Fat	5.1	4.2	5-0 (max.)
Protein	6.2	6.4	5-0 (min.)
Sugar	37.7	36.1	45.0 (max)
Ash	1.0	0.5	---

Source : I.D.A. Publication

The method of manufacture of Rasogolla has been given in flowchart 9.12.

Flow Chart 9.12 Production of Rasogolla

Chhena

↓

Kneading (by tapered screw extruder)

↓

Ball making

↓ Chhena balls

Cooking at atmospheric pressure in sugar syrup for 25 min.

↓ Cooked chhena balls

Soaking

Swelling of cooked chhena balls

↓

Rasogolla

Modification of Buffalo milk for Rasogolla Making: Generally cow milk is preferred for chhana based rasogolla production because it gives a soft textured body. Buffalo milk produces chhena with hard body and coarse texture besides greasy surface due to high calcium content thus making it unfit for rasogolla preparation. A mixture of buffalo and cow milk in the ratio of 1:3 produces chhana with moist surface, light yellow colour, soft body and smooth texture thus making it suitable for rasogolla and sandesh.

9.9.7.2 Sandesh

It is very popular for its palatability and aroma and is a good source of milk protein and fat. Chhena and sugar are the raw materials in the preparation of sandesh. Kneading and mixing of chhana sugar mixture forms the base and an important unit of operation in the preparation of sandesh. The chemical composition of the sandesh as well as the manufacturing process has been given in the table 9.29 and the flow diagram 9.13 respectively.

Table 9.29 Average chemical Composition of Sandesh (%)

Constituent	Sandesh	
	Cow Milk	Buffalow Milk
Moisture	25.50	27.14
Fat	19.89	18.42
Protein	18.48	18.71
Sugar	34.47	33.83
Ash	1.66	1.90

Source : I.D.A. Publication

9.9.7.3 Chhana Podo

It is a popular milk sweet in the eastern parts of India, which is baked sweet with caramelized colour and flavours. A mechanized production line of chhana podo has been developed and standardised by the NDDB. In this process the chhana is mixed with the required amount of sugar, suji and water a dough mixer. The chhana podo premix is transferred to rectangular aluminium pan previously smeared with ghee. The pan is then transferred to a hot air oven maintained at a temperature of 200 + 5 °C for 50 minutes. The baked product is cooled at ambient temperature to 40 - 45 °C leading to the development of puffed, uniformly brown product with desirable body and textural attributes. The pan is then reheated to 50-55 °C for the melting of the ghee which facilitates de-panning. The product is cut into pieces, packed in polythene pouches and stored under refrigeration. A flow chart 9.14 for the commercial production process of chhana-podo is presented.

Flow Chart 9.13 for production of Sandesh

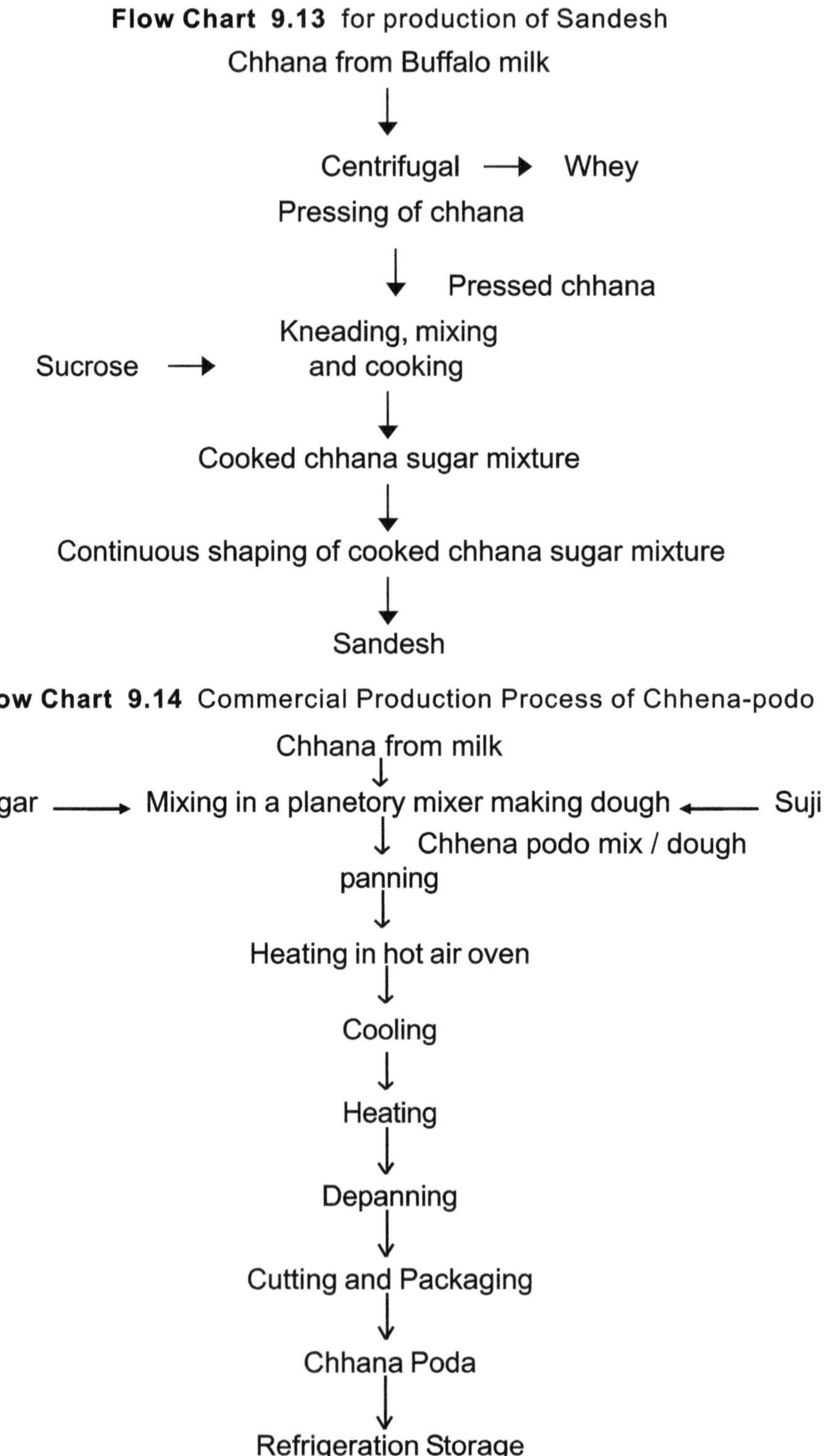

The emergences of new avenues for the production and marketing of Indian milk products in the western countries, chhana based products have gained much importance. The following steps need to be taken for the global marketing of chhana based products.

1. The procurement of milk with improved efficiency and quality control methods is essential in order to ensure the physico-chemical and microbiological quality of the products with regard to freshness, appearance, colour, nutritive value, wholesomeness composition as well as human safety.
2. Good manufacturing practices (GMP) is to be adopted in plant design, process control and hygienic processing of milk.
3. Packaging and storage of chhana for its use and conversion into chhana based products is to be aimed through modern packaging systems.
4. The labeling and quality assurance programmes is to be in conformity with the international standards.
5. Indian manufacturers must adopt the quality and safety standards of chhana and chhana based products conforming to the FDA standards by keeping pace with international developments.

9.10 Dahi/Curd

The conversion of milk into dahi/curd is an intermediary step in the production of indigenous butter/makkhan and ghee. The curd in our country is also called dahi (Sanskrit) is a well known fermented milk product, which is consumed by the majority of the population belonging to all sections throughout the length and breadth of the country irrespective of the relegion either as a part of the daily diet or as a refreshing beverage in the form of lassi. There are two types of dahi prevalent in the country namely sweet/mildly sour variety with a pleasant flavour and a sour variety with a sharp and strong acid flavour for direct consumption.

According to PFA Rules (1976) dahi or curd is the product obtained from pasteurized or boiled milk by souring natural or otherwise, with a harmless lactic acid or other bacterial culture. Dahi may contain additional cane sugar and should have the same percentage of fat and solids not fat as the milk from which it is prepared. When dahi or curd, other than skimmed milk dahi, is sold or offered for sale without any indication of the type or class of milk, the standards prescribed for dahi prepared from buffaloe milk shall apply.

The FSSAI (2017) standards/specifications for designation of fermented milk products based on the types of culture used and the requirements have been given in the tables 9.30 and 9.31.

Table 9.30 FSSAI (2017) standards for fermented milk products

Sl No.	Product	Starter Cultures
1.	Dahi (curd)	Lactic acid bacteria
2.	Yoghurt	Symbiotic cultures of strept coccus thermophiles and lactobacillus delbrueckii sub sp bulgarius
3.	Alternate culture yoghurt	Cultures of streptococcus acidophilus and lactobacillus species
4.	Acidophilus milk	Lactobacillus acidophilus

Note: Other harmless microorganisms than those constituting the specific starter cultures specified above may also be added.

The average composition of dahi prepared from the whole milk is given below on percentage basis.

Water	-	85-88
Fat	-	5-8
Protein	-	3.2 -3.4
Lactose	-	4.6-0.72
Ash	-	0.70-0.72
Lactic Acid	-	0.50-0.11

Dahi may be classified into two types of categories.

1. For the preparation of indigenous butter or makkhan by churning.
2. For direct consumption, which is further classified into
   - (a) (i) Whole milk dahi
     - (ii) Skimmed milk dahi
   - (b) (i) Sweet or mildly sour dahi
     - (ii) Sour dahi
     - (iii) Sweetened dahi

It is beyond doubt that the fermented milk products including dahi increase in their food and nutritive value as compared to the milk from which they are prepared due to.

1. People who do not like to consume milk readily accepts dahi because it is more palatable.
2. It is more easily digested and assimilated.
3. It appears that dahi exerts a therapeutic value in the stomach during intestinal disorders, possibly due to its content of antibiotics.

The method of manufacture of dahi on commercial basis has been given in flow chart 9.15.

Flow Chart 9.15 for the Manufacture of Dahi

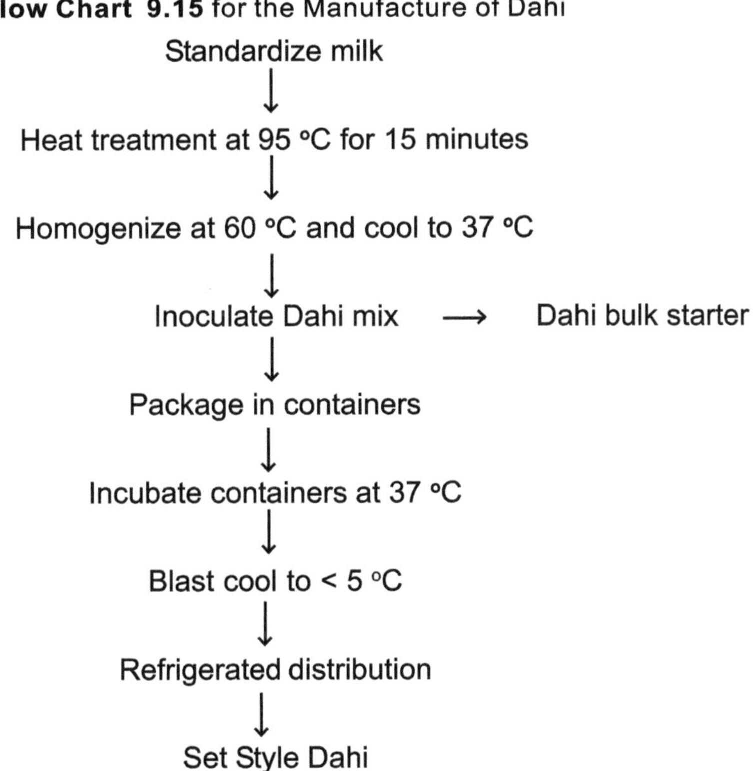

9.10.1 Production Methods

Dahi is prepared/produced either by traditional method or by an improved standardised method in our country to meet the requirements of the following namely:

1. For the production of desi butter/makhan which in turn is being used in the production of ghee.
2. For direct consumption.

9.10.1.1 Traditional Method

The preparation of curd on a small scale basis either in the consumer's household or in the sweet maker's shop in the urban areas, invariably involves the traditional method. In the household the milk is boiled and cooled to the room temperature inoculated with 0.5 to 1.0% starter, generally consists of previous days dahi or buttermilk and is allowed to settle undisturbed for overnight. In cool weather conditions, the dahi setting vessel is generally wraped up with a woollen cloth in order to maintain the warmthness. In the shops, more or less the same procedure is followed except that the milk is concentrated to a little extent, before inoculation and the dahi is allowed to set in a circular earthenware mould.

9.10.1.2 Standardised Method

Generally, the quality of the dahi sold in the market by the halwai shops in our country is substandard and variable in its composition and the factors that are responsible for this kind of solution/condition are as under:

1. Use of low quality milk.
2. Use of unsuitable starter culture.
3. Unfavourable temperatures of incubation.
4. Contamination from the uncleaned utensils and unsanitary conditions/ premises of the halwai shops.

In view of the factors mentioned above as well as the nutritional and economic importance of dahi, there is an urgent need to produce the product on scientific basis in the organised dairy sector to ensure the supply of an uniformly good quality product every day.

The method of production of dahi by the use of the standardised method has been given in a flow diagram/chart 9.16.

9.10.2 Quality Control of Dahi

These programmes focus on product viscosity, flavour, body and texture, colour fermentation process and composition. The ingredients used may impart undesirable flavours such as stale, metallic, oxidized, rancid or unclean, lack of control in processing procedures may cause over cooked, caramelized or excessively sour flavours in the products. Proper control of processing parameters and ingredient quality ensures good flavour. Some of the common quality problems of Dahi and the steps that are needed to rectify them have been summerized in the table 9.31.

Flow Chart 9.16 for Mass Production of Dahi

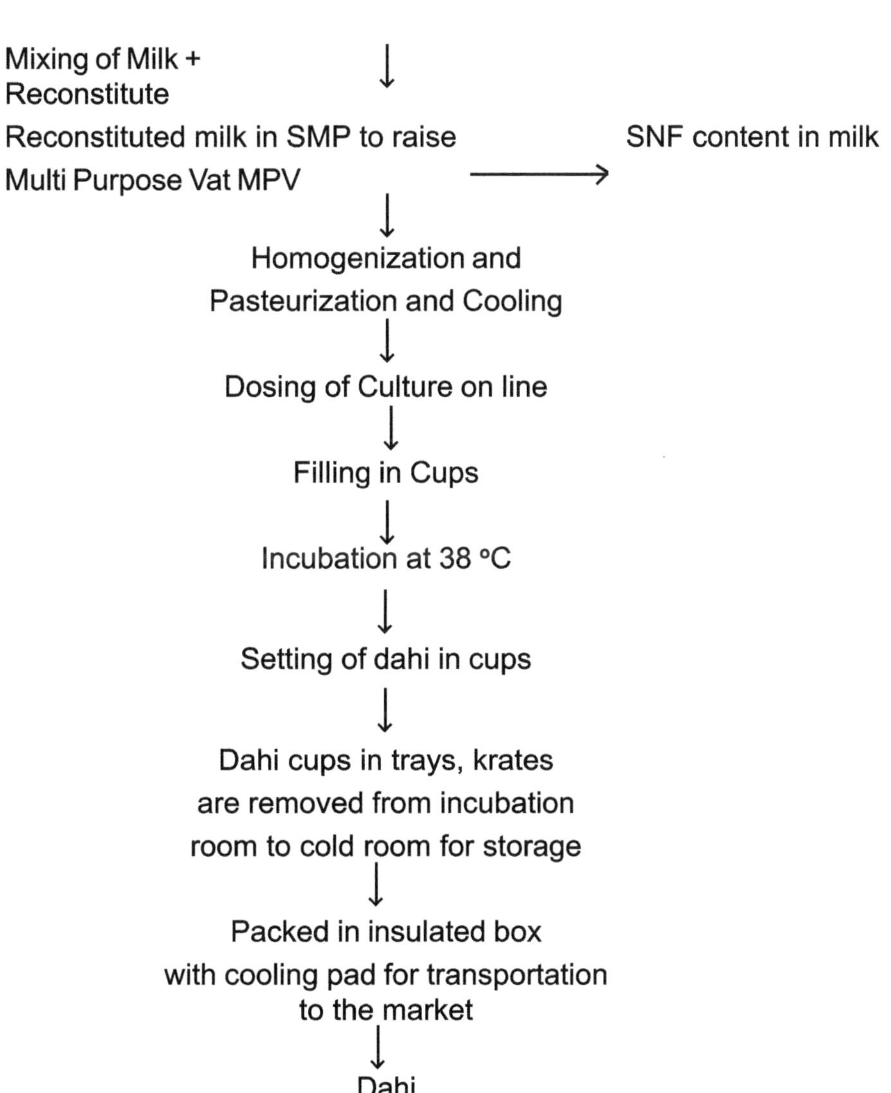

Table 9.31 Common quality Problems in Dahi

Defect	Probable cause	Remedy
Insufficient flavour	1. Low citrate level in milk 2. Diacetyl destroyed	1. Add 0.02 to 0.05% (upto 0.1%) sodium citrate prior to the mix 2. Cool rapidly after culturing 3. Agitate gently after breaking to incorporate oxygen in the product 4. Avoid the usage of lactococcus diacetyl lactis
Green/Yoghurt flavour	Accumulation of accetal dehyde	Avoid the usage of lactococcus diacetyl lactis
Oxidized (card board) flavour	Copper contamination and/or exposure to flouroscent light or Sunlight	1. Avoid exposure to copper utensils 2. Protect product from direct sun light / UV light exposure
Yeast like/ cheesy	Contaminating yeast growth	1. Santation check 2. Avoid return milk
Rancid flavour	Hypolytic activity	Do not mix pasteurized and raw dairy ingredients prior to homogenization.
Weak body	1. Insufficient heat treatment of the mix 2. Too low content of milk SNF 3. Too severe agitation after fermentation	1. Heat treatment should not be less than 85 °C/30 minutes for Dahi/Lassi and 95 °C/30 minutes for Yoghurt and sour cream 2. Fortify with 0.5-1% skim milk powder for dahi and lassi and 3-4% for Yoghurt 3. Homogenize the dahi mix prior fermentation.
Grainy texture	1. Too high acidity 2. Skim milk powder is not dispensed properly.	1. Exercise rigid acidity control 2. Check the dispersion equipment 3. Use in-line screen
Chalky / Pardoxy texture	Too much of skim milk powder	Check on the quality and quantity of milk powder

Sweetened Dahi

In West Bengal and other eastern parts of the counrty, a sweetened variety of dahi known as misti dahi, lal dahi or payodhi is quite popular, which has a characteristic brown colour, a cooked and caramelized flavour with a firm body. Commercially it is prepared by adding 6.25% cane sugar to the milk of either Cow, Buffalo or mixed milk before boiling or at the time of setting. The pronounced and intense heating causes the milk to brown and get partially concentrated, reducing the volume to about (3/4) three fourths of the original. Artificial, colour, sugar caramel and gur (jaggery) are also added during production. After heat treatment the milk is cooled to room temperature and then seeded/inoculated with the previous days product. Generally it is set in earthernware pots/basins and the finished product is obtained after 15-16 hours.

Dahi which is prepared in the conventional manner, has a short keeping quality at room temperature, however on prolonged storage it becomes highly acidic, which is accompanied by the formation of whey, thus making the product unfit for human consumption. Under refrigerated conditions of storage (5-10 °C) dahi is usually keeps well for a maximum period of one week.

9.11 Srikhand

It is defined as a semi soft, sweetish, sour, whole milk product prepared from lactic fermented curd. The curd/dahi is partially strained through a cloth in order to remove the water or whey resulting in a solid mass called "Chakka", which is the basic material for the preparation of Srikhand.

The chakka is then mixed with the required quantity of sugar and condiments to yeild srikhand, which is further desiccated over an open pan to make the Srikhand wadi sweet. These products are very popular in Gujarat and Maharashtra States.

The composition of chakka will depend on the intial quantity and composition of milk, degree of fermentation and the extent of whey removed along with the amount of sugar added. Further, the composition also depends upon the extent to which the srikhand is desiccated in the case of Srikhand wadi.

The authors have made an attempt to update the standards for chakka and srikhand wadi as prescribed by the FSSAI act (2017) and published in the Gazette govt. of India have been presented in table 9.32 and 9.33 for the better understranding by the students of dairy science as well as by the dairy entrepreneurs.

Table 9.32 FSSAI (2017) standards for chakka

Parameter	Chakka	Skimmed milk chakka	Full cream chakka
Total solids minimum % (m/m)	30.0	20.0	28.0
Milk fat % (m/m) on dry bases	33.0 (minimum)	5.0 (maximum)	38.0 (minimum)
Milk protein ®minimum % (m/m) on dry basis	30.0	60.0	30.0
Titrible acidity max % as lactic acid	2.5	2.5	2.5
Total ash max % (m/m) on dry basis	3.5	5.0	3.5

®Protein content is 6.38

Table 9.33 Standards for Srikhand FSSAI (2017)

Parameter	Srikhand	Full cream Srikhand	Fruit Srikhand
Toal solids minium % F (m/m) on dry basis	58.0	58.0	58.0
Milk fat minium % (m/m) on dry bases	8.5	10.0	7.0
Milk protein ®minimum % (m/m) on dry basis	9.0	7.0	6.0
Titrible acidity max % (as lactic acid)	1.4	1.4	1.4
Sugar source max % (m/m) on dry basis	72.5	72.5	72.5
Total ash max % (m/m) on dry basis	0.9	0.9	0.9

®Proteine content is 6.38

Standardised, fresh, sweet buffalo milk with 6 percent fat is pasteurised at 71°C for 10 minutes and then cooled to 28-30 °C. It is then inoculated at the rate of 1 percent with lactic culture, mixed well and incubated at 28-30 °C for 15-16 hr preferably over night. The curd when set firmly having an acidity of 0.7 to 0.8 percent lactic acid, etc. is broken and placed in a muslin cloth bag and hung on a peg or hanger for removal of the whey for about 8-10 hr. During this period the position of the curd may be altered or the curd may be squeezed gently in order to facilitate the drainage of the whey. The solid mass thus obtained is called as Chakka which is the base for the prepation of the Srikhand, which is admixed with sugar either in crystalline or ground form and kneaded well for the uniform mixing, colour, flavouring substances and condiments may also be added. The final product thus obtained

Flow Diagram 9.17 for Production of Shrikhand

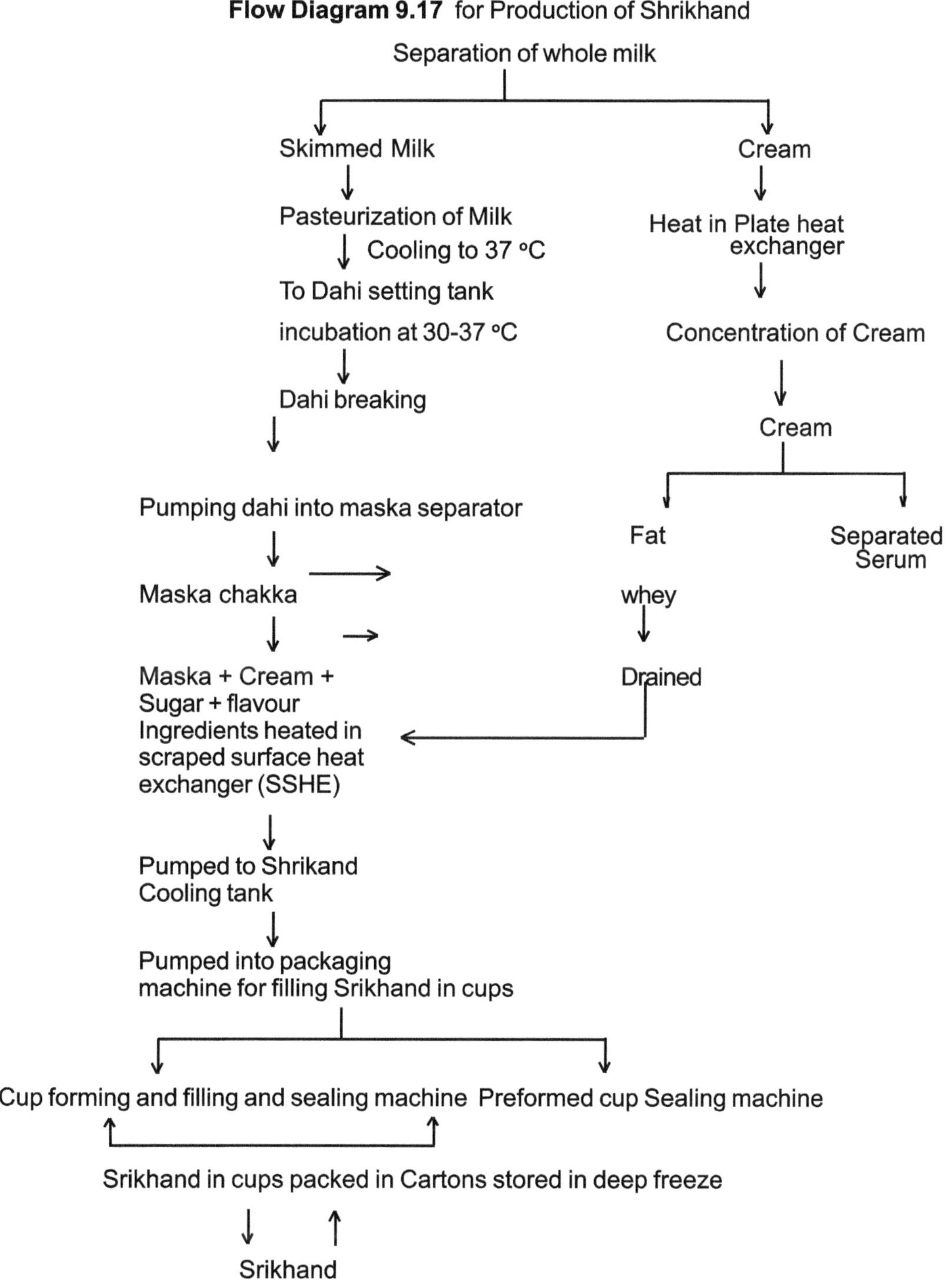

now is called Srikhand. This srikhand, when further desicated to a hard mass by heating in a open pan under direct flame or fire, srikhand wadi is obtained. The method of manufacture of srikhand commercially has been given in flow chart 9.17.

Sensory Profile of Traditional Dairy Products

For millennia, traditional dairy products have been contributing to the health, nutrition and delight of discriminating consumers in the Indian sub-continent. To establish an efficient, economic and profitable system for their industrial scale production, a proper understanding of these products' sensory profile is essential. This knowledge will help to design and develop product innovations and variations to meet the new demands of the market. The desirable sensory attributes of major traditional products are tabulated in Table 9.34.

Table 9.34 Sensory Attributes of Traditional Dairy Products

Product	Flavour	Body and texture	Colour and appearance
Khoa	A typical mildly cooked flavour similar to that perceived from boiled milk is most acceptable. The taste should be pleasantly sweet.	Soft and uniform body and granular texture is most desirable. Pindi khoa shall have smooth, compact, homogeneous texture with very fine grains. Dhap khoa shall have granular texture and slightly soft body. In Danedar khoa, presence of big grains with brown colour is desirable.	Cow milk khoa shall be pale yellow with a tinge of brown having moisture surface, whereas buffalo milk khoa shall be white with a tinge of brown having slightly greasy/oily surface.
Peda	Cooked to slightly heated flavour with sweet taste	Soft, greasy to dry body with grainy texture	White to brown colour with absence of burnt particle
Burfi	Mildly caramelised and pleasant flavour with sweet taste	The body characteristics of burfi may range from very loosely compacted to close-knit body. The texture could also vary from smooth granular and crisp to chewy	The colour may range from off white to creamy or light caramel, depending upon the type of milk solids used as base material and also the extent of heat desiccation during preparation. It should be free from burnt particles
Kalakand	Fresh, clean, pleasant caramel sweet flavour	Cohesive body with granular close knit texture	Off white to light brown with absence of burnt particles
Gulabjamun	Typical heated fresh aroma, taste moderately sweet, free from doughy feel and fully saturated with syrup.	Soft and thin crust, smooth granular texture, soft and spongy, free from lumps and hard central core.	Light to yellowish brown, uniform, round/elongated shape, smooth, glossy, moderate size.

Table 9.34 ***Contd...***

Product	Flavour	Body and texture	Colour and appearance
Basundi	Pleasant caramelised flavour	Creamy consistency and viscous body with soft textured flakes uniformly suspended throughout the product.	Cream to light caramel colour
Rabdi/Rabri	Pleasant caramelised flavour	Creamy consistency and viscous body containing several layers of clotted cream with a chewy texture	Creamy white to light caramel in colour
Kheer	Sweet, nutty, pleasant flavour	Thick, viscous mass with uniformly distributed rice	White to slightly brownish. A rich creamy shade is preferred
Chhana	Mildly acidic smell and pleasnat sweetish taste are considered desirable flavour attributes.	For rasogolla, chhana should have moderately soft and uniform texture and shall show slight springiness, when pressed with thumb ; upon rolling between palms, it should yield round ball of even surface and no cracks. It should not release fat on kneading/working.	It should have uniform yellow colour (cow milk) to whitish colour (buffalo milk, calcium lactate as coagulant). It should be neither too moist nor dry, and absence of visible soil of burnt particles.
Paneer	A characteristic blend of flavour of heated milk and acid, i.e. pleasant mildly acidic and sweet (nutty).	It should be sufficiently firm to hold its shape during cutting/slicing yet tender enough not to resist crushing during mastication. The texture should be compact, smooth and velvety.	Uniform, pleasing white colour and greenish tinge (buffalo) and light yellow for cow milk paneer.
Sandesh	Typical cooked and heated aroma, sweet, fresh creamy taste.	Soft grade is cohesive, smooth and little grainy. Hard grade is crumbly, smooth and has fine grains.	Yellowish white and uniform in colour. Attractive size and shape, free from mechanical holes.
Rasogolla	Pleasant flavour, moderately sweet, free from doughy feel and fully saturated with syrup.	Soft body and maximum sponginess, free from lumps and hard centre.	White colour, round shape, smooth, moderate size.
Dahi	Pleasant sweetish aroma of diacetly and clean acid taste is desired. Skim milk dahi lacks this.	A weak gel-like junket, a creamy layer of fat if whole milk is used. Homogenous body, while cut surface is trim and free from cracks and gas bubbles.	Creamish yellow for cow milk and creamish white for buffalo. It should have smooth and glossy surface and no free whey separation.
Shrikhand	A clean, pleasant sweet-sour flavour representing blend of sugar and fermented milk solids.	Typical semi-solid uniform consistency showing a characteristic firmness and pliability and shall show a smooth texture and no graininess.	Uniform colour and glossy appearance and devoid of free fat and syrup separation.
Mishti doi	Pleasant sweet fermented flavour	Firm consistency with smooth texture	Uniform cream to light brown colour

Table 9.34 *Contd...*

Product	Flavour	Body and texture	Colour and appearance
Ghee	A natural sweet, pleasant nutty, slightly cooked / caramelized aroma and an aggreable taste	A good textured ghee requires large and uniform grains with very little liquid fat. A greasy texture is objectionable.	When melted, it should be clear, transparent and free from sediment and uniform throughout. It should be bright yellow for cow and white with greenish tinge for buffaloe milk.
Lassi	Sweetish rich aroma and mild to high acidic taste, flavoured either with salt or sugar depending on regional preference	Homogenous and viscous liquid	White to creamy white

Source : Dr. Suneeta Pinto, Professor, Dairy Technology, SMC College of Dairy Science, Anand, Gujarat.

Chapter 10
Dairy by-products

In any food industry, the disposal or the proper utilization of the by-products invariably poses a major problem during the manufacture of the main products. Similarly, during the manufacture of the dairy products, the utilization of the by-products also creates a problem, because of their unique and great nutritional value. The most profitable and the most rational method is the utilisation of these by-products in the food products. It has been well recognized that the economic disposal of the by-products is an essential pre-requisite to the profitable dairying. It is not justifiable to convert the edible substances into non-edible food items, in countries where there is an overall shortage of liquid milk supplies. It appears that an increasing interest has been developed in a wide spread manner throughout the world in creating newer channels or finding new methods for the utilization of the by-products of the dairy industry.

A dairy by-product may be defined as a product of commercial value produced during the manufacture of a main product. However, it should also be borne in mind that there is a possibility for a by-product of the present dairy may become the main product in the future.

The most important dairy by-products of the developed dairying and those of the Indian dairy industry have been given in the table10.1 for a comparative study.

Table 10.1 By-products of Western and Indian dairy industry

Main product	Western By-product	Indian By-product
Cream	Skim milk	Skim milk
Butter	Butter milk	Butter milk and Lassi
Ghee	--	Ghee residue
Cheese/Casein	Whey	Whey
Channa/Paneer	--	Whey

Source : Outlines of Dairy Technology - S.K. De

The average chemical composition of some of the by-products has been given in the table 10.2.

Table 10.2 Average composition of some by-products (percentage)

Name of the by-product (from Cow Milk)	Composition				
	Water	Fat	Protein	Lactose	Ash
Skim milk	90.6	0.1	3.6	5.0	0.7
Butter Milk (Sweet cream)	91.0	0.4	3.4	4.5	0.7
Lassi	96.2	0.8	1.4	1.2	0.4
Ghee residue	9.7	61.4	24.8	--	4.1
Chhana whey (Sweet Cream Butter)	93.6	0.5	0.4	5.1	0.4
Cheese whey	93.1	0.3	0.9	4.9	0.6
Acid casein whey	93.1	0.1	1.0	5.1	0.7

Source : Outlines of dairy Technology S.K. De

An attempt has been made to present the broad principles pertaining to the utilization of the various dairy products, both foreign and Indian, together with the names of the commonly made food products has been given in the table 10.3.

Table 10.3 By-products, their utilization and food products manufactured

By product	Principle of Utilization	Food products manufactured
	Pasteurisation	Flavoured milks
Skim milk	Sterilization Fermentation Fermentation and Concentration	Sterilised, flavoured milk Cultured Butter milk Acidophilus milk Bulgarian butter milk concentrated sour skim milk
	Concentration	Plain condensed skim milk, sweetened condensed skim milk, Low-lactose condensed skim milk (Frozen condensed skim milk)
	Drying	Dried skim milk
	Coagulation	cottage cheese, Baker's cheese quarg, Gammelost sap sago, casein (edible)
Butter milk	Fermentation and concentration	condensed butter milk.
	Drying	Dried butter milk.

Table 10.3 *Contd...*

By product	Principle of Utilization	Food products manufactured
Whey	Coagulation	soft cheese, whey beverage, yeast whey, plain condensed whey
	Fermentation	Whey beverage Yeast whey
	Concentration	Plain condensed whey Sweetened condensed whey protein concentrate whey, paste-lactose
	Drying	Dried whey
	Coagulation	Ricotta Cheese Mysoft, Greatest and Primost cheese
Lassi	Icing and seasoning	Beverage
Ghee residue	processing	Toffee or Paste or Indigenous sweet meat

Source : Outlines of dairy Technology, S.K. De

10.1 Skim Milk - its Utilization

The utilization of the various by-products resulting from the manufacture of different products has been described briefly under appropriate headings. The use of skim milk has been discussed already in the standardisation of milk, cream etc. The other uses of skim milk are as under:

1. **Concentrated Sour Skim Milk:** It is primarily used as an animal feed especially for the poultry and pigs in the developed dairying countries. Lactic acid is used as a preservating agent in its preparation/manufacture. The skim milk is fermented with the addition of *Lactobacillus bulgaricus* and an *Acetobactor*, to produce the required acidity of about 2 percent, which is then concentrated to about (1/3) one third of its actual weight. Fresh, sweet skim milk is pasteurised at no hold time (77-82 °C) and then immediately cooled to 46 °C (115 °F) in actual preparation/practice. A starter having 2 percent is then added and properly mixed, which is then incubated at the same temperature of 46 °F for 16-18 hours, after which the acidity is expected to be about 1.7 to 2 percent lactic acid. The curd is then broken up thoroughly and concentrated in a single effect evaporator. For warming or preheating is not necessary. The final concentrated product should have an acidity of 5-6 percent lactic acid in order to ensure not only a good keeping quality but also a smoother texture and a more stable product.
2. **Sweetened Condensed Skim Milk:** The procedure of manufacture of this milk is similar to that of condensed milk, except that the standardisation is not necessary in this case.

3. **Low Lactose Condensed Skim Milk:** These are the derivatives of the skim milk in which a part/portion of the lactose is removed or hydrolysed. They were initially developed to prevent the sandiness in ice-cream and to improve the body and texture.

10.2 Casein-Industrial

The production/manufacture of industrial casein which is a non-food item is the most important utilization of the skim milk, due to its wide spread demand it is being produced extensively throughout the world. There are two types of industrial casein namely acid casein and rennet casein. The casein precipitated by various acids is called acid casein, free casein or simply casein, whereas that is precipitated by rennet is called rennet casein and these two types of casein have different uses. An attempt has been made to classify the casein in a flow chart 10.1 form.

Flow Chart 10.1

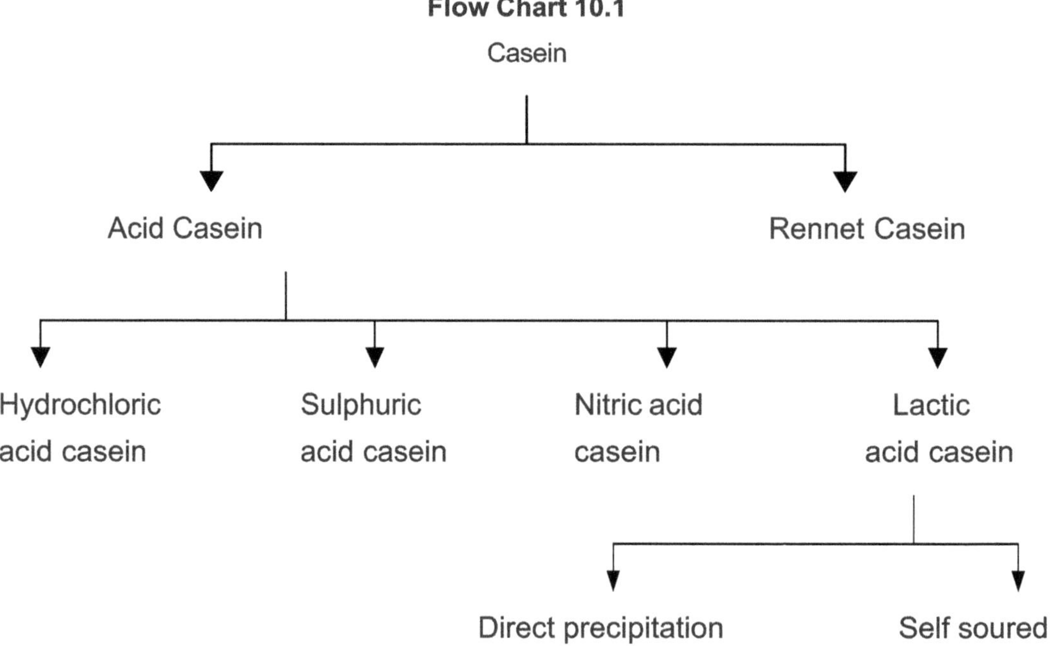

(a) Based on the coagulating agent used
   1. Acid casein
   2. Rennet casein

(b) Based on the type of acid used
   1. Hydrochloric acid casein
   2. Sulphuric acid casein

3. Nitric acid casein
4. Lactic acid casein
    (i) Direct precipitation
    (ii) Self Soured

(c) On the basis of the texture of the curd.
    1. Grain curd casein
    2. Cooked curd casein
    3. Pressed curd casein

10.2.1 Manufacture of Casein-Industrial

The manufacture of industrial casein from the skim milk has been presented in a flow diagram 10.2 for the better understanding by the students of dairy science.

Flow Chart 10.2 Manufacture of Industrial Casein

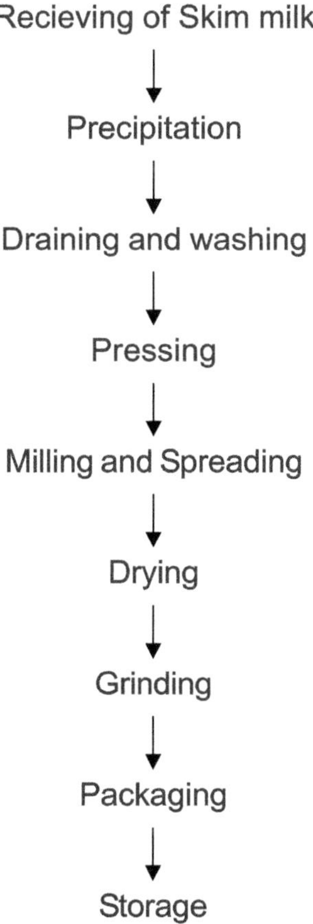

Sweet skim milk which is fresh, clean and fat free as far as possible should be used in the manufacture of the casein. It has already been stated that casein exists in milk as calcium caseinate calcium phosphate complex. Addition of acid to milk results in the dissociation of this complex. The casein begins to precipitate from the solution at a pH of about 5.3 and at the pH of 4.6, the maximum precipitation occurs, which is considered as the iso-electric point of casein, which coagulates as a watery, gel like curd. The gel is broken to allow the expulsion of water from it. There are three considerations which are to be borne in mind during the process of precipitation of casein namely:

1. acidity and pH
2. temperature and
3. the speed at which the skim milk is stirred/mixed while adding the acid.

The casein does not dissolve appreciably either in a slightly more acid or alkaline solution once it is out of solution as precipitate. However, in practice a little more acid is added after precipitation till a pH of 4.1 is reached in order to dissolve the calcium which flows out with the whey, thereby improving the texture of the curd which becomes coarse in nature. Removal of impurities viz. whey, proteins lactose, salts and acid by washing which is normally carried out becomes easier. The temperature should be carefully controlled during the precipitation, because of the fact that the casein curd is sensitive to heat. The effect of temperature on the nature and character of the casein curd has been given in the table 10.4.

Table 10.4 Effect of temperature on Character of casein curd

Temperature of Precipitation	Character of casein curd
Less than 35 °C / 95 °F	Soft and fine, slow to settle and difficult to wash.
35 - 38 °C / 95 - 100 °F	Coarse, provided it is not stirred too quickly/rapidly settles quickly and easy to wash
More than 38 °C / 100 °F	Chewing gum texture i.e rubbery, stringy, lumpy and coarse.

Source : Text Book of Dairy Chemistry - M.P. Mathur, D.D. Roy & P. Dinakar

Stirring is also essential to distribute the acid uniformly, which should be carried out at a very slow speed. Rapid or vigorous stirring at less than 38 °C (100 °F) results in the production of a fine curd which settles at a very slow rate when being drained and washed, and is likely to be partly lost in the whey and also during the subsequent washings.

It is very important to be kept in mind that when hydrochloric or sulphuric acids are used for the precipitation of casein, they must be diluted with water in order to prevent any local injury to the curd, either through the action of concentrated acid even though the agitation may be rapid. The dilution/strength of the acids and the proportion of acid and water to be used for the precipitation is given in the table10.5 for ready reference.

Table 10.5 Dilution of acids

Proportion (By weight)		
Name of acid	**Water**	**Remarks**
Hydrochloric 1	3 - 9	-
Sulfuric 1	6 - 30	Add acid to water very slowly with proper care for dilution

The whey should be removed immediately and promptly because it becomes more difficult to remove the impurities like, whey proteins, lactose salts and acid etc., when the casein curd stays for a longer period in the whey. Initially as the first step the curd must be allowed to settle and should be pushed to the back from the outlet valve in order to recover the small curd particles, a strainer along with another cloth lined strainer is fixed in position at the discharge valve thus allowing the whey to be drained off promptly in a perfect manner. As soon as the whey is drained off or ceases to run freely from the outlet valve, cold wash water which is made slightly acidic pH of 4.1 is used for the first washing by adding approximately equal to the quantity of whey that has been removed. The curd should be well stirred or mixed in the wash water taking proper care not to break the curd into smaller particles, which is then allowed to settle and the wash water is drained off in the similar manner. This process should be repeated at least twice in order to wash the casein curd to remove all the impurities. Inadequately washed casein is less soluble when compared to the one which is washed thouroughly, further it has less adhesive strength and develops more colour when dry.

Under normal conditions it is not possible to reduce the level of moisture in the pressed curd to less than 55 percent and with an intention to reduce the cost of drying the wet curd is pressed to remove as much water as possible. The pressed curd is then milled to produce particles of uniform size and surface for quick drying, without which it dries in an uneven manner. Proper drying of the milled product is essential, because of the fact that improper drying may result in the spoilage by mould and bacterial action. The curd which was milled into a shredding variety should be dried efficiently and to achieve the same there must be an effective control measure of the temperature and humidity of the incoming air. The process of drying once commenced should not be interrupted till the moisture content has been reduced to less than 8 percent. The temperature of the incoming air may range from 71-77 °C (160-170 °F) whereas the out going air temperature should be between 52-57 °C (125-135 °F) so as to avoid the risk of discolouring and impairing the solubility of casein.

The casein which is warm in nature is not suitable for grinding because it sticks to the surface of the grinder and become plastic or elastic, therefore it should be cooled properly before it is ground. The fineness of grinding is dependant on the requirements of the user and the size and uniformity of the ground casein is ensured by using screens having different mesh sizes and passing it through them. The dried casein, which ground or unground is normally packaged in either gunny bags lined with cloth or in a heavy paper bag with polythene liners. The casein filled packages/bags, should be closed air tight and stored in a dry storage room, which is maintained at a uniform temperature. Any variation in the storage temperature may result in the 'sweet' and mould in the casein. If the casein is to be stored for a longer periods it is advised to store it in a dark place away from the sun such as basement or cellar rooms and when stored in a damp atmospheric conditions there is a possibility that the casein may absorb moisture/water resulting in the deterioration of its quality.

10.2.2 Grain-Curd Casein

Acid casein of a high grade quality, which is low in ash and readily soluble is made by the grain-curd process provided the pH and temperature are monitored and controlled very closely using a jacketed vat for an accurate control of the temperature. The best product is obtained by direct precipitation using hydrochloric acid at a temperature of 35 °C (95 °F), lactic acid at 38 °C (100 °F) and sulphuric acid 37 °C (98 °F) at a pH of 4.1 by the addition of the dilute acids with gentle stirring of the skim milk.

Similarly in the manufacture or the preparation of casein by the self soured process, the skim milk is held at 38 °C (100 °F) and an active, high producing lactic starter culture is added at 10 percent by thoroughly mixing into it. The acidity is allowed to develop to reach up to 0.55 percent and then the temperature is raised to 40-41°C (104-105 °F) stirring the curd until a clear whey is separated, allowing the curd to settle and draining off the whey as quickly as possible. The speed of removal of the whey is most important than the thouroughness of the process/procedure. Then lactic whey containing about 2 percent acid which is previously heated to a temperature of 38°C (100°F) is added and stirred until the pH reached to about 4.1. Then the granular curd is allowed to settle and washed twice or thrice with acidulated water having a pH of 4.1 having a temperature of 32 °C (90 °F) and finally washed once with cold neutral water. Thus the washed curd is pressed, dried and ground, packaged and stored till further use to under damp free conditions avoiding the direct sun light.

10.2.3 Rennet Casein-Manufacture

Fresh, sweet skim milk of very good and high quality is essential for the manufacture of rennet casein which is heated in a Jacketed vat up to 36-38 °C (96-100 °F). Sufficient quantity of rennet along with calcium chloride are added to the skim milk allowing a setting time of about 20-30 minutes. The rennet used should be diluted with about 15-20 times its weight in water (by volume). The calcium chloride which is being used in this process should be 1 percent solution. The coagulum is gently agitated for a period of about 2-5 minutes, during the process of beginning of the coagulation but before the coagulum reaches a solid clot in order to have an optimum size of curd particles for further processing. The temperature of the curd is raised to 54-66 °C (130-150 °F) with simultaneous agitation, then the curd is cooked for about 30 minutes at this temperature, which is then processed in the same manner as in the case of acid casein.

10.2.4 Manufacture of Casein from Butter Milk

Casein can also be prepared or manufactured from butter milk/Lassi and the method has been described briefly hereunder:

The casein which is present in the butter milk/Lassi is firt dissolved by adding enough sodium hydroxide solution at 32 °C (90 °F) in order to raise the pH to 7.5 which is then precipitated by adding dilute hydrochloric acid thus bringing the pH to 4.6. The precipitated curd is then allowed to settle, draining the whey, washing the curd and finally pressing it. The wet curd thus formed is dissolved once more in dilute alkali at 38 °C (100 °F) and allowing the solution to pass through the cream separator for the removal of fat. The casein thus separated is once again re-precipitated with the dilute acid allowing the curd to be settled, draining the whey and finally washing, pressing to drying, grinding and packaging the casein obtained by this process. Though a fairly acceptable, good amount of casein is obtained by this double alkali treatment in mechanical separation of process, the method has not gained much importance as it was found to be uneconomical. Further, the casein obtained from the butter milk tends to be soft finely devided state, thus making it difficult to wash the curd without excessive loss. Moreover it contains a greater amount/quantity of fat which makes it highly coloured during the process of drying. The yeild of casein is roughly equal to its percentage in the skim milk and may vary from 2.8 to 3.2 percent depending on the nature of its source namely whether it is cow or buffalo milk.

The characteristics or specifications of both acid and rennet casein have been presented in tabular form 10.6 for comparison.

Table 10.6 Characteristics or Specifications of both acid and rennet casein

Characteristics	Acid Casein		Rennet Casein	
	Cow	Buffaloe	Cow	Buffaloe
Colour	Yellowish white	Chalky white	Light Yellowish	Whitish
Odour	Pleasant	Pleasant	Faint	agreeable
Taste	Slightly acidic	Slightly acidic	tasteless	--
Total acidity	Equivalent to not more than 10.5 ml and N/10 alkali per gram	--	pH around 7.0	--
Moisture	Not more than 10 percent	--	Not more than 10 percent	--
Fat	Not more than 1 percent on moisture free basis	--	--	--
Ash	Not more than 4.0 percent on moisture free basis	--	Approximately 7.5 percent	
Nitrogen	Not less than 14.25 percent on moisture fat and ash free basis	--	--	--

Source : Text Book of Dairy Chemistry - Edgar. R Ling

10.2.5 Casein - Defects, Causes and Prevention

Generally, during the manufacture of any product, will be carried out at different stages and in batches. In this process defects are likely to occur in spite of the prompt care which may vary from batch to batch. Thus casein is not an exception and defects may arise from the usage of low grade material i.e, the skim milk, faulty methods in the production and drying etc. The important defects, their causes and prevention has been presented in table10.7.

Table 10.7 Casein Defects, Causes and Prevention

Name of the Defect	Causes	Prevention
Browning	(i) Over heating of casein when drying (ii) Presence of excess fat in casein during drying.	(i) Maintaining the optimum casein drying temperature (ii) using low fat skim milk for casein manufacture

Table 10.7 ***Contd...***

Name of the Defect	Causes	Prevention
	(iii) Presence of Sulphuric acid/sulfate in casein during drying.	(iii) Proper washing of the curd
Poor Solubility	(i) Excessively high temperature of cooling curd	(i) Maintaining the optimum temperature of cooling curd
	(ii) Over heating of casein during drying	(ii) Maintaining optimum casein drying temperature
	(iii) Moulding of curd before drying	(iii) Proper drying of the wet curd
Low adhesive strength	(i) High ash content in casein	(i) Adequate washing of the curd
	(ii) Moulding of curd before drying	(ii) Prompt drying of the wet curd.

Source : Outlines of Dairy Technology - S.K. De

Uses

The two types of casein namely acid and rennet casein have separate and different type of uses and cannot be used as a substitute for the other and their uses have been given hereunder.

Rennet Casein

In the manufacture of plastics such as buckles, brush backs, buttons, costume, jewellery, umbrella handlers etc.

Acid Casein

1. Adhesive or glue
2. Paper coating
3. Paint Industry
4. Fibre industry
5. Bristles
6. Films
7. Leather finish
8. Textile finish
9. Spreader for insecticides
10. In medicines (silver and mercury caseinate)

Casein Edible

The edible casein may be defined as the casein, which has been isolated from the skim milk by taking certain special precautions in order to ensure its suitability for usage in patented foods and pharmaceutical preparations. These precautions are chiefly concerned with:

1. Strict quality control of the raw material namely the skim milk.
2. Use of standard equipment and techniques for production.
3. Maintenance of strict hygienic conditions of production and
4. Approved packaging and storage conditions.

The FAASI (2017) has prescribed certain specifications for the edible casein which are given in the table 10.8.

Table 10.8 FSSAI (2017) standards for edible casein in products

Sl.no	Parameter	Edible acid casein	Edible rennet casein	Edible caseinate
1.	Moisture [(a)] Maximum % (m/m)	12.0	12.0	8.0
2.	Milk fat maximum % (m/m)	2.0	2.0	2.0
3.	Milk protein[(b)] minimum % (m/m) dry matter basis	90.0	84.0	88.0
4.	Casein protein minimum % (m/m)	95.0	95.0	95.0
5.	Lactose[(c)] Maximum % (m/m)	1.0	1.0	1.0
6.	Total ash including P_2O_5 % (m/m)	2.5 (maximum)	7.5 (minimum)	-
7.	Free acid maximum ml of 0.1N NaOH per g	0.27	-	-
8.	pH (in 10% solution) maximum	-	-	8.0

Notes: (a) The water content does not include water of crystallization of the lactose (b) Protein content is 6.38 multiplied by the total nitrogen determined (c) Although the powders may contain both anhydrous and lactose monohydrates, the lactose content is expressed as anhydrous lactose. 100 parts of lactose monohydrate contain 95 parts of anhydrous lactose.

The edible casein is prepared/manufactured by following the same procedure and steps similar to the production of acid casein, preferably grain curd hydrochloric by taking care and following the special precautions as stated earlier which include:

1. Precipitation of curd at pH 4.1 to 4.3.
2. A minimum of at least 3 separate washings of the curd with waters of proper pH and with a minimum contact time of 15-20 minutes each.
3. Last but one washing with hot water at 71-77 °C (160-170 °F) which gives the affect of pasteurization to the curd thus reducing the bacterial count.
4. Last washing with neutral water at 41°C (105 °F) etc.

Edible casein in its original form or as sodium caseinate is being used in various food products such as:

1. Ice cream
2. Coffee whiteners
3. Imitation milk
4. Whipping powders
5. Instant break fast
6. Water binders in sausages and
7. Protein hydrolysates.

10.3 Butter Milk-Condensed Butter Milk

Generally, the sweet cream butter milk which is ripened has been utilised and condenced approximately in ratio of 3:1 by following the similar procedure as in the preparation/manufacture of concentrated sour skim milk. It is also known as semi-solid butter milk. As an alternative sour-cream butter milk may be used without ripening, however, the lactic acidity of the butter milk prior to its condensation should be greater than 1.6 percent, thus it will ensure a 5-6 percent acidity in the finished product, which is essential and necessary for the long keeping quality of the product. In order to increase the acidity of the butter milk either sour whey or an organic acid such as lactic, acetic, citric or tartaric may be used. In olden days, the packages consisted of air tight wooden barrels, while in the present the modern packages consist of an acid resistant tight plastic film with a suitable fibre board container. It is being used in the feeding of pigs and poultry in the developed dairying countries.

10.4 Whey

Whey is a dilute highly perishable greenish yellow fluid and the largest by-product of the dairy world produced during the manufacture of cheese, casein, chhana, paneer, chakka and other co-precipitates. The continuing and ever increasing annual growth in the production and consumption of cheese as well as the coagulated milk products represents the generation of extremely large quantities of whey because of its huge base. About 68-70% of whey is produced in European countries and about 24-25% in North America, whereas the total world production of whey is estimated to be about 170 million tonnes and whey produced from cheese alone accounts for about 95%. In India the major source of whey production is from the manufacture of chhana and paneer. In the absence of systematic statistics or surveys the predicted value of whey production is around 5-8 million tonnes per annum in India. There has been a gross lack of interest in the utilization of whey due to the low solid content, however, the presence of high organic matter causes environmental pollution in its disposal.

Composition of Different Types of Whey

Whey produced during the manufacture of different products like cheese, casein, chhana, paneer, chakka, and co-precipitates vary in its chemical composition and acidity, and has been presented in the table 10.9.

Table 10.9 Percentage Composition of different whey systems

Constituents	Cheddar cheese	Acid Casein	Rennet Casein	Chhana & Paneer	Co-precipitates
Total solids	6.7	6.9	6.7	6.4	6.1
Fat	0.3	0.1	0.1	0.5	0.1
Protein	0.9	1.0	1.0	0.4	0.3
Lactose	4.9	5.1	5.1	5.0	5.1
Ash	0.6	0.7	0.5	0.5	0.6
T.A.	0.2	0.4	0.2	0.4	0.3

Source : IDA Publication

In general, the whey contains about half of the total solids of the milk and is a source of precious nutrients like lactose, whey proteins, minerals and vitamins.

10.4.1 Utilization of Whey

In the dairy industry, which is engaged in the manufacture of cheese and coagulated milk products, the disposal or utilization of whey has been the great concern/problem. The techno-economic problems that have been associated with the utilization of whey have been receiving considerable attention and made remarkable advances. The modern industrial processing technologies that are available today such as ultra filtration (UF), reverse osmosis (RO), new drying methods, hydrolysis, electrodialyses, ion-exchange fermentation and protein fractionation, are some of the methods, which have converted whey into a major source of ingredients with differing functional and nutritional properties, which could be used in food and dairy industry. The global market for whey derivatives is growing at an average of about 10% annually, due to the stringent regulations imposed by the environmental pollution agencies all over the world, as well as the economic returns from whey, which contains almost economic returns from whey, which contains almost half of the solids of original milk. The major models/sources of whey utilization has been detailed diagramatically as shown in Flowchart 10.3.

Whey is a good fermentation medium, because it is a rich source of lactose, which in many applications either as whole or deproteinized is hydrolysed to glucose and galactose thus resulting in the increase of sweetness. Such lactose, hydrolysed syrups, after condensing are generally utilized in the confectionary and ice cream products.

Flow Chart 10.3 Whey and its Utilization

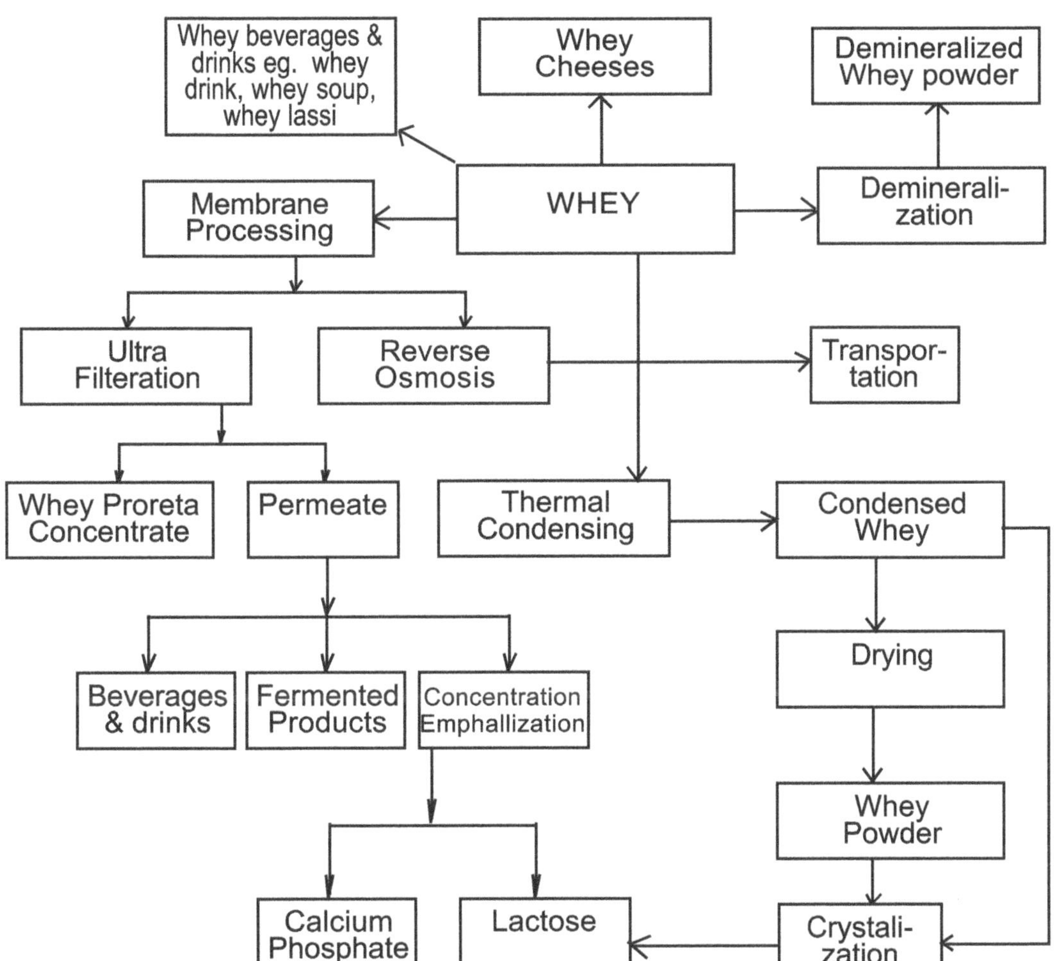

In advanced dairy countries, the production of products from whey such as whey powder, whey beverages, whey protein concentrates, lactose and many others have prominence, whereas in India the dairy industry is still a novice in this field. The economic liberalization and de-licensing of the dairy industry which began 15 years ago, rapid changes have been witnessed in the dairy industry with the setting up of many manufacturing plants with large automatic and continuous process for the utilization of the whey.

The major problem with whey based products is their salty flavour due to the high mineral content. Despite significant gains in the amount of whey being processed, still a large amount is disposed of as raw whey. Much of this represents, the whey produced from small plants, where the cost of purchasing, processing as well as the subsequent transportation and handling, clearly exceeds the value of any product that might be produced from it.

In view of this it can be concluded that the utilization of whey can be undertaken at two levels i.e. small and large scale.

10.4.2 Small Scale Utilization of Whey

In small dairy plants, the choice of disposal of whey may be in the form of municipal treatments, spreading on local farm lands, or feeding to live stock. The acid whey because of high mineral content and low pH, poses considerable difficulties in its utilization hence mostly remains un utilized. However, utilization of whey in the form of different whey beverages and drinks at a small scale appears to be more economical, viable and promising.

Whey beverages and drinks such as:

1. Whevit
2. Acido whey
3. Whey based fruit beverages
4. Whey based soups
5. Whey based Lassi
6. Whey based Spoil beverage

10.4.3 Large Scale Utilization of Whey

The single largest use of whey solids on global basis is in the form of whole dry whey, which is growing continuously. This is whole whey that has been condensed and spray dried as such or after blending with certain other liquid ingredients. The concentration and drying of whey assumes a special significance because of its highly perishable nature, whey can be preserved as plain, or sweetened condensed whey. The development of the membrane technology (RO), and its economical feasibility to manufacture R.O. Whey concentrate with 28% T.S. retaining functional properties of whey proteins. The potential use of R.O. can be a pre-concentration step prior to transportation of whey concentrate to a central processing facility for further processing for the manufacture of whey powder, whey protein powder, de-mineralised whey powder and lactose.

10.4.4 Concentration of Whey

Whey concentration is carried out with the following objectives:

(a) Reduction of volume for times post to other processing facilities.

(b) Pre-concentration for drying into whey powder.

(c) Manufacture of condensed whey plain or sweetened.

The various steps involved in the production of concentrated whey are as follows:

(a) Clarification

(b) Separation and Pasteurization

(c) Concentration

Plain Condensed Whey

It is prepared as an intermediate product for whey powder making the whey is concentrated to about 40-60% solids and pre-crystallized before drying. The desirable degree of condensation of plain whey is 35-50%. Whey when condensed to contains higher solids more than 55% terms of gel after cooling hence not recommended to be used in any food or feed products due to its coarser texture and low solubility.

Sweetened Condensed Whey

Sweet cheese whey is mixed with sugar equal to the weight of the solids present in the whey for the preparation of the sweetened condensed whey. The mixture is concentrated using multistage evaporators to at least 76% T.S. The concentrate is cooled to 30 °C used flash coolers and seeded with lactose crystals. The seeded mixture is stirred for 1-3 hr to crystallize lactose. It may then be packed in barrels or in laws. The product does not require any refrigeration for storage. The sweetened condensed whey darkens and thickens in storage, however these changes do not develop to an objectionable level during the first year. The composition of different types of whey is given in the table 10.10.

Table 10.10 Proximate Composition of Condensed Whey

	Constituent %					
Product	**Sugar** Lactose	**Sugar** Sucrose	**Water**	**Ash**	**Fat**	**Protein**
Plain condensed Whey	51.3	--	32.0	6.0	0.6	10.1
Sweetened Condensed Whey	28.7	36.0	24.0	3.4	0.3	5.6
Condensed acid Whey	49.0	--	33.5	8.2	0.6	7.7

Source : IDA Publication

10.4.5 Uses of Condensed Whey

Whey solids in the form of condensed whey are used in many dairy products, bakery goods, baby foods, etc. Concentrated whey has world wide industrial application in bread and other bakery products, condensed whey is also used with fruits and jams, in various tippings and spreads in confectionary. In Norway "My Sost" and "Primost" cheeses are produced from concentrated whey with 84% total solids.

Lactose Crystallization

It is of great industrial importance to get the major part of the lactose content in a crystalline form, it is very essential to avoid the very undesirable cacking properties of ordinary whey powder. The advantage of this lies both in energy savings and in improved powder properties. Controlled crystallization can be initiated by immediate flash cooling of condensed whey after evaporation to about 30 °C.

Spray Drying

Recent trend in drying of milk and whey is expensive use of spray drier that may be single stage, double stage and more recently the three stage drying.

Chemical Composition of Whey Powder

The composition of whey powder varies depending upon the type of whey from which it has been prepared, pretreatment given to the whey and the various processing steps involved in its manufacture/production. The average chemical composition properties of different varieties of whey powder have been given in the table 10.11. An attempt has been made to present the FSSAI (2017) standards as published in the Gazette of Govt. of India in table 10.12 for the benefit of the students of dairy science.

Table 10.11 Properties of the whey powder (Percentage)

Properties	Ordinary Whey Powder	Pre-Crystallized Whey powder	Non Whey Powder	Non-coding whey Powder
Total moisture	3-4	3-5	5-6	4-5.5
Free Moisture	3-4	1-2	1.5-2.5	1-2
Bulk density	06-07	06-07	0.55-0.65	0.4-0.5
Degrees Celcius	100	40-100	0.5	0.5
Degree of crystallization	0.1	50-75	75-85	85-95

Source : IDA Publication

10.4.6 Storage of Whey Powder

The shelf life of the whey powder is limited by the non-enzymatic browning by the Maillard reaction, which is one of the important mode of deterioration. Whey powders contain relatively high concentration of lactose and protein. In the presence of moisture these components readily participate in the Maillard reaction. This interaction may result in a decrease in protein quality which is accompanied or followed by undesirable colour changes.

Table 10.12 FSSAI (2017) standards for whey powder

Sl. No.	Parameter	Whey powder	Acid whey powder
1.	Moisture [(a)] maximum % (m/m)	5.0	4.5
2.	Milk fat maximum % (m/m)	2.0	2.0
3.	Milk protein [(b)] minimum % (m/m)	10.0	7.0
4.	Lactose content[(c)] as anhydrous lactose minimum % (m/m)	61.0	61.0
5.	pH (in 10% solution)	more than 5.1 (d)	5.1 (e) maximum
6.	Total ash maximum % (m/m) on dry basis	9.5	15.0

Notes: (a) The water content does not include water of crystallization of the lactose (b) Protein content 6.38 multiplied by the total nitrogen determined (c) Although the powders may contain both anhydrous lactose and lactose monohydrates the lactose content is expressd as anhydrouslactose, 100 parts of lactose of monohydrate contain 95 parts of anhydrous lactose. (d) Or titrable acidity (calculated as lactic acid) < 0.35% (e) Or titrable acidity (calculated as lactic acid) > 0.35%.

10.4.7 Whey Protein Concentrate

The introduction of the Ultra Filtration (U.F.) process, there has been a continuous increase in the production of whey protein concentrate (WPC). Clarification of cheese whey, prior to U.F. is an essential step for the effective operation of the U.F. plant. The UF permeation rates are greater with higher temperature and with pre holding of the whey for 30-40 min at temperatures (68-72 °C) higher than the UF temperature (50 °C). Pre-heating and UF temperature treatments affect the constituents of concentrates. Protein purity of WPC improves greately with increased UF concentration but at diminishing rate during later UF stages.

WPC even though soluble has been found to be technically suited to a wide range of products, its use is not cost effective in all cases. The largest potential use of WPC is as a replacement for non fat dry milk (NFDM) in the food industry. WPC with 35% protein is perceived to be a universal substitute for NFDM, because of the similarity in gross composition and its dairy character as well as the cost advantage over NFDM.

Buffalo and cow milk are being humanized, due to various reasons and are being used either partly or exclusively for the feeding of the human infants throughout the world. The whey protein proportions needs to be increased in milk for humanization in addition to making certain other modifications. In view of this there is a great potential in the application of WPC.

10.4.8 Manufacture of Whey Protein Concentrate (WPC)

The manufacture of whey protein products are based on known behaviour of whey components under defined conditions. The properties that have been exploited commercially include molecular size differences (ultra filtration, gel filtration) insolubility of protein at high temperature, charge characteristics (demineralization, protein removal by ion exchange), aggregation by poly phosphates and crystallization of lactose.

The different methods/processes that have been employed in the manufacture of WPC are:

1. Ultra filtration process
2. Gel filtration process
3. Heat precipitation process and
4. Precipitation by complexing agents.

There has been many innovative uses of whey through microbial fermentation. Many products of industrial importance and food value have been produced through microbial fermentation of whole whey and UF permeate. Whey being a rich source of lactose, there is a good potential for the production of fermented products like lactic acid, other organic acids, alcohol, microbial biomass, flavour concentrates and vitamins.

10.5 Lactose

Lactose is the characteristic carbohydrate and is the only sugar of animal origin. Lactose in its pure form is a white water soluble crystalline powder of moderately sweet taste with no odour. Lactose is extremely used in food and pharmaceutical industries due to its nutritional importance and multiple functional properties. Low sweetness consistency and colour improvement in confectionary and bakery products as a carrier and anti-caking agent in powder foods, filler for tablets and capsules, flavour enhancer in sauces and dressings, nutritional importance in infant formulas, an additive in culture media etc. or some of the desirable functional properties and applications of crystalline lactose powder. Lactose is also an important fermentation substrate for the production of a variety of fermented products. As a rough estimate, the quantity of lactose produced on a world wide scale amounts to about 50 lakh metric tonnes per annum.

Lactose is produced commercially from whole whey or from ultra filtrate permeate. The protein and mineral contents of the whey are the limiting factors for the crystallisation of lactose and this is the reason why the U.F. permeate of whey is preferred as the starting material in the production of lactose. Because, it gives a relatively higher yield and purity of lactose as compared to whey lactose.

Normally lactose occurs in one of its two crystalline forms or as a non-crystalline glass. The more commonly present crystalline form of lactose is that of α-lactose hydrate which is considered as the commercial lactose. Generally the most important source of lactose in all the major producing countries is cheese-whey although hydrochloric acid casein whey is still considered as the best source for its production. In India, chhana whey is produced in larger quantities than any other whey, therefore, should be considered as the main source for the manufacture of lactose in our counrty. However, the collection of whey from the small scale and scattered producers/manufacturers needs to be organised systematically as in the case of milk collection/procurement by the dairy industry.

10.5.1 Grades of Lactose

Three common grades of lactose are available in the market namely Crude, Edible and USP (United States Pharmacopea). The crude lactose is the product obtained in the first instance during the common process of manufacture which may contain many impurities and in general being used in the refining of lactose to the USP grade. The edible grade contains less impurities as compared to the crude variety but does not meet the requirements of USP grade as it is below the standards, but it is suitable for use in infant foods. The USP grade lactose is obtained by refining the edible lactose grade to meet the highest standards of purity and is being used in the pharmaceutical preparations as well as for infant feeding. The specifications for the above grades of lactose has been presented in the table10.13.

Table 10.13 Specifications for Various Grades of Lactose

Particulars	Requirements		
	Crude	Edible	USP
Lactose percentage	98.4	99.0	98.85
Moisture non-hydrate percentage	0.3	0.5	0.1
Protein (N × 6.38) percentage	0.8	0.1	0.01
Ash percentage	0.4	0.2	0.03
Lipids percentage	0.1	0.1	0.01
Acidity as Lactic percentage	0.4	0.06	0.04
Standard plate count (Per g)	*	Less than	less than
Coliforms, Spore formers	*	30	30
Moulds, Yeasts (Per g)	*	–ve	–ve
*Not determined routinely for this grade.			

Source : Outlines of Dairy Technology - S.K. De

The International trade recognises the following commercial forms of lactose has been given in the table 10.14.

Table 10.14 Grades of Lactose

S.No.	Grade	Lactose (%)
1.	Technical Lactose	90 - 92
2.	Crude Lactose	95 - 99
3.	Food grade Lactose	98 - 99
4.	Pharmaceutical Lactose	99.5 - 99.9

Source : I.D.A. Publication

An attempt has been made to present the FSSAI (2017) standards as published in the Gazette Govt. of India in the table 10.15 for the benefit of the students of dairy science.

Table 10.15 FSSAI Standards for edible lactose (2017)

Parameter	Limits
Total moisture max % (m/m)	6.0
Lactose minimum % (m/m) on dry basis	99.0
Sulphide ash max % (m/m)	0.3
pH (10% solution)	4.5 to 7.0
Scorched particle Max	Disc B.

10.5.2 Method of Manufacture

Generally, the lactose is separated from the whey obtained either by rennet or acid precipitation by the process of concentration and crystallization. The whey is pretreated for the removal of the impurities such as whey proteins, salts and acid primarily for two reasons namely:

1. to reduce the viscosity of the concentrated whey in order to make the possible separation of lactose crystals and
2. to increase the purity of the recovered lactose.

Most of the whey proteins and salts pass on to the remaining liquid which is called as the mother liquor, whereas the lactose crystals are removed from the concentrate in a centrifuge. The mother liquor is usually utilized in the animal feed. The manufacture of lactose is represented in the form a flow diagram 10.4 and the various steps involved in the process for the better understanding by the students.

Flow Chart 10.4 Manufacture of lactose

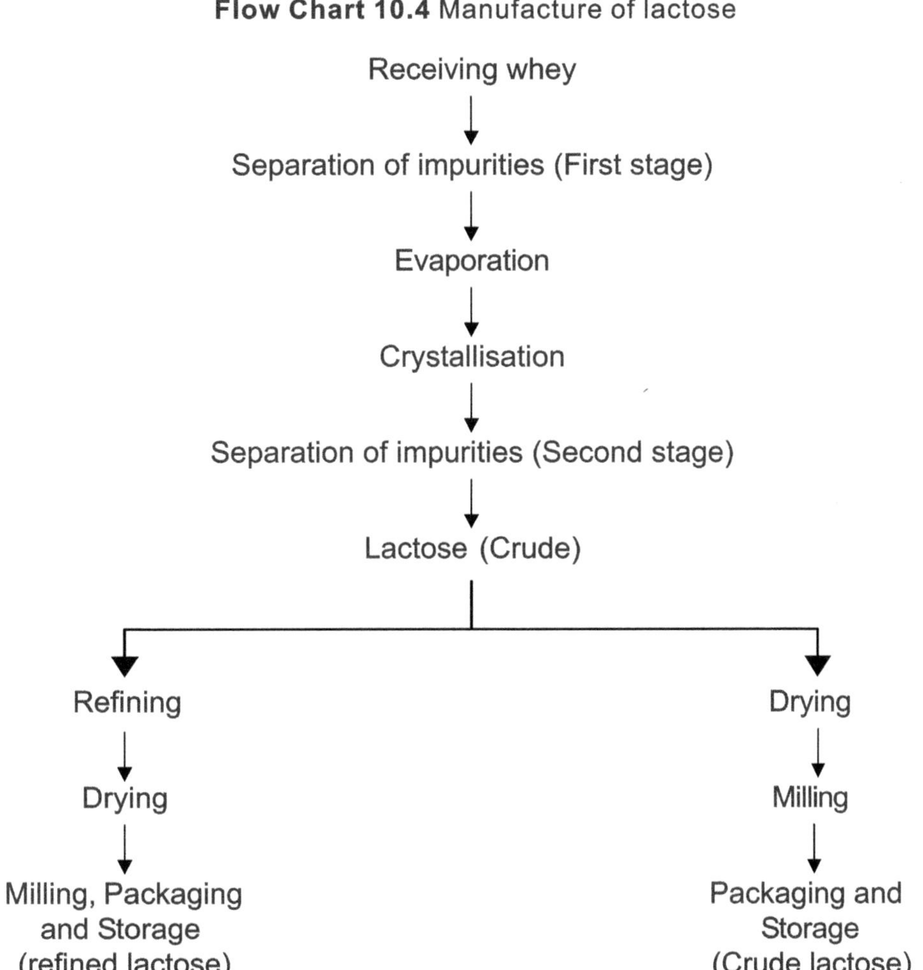

The whey which has been received is collected in a tank/vat and boiled with the help of the steam through the perforated pipes, and during the process milk of lime is added reducing the acidity to about 0.05 percent or a pH of 6.2. The steam boiling is completely stopped, when the milk proteins are coagulated and the whey is allowed to settle undisturbed and the impurities are settled as a sludge. Then the clear whey is transferred to a storage tank, which supplies it to a double or tripple effect evaporator, leaving the sludge in the vat itself, thus concentrating the whey to 20° Baume which is equivalent to a lactose content of 30 percent. The impurities are removed in the first stage by filtering the whey through the filter press and subsequently the sludge in the coagulating vat also is passed through the same filter press, thus mixing the partially purified whey with the earlier whey and the pressed cake which was obtained is used in the feeding of poultry.

The partially concentrated and filtered whey is then transferred into a single effect vacuum evaporator for the second stage, which is generally carried out at 40° Baume. Crystallization of lactose occurs partially during the process of the concentration in the pan and when the graining has progressed sufficiently it is "Struck" or stopped. The hot mats of crystals and the syrup (51-60 °C/125-140 °F) is poured into the crystallization vats, which are provided with slow speed spiral screw agitators and with jackets with a provision for the circulation of cold water. The lactose crystals must be kept in motion in order to prevent the formation of "setting up" or formation of solid mass. By regulating the speed of agitation and rate of cooling, uniform size of the crystals can be obtained. Further, prolonged holding with reduced temperature the yeild of less soluble lactose may also be increased. Lactose crystals are separated from the mother liquor and washed at an rpm of 600 during the initial stage. When the mother liquor ceases to run freely from the sugar crystals, it is then washed off with a small stream of cold water. Then the speed of the basket is increased to 1200 rpm in order to remove the wash water as much as possible. Finally the crude sugar which emerges from the centrifuge is either refined promptly or dried at 79 °C (175 °F) in a rotary drum drier.

The refining of crude lactose sugar to produce the USP grade requires that it should be refined further in order to remove the colour, protein and salts by treating the lactose solution. The crude sugar is either dissolved in water or the washings of the previous batch in a stainless steel tank with the help of the live steam by adjusting the volume or the proportions of the sugar and water so that a reading of 20° Baume is obtained indicating that approximately 30 percent lactose solution is obtained. At this stage, for every 100 kg of the sugar in the batch (1/4) one-fourth kg of filtering aid and 1 kg of decolonizing paste consisting of 3 parts of bone black, 1 part of activated carbon, 1 part of 36 percent hydrochloric acid and enough water making the paste easy to handle. The treated solution is then boiled and the acidity is adjusted to 0.09 percent of lactic acid with the help of hydrochloric acid. Before filtering the acidity is lowered to 0.05 percent with milk of lime avoiding the over neutralization because, this will result in the caramelization of the sugar thus giving a discoloured appearance. After lime treatment the solution is boiled vigorously for a few minutes and then allowed to stand undisturbed such that the flocculated mixture of carbon, protein and insoluble salts have settled. Then it is filtered through the cloth in a filter press using the filter aid such as diatomaceous earth in the first instance followed by some fine textured rag paper. Finally, the clear solution is then transferred into a single effect evaporation and then concentrated to about 40° Baume. This is followed by crystallization (initial seeding), centrifuging, washing, drying, milling, packaging and storage following the same procedure as that of crude sugar, by giving special attention to the details so as to meet the standards that have been laid down or prescribed for the USP grade lactose.

10.5.3 Yeild

The crude lactose which is obtained as the first crop/yeild should contain about 75 percent of sugar present in the whey and between 85-90 percent of lactose. The conversion of crude to refined lactose/sugar gives an approximate yeild of about 90 percent however, the overall yield is about nearly 100 percent, depending upon the lactose content of the crude since the mother liquor and washings are reprocessed.

Specifications

The Bureau of Indian Standards specifications for the commercial lactose have been given in the table 10.16.

Table 10.16 Specifications for Commercial lactose

S.No.	Specifications	Requirement
1.	Lactose on dry basis % by mass min.	99.0
2.	Moisture % by mass max.	
	(a) For lactose Mono-hydrate	5.5
	(b) For lactose anhydrous	1.0
3.	Sulphated Ash (on dry basis)	
	% by mass max.	0.2
4.	pH (of 10% solution)	4.0 - 6.5
5.	Specific Rotation	52.0 - 52.6
6.	Nitrogen % by mass max.	0 - 0.6
7.	Arsenic (As) mg/kg max.	1.0
8.	Lead (Pb) mg/kg max.	2.0
9.	E.Coli per 0.1 g	Absent
10.	Solmonella per 0.1 g	Absent

Source : IDA Publication.

10.5.4 Uses

The various uses in which the lactose is being utilized are:

(i) In the preparation of humanised milk
(ii) In Infant foods
(iii) In Caramels or fudges
(iv) Solid pharmaceutical preparations such as pills and tablets.
(v) In silvering of mirrors
(vi) In bakery products.

10.6 Lassi

Synonyms : Chhas, Matha, Desi butter milk

Flow Chart 10.5 for Mechanised Production of Lassi

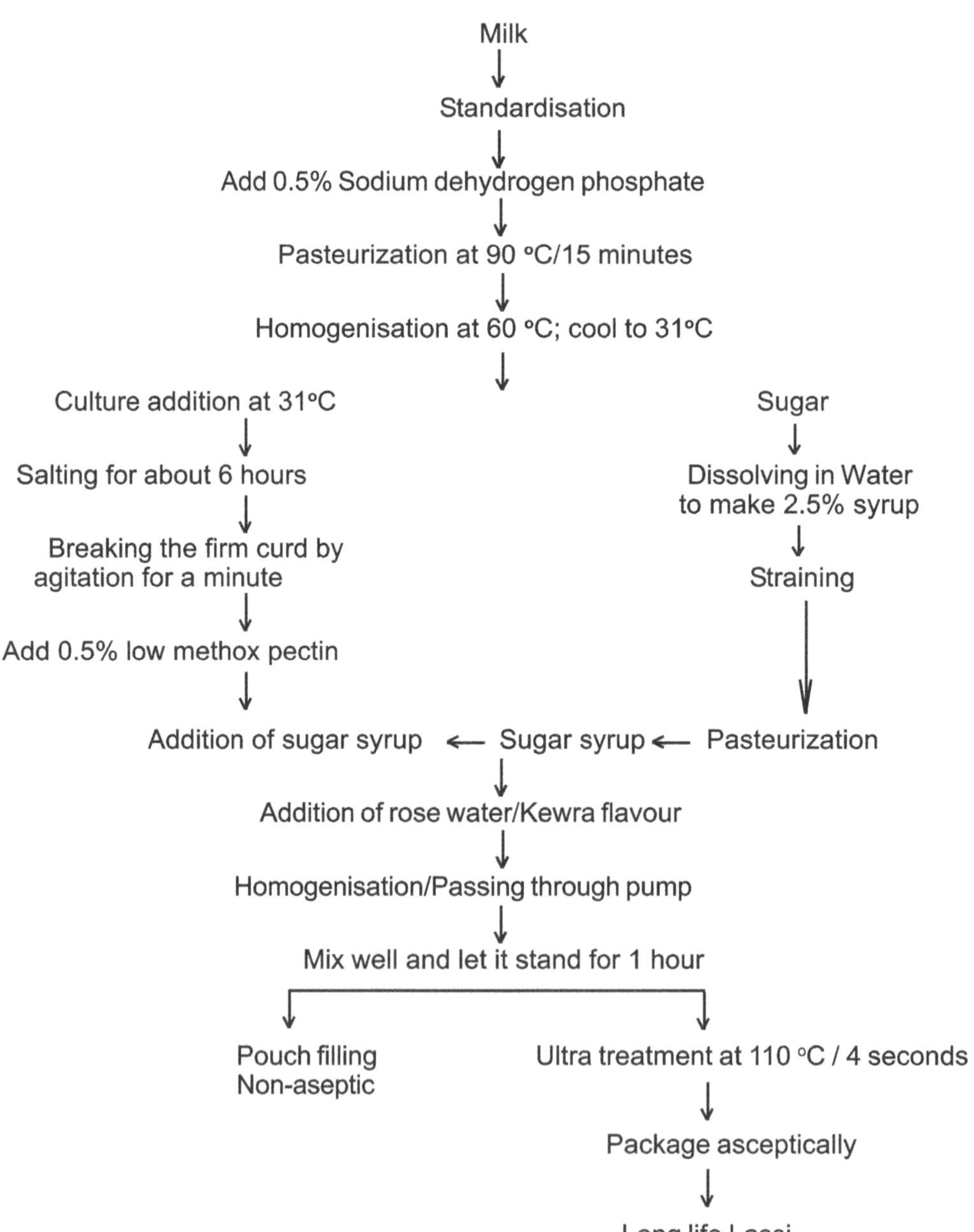

It is the by-product obtained by churning the curdled whole milk, with crude indigenous devices for the production of desi butter/makkhan. The composition of Lassi varies considerably depending upon the composition of dahi, the extent to which the curd is diluted when churning and the efficiency of fat removal. Lassi contains appreciable amounts of milk proteins and phospholipids and is an excellent beverage for quenching the thirst due to its lactic acidity. The yeild of lassi mainly depends on the extent to which the curd is diluted during churning. The greater the dilution the greater the yield and vice versa.

The uses of lassi are as under:

1. As already stated it is used as a beverage, generally with the addition of ice, sugar, or salt, with or without additional flavours in hot weather.
2. As a starter culture in almost every household to prepare the curd for domestic consumption.
3. It is also used as a coagulant.

10.7 Ghee Residue

Generally it is wasted in the large ghee making/manufacturing dairy industries even though it is usually consumed along with sugar in the farmers household to a very limited extent. This is a highly nutritious food which contains fat, denatured protein burnt lactose and minerals. The research work carried out in India has revealed that it can be utilized in the preparation of candies and chocolate industry. It can also be used as a part in sandwiches and alone with Indian foods such as idli, dosa and samosa. However, before it is being utilized, it is desirable and essential to remove the excess fat from the ghee residue. This can be done by keeping the material in a fine muslin cloth and covering it from all the sides with the muslin cloth in the boiling water bath for 30 to 40 minutes. The bundle is then taken out allowing the water to drain out by itself in the first instance, followed by gentle squeezing when it is cooled to a reasonable temperature. This residue is then given a pre-treatment in the similar manner in a 10 percent sodium bicarbonate solution, with constant stirring during the process of cooking.

The processed ghee residue thus cooked can be made into candy and chocolate as under:

Candy

Prepare a 50 percent sugar syrup in the first instance and mix the processed cooked ghee residue thouroughly in a 'Karahi' with the help of the 'Khunti'. Simultaneously heating the mixture on a low flame until a good amount of water is evaporated. Then dry coconut powder may be added to this sticky mixture and the candy is evenly spread on a plate at 5-10 °C for an hour or so allowing it to cool. Then it is cut into small cubes and wrapped in a parchment paper for storage at room temperature till it is marketed. For every 1 kg of cooked and processed ghee residue about 500-625 gm of sugar and 125-250 gm of dry coconut powder is required.

Fig. 10.1 Candy Plant

Chocolate

The processed cooked ghee residue is mixed with 50 percent sugar syrup in the same manner as explained above and to this sticky mass, cocoa powder and skim milk powder are added and stirred vigorously till the formation of a pat. Then the finished product is spread on a plate and allowed to cool to a temperature of 5-10 °C overnight. Then it is cut into cubes and wrapped in parchment paper for storage at room temperature till it is being marketed. For every 1 kg of cooked residue about 500-625 gm of sugar, 60-90 gm Cocoa powder and 250 gm of skim milk powder is required for the preparation of the chocolates.

Preparation of Paste

The cooked residue is mixed with salt at the rate of 2.5-3 percent followed by 'marmite' an yeast product at the rate of 0.1 to 0.5 percent and heated on a low flame for about 5 minutes till a paste is formed, which can be used in sandwitches. As an alternative the salted residue is mixed with chutney powder at the rate of 2-4 percent ground thouroughly and then heated as above to yield a paste for use with idli, dosa and samosa.

In the recent years, a method has been developed for the manufacture of a burfi type of sweet and the method is as follows.

The processed sweet ghee residue is mixed with standard quality khoa, which is prepared from 5 percent standardised buffalo milk having a moisture content of 30 percent in the proportion of 1 : 1 of total solids content. The mxture is then heated in a steam jacketted vessel/kettle, agitating thouroughly with a 'Khunti' in order to ensure proper mixing. The total quantity of crystal sugar calculated at the rate of 45 percent of total solids in khoa and 75 percent of total solids in the processed ghee residue is added to the mixture, which is then worked rigorously for 10-15 minutes in order to dissolve the sugar completely. About one third of the sweetened mass is separated and chocolate at the rate of 8 percent of the total solids content of the processed residue Khoa and well mixed into it. This portion, containing the dissolved chocolate, is applied as a thin layer over the remaining two thirds of the mixture, which has already been spread out as a thick layer on a well greased tray. This combined mass is allowed to cool and set and then cut into pieces of a uniform size and shape for packaging and sale. This product has been called 'Choco' burfi.

CHAPTER 11
Milk–Different Varieties/Forms

Generally, the milk is sold by the dairy plants in the liquid form for the consumers after it is processed/pasteurised because it is perishable and has very limited shelf life. In order to increase the shelf life of the milk it is processed, fermented and converted into different categories which physically resemble and behave like liquid milk and are being marketed. These include:

11.1 Sterilized Milk

11.1.1 Definition

Sterilized milk may be defined as the milk which has been homogenised and has been heated to a temperature of 100 °C or above for such a length or period of time that it remains fit for human consumption for at least 7 days at room temperature. Commercially sterilised milk is rarely sterile in the strict bacteriological sense, because, the requirements of complete sterility conflicts with the consumer's preference for normal colour and flavour in the sterilized product. The spore forming bacteria which are present in the raw milk, are highly heat resistant and survive the sterilization temperature time employed in the dairy and ultimately lead to the deterioration of the sterilized milk.

The sterilized milk must meet the following requirements which include:

1. The milk should remain stable without any deterioration having good commercial value for sufficient period to satisfy the commercial requirements.
2. It should be free from microorganisms which are harmful to the health of the consumers namely the pathogenic toxinogenic germs and toxins.
3. It should also be free of any microorganisms which are liable to purify and should not show any signs of bacterial growth that leads interalia to an absence of deterioration.

The sterilised milk has both the advantages and disadvantages which include:

1. It has a remarkable keeping quality and does not need any refrigerated storage.
2. There is no cream layer/plug.
3. It forms a soft digestible curd hence useful for the feeding of infants and invalids.
4. It has a distinctive rich flavour due to homogenisation.
5. It is economical in usage due to its long shelf life.
6. It is less prone/liable to the oxidised taints/flavours and,

the disadvantages are:

(a) There is an increased cost involved in its production.

(b) There are more losses in its nutritive value when compared to the pasteurised milk. About 50 percent of the Vitamin C and 33 percent of Vitamin B Complex, which are originally present are distroyed and also there is a slight reduction in the biological value of the milk proteins.

(c) The normal 'Gerbers' fat test procedure is not so accurate.

The method of manufacture of the sterilised milk has been given as flow diagram 11.1 for better understanding.

Flow Diagram 11.1 Manufacture of Sterilised milk

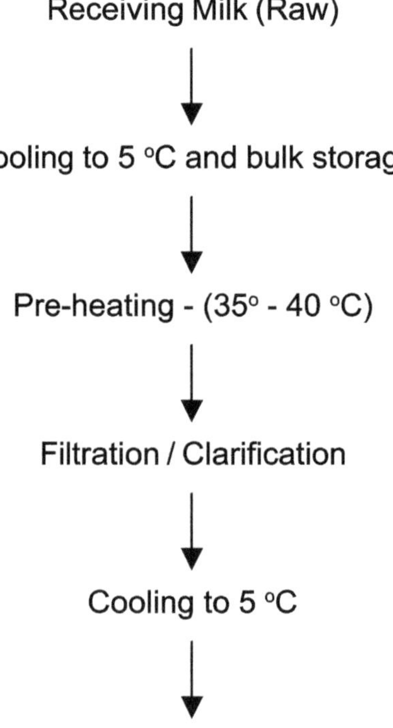

Flow Diagram 11.1 *contd....*

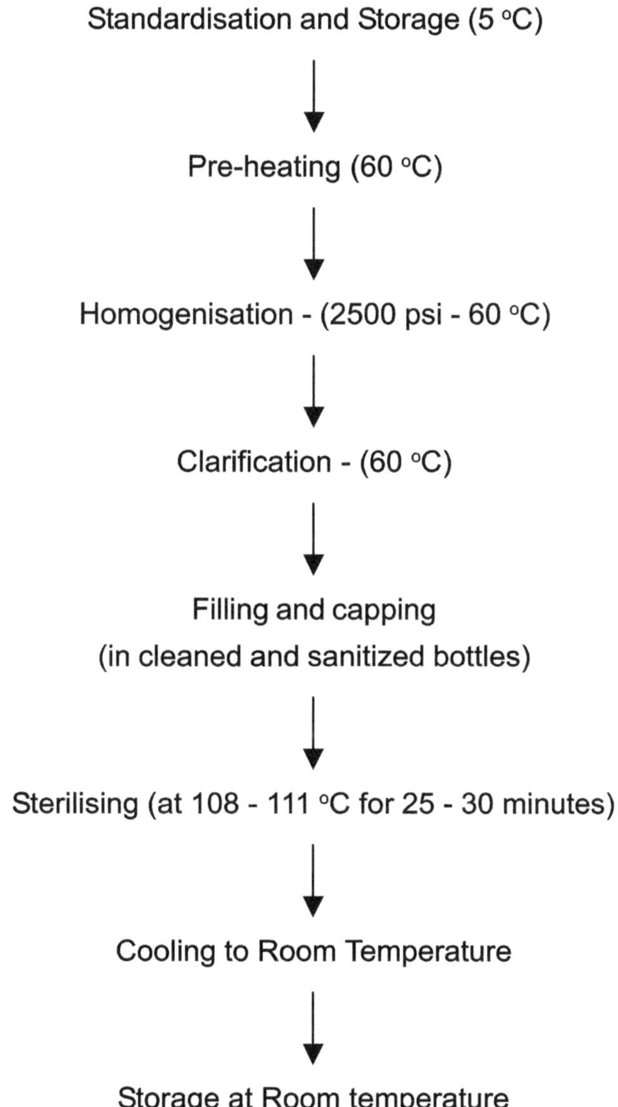

11.1.2 Manufacturing Details

The raw milk which has been received from the different procurement milk centres should be examined strictly by the established and prescribed physico-chemical and bacteriological tests and the high quality milk only should be used for the production of the sterilized milk. Enough care should be taken to accept the milk supplies which do not have any developed acidity and containing the least number of spore forming bacteria. The intake milk should be promptly cooled to 5 °C for bulk storage in order to check the bacterial growth if any. Then the milk should be pre-heated to 35-40 °C for an efficient filtration/or clarification so as to remove the visible dirt etc and to increase the aesthetic quality. The milk again should be cooled to 5 °C in order to preserve its quality and it should be standardised to the desired percentages of fat and SNF content in order to confirm to the legal standards in accordance with the PFA rules 1976 which varies from state to state for both cow and buffaloe milk and the milk must be cooled and stored at 5 °C until it is processed in the storage tank. The milk should be promptly pre-heated to 60 °C for efficient homogenisation and to prevent any subsequent cream layer formation. Generally the single-stage homogenization is carried out at 2500 psi pressure. The milk which has been homogenised must be clarified in order to remove the sediment formed during the process of homogenisation. The hot milk from the homogeniser should be filled into the cleaned and sanitized bottles which are coming from the bottle washing machine and then sealed with special caps of the crown seal type. The filled and capped botles should then be placed in the metal crates for sterilization by the 'batch process' or fed into conveyor for continuous process. Usually the milk is sterilized at 108-111°C (225 - 230 °F) for 25-30 minutes. The sterilised milk bottles should be gradually cooled to room temperature. Any sudden cooling may lead to the breakage of the bottles and finally the milk in bottles should be stored in a cool place until they are despatched for marketing.

The sterilizers are two types namely (1) Batch and (2) Continuous.

The batch type of sterilizers may be either rotary or non-rotary types. The batch or tank sterilizers are rectangular horizontal, boiler or cylindrical shaped, which retorts with a steam inlet and condensate outlet fitted with clamp down covers, into which the steam is adjusted for the required temperature and time for sterilisation. The batch sterilizers are simple and flexible in its operation and less in its initial cost and recurring expenditure. The sterilized milk generally produces a brownish appearance and gives a cooked flavour to the finished product. Sometimes the sterilization may be faulty and the cooling process has to be very slow in order to avoid the breakage of the bottles. During the large scale processing no economic advantages have been obtained. In the batch rotary type, the filled bottles are put into holders which are rotated at 6-7 rpm. The sterilized milk is of a slightly better quality in rotary type sterilizers than in non-ratary ones.

In the continuous type the filled and sealed milk bottles are automatically placed by means of a slat conveyor into the pockets of carrier cages. Then they pass into water at near boiling temperature and from there they enter into the sterilizing zone, consisting of a steam chamber at 108-111°C (225-230 °F) where the bottles remain for a predetermined time namely 25-30 minutes for sterilization of milk. After completion of the heat treatment in the batch or tank sterilizers the milk bottles may be cooled in air or water. In case the cooling is too rapid there is a danger of breakage of the bottles and if it is too slow then browning due to caramelisation may occur. In the continuous system, after leaving the sterilisation zone the bottles enter a column of hot water where the cooling process begins. This is followed by their passage through another tank of water where a lower temperature is maintained than the previous one for further cooling, and lastly through a shallow tank of cold water for final cooling. The bottles are then discharged automatcially and conveyed to a point where they are placed in crates in which they are transferred to the storage room.

11.1.3 UHT Method of Sterilization

In this process, the milk is heated to 135-150 °C for few seconds, generally in a plate or tubular heat exchanger. The milk which is almost sterile has to be filled into containers for distribution, where the filling is carried out asceptically. In many cases pre-sterilization, as above, is followed by in bottle, sterilization. The distribution of milk is carried out once a week; due to this reason the sterilized milk has great scope in warm countries as long as household refrigerators are not in common use.

The common tests that are generally carried out for the confirmation of the sterilization are:

1. Turbidity test which is not an official test.
2. The bacterial count. The phosphatase test is not applicable to the sterilized milk. Browning is the most common feature in the sterilized milk and due to this reason or fault, the plain sterilized milk is not so popular. Flavoured and simultaneously coloured sterilized milk is more popular.

11.2 Homogenized Milk

Homogenized Milk according to the United States Public Health Service is defined as the milk which has been treated in such a manner as to ensure the break up of the fat glabules to such an extent that even after 48 hours of quiescent storage no visible cream separation occurs on the milk and the fat percentage of the milk in the top 100 ml of milk in a quart bottle or of proportionate volumes in containers of other sizes, does not differ by more than 10 percent of itself from the fat percentage of the remaining milk as determined after thourough mixing. Inefficiently homogenised milk the fat globules are subdevided into 2 microns or less in diameter.

For preparation of homogenised milk the students are advised to refer to the Homogenisation process of milk.

11.3 Soft Curd Milk

It is defined as the milk which forms a soft curd, when coagulated with rennet or pepsin under standard procedure, which has a curd tension (CT) of less than 25 g. The standard procedure such as the Hill Curd Tension test consists in coagulating milk by means of a pepsin-calcium chloride solution and then measuring the resistance in which a special multi bladed curd knife encounters in its passage through the coagulated milk.

The characteristics of the soft curd milk are:

1. Low casein content and
2. low calcium content. The average casein and calcium contents of Indian cow and buffaloe milks have been compared to human milk and presented in the table11.1.

Table 11.1 Casein and calcium content of different milks

Type of Milk	Casein (%)	Calcium (%)
Indian Cow	2.5	0.13
Indian Buffalow	3.0	0.19
Human	0.4	0.03

Source : Outlines of Dairy Technology - S.K. De

A close examination of the above table reveals that the human milk is very low in casein and calcium. Therefore, the mothers milk is best suited for the infants feeding because it forms a soft curd when coagulated in the stomach, which is apparently more quickly digested as compared to the cow or buffaloe milks.

Several methods are being employed in the preparation/manufacture of soft curd milk. These include

1. **Natural soft curd milk:** This is possible by segregating the cows that produce milk with a curd tension of 30 gm or less from the herd by closely watching them over a period of time.
2. **Dilution with water:** The casein and calcium contents of the milk are reduced with the addition/dilution of water.
3. **Heat Treatment:** This may be carried out either by boiling, sterilization or super heating. Of these the boiling of milk is the most common practice adopted by every housewife, which is fairly effective as it is the standard practice in India, whereas the sterilisation and super heating are markedly effective.
4. **Homogenisation:** A pressure of 2500-3000 PSI markedly reduces the curd tension.

5. **Acidification:** Addition of lactic acid facilitates the digestive action which is brought about by the fine coagulation of the casein.
6. **Enzyme treatment:** The curd tension is reduced, when the milk is acted upon under controlled conditions by certain proteolytic enzymes such as Trypsin.
7. **Addition of salts:** The following salts like sodium citrate, sodium pyrophosphate, sodium hexameta phosphate etc when added to the milk reduces the curd tension.
8. **Base Exchange Treatment:** In this process the milk acidified with citric acid is allowed to pass through a Zeolite filter bed which contain about 20 percent of Ionic Calcium.

Different types of milk that are suitable for the feeding of infants, have been arranged in the descending order of their digestibility and/or curd tension are presented in the table 11.2.

Table 11.2 Milks suitable for Infant use

Type of Milk	CT in gm (Average)	Remarks
Human	0	Breast milk is the best possible food for human infant and used as a standard for comparison
Boiled Cow Milk	3	Suitable for infant feeding.
Trypsin-treated Cow milk	10	- do -
Homogenised Cow milk	14.5	- do -
Pasteurised Cow milk	44.5	Unsuitable for infant feeding
Raw Cow milk	55	- do -

Source : Text Book of Dairy Chemistry - M.P. Mathur, D.D. Roy & P. Dinakar

11.4 Flavoured Milks

Flavoured milks are defined as the milks to which some flavours have been added. When the milk is used, the product should contain a milk fat percentage which is equal to the minimum legal requirement for the market milk. However, when the fat level lower (1-2%) then the term 'drink' is used instead of milk.

The main purpose in the manufacture of these flavoured milks/drinks is

1. to make the milk more palate to those who do not relish it as such.
2. To stimulate the sale of liquid milk and
3. To put the skim milk for more profitable use.

There are different types of flavoured milks and these are:

1. Chocolate milks/drinks
2. Fruit flavoured milks/drinks and
3. Sterilised flavoured milks/drinks.

The method of manufacture of these flavoured milks or drinks has been given in the form of a flow diagram.11.2

Flow Diagram 11.2 Manufacture of Flavoured milks

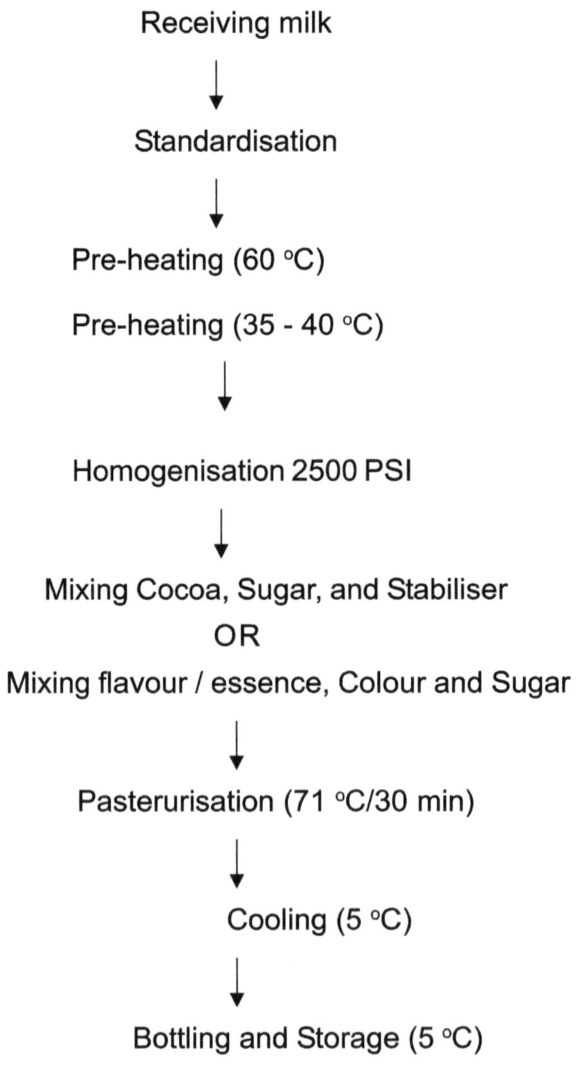

The details of manufacture of flavoured milks and/or drinks has been given as under:

1. *Chocolate milks/drinks*: In the preparation of these products the following formula may be used and the ingredients are:

(a) Cocoa Powder	1 to 1.5%
(b) Sugar	5 to 7.0%
(c) Stabilizer (Sodium alginate)	0.2%
(d) Fat level in milk/Drink	As per the minimum legal standards - 2%.

The milk on receipt is standardised, pre heated to 35-40 °C and filtered or as an alternative, after standardisation it is pre-heated to 60 °C, homogenised at 2500 PSI and then clarified. The desired amounts of cocoa-mix, sugar and stabilizer are slowly added and stirred so as to dissolve them properly to the warm milk. The cocoa powder may also be added in the form of a syrup and the stabilizer in the form of solution. The mixture is then pasteurized at 71 °C for 30 minutes cooled rapidly to 5 °C, bottled and kept under refrigeration (5 °C) until used. The bottles are invariably inverted up and down or shaken for few minutes/times before consuption. Standardised milk is homogenised to prevent or delay the rising or cream. It may be homogenised after the addition of cocoa and sugar, however this has the effect of increasing the sedimentation. Stabilizer is usually added to delay or prevent settling of cocoa particles, which also ends in the prevention of the cream rising.

11.4.1 Fruit Flavoured Milks/Drinks

In the preparation of these milks/drinks, permitted fruit flavours or essences along with the permitted matching colours and sugars are only used. The method of preparation is very similar to that followed for the preparation of Chocolate milks/drinks. The common flavours used are Strawberry, orange, lemon, pine-apple, banana, vanilla, rose etc. Pure fruit juices or syrups can be satisfactorily combined with milk to form "Milk shakes". However, in order to obtain good results the following precautions must be taken care.

1. No acid either citric or tartaric should be added to the fruit syrup, because it may cause curdling. The pH of the milk syrup mixture should be about 5, which is safe from causing curdling.
2. Avoid excessively sweet syrups. The best sugar content of the syrup is around 45-55 percent.
3. Add 1 part of fruit syrup to 5 parts of milk. The fruits which give good results particularly are strawberry, lemon, orange, pineapple, blackberry, raspberry and blackcurrant.
4. Care should be taken to see that there is a pleasant blend of sweet, fruity, and milky flavours together with an appealing colour.

11.4.2 Sterilized Flavoured Milks/Drinks

These milks have the combined advantages of both sterilized and flavoured milks or drinks. The method of preparation has been given in the form of a flow diagram 11.3 and the preparation is similar to the preparation of chocolate and flavoured milks/drinks as explained earlier. In addition, in between clarification and filling flavour or essence and the permitted matching colours and sugar are added and mixed into the milk.

Flow Diagram 11.3 Manufacture of Sterilised flavoured milks

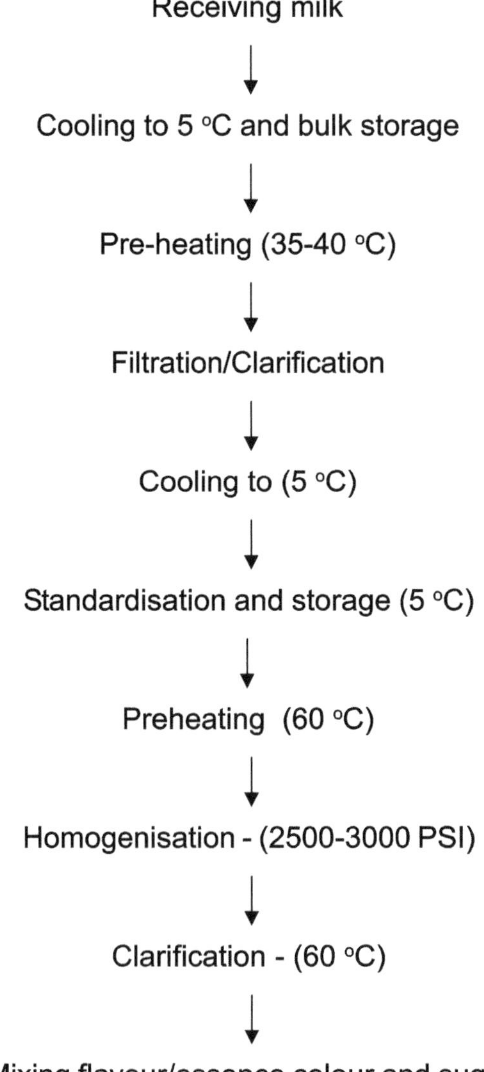

Flow Diagram 11.3 *contd....*

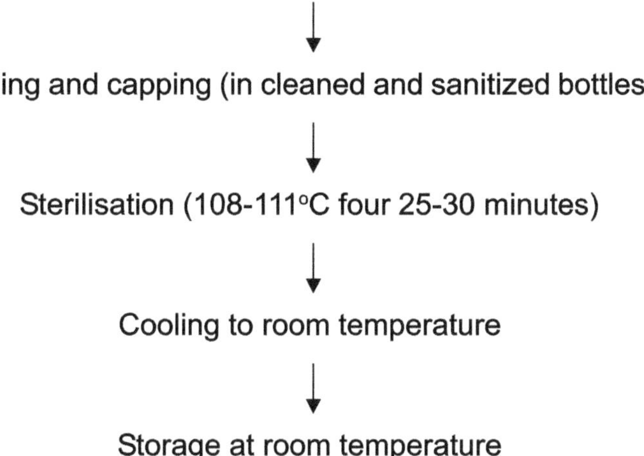

11.4.3 Vitaminized/Irradiated Milk

Vitaminized milk is defined as that milk to which one or more Vitamins are added. Whereas the Irradiated milk is defined as that milk in which the Vitamin D content has been increased by exposure to Ultra Violet rays. Mineralised milk is the milk to which the minerals have been added. It is a well known fact that lack or deficiency of Vitamins in the diet causes specific deficiency diseases which in turn, can be cured by the administration or intake of those particular Vitamins. Addition of vitamins and minerals to milk is called as fortification and such milk is called as fortified milk. The vitamins and minerals may be added either simply or in combination more commonly as multi vitamin or mineral preparations. Commercial vitamin-D milks can be prepared by any of the following methods.

1. **Metabolized:** In this case the milch animals are fed with irradiated yeast, thereby increasing the Vitamin-D content of the milk.
2. **Irradiated:** Rapidly flowing thin film of milk is exposed to an intensive source of Ultra-Violet radiation such as carbon arc lamp or a quartz mercury lamp, by which the milk develops an increased Vitamin-D potency due to the conversion of milk cholesterol to Vitamin-D by rays of wavelength at 2700 - 2900 Å.
3. **Fortified:** As already explained a definite amount of standardised Vitamin-D concentrate is added to the milk.

11.4.4 Frozen Concentrated Milk

This refers to the milk which has been partially concentrated and then solidified by means of freezing. The main advantage of this method is that the milk has a long keeping quality and is easy to transport for even long distances whereas the main drawback in this system is the destabilization of the fat and casein, resulting "in oiling off" and sediment formation. The Delhi Milk Supply scheme for the past several years has received the un-concentrated frozen cow milk regularly from Rajasthan. In this the raw milk is filled in metallic cylinders/containers which are covered tightly and then lowered into the refrigerated brine tanks and kept for 10-12 hours, whereby the milk in the cylinders is frozen. These cylinders containing frozen milk are then transported overnight in insulated railway wagons.

11.5 Fermented Milk

Fermentation has been defined as the metabolic process in which chemical changes are brought about on an organic substratum where the protein, carbohydrate or fat, through the action of enzymes liberated by specific living microorganisms. In the dairy industry, the most important fermentation is the lactic acid fermentation or the souring of milk. Fermented milks refer to those milks, which have been made by employing the selected microorganisms to develop the characteristic flavour and/or body and texture. The various types of fermented milks which will be discussed here are:

(a) Natural butter milk
(b) Cultured butter milk
(c) Acidophilus milk
(d) bulgarian butter milk
(e) Kumiss
(f) Kefir
(g) Yoghurt and
(h) Dahi

11.5.1 The Merits of the Fermented Milk are

1. It is more palatable than milk itself.
2. The nutritive value is generally higher as compared to the milk.
3. It is more easily assimilated by the human system than the milk.
4. It may contains the beneficial antibiotics.
5. It may possess some therapeutic properties.
6. Regular consumption of some varieties/types of fermented milk claimed to increase the longevity of human beings.

11.6 Natural Butter Milk

It is a by-product obtained by churning cream in the dairy plant and dahi in the common households for butter making. Ripened cream, which has undergone a clean lactic fermentation is usually preferred. Under Indian conditions butter milk is obtained in the villages as a by-product of churning the whole milk curd/dahi for the production of butter in the house.

11.6.1 Cultured Butter Milk

The cultured butter milk is obtained by inoculation and incubation of pasteurised skim milk with lactic starter and its manufacture has been given in the form of a flow diagram.11.4.

Flow Diagram 11.4 Manufacture of Cultured butter milk

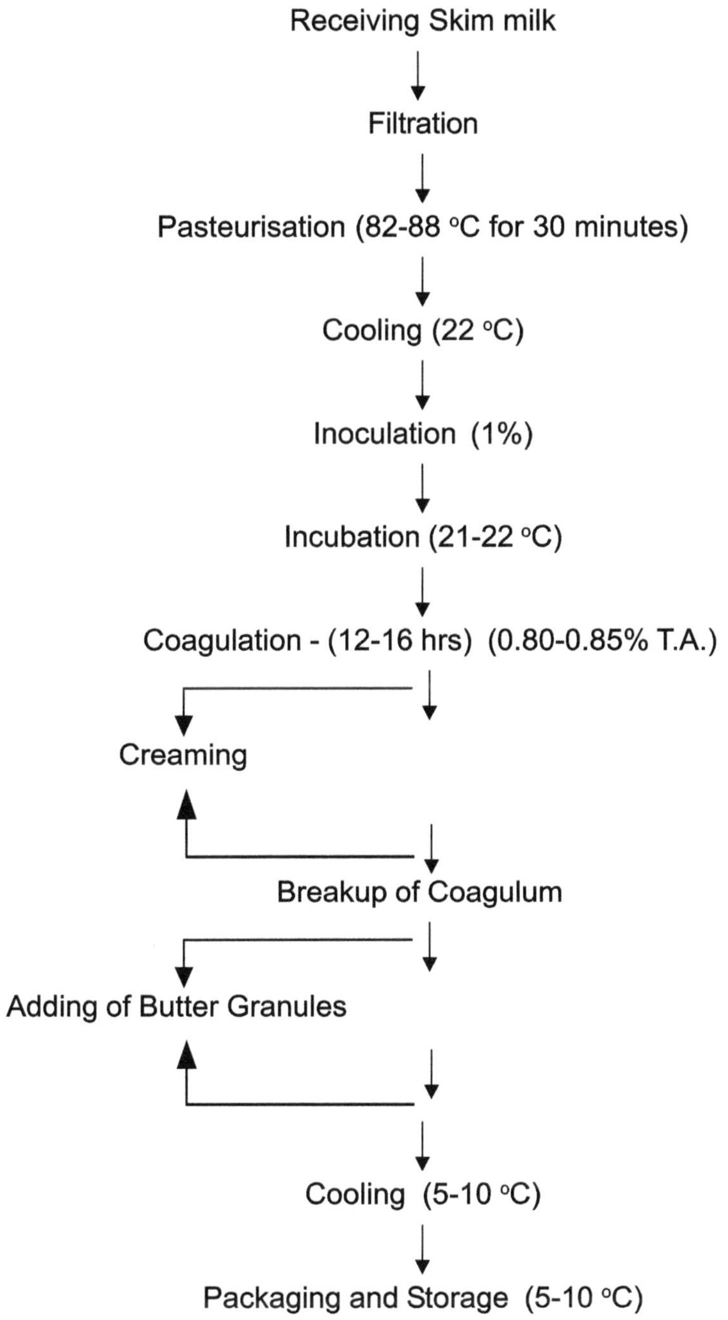

Receiving Skim milk

↓

Filtration

↓

Pasteurisation (82-88 °C for 30 minutes)

↓

Cooling (22 °C)

↓

Inoculation (1%)

↓

Incubation (21-22 °C)

↓

Coagulation - (12-16 hrs) (0.80-0.85% T.A.)

↓

Creaming

↓

Breakup of Coagulum

↓

Adding of Butter Granules

↓

Cooling (5-10 °C)

↓

Packaging and Storage (5-10 °C)

11.6.2 Manufacturing Procedure/Preparation

Fresh skim milk containing solids not fat content in a range of 9-10 percent, which is free from antibiotics and inhibitory substances is most preferable. Milk containing an appreciable amount of fat is used, it should be homogenised and in case if cream or butter fat is desired in the product it may be added to the cold butter milk when the curd is broken up. The skim milk which is being used for the manufacture of butter milk is heated in a holder pasteurizer vat to a temperature of 82-88 °C (180-190 °F) for 30 minutes with frequent stirring and then cooled to 22°C (72 °F). Bulk lactic starter at the rate of one (1) percent is added and mixed thouroughly in the milk and then allowed to incubate at 21-22°C (70-72 °F) without any further agitation until coagulation occurs with an acidity ranging between 0.80-0.85 percent is achived/reached. Then the curd is broken up by slow and gentle agitation to give a smooth and uniform consistency and cooled to 5-10 °C which is then packaged and stored at the same temperature until it is distributed. Proper care should be taken not to incorporate any air in the product during agitation and filling and the acidity at the time of packaging should be in the range of 0.85 to 0.90 percent.

11.6.3 Creaming

Addition of cream imparts a rich flavour to the butter milk which has the effect of decreasing the viscosity. Viscosity can also be decreased by adding 20-25 percent of pasteurised sweet whole milk during the process of break up of the curd. Further, salt at the rate of 0.1 percent may be added to the finished product in order to increase its flavour. Sometimes butter granules or flakes are added to the cultured butter milk made from skim milk to simulate and give an appearance of an old fashioned churned butter milk which is produced in the villages or village conditions. Churned butter milk may be prepared by culturing milk containing 1-2 percent butter fat. Cooling the coagulated milk to 13 °C (50 °F) and then churning it until butter granules appear or look like rice grains. Butter colour may be added to increase the appearance and colour of the granules and the product is finally cooled, packaged, and stored until it is distributed.

11.7 Acidophilus Milk

This kind of fermented milk is produced by the development of a culture of *Lactobacillus acidophilus* in the milk. It is claimed that the acidophilus milk has the therapeutic and health promoting properties. It is also claimed that the growth of lactic acidophilus under the conditions existing in the intestinal track will replace the undesirable putrefactive fermentations with a beneficial lactic fermentation.

The manufacture of acidophilus milk has been given in the flow chart 11.5.

Flow Diagram 11.5 Manufacture of Acidophilus milk

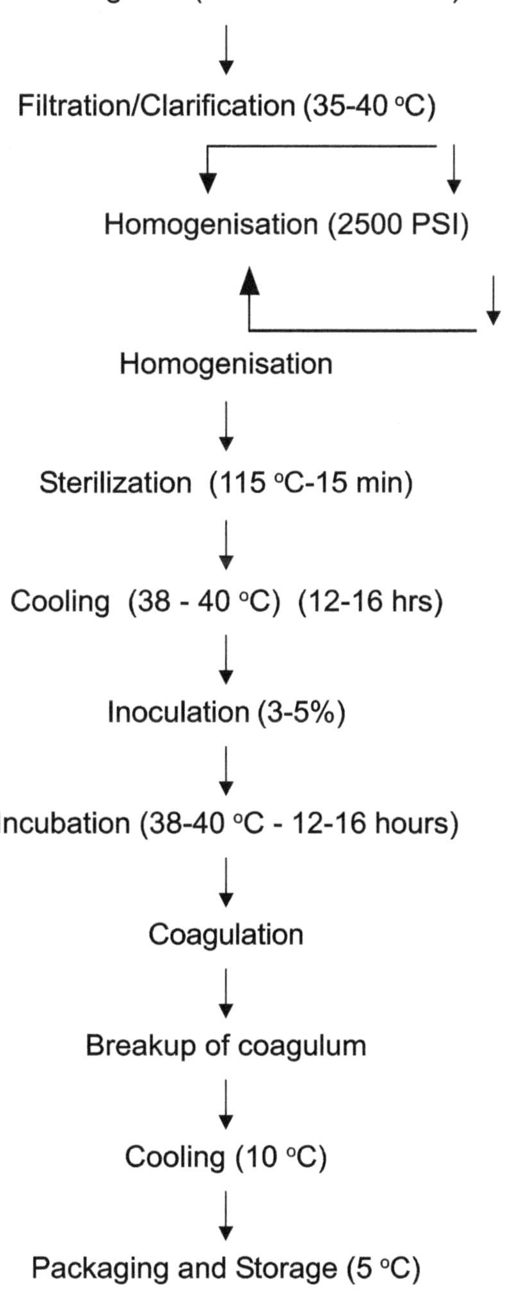

11.7.1 Procedure

In the manufacture of the acidophilus milk, either skim, whole or defatted milk may be used, however, if the milk which is being used contains more than 1 percent fat should be homogenised. Acidophilus milk can be easily and simply be made with added sugar, or honey or with tomato or carrot juice. Fresh milk is heated to 115 °C (240 °F) for 15 minutes or may be heated intermittently in order to obtain sterile milk. Addition of 1 percent glucose, or honey or 5 percent tomato juice helps to hasten the fermentation, but is not essential. The milk is then cooled to 38-40 °C (100-104 °F) and inoculated with 3-5 percent inoculum or mother starter. The inoculated milk is then mixed thouroughly and incubated at the same temperature i.e. 38 - 40 °C until the milk coagulates at least for 12-16 hr. Then the coagulum is slowly broken up and cooled to 10 °C (50 °F). Five to ten percent lactose or dextrin may be added if necessary. The milk is stirred until it becomes smooth, packaged and stored at 5 °C (40 °F) until used. The strains of *lact.acidophilus* are difficult to maintain and propagate because of the ease with which the contaminating bacteria outnumber them.

11.8 Bulgarian Butter Milk

In the preparation of this kind of milk a pure single strain starter of *Lacto bulgaricus* is propagated and the incubation temperature ranges from 38-43 °C (100-110 °F). The finished product is characterised by a higher acidity of about 1.2 to 1.5 percent.

11.9 Kumiss

In the preparation of this milk, in earlier days mares milk has been used whereas in the present days cows milk is being used and is fermented with lactic acid-alcohol and is originated in Russia. The culture consists of *Lack acidophilus or bulgaricus*. The first incubation period is followed by the addition of yeast and a second incubation period. The finished product may contain up to 2-5 percent alcohol.

11.10 Kefir

This milk has its origin in the Caucasian Mountains. It is a self carbonated beverage containing 1 percent lactic acid and 1 percent alcohol. It is made with a fermenting agent called "Kefir grains" which contains *Str. Lactics. Betabacterium caurasicum* Kefir bacilli and lactose fermenting yeasts.

11.11 Yoghurt

This has got many synonyms. In turkey its called as Yoghurt. In Egypt it is known as leben. In Armenia it is called as Metzoon and in India it is popularly called as Dahi which are all similar in their nature. Yoghurt was originally prepared by using the boiled concentrated wholemark, however, in the present modern days with the introduction of modern methods of manufacture use, whole or partly defatted milk containing small amounts of skim milk powder or concentrate. The fat content in Yoghurt may vary from 0 to 5 percent and the solids content from 9 to 20 percent.

The method of manufacture of Yoghurt has been given in the form of a flow diagram 11.6 for better understanding.

Flow Diagram 11.6 Manufacture of Yoghurt

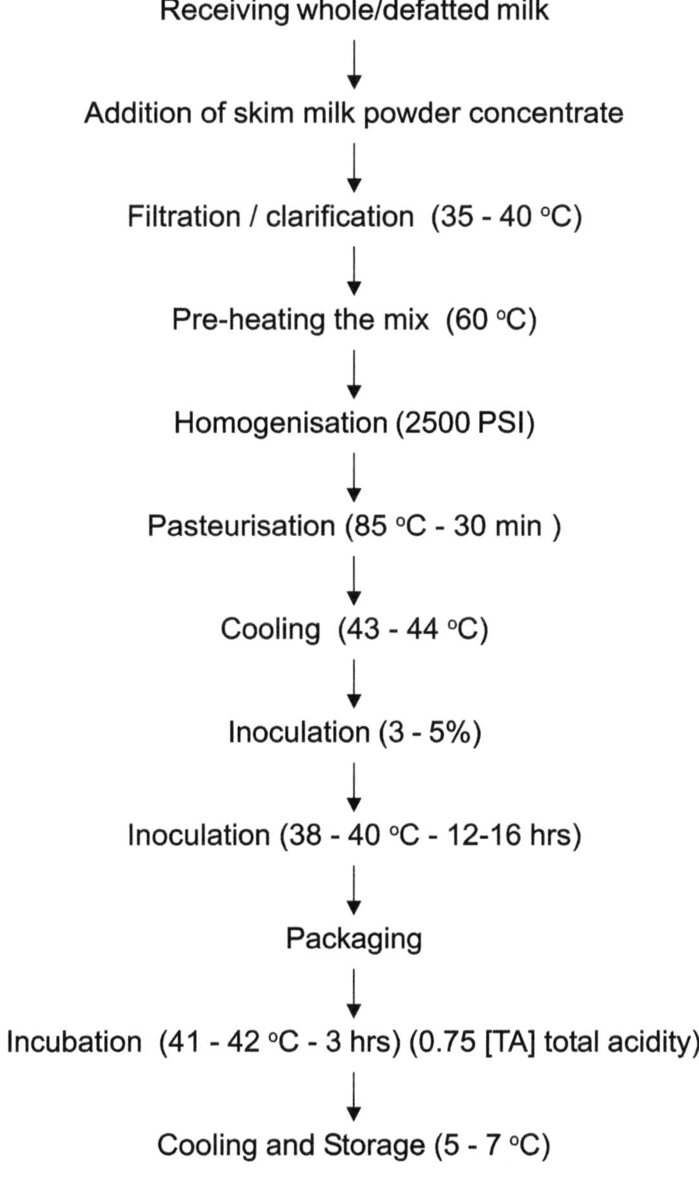

11.11.1 Procedure

The solids not fat content is increased by about 2-3 percent, thus bringing the total to approximately 12 percent by adding skim milk powder/concentrate to whole or partly defatted milk. Then the mix is pre-heated to 60 °C (140 °F) and homogenised in a single stage homogeniser at 2000-2500 PSI. In general stabilisers are added frequently to the mix, however, it is not essential and avoid excessive amounts. The mix is now heated in a vat to 85 °C (185 °F), inoculated with 2 percent bulk starter and stirred briefly to ensure proper mixing. The mix is then packaged taking proper care that the temperature does not fall below 41°C (106 °F) during the filling operation. It is most important to be borne in mind that the time interval between the inoculation and filling should not exceed 45 minutes. The yoghurt is incubated in the package without any further agitation at 41-42 °C (106-108 °F) for about 3 hours till a titrable acidity of 0.75 percent is reached. Then it is placed under refrigeration to cool to 5 - 7 °C (40-45 °F) and approximately about 8 hours is required for this purpose, and at this stage the product is ready for distribution with the final acidity of 0.9 percent is optimum and desired to this product at the time of distribution.

An attempt has been made to present the standards as prescribed by the FSSAI (2017) and published in the Gazette Govt. of India for the yoghurt and flavoured Dahi has been presented in table 11.3 for the benefit of the students of dairy sceince.

Table 11.3 FSSAI (2017) Standards for Yoghurt and Flavoured Dahi

Sl.No	Parameter	Yoghurt and flavoured Dahi	Partly skimmed yoghurt and partly skimmed flavored Dahi	Skimmed yoghurt and flavoured skimmed Dahi
1.	Milk fat % (m/m)	Not less than 3.0 and not more than 15.0	More than 0.5 and less than 3.0	0.5 (maximum)
2.	Milk S.N.F minimum % (m/m)	8.5	8.5	8.5
3.	Milk protein®minimum % (m/m)	2.9	2.9	2.9
4.	Titrable acidity minimum % (as lactic acid)	0.6	0.6	0.6

Note: ®Protein content is 6.38 multiplied by the total nitrogen determined when sold without any indication the product shall conform to the standards of yoghurt or flavoured Dahi as appropriate.

11.12 Flavoured Yoghurt Preparation

In the recent years the flavoured yoghurt has become more popular and the flavours used are of three types namely Synthetic, Artificial, or Natural. Synthetics are sometimes supplemented or added with natural fruits. These flavours may be added to Yoghurt in one of the following ways.

1. **By mix blending:** In this case the flavouring material, sugar and colour are added to the mix prior to fermentation.
2. **Fruit on bottom:** Preservatives or fruits are added to the package prior to filling or setting.
3. **Bulk Mixing:** Yoghurt is prepared, bulk chilled and mixed with a fruit dressing and packaged preservatives, flavours, sugar and colour are blended in after fermentation.

The flavoured Yoghurt has a definite advantage over the plain in that the harsh acidity in the product is less pronounced, and the incidence of objectionable off flavours is reduced and the need for the concentration of the milk is eliminated.

11.13 Standardised Milk

The standardised milk for liquid consumption, should contain a minimum of 4.5 percent fat and 8.5 percent solids not fat according to PFA Rules (1976) throughout the country.

Standardised milk is defined as that milk whose fat and/or solids not fat content have been adjusted to a certain predetermined level. The standardisation can be done by partially skimming the fat in the milk with a cream separator or by admixture with fresh or reconstituted skim milk in proper proportions. Standardised milk may be marketed as such or used for making certain products. In the latter case the fat and solids not fat contents may be varied according to the product requirement.

The advantages of the standardised milk are:

1. The consumer is assured if getting a milk which ensures a practically uniform and constant composition having a good nutritive value.
2. The surplus fat can be converted into butter and ghee.
3. It is possible to supply milk at cheaper rate.
4. It is more easily digestible due to the reduced fat content.

It is important to kept in mind that the initial reaction of the consumer may be that the standardised milk appears to be thinner than the whole milk.

The method of preparation of the standardised milk has already been discussed under standardisation.

11.14 Reconstituted/Rehydrated Milk

This is the milk which is prepared by dispersing or dissolving, whole milk powder, which is also called as dried whole milk in water approximately in the proportion (1: 7 or 8) 1 part of powder to 7-8 parts of water. At present, the spray dried milk powder is used because it is more readily soluble resulting in the production of less sediment. The main advantage of this milk is that it helps in making up the shortage of fresh milk supplies in developing countries. It is also used by the Armed forces in our country as well as in other countries also. The method of manufacture of reconstituted milk has been given in flow chart 11.7.

Flow Diagram 11.7 Method of Preparation of the Re-constituted milk

Receiving water in pasteurising vat

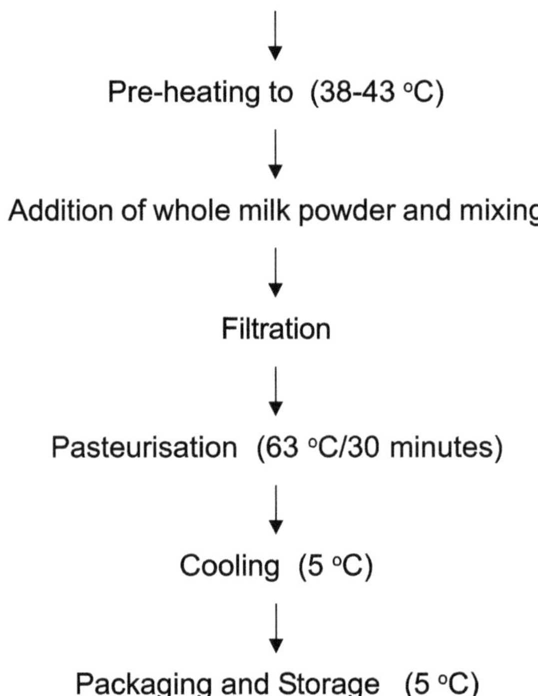

11.14.1 Procedure

Calculated amount of potable water is received in the pasteurizing vat/tank equipped with an agitator. The water is then heated, keeping the agitator in motion to 38- 43 °C. Then the calculated amount of spray dried whole milk powder is added very slowly, at the point of agitation and the mixture is thouroughly agitated till it dissolves completely. At times a special powder mixture equipment may be used for this purpose. The mixture is then pumped through the filter, pasteurized at 63 °C for 30 minutes and promptly cooled to 5 °C or below until distribution.

11.15 Recombined Milk

It is the milk or product obtained by the combination of butter oil which is also called as dry/anhydrous milk fat, skim milk powder and water in the correct and predetermined proportions to yeild fluid milk. The milk fat may also be obtained from other sources such as unsalted butter or plastic cream. The recombined milk should contain a minimum of 3.0 percent fat, and 8.5 percent solids not fat throughout the country according to PFA Rules (1976). The advantages of the recombined milk are that it helps in making up the shortage of fresh milk supplies in the developing countries. It also helps in preventing the price rise of liquid milk in the metropolitan cities. The method of manufacture of recombined milk has been given in flow chart 11.8.

Flow Diagram 11.8 Method of Manufacture of the Re-combined milk

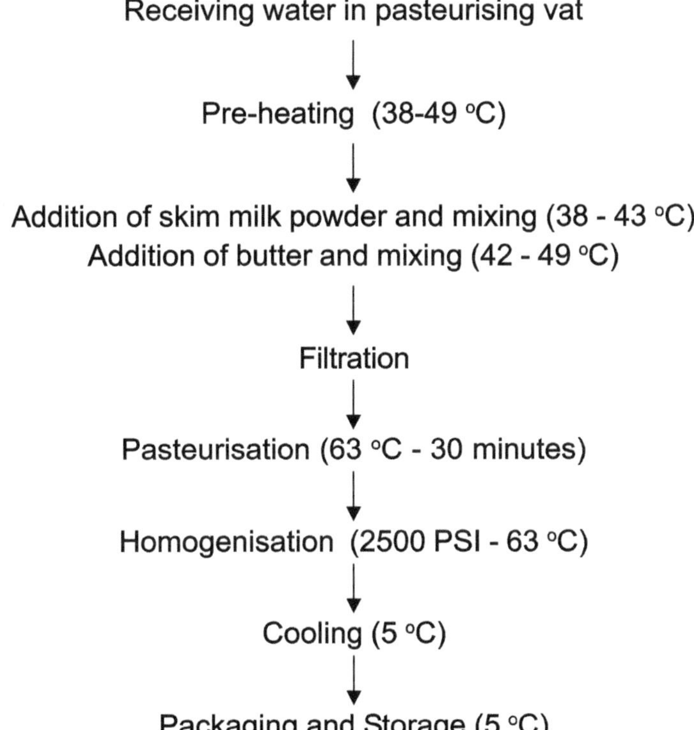

11.15.1 Procedure

As in the case of preparation of reconstituted milk, the initial steps are similar up to the stage of adding the whole milk spray dried powder at a temperature of 38-43 °C. In this case instead of spray dried powder, a proportionate amount of spray dried skim milk powder is slowly added at the point of agitation. When the water reaches a temperature of 43-49 °C another proportionate amount of butter oil is added. These constituents are then thouroughly agitated till a homogeneous mixture is obtained. The mixture is then pumped through a filter and then pasteurized at 63 °C for 30 minutes. It is then homogenised at the pasteurisation temperature of 63 °C at 2500 PSI single stage to ensure proper emulsification of the milk fat or butter oil. The product is then cooled quickly to 5 °C, packaged and stored at 5 °C or below until it is distributed.

It is very important to mention here that for the benefit of the students of dairy science that the re-combined milk was produced extensively under the operation flood scheme in the four major metropoliton cities of India viz. Bombay, Culcutta, Delhi and Madras during the years 1970-75, from the butter oil and skim milk powder donated under World Food Programme project by the Food and Agriculture Organisation of the United Nations Organisation.

11.16 Toned Milk

Shri D.N. Khurody, while working at the central dairy of the Aarey milk colony in the year 1946, has produced this milk in order to meet the every increasing demand for the supply of liquid milk by the citizens of Bombay. The production of this milk was his brain child and also credited with the coining of the name "Toned milk". In the opinion of Mr. Khurody, simply by adding water to the whole milk of the buffaloe, both the fat and solids-not-fat content are reduced. However by adding skim milk powder or skim milk to the whole buffalo milk, the soilids-not-fat content of the mixture is toned up or increased to the original level. The product thus obtained was neither whole milk, nor standardised and hence a new name "Toned milk" was given to it. In those days the skim milk powder was being imported from the European countries like, Denmark, Switzerland etc in order to meet the acute shortage of whole liquid milk as there are no organised dairies in the country. Soon other cities notably Calcutta, Madras and Delhi started producing and marketing toned milk which has become a permanent feature ever since of the market milk industry in India.

Toned milk or single toned milk is defined as that milk which is obtained by the addition of water and skim milk powder to whole milk. In practice, whole buffalo milk is admixed with the reconstituted spray dried skim milk for its production. Toned milk should contain a minimum of 3.0 percent fat and 8.5 percent solid not fat throughout the country according to PFA Rules (1976). The advantages of the toned milk are:

1. There is an increase in the supply of milk. The buffaloe milk used initially is increased by 100-150 percent.
2. It helps in reducing the price of the milk so as to reach the lower income groups of population. The method of manufacture of toned milk has been given in the flow chart 11.9.

Flow Diagram 11.9 the Method of Manufacture of Toned milk

Receiving water in pasteurising Vat

↓

Preheating (38-43 °C)

↓

Addition of skim milk powder and mixing and addition of whole buffaloe milk and mixing

↓

Filtration

↓

Pasteurisation (63 °C - 30 minutes)

↓

Cooling (5 °C)

↓

Packaging and Storage (5 °C)

11.16.1 Procedure

The calculated amount of potable water is received in the pasteurizing vat/tank equipped with an agitator. The water is heated while the agitator is kept in motion to 38-43 °C. Then a proportionate amount of spray dried skim milk powder is slowly added at the point of agitation and the mixture is thouroughly agitated till it get dissolved completely. A calculated quantity/amount of whole buffaloe milk is now added and the mixture is again agitated thouroughly till a homogenous mixture is obtained. Then the mixture is pumped through a filter and pasteurised at 63 °C for 30 minutes, rapidly cooled to 5 °C, packaged and kept at 5 °C or below till the milk is distributed.

11.16.2 Double Toned Milk

It is similar to the single toned milk in all aspects relating to the origin and other details except for the difference that according to the PFA Rules (1976) this “double toned milk” should contain a minimum of 1.5 percent fat and 9.0 percent solids not fat content throughout India.

11.16.3 Humanized Milk

Whole cow milk or buffaloe milk when so modified in its chemical composition that it resumbles human milk then it is called as humanized milk.

11.17 Miscellaneous Milks

11.17.1 Filled Milk

It resembles in all aspects the recombined milk except that for the difference that the source of fat in this case is derived from a vegetable source.

11.17.2 Vegetable Toned Milk

In this milk the protein of the skim milk powder is substituted by the Vegetable protein isolated from groundnut.

Example: MILTONE of the Central Food Technological Research Institute. Mysore.

11.17.3 Artificial Milk

Literally speaking artificial milks are prepared from substances/materials that are not found in the natural milk produced either by the cow or buffaloe and infact by any other mammal. However, these artificial milks are entirely different from the so called synthetic milk depending on the base substances that are included in their preparation/manufacture. The authors have made an effort to distinguish between these two milks with an intention to clarify and to remove the confusion that is prevailing in the minds of the students and the general public/consumer as well. The artificial milks may be classified as plant based milks which are then fortified with certain flavors, vitamins and minerals depending on the type of base material used by the manufacturers in their production. Further they may also be included in the category of milk substitutes. Synthetic milk even through prepared artificially by using the hazardous chemicals such as Urea, formalin, detergents, boric acid ammonium sulphate, caustic soda, benzoic acid, salysolic and hydrogen peroxide and melamine are being used as the base materials fradulently and scrupulously by the manufacturers of these milks mainly with an intention to satisfy their greed/ hunger with the sole object of financial gain and to become millionaires by illicit adulteration of the natural dairy milk. Thus there is a vast difference between the artificial milk and the synthetic milk in their preparation/manufacture and the purpose for which they are being utilized by the manufacturers.

W. Matlieu Williams (1871) while discussing and narrating the circumstances and the conditions that prevailed in France and some other European countries who have faced an acute shortage of milk and milk products mentioned about the artificial milk for the first time. According to him the concept of artificial milk was developed by M. Dubrunfant, who proceeded in a scientific manner after convincing and examining himself the theoretical aspects based on the Boussingaults analysis of the cows milk. For the first time the artificial milk was developed and prepared by Dubrufant by mixing water saccharine material such as cane sugar, glucose or sugar milk, dry albumin made from while egg and rub carbonate of soda along with olive oil or any other edible/consumable fatty material which are then together agitated till they form an emulsion. This may be carried out either with warm or cold water however water with 50 °C to 60 °C temperature is preferred to form a pasty liquid, which may be further diluted with its own bulk of water to assume the consistency and general appearance of milk. He further added little quantity of gelatin to give a rich cream by giving a slight viscosity to the liquid (assists in the formation and maintenance of the emulsion.

Andrew Miller (2018) in his article "artificial milk here to stay" has stated that the first stage was the acceptance that the artificial dairy is going to be the order of the day in near future and it was a real thing. He further stated that the scientists had also "grown" milk in the laboratory.

11.18 Synthetic Milk

The simplicity and rapidity with which milk can be adulterated always tempted the unscrupulous milk vendors to indulge in fradulant practices to adulterate the milk. The ever rising greed has given way to the development of a new type of adulterated milk known as synthetic milk. As the name suggests synthetic milk is not milk but it is entirely different component with high degree of adulteration to increase the volume of milk and thereby the profit. Generally it is a mixture of water pulverised detergent or soap. Sodium hydroxide, vegetable oil salt and urea. Similar to genuine milk production the practice of preparing the synthetic milk too starts at the village level. The places notorious for the production of synthetic milk includes parts of Rajasthan, Haryana, Uttarpradesh and in some areas of Andhra Pradesh and Telangana in India. Slowly and steadily the practice is spreading to other parts of India also.

11.18.1 Procedure

Vegetable refined oil of any brand whose butero refractometer reading is less than 42 is selected and taken in a wide mouthed container along with a suitable emulsifier and mixed thoroughly so that the entire content is made into a thick white paste. At this stage water is added slowly to this paste until the density of the liquid is achieved similar to that of milk. Then it is added with urea or sodium sulphate or glucose or maltose or sometimes commonly available any one of the fertilizers. These substances are usually dissolved in hot water and then added to the seemingly milk like solution. The refined oil in synthetic milk acts as a source of fat whereas the hot solution of any one of the substances mentioned above acts as a source of solids not fat (SNF). The ingredients that go into the making of synthetic milk are calculated in such a way that the fat and SNF percentage is similar to mixed milk. Hence it easily passes the platform tests generally carried out at the village level dairy coop society, but from the health point of view of the consumers it is highly dangerous. The taste is highly objectionable. An attempt is made to make a comparison between the Genuine and synthetic milk in a tabular form in table 11.4.

Table 11.4 Comparison between Genuine and Synthetic Milk

Differential tests	Genuine milk	Synthetic milk
1. Physical tests:		
(a) Colour	White	White
(b) Storage	It remains white on storage	It turns pale yellow after some time on storage
(c) Texture	When rubbed on the palm it does not form foam	when rubbed on palm foam formation is noticed
(d) pH	6.6 - 6.8	10 - 11 highly alkaline
(e) Fat	4.5 - 5.0%	4.5 - 5.0%
(f) SNF	8.0 - 9.0%	8.0 - 9.0%
2. Chemical		
(a) Heat	No change in colour on heating	It turns yellow on boiling
(b) Urea test	Pale yellow colour develops	Dark yellow colour develops

Source : Dairy for all.com

11.18.2 Harmful Effects and its Prevention

A peculiar type of uneasiness prevades the people these days pertaining to the matter of synthetic milk which has started its journey from the cities to the villages. India is considered to be the largest milk producing country attaining the No.1 position in the world on one side while on the other side it is becoming notoriously famous and alarming by becoming the first ever country to produce the synthetic milk. According to one estimated report every day production and sale of approximately 1.10 crores of liters of synthesized milk is taking place in different states of India. According to a study carried out by the Ahmedabad based consumer education and research society it has been reported from 31 popular milk brands in different parts of the country different types of harmful bacteria have been found which causes serious damaging consequences. There is no exception to say that even the Government controlled organizations are not exempted from this unprofessional conduct of mixing of the deadly chemicals which has become the order of the day.

Unfortunately despite becoming self dependent and the massive increase in the production of milk and attaining the No. 1 position in the world surpassing the records of Europe and United States of America, the illicit production of the synthetic milk has emerged as the biggest problem to the milk industry. Because of this reason the confidence of the general public consumers is warning from the dairy industry and are reluctant to consume the dairy products. Therefore it is of paramount importance to find the solutions very soon to this raging problem, because the dairy industry occupies an important place in the Indian rural economy and the millions of farmers who are producing the milk including the consumers as well.

11.18.3 Effect of Synthetic Milk

The usage of synthetic milk causes very serious damage to the human body causing swelling of the eyes, and damaging the liver and kidney. Further it is very deadly for the pregnant women and the patients suffering from the heart ailment with high B.P. Apart from this it is extremely poisonous for infants and young children. Therefore to safe-guard ourselves from the ill effects of this synthetic milk one should take care and follow the suggestions by taking into consideration.

1. The entire population of the country should be made aware the health hazards caused by the synthetic milk by holding consumer awareness camps.
2. The Govt. must be vigilent and the officials of the state or central food and civil supply department should be empowered with certain powers to procecute the persons involved in illegal practices.
3. The PFA act which remained totally in effective in checking and controlling the massive production of this illegal synthetic milk must be made more strict and stringent law should be made with capital punishment for those persons who are involved in the illegal practice.
4. In order to contain and block the loop holes in the legal aspects of the relevant acts made by the government a more stringent and perfect law must be enacted by giving additional powers to the executive authority and capital punishment may be the life imprisonment to those involved in the illegal practices of preparing the dangerous synthetic milk which cause many health hazards to the human beings.

The synthetic milk looks almost similar to the regular milk in all aspects. The differentiating factors would be the taste and the nutritional elements. It may not be easy to detect and check the milk at home but some simple ideas could be helpful.

1. The natural milk obtained from the cow or buffaloe will have a sweetish taste which is very unique. But the synthetic milk tastes little bitter.
2. Natural milk has a pH value around 6.8 which is slightly on the acidic side of neutrality. Synthetic milk is highly alkaline with a pH value of over 10.
3. Synthetic milk gives a soapy feeling when rubbed between the fingers and the hands because of the presence of the detergents.
4. Natural milk remains white even on boiling whereas the synthetic milk turns slightly yellow on heating/boiling.
5. The Sp. Gravity/density of the natural milk 1.032 whereas the synthetic milk is apprecidly higher than this value.

6. Synthetic milk turns dark yellow on storage for 3-6 hrs where the natural milk remains white.
7. Synthetic milk is more slippery in touch while natural milk is non-slippery.
8. Synthetic milk on keeping in freeze forms an oily layers on the top of the milk due to the presence of the oil which floats over water while no such appearance occurs in natural milk.
9. Synthetic milk gives a bad smell with dark yellow color after storage for long period/hours but no such bad smell appears in natural milk.

Mudgil and Barak (2013) while discussing the synthetic milk a threat to Indian dairy industry have narrated the pros and cons of the adulteration by the synthetic milk. They have also made an attempt to identify the most common differences in physical and chemical properties of natural and synthetic milk and presented in a tabular form in 11.5.

Table 11.5 Differences in physical and chemical properties of natural and synthetic milk.

Properties	Synthetic Milk	Natural Milk
Physical		
Color	White	White
Odor	When freshly prepared, the mixture has a 'soapy' smell, which disappears on overnight storage at 4 °C	Not distinctive
Taste	Bitter	Palatable
Density	1.025-1.035	1.025-1.035
Storage	Turns yellowish after sometime	Curdling but no change in colour
Texture	Give soapy feel when rubbed	No soapiness
Boiling	Becomes yellow	No change
Chemical		
Fat	4.5%	4.5%
pH	Highly alkalin, 10.5	Slightly acidic, 6.4-6.8
Urea test	Intense yellow color	Faint yellow color
Urea concentration	14 mg/ml	0.2-0.7 mg/ml
Sugar test (Resorcinol)	Positive	Negative
Neutralizers test (Rosalic acid)	Positive	Negative

Source: Mudgil D and Barak.S.

11.18.3.1 Milk Substitutes

A milk substitute is a liquid meant to replace the milk obtained from a mammal i.e the natural milk. It is a whitish liquid that resembles milk and is often packaged as powder to be mixed with water to yield a liquid that resembles milk. Generally milk substitutes contain milk powder in varying proportions of 30 to 80% with the remainder consisting of other ingredients such as whey, palm oil, coconut oil and similar other vegetable extracts. Due to their composition milk substitutes tend to have a longer shelf life and can withstand higher temperature than the natural milk without any spoilage. Milk substitutes include a variety of plant milks which is being used by those who are intending to avoid animal products for health or other ethical reasons or because of taste preference.

Some of the milk substitutes are marketed to the consumers that are healthier than cows milk because they may be-lower in saturated fat and of they are entirely free from animal products – cholesterol free. If the milk substitutes are lacking in vitamins or dietary minerals that are present in natural milk they are usually fortified.

11.18.3.2 Animal vs Plant Milk

In general all over the world the grocery stores are flooded with non-dairy milk aisle which is expanding to include everything from soy to Rice to peanut milk. Though non-dairy milks have been around for a while they have recently experienced a spurt with soaring sales and enormous popularity partly due to environmental concerned dairy issues and also to the growing general interest in the consumption of dairy alternatives. From the variety of base ingredients to the assortment of flavours, they have never been more kick offs to choose from. In case if you are confused by the number of choices here is a quick guide to help you to navigate the dairy and non dairy aisle.

Generally a doubt/question arises in the minds of the consumers that which milk is best to consumer which again rests on the assumption that people need to drink. Strictly speaking plant milk is a mildly processed food hence a 100% unprocessed product which may not be included in the diet. The process of making plant milks varies from variety to variety. However the processing is minimal in general while milk pulverizing the plant material and the consumer must aware that it is unsweetened. The desire to drink a milk beverage is entirely a choice. It is also not necessary to consumer plant based milk beverages. It is also possible to obtain all the essential nutrients by eating a plant base whole food. Consumption of conventional dairy milk produced by the animals sometimes may carry health risks associated with the animal protein source.

An attempt has been to compare the macro nutrient profiles for the variety of milk listed and discussed above and presented in a tabular form. The animal milks listed in the table 11.6 are for full fat milks. It is essential to remember here that as fat is removed the protein becomes a greater part of the whole which further increases the excess of protein the coconut milk listed in the table is that sold in stores as a substitute for cows milk but not the coconut milk sold in cans as a cooling ingredient sources for the nutrient composition of the animal milks were the USDA nutrient data base. Sources for the plant based milks were the manufacturers published nutrient content. All the plant milks listed here are unsweetened. Small discrepancies in the % of kcal may be due to rounding errors. Fat has 9 kcal protein and carbohydrates have 4 kcal per gram of milk.

Table 11.6 Nutrient Profile of animal and plant based milks

Per 1 fluid cup energy (kcal)	Flat (g) -- %fat	Protein (g) -- % protein	Carbohydrate (g) -- % carbohydrate
Cow's milk 152	8.11 -- 48	7.86 -- 21	11.96 -- 31
Sheep's milk 265	17.15--58	14.65 -- 22	13.13 -- 20
Goat's milk 168	10.10 -- 54	8.69 -- 21	10.86 -- 26
Almond milk (Blue Diamond) 30 (larger rounding errors due to small volume from manufacture)	2.5 -- 75	1 -- 13	2 -- 26
Oat milk (Pacific foods) 130	2.5 -- 17	4 -- 12	24 -- 74
Hemp milk (Pacific Food) 140	5 --32	3 -- 8	20 -- 57
Soy milk (Edensoy) 140	5 -- 32	11 -- 31	14 -- 40
Rice milk (Rice Dream) 70	2.5 -- 32	0 -- 0	11 -- 63
Coconut milk (So Delicious) 70	4.5 -- 58	0--0	8 -- 46

Source: Micnela karisen. MSPH

11.18.3.3 Dairy Milk

Most of the non-dairy milks are generally compared with the cows milk which has a strong nutritional package. It is hard to argue with the spectrum of nutrients in milk unless and of course one has lactose in tolerance or a milk protein allergy. Milk itself comes in many varieties from fat free (skim) to whole milk. Organic and lactose free. As far as organic milk is concerned, it is a term that refers to the sustainability and manage mental practices in the dairy farm. Organic and conventional milk have the same nutritional and safety profile therefore deciding between the two comes down to a personal choice. Recognising the cows milk for infant feeding is deficient in lactose, although containing excessive amounts of protein and ash. It has been customary to add lactose, water and cream in such a way/proportions as to simulate approximately human milks. In order to prevent the formation of hard lumps of curds in the stomach and to secure a flocculant curd resembling that of human milk lime water, sodium citrate or a gelatinous substance such as cereal paste is often added to the milk or the milk is heated or curdled by acid.

11.18.3.4 Soymilk

Soymilk is possibly one of the most popular non-dairy milk products which was declared as a fore runner amongst the non-dairy milks that are available in the market. Unlike other plant based milks soy milk contains the same amount of protein as cow milk giving it the edge. It is made from soybeans. Although it is clained by some people that it contain twice as much sugar and saturated fats, when enriched by the manufacturer it may be a source of calcium and Vitamin D and B.Vitamins such as B_{12}. However those who are allergic to soy or who have other soy related concerns this milk is a good protein rich option. The most recent evidence does not support the association of soy foods rise the risk of certain concerns as suggested by some people earlier.

Soya beans are soaked and crushed and the liquid strained diluted boiled and again strain. The product is made at night and delivered the following day in bottles. It has a specific gravity of 1.02 and contains over 3% Fat. Another soya bean preparation is said to contain solids 12%, protein 3.5, Fat 3.5, carbohydrates 4.25 and mineral matter 0.75% respectively. It is low in acidity and keeps better than milk.

Vegetable Milk has been prepared by grinding oil seeds notably soya bean and pea nuts with water in the proportion of 100 gms / ltr. The analysis gives solids 10.0, protein 2.0, fat 4.5, carbohydrates 2.5 and ash 1.0 percent respectively. By the action of certain organisms the taste of cow milk is simulated.

A norwagian patent has been issued for preparing artificial milk from soya beans, the natural oil being replaced by an emulsion of sesame oil. Methods of emulsifying fats and collesterol with skim milk or other liquids containing proteins have been patented in England and Sweden. Purely animal substitutes for whole milk may be prepared homogenizing oleo oil with skim milk. Using both animal and vegetable fats in the artificial milk prepared by Haskins, McGregor and Ruh was closely resembles the constitution of human milk especially with regards to vitamins and fatty acids of low volatility. It was reported that a mixture of 14 parts of coconut oil and 80 parts of lard has about the same composition as the fact of the human milk.

11.18.3.5 Pea Protein Milk

With less impact on the land than almond milk and more protein than soy or cows milk the milk made from Pea protein have a lot love by the consumers.

11.18.3.6 Rice Milk

Unlike soy milk and nuts which are among the top eight food allergies a grain milk processed and made with brown rice may be preferable and well tolerated and great option for those whose choices are limited. It is mostly used for baking because of its sweet taste. When footfield this milk can be a source of calcium vitamin B_{12} and D_2. However, one must be careful and concerned about the presence of arsenic in rise particularly with infants and pregnant women.

11.18.3.7 Oat Milk

Like rice milk oat milk is generally well tolerated, making it suitable for the people with food allergies and in tolerances. Unlike other non-dairy contenders, this beverage boasts soluble fibre. The kind that helps lower cholesterol. Yet one may get more of these beta, glucons.

11.18.3.8 Peanut Milk

The latest nut based milk is made from peanuts and is in fact labeled as "milked peanuts". According to the manufacturer the distinction is in the minimal processing a fact that allows this beverage to keep its creamy consistency without added thickness and stabilizers. According to a report that almond milk is made with very few almonds but the newest nut milk boasts 31 peanuts per cup. These sips are obviously not suitable for any one with a peanut allergy.

11.18.3.9 Almond Milk

It is produced from almonds by grinding with water, then straining the pulp from the liquid. This procedure can be followed even at home. This is the darling of the non-dairy milk world. Probably because almonds themselves are deserving the nutritional praise with protein, fibre calcium, vitamin E and healthy mono unsaturated fats (MUFAS). Almonds have a lot going in favour of them. However, the miling and processing means poses a problem and that a lot of goodness does not end up in the drinkable version. Almond milk does not have the healthy fats, protein and fibre that are expected from this nut based beverages. It is not appropriate for those who suffer from nut allergies.

11.18.3.10 Coconut Milk

It is made by mixing water with the freshly grated, white side pulp of a ripe coconut. It is also a source of calcium and vitamins C, E, B_1, B_3, B_5 and B_6. Coconut milk is usually very high and rich in fat and calories but low in protein. There are different opinions and schools of thought regarding coconut milk because of its highly saturated fat content. However the type of saturated fat content in coconut milk is probably more neutral than harmful provided that if one avoids the other prime sources of saturated fat and eating lots of produce adding coconut milk. Due to its low protein content it makes a good substitute for cream because it can be whipped up in a similar fusion to decorate baked goods or deserts.

11.18.3.11 Hempmilk

It is made by grinding hemp seeds with water, which are then strained to yield a nutty creamy flavoured milk. Hempmilk is naturally rich in protein and amino acids.

11.18.3.12 Carrot Milk

The rich nutrient content of carrot especially the Vitamin-A and its precursor carotene cannot be questioned. It is also rich in health giving fibre. Some people relish carrot as such in its raw form. And some cook and use it. Another form of utilization is extraction of juice from carrot. But carrot juice when consumed as such will not be tasty. It has to be added with sugar just to make the carrot juice tasty and at the same time utilize the nutrients present in the juice in its natural form. Carrot juice can be added with milk to make "Carrot Milk".

11.18.3.13 Organic Milk

It is slowly catching up with the health conscious population for its health promoting properties and it is produced in the farms where the routine practices are strictly controlled. Organic foods are coming up in a big way in lebanon, Rumania as well as in India. The Indian organic farming is expected to grow mainly due to the increasing awareness among people regarding these products and raising income levels.

11.18.3.14 Some of the other plant based milks include

1. Mecadamia milk
2. Tiger nut milk
3. Cashew milk
4. Hazel nut milk
5. Flex milk and
6. Sesame milk

11.18.4 Yeast Milk

Two bio engineers Perumal Gandhi and Ryan Pandya who are working in California developed the world first artificial cows milk made from genetically engineered yeast in an effort to put a more environmentally sustainable option in the milk market. Earlier they produced a synthetic startup called 'Muufri' pronounced as Moo-free.

Nick named as "out of body udder" their system is designed to produce milk that retains the taste and health benefits of real natural milk setting it apart from, Soy, rice and almond milk varieties. If we want the world to change its diet plan a product that is not sustainable to something that is it has to be identical (to) or better than the original product. The world will not switch from milk of a cow to the plant based milks. But of the cowless milk is identical and priced right they just might change their habits. The vegan pair Gandhi and Pandya were inspired to invent the milk to reduce the need for the over crowded dairy barns, where the cows are fed a constant cocktail growth hormones and antibiotics have their tails docked and with the removal of their horns. According to FAO of the United Nations dairy production is responsible for 3 prevent of the worlds annual green house gas emissions.

Fortunately for them synthesizing cows milk a relatively simple process it has got less than 20 components and consists of about 87% water. Muufri milk contains six proteins to help from its structure and eight different fatty acids to give it its rich flavours. The milk is made using the same process that pharmaceutical companies used to produce insulin. DNA is extracted from dairy cows and certain sequences are inserted into the yeast cells. The yeast culture is then grown in the industrial sized Retor dishes at just the right temperature and concentrations and within few days the yeast will have produced enough milk for harvesting.

Although the proteins in Muufri milk come from yeast, the fats come from vegetables and are tweaked at molecular level to miss or the structure and flavour of milk fat. Minerals like calcium and potassium and sugars are purchased separately and added to the mix. Once the composition is fine tuned and the ingredients emulse naturally into milk. This of course means that all the nutritional values can be tweaked and the so prepared artificial milk could potentially be even better for the consumer than the regular original milk. Initially the "Muufri milk' even though is more expensive to buy than the regular cows milk it is hoped that eventually it can be made cheaper as the production of this milk is scaled up. Because of the fact that this milk wont contain any bacteria like that of regular cows milk the shelf life will be much longer which can be used for the cheese making like regular milk. The milk substitutes produced in this way do not required the use of animals therefore it is more efficient produces fever green house gases and utilizes less land because of the fact that no animal is to be fed and milked for obtaining the natural milk. (Source National Geographic-Gazmag)

The inadequate supply to meet the demand for the ever increasing population allover the world coupled with the hunger/greed for the financial gains is considered to be one of the major reasons that paved the way for the scrupulous fraudulent adulteration methods of milk by the people who are involved in the trading of milk in illicit manner/ways. This problem is more acute in the developing and under developed countries, than in the developed countries due to lack of adequate monitoring and stringent laws with capital punishment for enforcement. Existing common detection techniques are not always convenient and accessible in these countries making much more difficult to address the diverse ways fraudulent scrupulous and illicit ways/methods of adulteration of milk. Therefore it is essential and calls for combined efforts from the scientific community and the administrative enforcing executive regulatory authorities for the implementation of the stringent laws and dissemination of better techniques for the detection of adulteration of milk. In addition consumer awareness and access to the information can play a vital role in order to over come the issue/problem. Some of these easy detection methods at the consumer level and the state of art techniques of the executive authority level will be able to bring this problem to an end for the victims including millions of children and pregnant women in the developing under developed countries.

The major drawbacks of these techniques developed to detect the adulteration is due to the fact that these are valid for a limited range of concentrations and are not sufficiently precise. However qualitative detections are advantages because these are simple rapid and very easy to perform. Some of the edible compounds are often used as adulterants to improve the taste milk. Presence of these in milk can be detected rapidly and presented in the Table 11.7.

Table 11.7 Rapid qualitative detection of different edible adulterants in milk

Adul-terant	Procedure	Observation	Limit of detection (v/v) (R. Sharma Rajput, Barui, & N., 2012)	References
Sugar	Take 5mL milk sample in a test tube. Add 1 mL conc. HCl and 0.1 g resorcinol solution. Place the test tube in water bath for 5 min.	Appearance of red color indicates he presence of added sugar.	0.2% (w/v)	(Sharma et al. 2012); (Kamthania et al. 2014); (Arvind singh et al. 2012)
Starch	Take 3 mL sample in a test tube. After boiling it thoroughly, cool it to room temperature. Add 1 drop of 1% iodine solution.	Appearance of blue color indicates he presence of starch.	0.02% (w/v)	(Sharma et al. 2012); (Arvind singh et al. 2012); (Kumar et al. 1998);
Glucose	Take 1 ml of milk sample in a test tube. Add 1 ml of modified Barfoed's regent. Heat the mixture for exact 3 min in a boiling water bath. Rapidly cool under tap water.	Immediate appearance of deep blue color indicates the presence of glucose.	0.1% (w/v)	(Sharma et al. 2011)
Common salt	Take 5 ml of milk sample into a test tube. Add 1 ml of 0.1 N silver nitrate solution. Mix the content thoroughly and add 0.5 ml of 10% potassium chromate solution.	Appearance of yellow color indicates the presence of added salts, whereas, brick red color indicates the milk free from added salt.	0.02% (w/v)	(Sharma et al. 2012)
Buffalo milk	Dilute the milk 1/10. Put a drop of diluted milk on the centre of a glass slide. Now place a drops of Hansa test serum (duly preserved) on the drop of milk and mix together with a glass rod or clean tooth pick.	Curdy particles develop within half a minute in milk containing buffalo milk,		(Kamthania et al. 2014); (Arvind Singh et al. 2012)

Source: Tanzina Azad and Shoeb Ahmed (2016)

There are some hazardous chemicals that are being added to the milk with an intention to improve the physical appearance and shelf life. These chemicals are very hazardous and leads to fatal diseases. A simple yet fast detection techniques of these hazardous chemical have been presented Table 11.8.

Table 11.8 Rapid qualitative detection of different hazardous chemical in milk

Adulterant	Procedure	Observation	Limit of detection (v/v) (Sharma et al. 2012)	Reference
Hydrogen peroxide	A. Add to 5 mL of suspected milk sample in attest tube an equal volume of raw milk and 5 drops of 2% solution of paraphenylenedi-amine. B. Take 1 mL milk sample in a test tube. Add 1 mL of potassium iodide-starch reagent solution and mix well.	Appearance of blue color indicates the presence of hydrogen peroxide as adulterant. Appearance of blue color indicates the presence of hydrogen peroxide as adulterant.	0.025% 0.004%	(Arvind singh et al. 2012); (Kamthania et al. 2014); (Sharma et al. 2012)
Formalin	A. Take 10 mL milk sample in attest tube. Add 5 mL conc. sulfuric acid with a little amount of ferric chloride without shaking. B. Take about 5 ml of milk in a test tube. Take 1 mL of 10% ferric chloride solution in a 500 mL volumetric flask and make up the volume using concentrated hydrochloric acid. Add 5 mL from this solution to the sample in test tube. Keep the tube in boiling water bath for about 3-4 min.	Appearance of violet or blue color at the junction of two liquid layers indicates the presence of formalin. Appearance of brownish pink color indicates the presence of formalin.	0.1%	(Arvind singh et al. 2012); (Kamthania et al. 2014) (Sharma et al. 2012)

Table 11.8 ***Contd...***

Adulterant	Procedure	Observation	Limit of detection (v/v) (Sharma et al. 2012)	Reference
	A. Take 1 mL of sample milk in a test tube. Take saturated solution of 1, 8-dihyroxynaphthalene-3, 6- disulphonic acid in about 72% sulfuric acid to make chromotropic acid solution. Add 5 mL from this solution to the sample in test tube.	Appearance of brownish pink color indicates the presence of formalin.	0.05%	(Sharma et al. 2012)
Ammonium sulfate	A. Take 2 ml. milk in a test tube and add 0.5 ml NaOH (2%) 0.5 ml sodium hypochlorite (2%) and 0.5 ml phenol (5%) Heat in boiling water bath for 20 sec. B. Take 10 ml of milk in a 50 ml stopped test tube. Add 10 ml of TCA solution. Filter the coagulated milk through Whatman filter paper Grade 42. Take 5 ml of clear filtrate. Add few drops of barium chloride solution.	A bluish colour forms immediately, which turns deep blue afterward. Pure milk shows salmon pink colour which gradually changes to bluish after 2 hours. Formation of milk-white precipitates indicates the presence of added sulfates like ammonium sulfate, sodium sulfate etc. zinc sulfate and magnesium sulfate to milk.	0.05% (w/v)	(Kumar et al. 2002) (Sharma et al. 2012)
Urea	A. Take 5 mL milk sample in a test tube. Add equal volume of 24% TCA to precipitate fat and proteins of milk. Take 1 mL filtrate and add 0.5 mL 2% sodium hypochlorite, 0.5 mL 2% sodium hydrxide and add 0.5 mL 5% phenol solution, then mix.	A characteristic blue or bluish green colour develops in presence of added area wheras pure milk remains colourless. Appearance of blue color after 10-15 min. indicates the presence of urea in milk. Normal milk shows faint blue colur due to natural urea present in milk.		(Meisel 1995) (Sharma et al. 1993); (arvind singh et al. 2012)

Table 11.8 ***Contd...***

Adulterant	Procedure	Observation	Limit of detection (v/v) (Sharma et al. 2012)	Reference
	A. Take 5 ml milk in a test tube, add 0.2 ml urease (20 mg/ml) shake well at room temperature and then add 0.1 ml. Bromothymol Blue (BTB) solution (0.5%) B. Take 5 mL milk sample in a test tube. Add 5 mL p-Dimethyl amino Benzaldehyde reagent.	Appearance of distincy yellow color indicates presence of added urea wheras formation of slight yellow color indicates natural urea in milk.	0.2% (w/v)	(Sharma et al. 1993); (Arvind Singh et al. 2012); (Bector et al. 1998); (Kavita, 2000)
Nitrate	Take 10 ml sample milk in a beaker. Add 10 ml mercuric chloride solution to it. After mixing filter through what man No 42 filter paper. Take 1 ml filtrate in a test tube and add 4 ml of diphenyl amine sulphate or diphenylbenzidine reagent.	Appearance of blue colour indicates the presence of nitrates pure milk sample will not develop any color.	0.2%	(Sharma et al. 2011)
Benzoic and salicylic acid	Take 5 mL milk sample in a test tube. Upon acidification with sulfuric acid, 0.5% ferric chloride solution is added to it drop by drop. Mix it. Five ml of milk is taken in a test tube and acidified with concentrated sulphuric acid. 0.5% ferric chloride solution is added drop by drop and mixed well. Development of buff colour indicates presence of benzoic acid and violet colour indicates salicylic acid.	Appearance of buff color indicates the presence of benzoic acid whereas that of violet color indicates salicylic acid.		(Arvind Singh et al. 2012)

Table 11.8 *Contd...*

Adulterant	Procedure	Observation	Limit of detection (v/v) (Sharma et al. 2012)	Reference
Borax and boric acid	Take 5 mL milk sample in a test tube. Add 1 mL conc. HCl to it. A turmeric paper is dipped and it is dried in a watch glass at 100 °C.	If the turmeric paper turns red, it indicates the presence of borax or boric acid.		(Arvind Singh et al. 2012)

Source: Tanzina Azad and Shoeb Ahmed (2016)

In addition some other mixed chemical such as soaps, detergents and coloring compounds are being added to the milk in order to improve the appearance. Qualitative detection of some these common adulterants in milk have been presented in the table 11.9.

Table 11.9 Rapid qualitative detection of different mixed adulterants in milk

Adulterant	Procedure	Observation	Limit of detection (v/v) (Sharma et al. 2012)	References
Detergent	A. Take 5 ml in a test tube and add 0.1 ml 0.5% Bromocressol Purple (BCP) solution. B. Take 5 mL of milk sample into a 15 mL test tube. Add 1 ml of Methyene blue dye solution and 2 ml chloroform. Vortex the the contents for about 15 sec and centrifuge at about 1100 rpm for 3 min.	Appearance of violet colour indicates the presence of detergent. Unadulterated milk shows faint violet color. Relatively, more intense blue color in lower layer indicates presence of detergent in milk. Relatively more intense blue color in upper layer indicates absence of detergent in milk.	0.0125%	(singhal, 1980); (Arvind Singh et al. 2012) (Rajput, Sharma & kaur)

Table 11.9 ***Contd...***

Adulterant	Procedure	Observation	Limit of detection (v/v) (Sharma et al. 2012)	References
Pulverized soap	Take 10 ml milk sample in a test tube. Add equal quantity of hot water to it, then add 1-2 drops of phenolphthalein indicator.	Appearance of pink color indicates presence of soap.		(Arvind singh et al. 2012); (Kamthania et al. 2014); (Ghodekar 1974)
Coloring matter	A. Take 10 mL milk sample in at test tube. Add 10 ml diethyl ether. After shaking allow it to stand. B. Make the milk sample alkaline with sodium bicarbonate. Dip a strip of filter paper for 2 hours. C. Add a few drops of hydrochloric acid to milk sample.	Appearance of yellow color in ethereal layer indicates the presence of added color. Appearance of red color on filter paper indicates the presence of annatto. Treatment of this paper with stannous chloride gives pink color. Appearance of pink color indicates azo dyes.		(Batis et al. 1981) (Lechner and Klostermeyer 1981) (DE. Souza) et al. 2000

Source: Tanzina Azad and Shoeb Ahmed (2016)

CHAPTER 12

Quality Control of Milk and Milk Products

12.1 Quality of Milk

The quality assurance is of paramount importance in all food manufacture and handling industries. It involves a coherent system of activities that assures/ guarantees that the products made meet a set of defined quality of marks.Quality can be defined in various ways: quality is fitness for use. The quality must be controlled/enforced and is thus a management function which involves an approach of integrated or total quality management. The quality concept has to be built in from the begining and the greatest concern is the safety of the product for the consumer. In practice, it is almost never possible to check energy unit of the product, therefore, a rigorous inspection and control system must be established.

Health hazards due to toxic or carcinogenic concentrations of substances in the product are very rare in milk products. Most contaminations with hazardous substances are restricted to one form and the dilution achieved by mixing the milk with that from several other forms generally causes the concentration in the final product to be far below the toxic level. Sampling of the final product will establish the contamination, whereas the samples taken from road tankers, and individual deliveries can readily locate the source of contamination.

The concept of quality control varies as per the situation and when viewed at from different perspectives. The consumer views quality control as a watch dog which ensures that he gets a good product for the money he is spending/paying. The processing/production people considers that the quality control as a necessary evil, which they have to live with, whereas the management views the quality control as a watch dog that keeps the consumers as well as the legal standards are complied or satisfied with and at the same time they ensure that the dairy does not incur any losses by exceeding or not meeting the prescribed standards for the different products. Therefore, the quality control plays an important role interacting with different agencies, departments and the public as well.

In the dairy industry, the functions of the quality control are manifold whose description has been given briefly here under.

12.1.1 Raw Milk and its Quality Control

The raw milk which is being received at the dairy plant must undergo stringent testing and be approved for its usage. It has to be borne in mind, very clearly understood and must be emphasized that good quality finished product can not be produced from poor quality raw material. All the ingredients that are being used in the milk product manufacture have to be vigourously inspected before purchase. These include milk, water, skim milk powder, butter/butter oil, sugar, salt etc. The milk supplies procured by the dairy have to be first tested for their acceptability by the organoleptic tests and the platform tests which include titrable acidity, the methylene blue reduction test, clot on boiling test, detection of addetives like adulterants, preservatives or neutralisers and so on. If the milk is found to be acceptable a small representative sample is taken for further detailed chemical analysis.

Milk which is found to be unsuitable should be rejected summarily or out rightly and should not be allowed to be entered into the dairy. In case if a small quantity of the poor quality milk is accepted and blended with good quality milk, it will lead to a lot of subsequent problems in the processing, storage and marketing.

In the same manner all other consumable ingredients have to be tested and approved before procurement.

12.1.2 Processing/Production - its Quality Control

The quality control plays an active supporting role to the processing or manufacture of dairy products. The raw material once accepted and stored, the quality control check at regular intervals provides its nature of composition and quality, which will help the production manager to decide, how it should be used in the product mix and diversification. The quality control reports are essential for the standardisation of the mixes and in the control of various parameters during production. An efficient quality control laboratory should give quick accurate reports, in order to monitor the production in an effective manner, which will help in avoiding the rejection of the product during the final quality control checking.

Quality control must also keep an eye on the cleanliness of the plant, machinery, premises etc., which will help in controlling the post pasteurisation contamination, to a large extent and in maintaining a good micro-biological quality check on the finished product. The quality control department when functions in an efficient manner can advise the management on the different avenues which will lead to losses thus enabling them to check the losses.

12.2 Finished Product - Quality Control

In order to meet the minimum legal requirements or the standards, the finished final product either it may be milk or its different products must undergo an exhaustive analysis for its worthiness before being released into the market. Any product which is suspected to be substandard either by way of its composition, keeping quality or defects in packaging etc should be rejected strictly and should not be allowed to be put up for sale. This will prevent the economic lossess, avoid the costly leagal battles and also ensures that the customer is satisfied and keep up the goodwill of the dairy plant, which is the most important factor/aspect in the quality control measures that are being taken up by the quality control personnel.

There are some other important functions to be performed by the quality control personnel in addition to the above stated essential functions. Standardisation of the testing equipment has to be carried out by quality control in order to ensure that the test results reported are accurate. Milk production centres, cooperatives and producers have their own milk testing equipment and evaluate their milk quality before supplying the milk to the dairy plant. Quality control has to help them in standardising their equipment and also train the personnel to ensure that there is co-ordination between their results and those of the dairy. Quality control also plays a major role in pollution control. The effluent water leaving the dairy premises has to be tested to ensure that it meets with the minimum legal standards/requirements. Otherwise, this could lead to trouble with the pollution control board and other agencies. Periodic checking of the maintenance of the effluent treatment plant helps in solving these problems. A close liaison with the P.F.A. authorities and other standard agencies such as Agmark and BIS is to be carried out by the quality control department. The PFA nominations, BIS and Agmark forms, undertakings etc., have to be signed and maintained by the quality control department. Thus the quality control in a dairy plant has to perform multifarious activities and keep track of different aspects while maintaining a strict vigil on the quality of every packet of every product originating from the dairy.

12.2.1 Quality Control Laboratory

The quality control laboratory in a dairy plant should essentially be set up in a clean well ventilated and well lighted place and should be located preferably in a corner place away from all the activities and hustle and bustle of the dairy plant. The laboratory personnel will then be able to work without any distraction and the laboratory will not became a thorough fare. In most modern dairies the quality control laboratory is located on the first floor over looking the processing hall below. This has the added advantage such that the quality control personnel can keep an eye on the processing activities. Another advantage is that the milk storage tanks are normally situated on the first floor and it will be convenient for drawing samples regularly during the standardisation and storage. At the beginning and end of each shift all the storage tanks etc. should be sampled and a report maintaining the temperature, acidity, chemical quality, history of processing etc should be submitted to the incoming staff of the next shift.

Sufficient care and thought/planning should be given/made in organising the equipment and the floor space, while setting up the quality control laboratory. The microbiological section should be separated from the chemical laboratory as far as possible. The microbiology section/room should be at the back such that only the qualified and essential people are allowed to enter the room. Provision of an air curtain helps in controling the contamination. The microbiology room should be self sufficient with all the necessary modern equipment, glassware, U.V. chamber counters, incubators and autoclaves etc.

The chemical laboratory room should be setup and organised properly. Routine tests like fat, acidity, S.N.F etc are done on hundreds of sample daily and those tests should be setup to handle large number of samples conveniently. Proper use should be made of the available wash basin, sink, etc. All samples should be properly labelled and after the sequential analysis the report should be recorded and dispatched as soon as possible. Normally, a few samples from each batch are retained to monitor the keeping quality and also for checking in case if there are any market failures or complaints. All the solutions that are being used for analysis in the quality control laboratory should be prepared taking utmost care and standardised properly before use. In the same manner all the equipment and glassware should be tested to verify that they confirm to the standards. One set of standardised glassware like milk pipettee, butyrometer, lactometer, thermometer etc should be kept on the reference against which all the other glassware can be standardised.

A proper record of maintenance system should be established and a copy of all the reports should be maintained either in the form of a permanent register or filed properly taking utmost care. An efficient filing system is very essential for a quality control laboratory since it has to keep track of so many diverse/different aspects. Thus the quality control laboratory is the nerve centre of the dairy and due importance should be given for setting it up properly.

12.2.2 Sampling - Quality Control

Milk which has been received either at a dairy or chilling plant is always tested for its quality, cleanliness and acceptability. The quality control personnel visits the raw milk reception dock and examines the milk for its acceptability, before it is allowed to be un loaded and processed. In order to carry out these activities the quality control personnel are to be trained and experienced adequately. The acceptability or rejection of the milk has to be decided by them by mere organoleptic examination only and the doubt samples are tested in the laboratory before taking the final decision. Samples which are doubtful should be rejected summarily. The sampling should be proper in order to have the testing to be accurate and efficient. The sample drawn should be representative of the bulk milk in all respects. A sufficient quantity of the sample has to be drawn for the various chemical and microbiological test. The quality and comparison of the sample should be identical with the rest of the milk and for this an efficient system of sampling has to be employed. Different methods of sampling are adopted to draw samples from different types of containers.

Sampling of milk from a single container is easily taken after proper mixing or if possible by shaking or inverting the container. Milk from small containers can be poured from one vessel to another for mixing and a small portion can be drawn as a representative sample. Very often a simple composite sample has to be drawn from different containers. In such circumstances, wherever possible the milk from different cans can be poured into a single large container like the dump tank or a weighing bowl and mixed properly with the plunger. From this pooled milk a single sample can be drawn. The level of milk in cans of the same height and diameter can be equalised by pouring from one to another. The same quantity of sample from each of these cans can be taken after proper mixing and pooled to get a single representative sample. Thus it will ensure that a proper sample has been taken/ obtained.

Tube method of sampling can be employed where the different cans of the same diameter contain different levels of milk. A sample tube of uniform bore which is open at one end and having a provision for closing the other end. After proper mixing of the milk the tube is introduced in a vertical position, and the top end of the tube is closed and withdrawn to deliver a proportionate quantity of milk. Similar samples are drawn from each can and pooled to obtain a single representative sample. This method also ensures that even if the different cans contain different quantity/level and quality of milk, a proprtionate sample is obtained which is representative of the bulk or pooled milk. Milk is received at the dairy in road tankers, samples here are to be drawn from storage tanks at periodic intervals. If the tank is equipped with a mechanical agitator, the milk should be agitated for not less than 15 minutes before taking the sample. In road tankers, the manhole at the top is opened and after reading the temperature, the milk should be agitated thoroughly with a plunger before sampling.

All the samples which have been taken should be stored in clean sample bottles. The bottle should have a wide mouth to facilitate proper cleaning. It should have stopping side and well fitting caps. There should be no head space left after the sample is taken and finally the sample bottles should be labelled properly for its contents.

12.2.3 Quality Control of Foods and Legislative Criteria

It is necessary for any country to have some quality control food laws and regulations, while dealing with the food safety measures. Food contamination and adulteration has created great concern and resulted in the passing of food laws in different countries of the world thus penalizing the sellers of unholesome and corrupted food products. The laws define foods, food preservatives and artificial colours and bar the sale of foods prepared under unsanitary conditions, or containing any deleterious toxic or unsafe substances. A food is called as adulterated food when it contains a poisonous or deleterious substance that may render it injurious to health.

12.2.4 International Food Safety Laws and Standard

The U.S. Government has enacted a comprehensive federal food law, the Food and Drug Act in 1906 to increase the food safety and decreasing fraud in the food market place. Another law the Food and Cosmetic Act of 1938 included all the provisions of the 1906 law but also fixed the problems that made the former law unforceable. This law ensured enforcement by assigning it to a federal agency, the Food and Drug Administration (FDA). Subsequently there are five major amendments to the Food Drug and Cosmetic Act 1938 enacted by the U.S. Government namely (1) The Pesticide Chemical Amendment -1954 (2) The Miller Food Additives Amendment - 1958 (3) The Colour Additive Amendment-1960 (4) The Fair Packaging and Labelling Act-1966 and (5) The Nutrition labelling and Education Act - 1990.

12.3 Food and Agriculture Organisation (FAO)

The Food and Agriculture Organisation (FAO) was founded in the year 1945 with a mandate to raise the levels of nutrition and standards of living, to improve the agricultural productivity and to improve the conditions of rural population. The FAO, since its inception has worked to alleviate poverty and hunger by promoting agricultural development, improved nutrition and the result of food security. The FAO is also active in land and water development, plant and animal production, forestry, fisheries, economic and social policy, investment, nutrition, food standard, commodities and trade. It works with governments to promote agricultural and rural development and to faster international cooperation on issues such as food standard fair trade, environmental management and conservation of genetic resources.

The FAO's role as a neutral forum is closely tied to its work as an adviser to governments. Member nations meeting at the council commit themselves to support the development initiatives such as (1) the world food security compact and (2) International code of conduct on the distribution and use of pesticides.

12.3.1 The Codex Alimentarius Commission and the FAO/WHO Food Standards Programme

The Codex Alimentarius Commission implements the joint FAO/WHO Food standard Programme, the purpose of which is to protect the health of consumers and to ensure fair practices in the food trade. The codex Alimentarius (Latin - meaning Food Law or code) is a collection of internationally adopted food standards, presented in a uniform manner. The publication of the codex Alimentarius is intended to guide and promote the elaboration and establishment of definitions and requirements for foods, assist in their harmonization and in doing so to facilitate the international trade. The codex general principles of food hygiene are aimed to identify the essential principles of food hygiene, applicable throughout the food chain, to achieve the goal of ensuring that the food is safe and suitable for human consumption, to recommend a hazard analysis and critical control point (HACCP) based approach as a means to enhance food safety, to indicate the procedure regarding the implementation of those principles and to provide guidance for specific codes which may be needed for sectors of the food chain, processes or commodities and to amplify the hygiene requirements specific to that area. The control described in this general principles document are internationally recognised as essential to ensure the safety and suitability of foods for consumption.

12.4 Hazard Analysis and Critical Control Point (HACCP)

In order to ensure the safety of food, the industry and regulators, traditionally have depended on spot checks of manufacturing conditions and random sampling of the final products. However, this approach tends to be reactive, rather than preventive and can be less efficient than the new system. The new system is known as Hazard Analysis and Critical Control Point or HACCP. Majority of its principles are already in place in the FDA, which regulated low acid canned food industry, and in 1995 a final rule that took effect in 1997, HACCP was established by the FDA for the sea food industry also. The FDA has also incorporated HACCP into its food code, on document that gives guidance to and serves as a model regulation for the state and territorial agencies that issues license and inspects food service establishments, retail food stores and vending operations in the United States.

FDA is now considering to develop regulations that would establish HACCP as the food safety standard throughout other areas of the food industry including both domestic and imported food products. The agency is conducting pilot HACCP programmes with volunteer food companies, to help and determine the degree to which such regulations would be feasible. HACCP has been endorsed by the National Academy of Sciences, the Codex Alimentarious Commission an international food standard setting organization and the National Advisory Committee on Microbiological Criteria for foods. A number of U.S. Food companies have already adopted the system in their manufacturing processes and it is in use in other countries also.

HACCP is a management system in which the food safety is addressed through the analysis and control of biological, chemical and physical hazard from raw material production procurement and handling to manufacturing, distribution and consumption of the finished products. The management must be committed strongly to the concept of HACCP for its successful implementation and such a firm commitment provides a sense of the importance of producing the safe food to the employees of the firm or company.

HACCP is designed for use in all segments of the food industry from growing harvesting, processing, manufacturing, distributing and merchandising to preparing food for consumption pre-requisite programmes such as Current Good Manufacturing Practices (cGMPs) are an essential foundation for the development and implementation of successful HACCP plans. The HACCP is based on seven principles, which have been accepted universally by the government agencies, trade associations and food industry around the world.

It is a method to establish the control measures that are essential to ensure the safety of the product for an existing production system. It can also be applied for other quality characteristics, however, the emphasis is generally on the safety. The method should be applied separately to every manufacturing process actually in operation. Thus indicating that a separate system for every produce or group of closely related products. The main features of the method are as the name indicates. Make an analysis of the potential hazards, identity the critical points in the process and establish the criteria for the control measures. A critical point is defined as one that must be controlled in order to ensure the safety and good quality. There must be several other points, but need not be controlled because control of another critical point will also detect the imperfection.

It is a control system applied after the analysis has been made involving the corrective measures where needed and its study may reveal that the process should be changed to allow an efficient control system. Instruction manuals should give more details of the product and its use must be described in detail. The manufacturing process should be described in a careful manner in a flow diagram, including the control points for adjusting the process (eg. temperature, flow rate, mixing intensity, and the rate at which a component is added). Every step in the process is analysed for its potential hazards, which are then eavaluated and quantified as to what measures can be taken to minimize the hazard, which is decided on the basis of systematic criteria, whether it is to be considered as a critical control point or not. In such a case a monitoring scheme is devised which is an essential part of the monitoring frequency. As an essential aspect is that HACCP system can not be copied. The development and repeated evaluation of the system by the people involved in its application is a prerequisite for its success. The system should regularly be evaluated and verified during its application and modified according to the need.

The procedure for developing a control system for the manufacture of a food product according to the European Hygienic Design Group has been presented hereunder for the better understanding of the HACCP by the students.

Stage	Action
1.	Define the terms of reference,
2.	Select the HACCP team,
3.	Describe the product,
4.	Identify the intended use of the product/products,
5.	Construct/draw the flow diagram of the product intended to be manufactured,
6.	On site verification of the flow diagram,
7.	List out all the hazards, with each process step and list all the control measures of the hazards,
8.	Apply HACCP decision to each process step in order to identify the critical control point (CCP),
9.	Establish the target level(s) and tolerance for each CCP,
10.	Establish the monitoring system for each CCP,
11.	Establish a corrective action plan,
12.	Establish record keeping and documentation,
13.	Verification and lastly,
14.	Review the HACCP Plan.

An essential aspect of the HACCP system is that it can not be copied because each and every manufacturer has his/her own particulars in the process applied and in outside conditions and constraints. The development and repeated evaluation of the system by the people involved in its application is a pre-requisite for its success.

12.4.1 Analyse Hazards

Potential hazards which are associated with foods and measures to control those hazards are identified. The hazard could be biological, such as a microbe, chemical such as toxin or physical such as ground glass or metal fragments.

12.4.2 Identification of Critical Control Points

In the production of food from its raw state through processing and shipping to consumption by the consumer at which the potential hazard can be controlled or eliminated, for example, cooking, cooling, packaging and metal detection.

1. Establish preventive measures with critical limits for each control, for example, in the case of a cooked food, this might include setting up of the minimum cooking temperature and time required to ensure the elimination of any harmful microns.

2. Establish procedures to monitor the critical control points.

   These include procedures which might be determining how and by whom the cooking time and temperature should be monitored.
3. Establish corrective actions to be taken when monitoring shows that a critical limit has been met for example, re-processing or disposing of food that has not met with the standard of minimum cooking temperature.
4. Establish procedures to verify that the system is working properly.

   For example, testing time and temperature recording devices to verify that the cooking unit is functioning properly.
5. Establish effective record keeping document - The HACCP System.

This include records of hazards and their control methods, the monitoring safety requirements and action taken to correct potential problems. Each of these principles must be backed up by sound scientific knowledge. E.g. published microbiological studies on time and temperature factors in order to control the food borne pathogens.

12.4.3 Advantages of HACCP

HACCP focuses on identifying and preventing hazards from contaminated food and permits more effective and efficient monitoring at government level, primarily, because of the fact that the record keeping allows the investigators to examine, how well a firm is complying with the food safety laws over a period of time rather than presently how well it is doing on any given day. It fixes the responsibility for ensuring the food safety appropriately on the manufacturer or the distributor of the food and helps the food companies to compete more effectively in the world market thus reducing the barriers to the international trade.

12.5 ISO 9000 Series and Other Standards

The International Organisation for Standardisation (ISO) is a non governmental body having membership of worldwide federation of national standard bodies from over 100 countries was founded in the year 1946. The ISO is responsible for the promotion, development of international standards and other related activities including confirmity of assessments such as testing, inspection, laboratory accrediation, certification, and quality assessments. This standardization leads to developing cooperation in the spheres of intellectual, scientific, technological and economic activity in various organisations including the food industry. The ISO standards have been adopted in 1978 by the organisation of International standards situated at Geneva, Switzerland, as a uniform standard of excellence for its use in the European Economic Community (EEC) which represents the latest developments of international standards for quality systems. The series have been recognised very widely and accepted Internationally, as national standards by about 90 countries. In the Indian context its equivalent standard is the Bureau of Indian standard (BIS) 14000 series.

Compliance with ISO 9000 by any company ensures that it has sound quality assurance system, which has resulted in a dramatic reduction of customer complaints against the registered companies, significant reduction operating costs with an increased demand for their services including the products. There is an ever growing trend towards the acceptance of ISO 9000 as an international standard universally. A quality system is a series of checks and balances which when introduced and followed scrupulously ensures quality output of the products. Thus the ISO 9000 establishes a standards framework for a quality system and identified about twenty (20) requirements, each of which affect the quality.

These standards themselves, are the progeny of the ISO 9000 family define the quality systems or models applicable to design, development, production, installation, servicing, final inspection and testing. The standards have been developed covering different aspects of combinations of these desciplines and a set of guidelines were issued to assist in choosing the correct standards for its application. These standards include:

12.5.1 ISO 9001

Quality systems: Models for quality assurance in design, development, production, installation and servicing and ISO 9001 is the super set of the contractual models for quality systems which contains 20 sections each of which specifies the requirements for a component of the quality system.

12.5.2 ISO 9002

Quality systems: Model for quality assurance in production, installation and servicing and ISO 9002 is the contractual model for quality systems, which include, production but do not include design. It is almost word to word equivalent to ISO 9001 except that it differs and does not include the requirements for design control and contains 19 sections or requirements.

12.5.3 ISO 9003

Quality systems: Model for quality assurance in final inspection and testing. This is the contractual model for quality systems which does not include, design or production. It contains about half of the requirements of ISO 9001 and modifies some of the requirement to suit to the needs of the inspection and final testing applications.

12.5.4 ISO 9004

It is a set of guidelines which includes quality management and quality systems.

All these systems require the development of a quality manual and well documented procedures, which define the organisation and its operation of the quality systems. It is the responsibility of the organisation or the company to create and maintain these documents so that they are relevant and appropriate to the specific business operation.

12.5.5 Differences between ISO 9001, 9002, and 9003

The organization which undertake design and development activities, manufacture those designs and complete final inspection and testing would choose and abide by ISO 9001 which is a quality assurance model, that is basically used by companies that design, produce, inspect, test, install and service items. Similarly those organisations, which include a stage of manufacture prior to final assembly and test would choose and follow the guidelines of ISO 9002, which is a quality assurance model, which is basically used by the organisations/companies that produce, inspect, test, install and service items. On the contrary those companies or organisations which are involved with the stages of product completion, final inspection and test would appropriately choose to abide by ISO 9003, which is a quality assurance model, that is basically used by the companies that must inspect and test items including an importer or distributor. Thus in short the ISO 9001 is considered as a super set standard while the ISO 9002 and 9003 are considered as progressively smaller subsetcs of that standard.

Organisations/firms or companies which have been registered with the ISO very often use to gain a definite advantage as compared to their competitors who are not the members of the ISO and did not register with them. In general the customers perceive that ISO registered firms as successful, competent industry leaders because, they follow the guidelines scrupulously in reducing the errors resulting for better systematic inspection and testing, increased employee participation involvement, awareness and systematic employee training. Better products resulting from better design control, improved productivity, resulting from planning and team work, reduction in cost associated with failures are some of the advantages of the ISO.

12.6 Indian Food Laws and Standards

The Government of India has promulgated several Acts and orders in order to ensure the safety and empowered several agencies or institutions to lay down the standards or regulations for the quality of foods, covering the processing and packaging. The Government of India has enacted an Act which is called the Prevention of Food Adulteration (PFA) Act pertaining to the selling of food and it defines what may be considered as adulteration and requires the foods must be pure, wholesome and labelled honestly. However, in our country, the consumer awareness about the various aspects the PFA act is lacking, and in the absence of the cooperation from the consumers, controlling of adulteration would prove to be an uphill task for the Government and now-a-days, there are consumer protection cells set up at all levels and the consumer awareness is growing/increasing at a slow but steady pace.

12.6.1 Prevention of Food Adulteration (PFA) Act

The Government of India has promulgated on the prevention of food adulteration Act in the year 1954 which is considered as the first of its kind and is in force since

1955 with the objective to ensure that the food articles, which are sold to the consumers are pure and wholesome and intended to prevent fraud or deception encouraging a fair trade. The Act was amended in 1964 and 1976 in the light of the experiences gained in order to plug the loop holes of escape in the act with an object to ensure stringent punishment to those traders who are indulging in malpractices in this trade. Standards have been laid under the PFA Act (1954) of the Govt. of India for various categories of foods. The International Codex Alimentation Commission is the principal organization under the joint auspices of FAO and WHO engaged in the preparation of standards for the quality of different categories of foods for international use. The PFA Act (1954) as amended in 1964 and 1976 prohibits the manufacture, sale and distribution of not only adulterated foods but also foods that are contaminated with toxins and mis branded foods. Some of the other acts promulgated by the Government of India in this direction are:

1. Fruit production Order (FPO) - 1955.
2. Meat Product Order (MPO).
3. Cold storage Order (CSO) 1980.
4. Promulgated under the essential commodities Act (1955).
5. Agmark standard.
6. MMPO - 1992

The word “Agmark” is derived from Agricultural Marketing. The standard laws are set up in the Agriculture Produce Act 1937, by the Govt. of India which provides a provision for the inspection by the Directorate of Marketing. The ‘Agmark’ seal ensures quality and purity. The quality of product is determined with reference to the size, variety, weight, colour, moisture, fat content and other factors. The Act defines the quality of cereals, spices, oil seeds, butter, ghee, legumes, and eggs and with a criteria provision for the categorization of the commodities into various grades depending upon the degree of purity in each case. The standards also specify the types of packaging to be used for different products. The central Agmark laboratory is located at Nagpur and is engaged in carrying out research and development work in this field. The certificate of authorization is granted only those who are engaged themselves in the trade having adequate experience and standing in the market.

12.6.2 Food Safety and Standards Authority of India (FSSAI)

The food safety and standards authority of India has been established under Food Safety and Standards Act in 2006 which consolidates different acts and orders that have higher to dealt with food related issues in various ministries and departments. FSSAI has been created for laying down science based standards for articles of food and to regulate their manufacturer, storage, distribution, sale and import to ensure availability of safe and wholesome food for human consumption.

The highlight of Food Safety and Standards Act (FSSA) 2006 is to emalgamate various central acts like Prevention of Food Adulteration Act 1954, Fruit Products order 1955, Meat Food Products Order 1073, Vegetable Oil Products (control) Order 1947, Edible Oil Packaging (regulation) Order 1968, Solvent Extraction Oil, De-oiled Meal and Edible Flour (control) Order 1967, Milk and Milk products Order 1992 etc will be repealed after the commencement of the FSS Act 2006.

The act also aims to establish a single reference point for all matters relating to food safety and standard by moving from multi level, multi departmental control to a single window line of control. To this effect the act establishes an independent statutory authority the food safety and standards authority of India having its head quarters/ office at Delhi. The food safety standards authority of India (FSSAI) and the state food safety authorities shall enforce various provisions of the act.

The ministry of Health and Family welfare Govt. of India is the administrative ministry for the implementation of FSSAI and the chair person is in the rank of Secy. to Govt. of India.

The FSSAI has been mandated by the FSS Act 2006 for performing the following functions:

(a) Framing of regulations to lay down the standards and guidelines in relation to articles of food and spaying appropriate system of enforcing various standards thus notified.

(b) Laying down mechanisms and guidelines for accreditation of certification bodies engaged in certification of food safety management system for food business.

(c) Laying down procedures and guidelines for accreditation of laboratories and notification of the accredited laboratories.

(d) To provide scientific advice and technical support to central government and state governments in the matters of forming the policy and rules in areas which have a direct or indirect bearing of food safety and nutrition.

(e) To collect and collate data regarding food consumption, incidents and prevalence of biological risk contaminants in foods, residues of various contaminants in food products, identification of emerging risks and introduction of rapid alert systems.

(f) Creating an information network across the country so that, the public, consumers, panchayaths etc., receive rapid, reliable and objective information about food safety and issues of concerned.

(g) Provide training programmes for persons who are involved or intend to get involved in food business.

(h) Contribute to the development of international technical standards for food sanitary and phytosanitary standards.

(i) Promote general awareness about food safety and food standards.

In order to perform these objectives the FSSAI has set up different sections/ branches under its administrative control namely

(a) Food import clearance system (FICS).

(b) Food licencing and registration system (FLRS) and

(c) Food safety reporting system (FSRS).

12.6.3 Export Inspection Council

The council has been set up with the sole objective to check the quality of a number of food materials that are meant for export. The council has vested with the powers to reject any food that does not meet the standards prescribed for food, canned food such as mango juice, pine apple juice, frozen foods etc are subjected to scrutiny by this body before export.

12.7 Bureau of Indian Standards (BIS)

Bureau of Indian Standards formerly known as the Indian Standards Institute is a national standards body of our country, is a statutory organization established under the Bureau of Indian Standards (BIS) Act 1986. It is a corporate body consisting of members representing industries, consumer organizations, scientific and research institutes, professional bodies, technical institutions, ministers of State Governments, and Members of Parliament. The bureau is responsible for facilitating harmonious development of standards, product certification, quality certification, quality system certification and Environmental Management systems certification and so on. The bureau is empowered by the legislative act of the parliament known as the Bureau of Indian Standards Act (1986) operates the product certification scheme. The certification allows the licences to use the popular ISI Mark, which has become synonymous with the quality of products for more than 50 years. The BIS product certification scheme basically is voluntary and is based on the ISO guide 28 which aims at providing quality, safety and dependability ultimately to the customer. Many of the operational elements of the BIS product certification scheme correspond with the requirements of the ISO guide 65. A good number Indian Standards have, however, been harmonised with ISO/IEC standards and some are dual numbered as IS/ISO or IS/IEC standard. All the BIS certification is carried out on Indian Standards, which have been found amendable to product certification.

12.8 Clean Milk Production

Cleanliness in the milk production means freedom from extraneous matter such as manure, dust etc, whereas "Safety" means freedom from pathogenic microorganisms for human consumption which is highly desirable. The sanitation of the milk supply can be safe guarded in two ways.

(A) Production and handling of raw milk in such a manner to prevent its contamination by the pathogenic organisms. This will require:

(a) Ensuring the health of the dairy cattle by various control measures

(b) Safeguarding the health of employees by regular medical examination

(c) Protection of the water supply from contamination by Pathogenic organisms and

(d) Control of flies in the milk barn etc.

(B) Pasteurization of milk, so as to kill all the pathogenic organims and avoidance of any post-pastereurization contamination.

In an endeavour to produce the clean milk certain standards have been fixed by the BIS, which are suggested as a guideline for grading the raw milk in our country (Table 12.1).

Table 12.1 Bacteriological Standards for raw milk

SPC / ml (or g)		Grade
Not exceeding	2,00,000	Very good
Between	2,00,000 - 1,000,000	Good
Between	1,000,000 - 5,000,000	Fair
Over and above	5,000,000	Poor

Source : BIS : 1479 (Part - III) 1962

Note: The pasteurized milk at the milk plant in its final container should have a SPC/ml (or g) not exceeding 30,000.

Further, the producer of milk must be fully aware of the various sources of contamination of the milk and their control measures in his attempt to produce the clean milk with good quality (Table 12.2).

Table 12.2 Sources of contamination of milk and their control

Sources of Contamination	Control measures
1. Interior of the Udder	(a) check for mastitis
	(b) discard fore milk
2. Exterior of the cow particularly Udder & Flanks.	(a) wash and wipe udder
	(b) clip the udder and flanks
	(c) dry milking
	(d) use of small top milk pail
3. Barn air and Dust	(a) keep milk covered
4. Flies and other vermin	(a) Elimination of breeding places
	(b) Fly control with fly sprays - fly trips etc.
5. The milker	(a) clean milk habits
	(b) dry milking
6. Utensils	(a) clean, sanitize and dry before use

Source : Outlines of Dairy Technology - S.K. De

In addition to the above said control measures the following methods are also being used namely:

Straining, filteration and cooling and these have been discussed in relevant chapters under dairy processes.

CHAPTER 13

Milk Residues

13.1 Pesticide Residue in Milk and Milk Products

Pesticides are the toxic chemicals which are generally used to destroy the organisms of several crops and the external parasites like ticks etc., of the animals. As many as 126 pesticides have been cleared by the Registration Committee for the usage in our country. Out of which 57 are being manufactured in our country. Nearly 40,000 tonnes of hazardous pesticides, which are either banned or highly restricted for use in developed countries are still being used every year in India. According to a study carried out on the usage of these pesticides, it was reported that DDT, BHC, Aldrin, Chlordana Heptachlor Sodium Cyanide 2.4 - D and Lindane are being used "indiscriminately", and are found to cause health hazards and environmental pollution. The study was carried out by the National Institute of Science and Technology and Development Studies (IVGSTADS), under the aegis of the Council for Scientific and Industrial Research (CSIR). The highly toxic or condemned pesticides carry no warning or caution. Besides, the study says the use of pesticides is concentrated in few areas only.

13.1.1 Foreign Substances

The usage/utilization of chemicals in agriculture in the form of insecticides and pesticides as well as the chemotherapeutics or drugs in fighting against the animal diseases results in residues of these compounds in milk. In addition, environmental contamination by the industries and other factors of civilization contributes to the occurrence of foreign material in milk. The foreign material/substances in milk/foods can be defined as those which are neither naturally do not present nor introduced by any normal physical treatment. The possible sources of contamination by these foreign substances in milk have been represented in a schematic flow chart 13.1.

Flow Chart 13.1 Contamination sources for foreign substances in milk

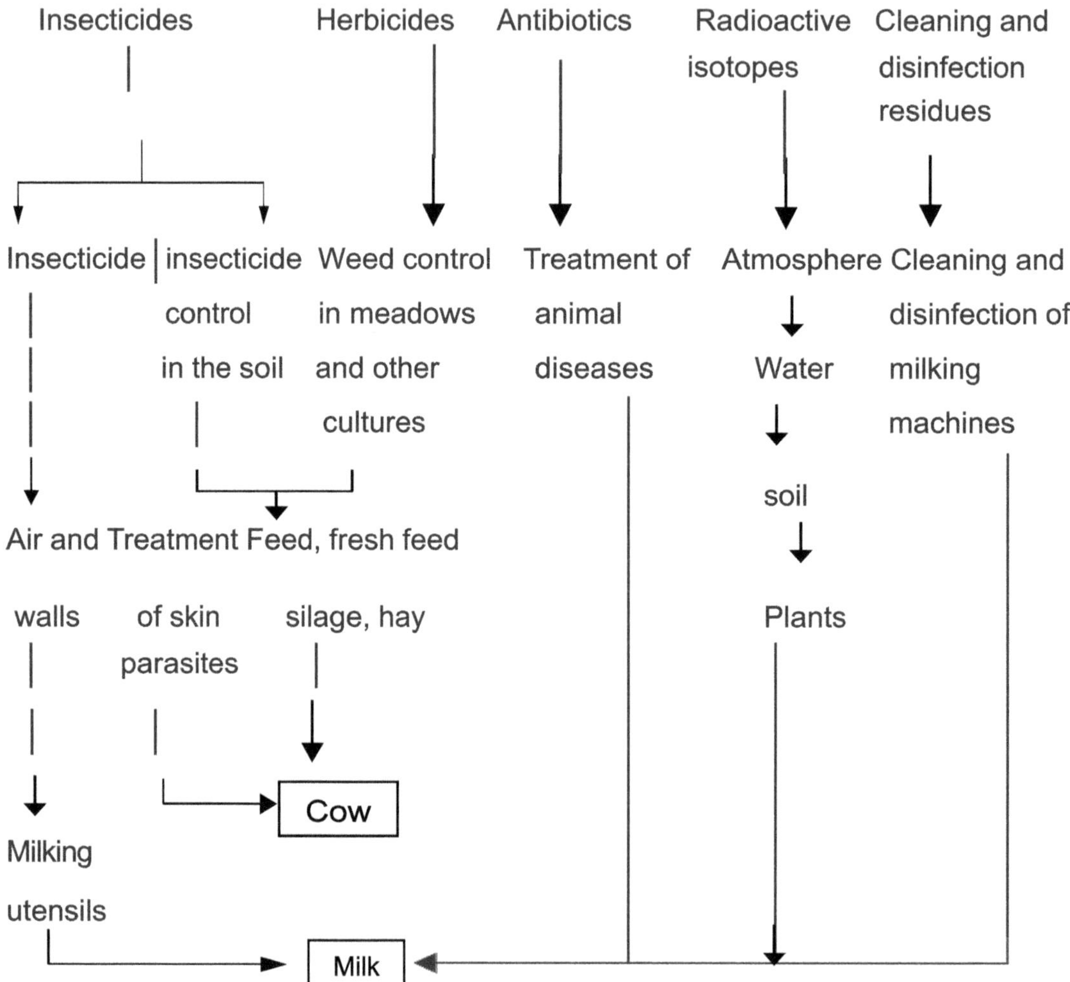

Source : Dairy Science Technology, Pieter Walstra et.al

The tolerance levels that have been set through the laws and regulations in the dairy products and other foods in order to avoid any damage/injury to the humans by the possible due to the consumption of these products have been given in table 13.1.

Table 13.1 Tolerance values for milk and milk products in mg/kg for selected foreign substances

Foreign Substance	Tolerance level/value for		
	Animal Fats	Dairy Products	Food and Baby food
DDT	1.0[a]	1.0[a]	0.02
Lindane	0.5	0.02	0.02
Trichlorphon	0.02	0.02	0.02
Dichlorphos	0.02	0.02	0.02
Butanol	0.5	0.02	0.02
Dimethoat	0.004	0.004	0.004
Carbaryl	0.02	0.02	0.02
Hexachlorobenzene	0.5[a]	0.5[a]	0.02
Lead	--	0.1	--
Zinc	--	5.0	--
Copper	--	0.5	--
Tin	--	250.0	--
Mercury	--	0.02	--
Nitrate	--	10.0	--

a - on fat basis

Source : Dairy Science Technology Pieter Walstra et.al

In addition to the possible damage to the human or animals due to the contamination with the foreign substances, defects in technological process have to be expected. The activity of butter or cheese cultures and yoghurt manufacture can be reduced significantly by the residues of pesticides, herbicides, insecticides and antibiotics.

The source of contamination of pesticide residue in milk and milk products are: contaminated utensils of milk, misuse of spray material as well as the contamination of feed. Extensive studies through the exhaustive analyses carried out in different laboratories in the country. It has been found, that the main channels of pesticide flow to milk are fodder and feed management practices and bad-farming husbandry. Fodder derive the pesticides from the soil and the commonly used fodders are leucern, berseem, green maize and wheat straw or paddy straw. Intensive agriculture forces or compells the farmer to use the pesticides to treat the crops in order to kill the white ants and other unwanted weeds which hamper the growth and affects the production of the crops. The quantity of these pesticides like DDT etc will increase every year as the farmers are using these pesticides injudiciously to save their crops from various types of diseases affecting them. Thus the crops are overdosed liberally with an idea to eliminate the insects completely. The pesticide dealers also advise the farmers wrongly to overdose the pesticides four times in order to push up their sales, unmindful of the resultant effect. The fodder thus produced contains excess of pesticide and such fodder which has the residual pesticide is harmful when fed to the animals, it passes into the milk thus making it hazardous to the health of the human beings.

Feed and fodders are apparently the by-products of agriculture produce such as wheat bran, rice polish, gram chuni, gram husk, sarson cake, cotton seed cake, groundnut cake, maize grain and molasses etc. Rice polish, cotton seed cake, groundnut cake, maize grain and molasses are supposed to carry the residual effect of the pesticides from the field i.e. pre-harvest period. Wheat bran, gram (chuni and husk), sarson cake, carry more with the post harvest practices, because these stored grains are mixed with heavy application of BHC and DDT and at the time of processing hardly any one leeds to wash the grains before milling. Thus wheat bran, gram chuni and gram husk have been seen with exceptionally high amounts of BHC or DDT. Thus, apparently it is very clear that the pesticide residue poses more serious problem where the milk production is solely obtained from the feeding of concentrates.

Another source of pesticide residue in milk may be due to direct spraying of insecticides in the barns and stables in order to control the menace of flies and ticks or direct application of the insecticide on the animal body. This problem is much less in the rural areas as compared to the urban dairies and stables. Neglecting the safe period, by overdosing the crops, between the last spray and the harvest time, directly fixing the cereals and pulses with pesticides are some of the bad farming husbandry practices to be mentioned. Milk is consumed by all people alike irrespective of their age and it is of paramount importance, that the milk they consume should be free from any chemical contamination. The defence mechanism of the new born is not powerful enough to withstand the threat posed by the ingestion of these harmful pesticide residues. Therefore, a regular analysis of fluid milk in different milk processing centres would help to assess the level of pollution or contamination. Scores of children are being fed by the infant foods manufactured by different companies, barring a few, none of them have the facilities to check the level of persistant pesticide residues in the infant formula feeds they are manufacturing. The effects of these pesticide residues on the new born are not yet clearly defined. Out of the various types of pesticides, only the organochlorine (non-polar) pesticides such as Aldrin, BHC, DDE, DDD (TDI) OP, and P.P. - D.D.T. Heptachlor Oxide, Lindane, Methoxy Chlor, Dieldrin, and Endrin find their way into milk through the animal's feed. These compounds are fat soluble and not readily excreted by the animal except into milk. However, other pesticides such as organo phosphates and carbonates are not fat soluble and they are broken down during metabolism and excreted by the animal.

Limited studies have been carried out in different regions of India on the samples collected (curd, ghee) by them show heavy contamination with DDT, BHC, DDE and DDD in some parts. The observed levels of these pesticides are above the recommended safety levels by the WHO. Therefore, there is an urgent need to monitor the milk and milk products for the pesticide residues in view of the ever increasing injudicious usage of these pesticides by the farmers in order to safeguard their crops against several crop diseases. In addition to the pesticides named above, the milk and milk products are also found to contain Lindane, Dieldrin, Heptachlor, Aldrin and HCH in our country.

According to a recent study carried out on the status of organo chlorine pesticide residues of milk in different zones of Haryana state by the research workers of NDRI Karnal have reported that the milk samples collected were found to contain the organochlorine pesticides residues such as Aldrin, Diendrin, Endosulfan - a, Endrin, Heptachlor. In an another study, the scientists of IVRI, Izatnagar, also reported that the organochlorine compounds pesticide residues are present in the milk. Samples collected from two blocks of Bareilly district, U.P., where the villagers are adopting improved agricultural practices. This indicates the indiscriminated usage of the organochlorine pesticides despite the prohibition on their usage in Agriculture.

The tables 13.2 and 13.3 gives an idea with reference to the contamination of milk, butter and ghee which differs from region to region.

Table 13.2 Pesticide Residue in butter and ghee

Pesticide	Range (ppm)
DDT (Butter)	0.30-8.10
DDT (Ghee)	2.0-13.0
BHC (Butter)	0.61-5.50
BHC (Ghee)	2.23-6.40
HCH (Butter)	2.85
HCH (Ghee)	2.01

Source : Text Book of Dairy Chemistry - M.P. Mathur, D.D. Roy & P. Dinakar

Table 13.3 Pesticide Residue in milk samples

Pesticide	Range (ppm)
DDT	Traces - 216
BHC	Traces - 513 (Human Milk)
Endrin	0.02 - 1.5
Heptachlor	Traces - 0.60
Aldrin	0.03 - 0.52
	0 - 44 (Human milk)
HCH	0.01 - 1.01
	115 - 325 µg/kg (Human Milk)
	0.01 - 1.01
	200 - 125 µg/kg (Human Milk)
Dieldrin	0.12 - 2.5

Source : Text Book of Dairy Chemistry - M.P. Mathur, D.D. Roy & P. Dinakar

Analyses of dairy products which are contaminated for the residual pesticides usually consists of following five steps.

(a) **Sampling:** The aim of sampling is to provide a reproducable portion of the environment on a scale that enables the sample to be handled in the laboratory. The results of analysis are meaningful only when the samples collected are truly representative.

(b) **Extraction of the residues from the sample matrix:** The residues are extracted with appropriate solvents from the sample substrate, the aim being to combine the maximum efficiency of extraction of the residue with minimum co-extraction of other constituents of the sample, which may interfere with the subsequent procedure.

(c) **Removal of interfering co-extractives (clean-up):** The clean up stage aims at the removal of the interfering substances from the extract in order to obtain a clean or pure solution of the extracted residue in a solvent suitable for quantitative examination by the choosen method of determination.

(d) Identification and estimation of the quantity of the residues in the cleaned up or pure extract usually at very low levels in the range of 10^{-9} to 10^{+12} gram. Estimation of the quantity of residues is normally done by the gas liquid chromatography with electron capture detector.

(e) **Confirmation of the presence and identity of the residues:** Chromatographic method is the most widely used method for the determination of pesticide residue, followed by spectrophotometry. The spectrophotometric methods are generally less sensitive and less selective than G.L.C.

Organochlorine pesticides present in milk and dairy products are usually analysed with a tritium or 63 N electron capture detector. The directly coupled chromatograph mass spectrometer is a powerful tool for positive identification of pesticides. HPLC has been found increasingly useful for the analysis of organochlorine pesticides as it requires less extensive clear cut than required with GLC. An attempt has been made to present the various standards prescribed for the milk products in accordance with the PFA Act. 1976 has been given in 13.4.

Table 13.4 Standards of Various Milk Products as per PFA

S.No.		Description
1.	Cream	Milk fat 25 percent (minimum)
2.	Dahi or curd	Dahi shall have the same minimum percentage of milk fat and milk solids not fat as the milk from which it is prepared.
3.	Paneer/Channa	Moisture 70.0 per cent (minimum); Fat 50.0 per cent of the dry matter (minimum)
4.	Skim Milk Paneer/Chhana	Moisture 70.0 per cent (maximum) Fat 13.0 per cent of the dry matter (maximum)
5.	Cheese (Hard)	Moisture 43.0 per cent (maximum) fat 42.0 per cent of the dry matter (minimum) Stabiliser/Emulsifier/NISIN 0.1 per cent
6.	Processed cheese	Moisture 47.0 per cent (maximum), Fat 40.0 per cent of the dry matter (minimum), Salt 3.0 per cent (maximum), Stabiliser/Emulsifier 4.0 per cent (Minimum) or sorbic acid 0.1 or Nisin 0.1 per cent
7.	Ice Cream	Fat 10.0 percent (minimum), 8.0 per cent (Minimum) if containing fruits or nuts. Protein 3.5 per cent (minimum) T.S 36.0% (maximum)
8.	Milk Ices/Lollies	Fat 2.0 percent (maximum), Protein 3.5 per cent (minimum), T.S. 20.0 percent (minimum), Stabilizer/emulsifier 0.5 percent (Maximum)
9.	Condensed milk unsweetened evaporated milk	Fat 8.0 per cent (minimum) Total milk solids 26 per cent (minimum)
10.	Condensed milk sweetened	Fat 9.0 per cent (minimum) Total milk solids 31.0 per cent (minimum), Cane Sugar 40.0% (minimum)
11.	Condensed skimmed milk unsweetened (evaporated skimmed milk)	Total milk solids 20.0 per cent (minimum) Milk fat-not more than 0.5 per cent by weight

Table 13.4 *Contd...*

S.No.		Description
12.	Condensed skimmed milk sweetened	Total milk solids 26.0 per cent (min.) Cane sugar 40.0 per cent (minimum) Milk fat Not more than 0.5 per cent by weight
13.	Whole milk powder	Fat 26.0 per cent, Acidity 12 percent,(L.A.) Maximum solubility 98.5 percent, (Spray) 85 per cent (Roller) S. Index 2.0 (ml), Spray 15.0 (ml) (Roller) Moisture 5.0 per cent (maximum), B.H.A. - 0.01 per cent, coliform 90/g, SPC - 50,000 (g) Sodium citrate 0.3 per cent
14.	Skimmed milk powder	Milk fat 1.5 per cent (maximum), Moisture 5.0 per cent (maximum), Acidity 1.5 per cent (maximum), Sodium citrate 0.3 percent Maximum, SPC - 50,000/g E, col. 90/g Coliform - 90/g
15.	Partially skimmed milk powder	Moisture - 5.0 per cent (maximum) Fat - It should be between 1.6 to 24.0 per cent, Acidity 1.5 per cent (L.A.) maximum, Sodium citrate - 0.3 per cent BHA - 0.01% The minimum solubility index of the product Roller dried / Spray dried solubility percent 85.0 / 98.5 solubility index 15.0 / 2.0 ml
16.	Khoa	Moisture 28.0 per cent (maximum), fat 20.0 per cent (minimum)
17.	Table Butter	Fat - 80.0 per cent (minimum), Salt 3.0 per cent (maximum) Curd 1.5 per cent (minimum), Diacetyl - 4.0 PPM (maximum)
18.	Desi Butter (Cooking)	Fat - 76.0 per cent (minimum)
19.	Ghee	Butyro-refractometer reading 40.0 to 43.0 reading at 40°C, Minimum Reichert value 28.0 per cent of free fatty acids as 3.0, oleic acid (maximum limit), per cent of moisture (maximum limit 0.5 per cent)
20.	Infant Milk Food	Moisture - 5.0 per cent (maximum), Fat - 18.0 per cent (maximum) 28.0 per cent (maximum), Total carbohydrate - 35.0 per cent (minimum) Milk proteins - 20.0 per cent (minimum), Ash - 8.5 per cent maximum SPC - 50,000/gm SI - 2.0 ml (spray dried) 15.0 ml (roller dried)
21.	Soy-whey Food	Fat - 12 per cent, Protein - 23 per cent, Carbohydrates - 50.5 per cent, Ash - 5.0 per cent, Moisture - 1.5 per cent (Fortification with vitamins and iron)

Table 13.4 *Contd...*

S.No.		Description		
22.	Coffee Complete	Fat	-	7.2 per cent
		SNF	-	20.0 per cent
		Sugar	-	39.0 per cent
		Water	-	25.0 per cent
		Total solids	-	75.0 per cent
		Milk fat	-	8.5 per cent
23.	Tea Complete	Fat	-	13.0 per cent
		SNF	-	31.5 per cent
		Sugar	-	39.5 per cent
		Tea Solids	-	14.0 per cent
		Moisture	-	2.0

Source : Text Book of Dairy Chemistry - M.P. Mathur, D.D. Roy & P. Dinakar

13.2 Oxytocin Residual Effect

Oxytocin (OT) is a harmone released from the anterior Pituitary gland in animals as well as in human beings. This harmone is protein in nature. It is transported through blood from the site of production/release to the site of action physiologically this harmone is important for milking. It causes contraction of smooth muscle cells surrounding milk producing cells in the mammary gland. Which are termed as alveoli and it also helps in the birth/delivery of the foetus causing the contractions of smooth muscles of the uterus thus expelling the foetus out side the body of mothers.

Oxytocin injections have been indiscrimately and rampantly used to increase the milk production in the lactating milch animals has become common practice by the dairy farmers of the dairy industry. There is no study/reports authentically available regarding surveillance of OT in market milk. However it is presumed that long term consumption of such milk leads to adverse health consequences. In a study conducted by Mid-Day a newspaper published in Mumbai in and around the existing cattle tables of Mumbai city has exposed how the owners of these tabels are using the OT indiscriminately injecting the cattle violating the ban imposed by the government thus resulting in the extra production of milk. Thus the drinking of the milk produced in this manners results in severe harmful effects in children. It was confirmed by the scientific personnel that the drug may decrease the reproductive ability of the cows/animals over a period of time eventually making them barren and also reducing the life span of the cattle. Studies also have confirmed that the consumption of milk contaminated OT results in the early onset of puberty in children which is alarming rise in these days.

Section 12 of the Prevention of Cruelty to Animal Act (1960), the Food and Consumable Substances Adulteration Act and the Drug control laws ban the sale of this drug without prescription from the Registered Medical practitioners. However these laws exists only on paper. The dairy farmers are using the drug with impurity and without any prescription having obtaining the drug from private suppliers. It is widely presumed and believed that the long time consumption of such type of milk leads to adverse health consequences even though there is no concrete evidence to support the views. There is no information on the effect of exogenous OT injections on milk OT content and its stability during pasteurization and gastrointestinal digestion.

Boiling milk before consumption is a good way/method to minimize the traces of OT in milk. Experts infact recommended boiling even the pasteurized milk, which refers to heating the milk to a specific temperature for a definite length of time and then cooling it immediately. This process slows down the specific spoilage due to microbial growth in the food. The enzyme of oxytocin breaks down into several small units which can limit the ill effects to a certain level.

In a study carried out by Sana Jawaid et al (2015) analysed a total of 84 milk samples for the occurance of OT residues by ELISA technique. The OT level was detected in the range of 39.29 – 456.55 pg/ml^{-1} with an average mean value of 138.53 μg/ml^{-1}. The average daily intake of OT was found to be 0.087 μg/day per person. They concluded that the OT concentration was relatively higher in the milk samples obtained from dairy farms, dairy shops and milk vendors than in UHT and powder milk samples. Further, they also stated that there are no reports subjected to standardizing a permissible OT level in dairy milk category although OT is considered as detrimental to both human and animal life (International food science technology Journal 50-(7) (2015).

There has been a controversy regarding the usage of exogenous OT by the dairy farmers in milking animals which may have some toxicological consequences during non-physiological exposure. A new sensitive extraction method for OT has been developed by Manjari Mishra et al (2014) followed by Enzyme Immuno Assay (EIA) or High Performance Liquid Chromatography (HPLC) analysis. The extraction of OT in milk involves two steps. (1) Tricholoacetic acid (TCA) precipitation of milk proteins and (2) Solid phase extraction (SPE) cleanup process. Without these steps the analysis of OT in milk was not possible. Utilization of EIA is a quantitative tool. The limit of detection (LOD) and limit of quantification (LOQ) were found to be 7.74 and 10.3 pg/ml^{-1}. Precision in terms of intra and intraday coefficients of variation was below 13% while precent recoveries were between 85 and 92%. OT was found to be stable against adverse temperature (up to 100 °C) and pH (2 – 10) and simulated gastric fluid digestibility assay. The intake assessment of OT through milk was assessed through the food frequency recal method employing a food frequency questionnaire. On the basis of milk consumption the values of OT in milk and the actual intake of OT was calculated. OT contamination in milk samples was found to be 21 pg/ml to 18.9 ng/ml with an average mean value of 89 ng/ml. The average daily intake OT in terms of μg/day/person was highest in 1-3 years of age group. Since there is no prescribed level of OT in milk and the intake of OT through this commodity is quite high and there is a need to implement regulatory laws in a stringent strict manner so that non-physiological OT exposure may not occur in children which may have deleterious effects.

(Journal of toxicology mechanisms and methods 24 (5) 2014).

CHAPTER 14
Packaging

Packaging is an art or technique which is an essential step which uses the most appropriate material or containers in order to protect, carry identify and merchandise any food item namely, the milk and milk products here in this context. The importance of packaging as a means of promoting exports has been well recognised and has become an essential and integral part of all the industrial activities in their approach to the modern marketing. Packaging, constitutes a vital link between the manufacturer and the ultimate consumer. Packaging has become an important integral part in the distribution of the products due to the changes in the socio-economic status of the valued customers/consumers. Attempts have been taken to produce the market milk or the dairy products of high grade by the processors or manufacturers with an aim to deliver in a fresh sound and convenient and attractive form to the consumer. However, these products sometimes are likely to be rejected, causing enormous wastage which can be minimised by adequate protective packaging to withstand the hazards pertaining to climate changes, transportation handling etc.

Today an array of packaging materials and systems such as aseptic and retort packaging, gas/vacuum packaging, thermoforming, bag-in inbox and lined carton systems, shrink wrapping is available. Development of higher barrier materials, controlled atmospheric packaging (CAP) or modified atmosphere packaging (MAP) systems, aseptic packaging system and the like has changed the food marketing system providing more "fresh food" to retail stores for a longer duration.

14.1 Importance of Packaging

Food must be made available throughout the year, irrespective of its season of production, and must be presented in such a way that it is convenient for its purchase and usage. The packaging technique and choice of a package with appropriate barrier properties, is designated to prevent the spoilage of food by the attack of the microorganisms, depending upon its physical nature and to preserve the quality and nutritive value of the food by the exclusion of oxygen and control of moisture. Several advancements have taken place in the packaging of dairy product. Packaging has become a socio-scientific discipline with the modern role of:

(a) Containing and safety, which is of paramount importance.
(b) Facilitating easy handling, storage and distribution.
(c) Protecting against biological, chemical and distribution damages.
(d) Provision of convenience and information through labeling.
(e) Security against tampering as evident through its design.
(f) Increasing the shelf life, ensuring longer availability.
(g) Contributing to the product image with structural, graphic designs which acts as a marketing and advertising tool.
(h) Environmental protection by taking care of the empty packages after its usage.

The efficiency and economy of distribution of milk and milk products have led to the innovation of newer packaging systems with the advancement of the technological progress and improvements. These innovations have provided greater convenience to consumers, extended the shelf life of the products, with lowering costs and increasing the sales, better hygienic conditions with ease in handling. Modified and controlled atmosphere packaging provides certain interesting features of long life, optimum quality of the product maintenance, good and reliable sealing and adaptability to various production programmes.

In India, the packaging industry is a heterogenous mix of both organised and unorganised sectors. The industry comprises of manufacturers of basic materials, converted package forms, ancillary materials and packaging machinery.

Development of suitable packaging system is essential for modernisation of the traditional dairy products with the greater participation of the industry and the organised sector. Already systems have been developed for assembly line packaging of ghee, shrikhand, dahi, Mistidoi etc. The high tech packaging technology has also extended the shelf life of paneer and facilitated its marketing even in the remote areas.

Packaging is defined as a means of placing a commodity in a protective wrapper or container for its easy transport or storage in a convenient manner. The most important aspects of packaging are:

1. Presentation (graphics and material)
2. Cost
3. Consumer friendliness (easy to open and use) and
4. Effect on environment.

The growing sophistication in marketing trend calls for consumer convenience in product buying and usage. Also from the environmental point of view emphasis is on developing such packaging material that can be recycled or reused. A food package has to satisfy many requirements. In essence they include:

1. Containment and portability
2. Preservation
3. Convenience and
4. Consumer information.

The cost of packaging is often seen in relation to the total selling price of the product. Selection of the packaging material must be in tune with the intended price of the product.

14.1.1 The Main Objectives of the Packaging are

1. **Containment :** In this the food is generally separated from the environment, involving the partitioning of the product into small units, which can be handled conveniently with ease during distribution, storage, transport and its final usage. It also prevents the contamination of the food material from the environment, which may result in hygienic problems and generally gives the guarantee to the integrity and the quantity of the contents.
2. Gives protection to the product from outside influences. Protection implies preventing or minimising the loss of components such as water and flavour substances to the environment. It also implies the prevention of contamination with microorganisms and chemical compounds or dust particles and exclusion of radiation especially light. Sometimes packaging forms as an essential part of the food preservation.
3. **Consumer Convenience:** The most important part /aspect in the packaging of any food item is that it should be easy to open and close it again. It is also very important to minimize the contamination, where a portion of it has been taken out of the package and is easy in stacking of the containers. Modern packaging systems often enable the distribution of foods that are ready to eat or that may need little preparation only. Packaging in different sizes is convenient for those consumers, who are in need or desire to purchase little or small quantities or lot of food per day.
4. **Providing information:** The packaging material or the wrapper must provide the factual information pertaining to the quality of the product, its composition, nutritive value, keeping quatity, the expiry date i.e. the day before which it should be consumed, method of storage precautions to be taken before handling etc. In addition to the above, several marketing messages can also be printed on the label, such as brand name, the possible application of the product (recipes) and potential benefits of its consumption.

The consumer is extremely packaging conscious, and in view of this the packaging industry has made tremendous advances and risen to greater heights in the developed countries, with the introduction of newer marketing systems such as departmental stores, super markets and self service stores etc. The development of packaging machinery and appliances, better packaging materials has improved the system or method of packaging and advanced in an integrated manner. On the contrary the packaging industry in the developing countries including India is still in its infant stage.

14.2 Selection of Packaging Materials

Selection or choice of an appropriate packaging material is governed by several factors, namely:

1. The specific sensitivities of the contents eg. moisture, oxygen etc.
2. Factors affecting the contents such as temperature, RH, pH, and the reaction machanism involved.
3. Weight and shape of the container and effect on filling and sealing speeds.
4. Contamination of food by constituents of the packaging material.
5. Storage conditions.
6. Biodegradability and recycling potential.
7. Transportation for long distances.
8. Due to the perishability of the dairy products, preservation during storage, handling and transportation utmost care is essential.
9. In the presence of oxygen and other contaminating agents that are present in atmosphere, the dairy products are subjected to spoilage at a faster rate at high temperatures.

In order to determine the packaging of dairy products, there are many peculiarities which could be identified under the following headings, namely:

14.2.1 Product Range

1. Milk based sweets such as - Peda, Burfi, Khoa, Basundi, Khurchan, Gulab jamun, Sandesh, Rasogolla, Rasmalai.
2. Frozen products like Kulfi and Ice cream.
3. Fat rich products - Ghee, Malai, Makhan.
4. Coagulated products - Channa and Paneer.
5. Fermented/cultured products - Dahi, Misti Dohi, Lassi and Srikhand.

Market: The indian dairy products market is huge with a domestic production of approximately 112 million tonnes of milk.

(a) Approximately 46% of the total milk produced in the country is consumed as fluid milk.
(b) Only 15% of the milk is packed.
(c) Approximately 15 to 20% of all the milk produced is processed. If we can create value added products with appropriate packaging, things will be quite different in our country.
(d) The dairy industry is highly competitive and the margin of profit per unit is low in some basic products where consumption is high.

14.2.2 Consumer Needs

Consumer trends will play a vital role in deciding the type of packaging. Some of the trends in the dairy industry in India are:

1. Well packed dairy products are associated with quality.
2. Packaging is expected to make dairy products to be more safe.
3. Increasing mobile life style.
4. Time has became a precious commodity.
5. Younger generation is looking for health and fitness conscious beverages.
6. The consumer is ready to try new products.
7. There is an increasing concern for freshness.
8. The traditional family meal no longer seems to exist. Seeking a home-meal dining experience.
9. Innovative or fun flavoured fortified milk for children.

14.2.3 The shift in the Consumer Perception of Packaging is Noticeable from:

(a) Commodity to performance.

(b) Value and the ease of usage which determine the cost of packaging.

(c) Package is not only the material, but it is a function for the product.

(d) The consumer has understood the effects on the environment and is looking for a balance with the environment and safety.

(e) Graphics have become a fashion. Industry with tremendous improvement in pre-press and printing capabilities.

14.2.4 There are several Attributes that Appeal to the Consumer such as:

(i) Product quality and protection with great emphasis on freshness.

(ii) Easy to open, dispense, reseal and store.

(iii) Appealing product presentation is gaining more prominence.

(iv) Durable and ecofriendly is being viewed together.

(v) Leak and spoilage proof is a must for the producer as well as to the consumer.

(vi) Reusable packaging.

(vii) Less hassels more convenience

(viii) Selection from a wider choice of available sizes.

Packaging for the growing dairy industry is an executing topic. However, it is very difficult to deal with because the dairy industry is highly competetive and the margin of profit per unit is very low.

With the increase in the consumption of the dairy products in our country, the future of packaging appears to be bright and increasing and must take into account the market need and other factors. In the dairy industry almost all types of packaging materials have been used. Developments and innovations are aimed at various alternatives and systems.

14.3 Packaging Materials

The production of milk in India is increasing and providing prospects for more entrepreneurs to get into the dairy packaging sector. Spoilage of milk is a constant troubling feature of the dairy industry which can be overcome by applying suitable and relevant effective packaging solutions/methods. The reasonable and cheep packaging options for milk and milk porducts can reduce their losses giving an opportunity to the manufactures to make these products more affordable to the consumers.

Wide variety of rigid plastics can be used as thermo formed like bottles cartons etc. Packaging of milk in HDPE may be a Declistic option and it was first introduced in 1964 which is more suitable for pasteurised UHT as well as sterilised milks. Polyethylene terephthalate (PET) as packaging material for dairy products using multi layer barrier technology can improve the shelf life up to two years.

Even though cartons are still the most popular form of aseptic packaging globally, but it is now trailing behind PET containers and pouches beacuse these became more user friendly from the consumers view point.

The public opinion of packaging as a waste that adds to environmental pollution is thought to be modern attitudes and media coverage towards environmental issues. Packaging of dairy products fulfills its major function of providing them in a safe and nutritions form to the consumers. Several research workers are carrying out for the development of newer materials that meet the eco friendly needs which is still in its infancy due to the breaches in the functionality.

Several widely varying packaging materials are in use and the extent to which they meet the various requirements and their preferences will be discussed briefly hereunder.

Packaging materials like Tin, Aluminium Plates, Foils, Glass etc., were extensively used during the formative years of food and dairy industry. These were replaced by a wide range of plastic materials because of the fact that they are relatively heavier and costly, as the packaging industry expanded. The developments in polymer chemistry has propelled the expansion because it was much possible to develop plastics of versatile properties. The advent of low cost poly laminated materials such as plastics in combination with paper/metal, the changes have been rapid.

The restriction on the use of plastics as packaging material, however, is imposed because of possibility of its components migrating into food products affecting their safety. In the context of safety two aspects have to be considered.

I. The toxicity level of the packaging material and
II. The extent of the package compounds that migrate into the food, which is mainly depends upon the type of food, temperature of packaging, storage conditions and the ratio of volume of food to weight of packaging material.

The packaging material should confirm to the following requirements.

1. It should not alter the colour, taste or character of the packaged product, protecting the organoleptic, physicochemical and microbiological attributes.
2. It should contain only the permitted ingredients and must not exceed the maximum permitted level.
3. It must provide barrier to light, temperature, gases, and moisture in order to protect the flavour and aroma of the packaged food material. It should also ensure requisite shelf life during distribution in marketing networks.
4. Migration of substance from the package to the food even though harmless, which is not perceptible organoleptically should not exceed the maximum possible level of 60 ppm.
5. Migration of a substance should not render the food unsafe for consumption.
6. The packaging material should have adequate tensile strength, elongation, streatchability, bursting strength, and heat sealability, in order to ensure the integrity of the package during transportation and handling.
7. The rigid container must have adequate impact strength and it should be suitable for capping and lid sealing.
8. The packaging material must be printable with or without treatment of the surface and
9. It must be environment friendly.

An overview of the packaging materials used in the dairy industries have been presented in the table 14.1 for information.

Table 14.1 Packaging Materials used in the Dairy Industry

Product		Packaging Material
Liquid Milk	:	• Glass bottles (obselete) • LDPE Film in combination with LLDPE or Octane/butane based films • Paper laminate for tetra packs
Milk Powder	:	• Tin plate containers, Nitrogen packed and lacquered from outside. • Flexible laminates such as metallized PET/BOPP/ Aluminium foil/Poly laminates. • Refill packs, lined cartons laminated with BOPP/PET varnished on the outside paper, laminated film also used. • Bag in box powder filled in laminate and packed in cartons.
Butter	:	• Duplex-board with vegetable parchment paper.
Cheese/Cheese Spread	:	• Tin plate containers lacquered from inside. • Packed in Aluminium foil and then in duplex board carton. • Injection moulded PP/HDPE container.
Ghee	:	• Tin plate containers lacquered from inside. • Glass Bottles
Ice Cream	:	• Thermo formed/injection moulded plastic containers. • Duplex board cartons which are poly laminated. • Laminates of BOPP or PET
Indigenous Products	:	• Injection moulded/Thermo formed containers (Shrikhand/ Gulab jamun) • Standup laminated pouches.

Source : Dairy Science & Technology-Pieter Walstra et.al

14.3.1 Packaging Materials - Range

The packaging industry in India is still in the infancy/evolutionary stage. The range of available packaging materials is very limited and their cost is relatively high.

1. Tin containers
2. Aluminium foil
3. **Paper:** Kraft paper, Brown paper, coated paper, parchment, glass line coated paper, wax.
4. **Paper board:** Pulp, Duplex, Triplex (Greese mill Solid, Corrugated, Moulded).

5. Glass bottles and Jars.
6. **Plastics/Metals:** As substrates or layer for application.
   - Poleofins, Low density polyethylene (LDPE)
   - Linear Low density Polyethylene (LLDPE)
   - Higher density polyethylene (HDPE)
   - High molecular high density poly ethylene (HMHD)
   - Biaxially oriented polypropylene films (BOPP)
   - Polyethylene terephthalate films (PET)
   - Metallized polyester (MPET) films.
   - Metallized poly propylene (MPP) films

Coexisting in the form of multilayer films may combine these packaging materials. Amenable properties of two or more materials can be combined in a synergistic manner to overcome their respective weaknesses as described below:

1. LDPE, LLDPE, HDE PP Ethyl, Vinyl Alcohol (EVA / EVOH).
2. Bonding agents like LXA or Primacor
3. Barrier resins like PA6 (Nylon)
4. Sealent layer like tonomers Primacor etc.

Flexible packaging materials (films) are also used in various other forms such as:

1. Monolayer - Blown, cast oriented, biaxially oriented, shrink.
2. Co-extruded - Blown, or Cast, Two, three, five and seven layers.
3. Metallized films - PET, BOPP and
4. Laminates - (Adhesive/extrusion) - Laminates of paper/polyester/Pol/Foil metallized polyester.

The various desired characteristics of the packaging materials are listed and presented in the table 14.2.

The data involved are highly approximate because they can vary widely according to the precise composition and the method of manufacture.

Processability: The processability of the packaging material depends upon the properties such as brittle, pliable or moldable and its availability in desired thickness and its suitability for sealing, lamination, sterilization and lastly its resistance to high temperatures.

Table 14.2 Properties of some packaging materials

Sl. No.	Material	Strength	Flexibility	Seala-bility	Resists sterili-zation	Resists freezing	Trans-perancy	H_2O 10^{-12} kg m^{-1} s^{-1}	Permeability O_2 10^{-18} kg m-1 s-1 Pa	CO_2	Fat
1.	Glass	Brittle	O	-	Yes	No	Clear	O	O	O	O
2.	Tinplate	Great	Small	-	Yes	Yes	O	O	O	O	O
3.	Aluminium Foil	+ + +	+ +	Not	Yes	Yes	O	< 0.1	0.002	0.003	O
4.	Paper Card board	+ +	+ +	Not	No	Yes/No	+	Great	great	great	great
5.	Cellophone	+ + +	+ + +	Good	No	Yes	Clear	100	1	10	tr
6.	Coated cellophone	+ +	+ + +	Good	No	Yes	Clear	1	0.1	0.1	tr
7.	Polyethylene L.D	+	+ + +	Good	No	Yes	+ + +	4	40	200	+ + +
8.	Polyethylene H.D.	+ +	+ +	Good	No	Yes	+ + +	1	10	50	+ +
9.	Poly Vinyl Chloride	+ +	+ +	Fair	No	No	Clear	10	1	10	+ +
10.	Polyamide	+ +	+ +	Poor	Yes	Yes	Clear	40	0.3	1	tr
11.	C.PET	+ + +	+ +	Good	Yes	Yes	Clear	5	0.3	2	tr
12.	Poly Carbonate	+ + +	+	Not	Yes	Yes	Clear	500	10	50	+ +
13.	Polypropylene	+ + +	depends	Depends	Yes	No	Clear	3	20	100	+ +
14.	Polystyrene	+ + +	+	Not	No	Yes	+ + +	30	20	100	+ +

Note : 10^{-12} kg m^{-1} S^{-1} Corresponds to 3.5 gr. M-2 d-1 at a layer thickness of 25 µm : 10^{-18} kg m^{-1} S^{-1} Pa^{-1} Corresponds to 0.35 g. M-2 d-1 at a pressure difference of 1 bar and a layer thickness of 25 µm. L.D. = Low Density H.D. = High Density

CPET = Crystalline Polyethylene terephthalate

Source : Dairy Science Technology-Pieter Walstra et. al

Resistance: The packaging material must be strong enough in order to resist the wearing and damage. It should be able to withstand the fluctuations in pressure and temperature such as sterilization, freezing and moist atmosphere.

Permeability: Passage of a substance through the packaging material may be by diffusion and consequently it greatly depends on the solubility of the substance in the material. The amount of substance permeating generally is proportional to the contact area, time and the difference in concentration and inversely proportional to the thickness of the material. When the closure of the package is perfect, generally the bacteria are not permitted to gain their entry through the packaging material. The permeability also depends on temperature. The permeability of aluminium to almost all substances is effectively zero, however, any perforation may lead to troubles. The permeability of the packaging material naturally depends on its thickness.

Heat Insulation: Most of the plastics have poor heat conductivity, because sometimes the layer is too thin for satisfactory insulation and in case the insulation is needed, expanded polystyrene can be applied. In general, a well insulating package is not desirable because after packaging heating and/or cooling are to be applied.

Light Transmission: In general for many foods, a transparent package is desirable such that the consumer will be able to see the contents. The drawback for milk products is that the light induced flavours may be developed.

Laminates: In many instances no single packaging material meets all the requirements, because of this reason, laminates are applied. In a milk carton for durable, aseptically packaged products, we may find (going from outside to inside)

Polyethylene	-	for water repellance
paper	-	for printing
card board	-	for firmness
Polyethylene	-	for making card board adhere to aluminium
Aluminium	-	against passage of light and all substances
Polyethylene	-	for good sealability. Sealing here means closing the filled package by pressing while heating.

14.3.2 Packaging Techniques

The first and foremost thing that attracts the consumer is the packaging and it ranges from 0.05 to almost 30 percent of the consumer price of the product. Good packaging lowers the product cost, reduces wastage of the product, helps the consumers make informed purchase decision, protects against tampering and spoilage, safety and comfort.

Innovations in materials, machinery designs, improvements in printing, labelling and computer software, barcoding etc., have revolutionized the packaging sector, which have resulted in the long shelf life, lower costs, and greater convince to the consumers, improvement in sales, market extension, better hygienic conditions, introduction of new product or product diversification and ease in handling. Some of the modern packaging systems have been listed below.

1. Vacuum packaging.
2. Modified atmosphere packaging (MAP) or
   - Controlled atmosphere packaging (CAP) / EMAP
   - Active packaging self heating and self chilling
   - Shelf life time temperature indicators
   - Shrink wrapping
   - Microovenable containers.
3. Oxygen absorbers/Scavengers.
4. Polyclip system
5. Aseptic in line dosing to test milk before the filler.
6. Computer aided designing and
7. Edible films and coatings biodegradable materials.

The different methods/classification of packaging systems has been presented in folw chart 14.1.

The extent of contamination by bacteria during packaging is essential for the keeping quality of the milk product. Strict aseptic packaging is far more difficult to achieve, but relatively simple measures may give substancial results. Satisfactory standards of hygiene during manufacture, transport and storage of the packaging material will control and prevent many problems, because of the fact that the materials involved are very poor substrates for the microorganisms. In the aseptic packaging of durable products, spoilage of fewer than 1 in 10^{-5} packages and even preferably less may be considered as acceptable.

Flow Chart 14.1 Classification of Packaging Systems

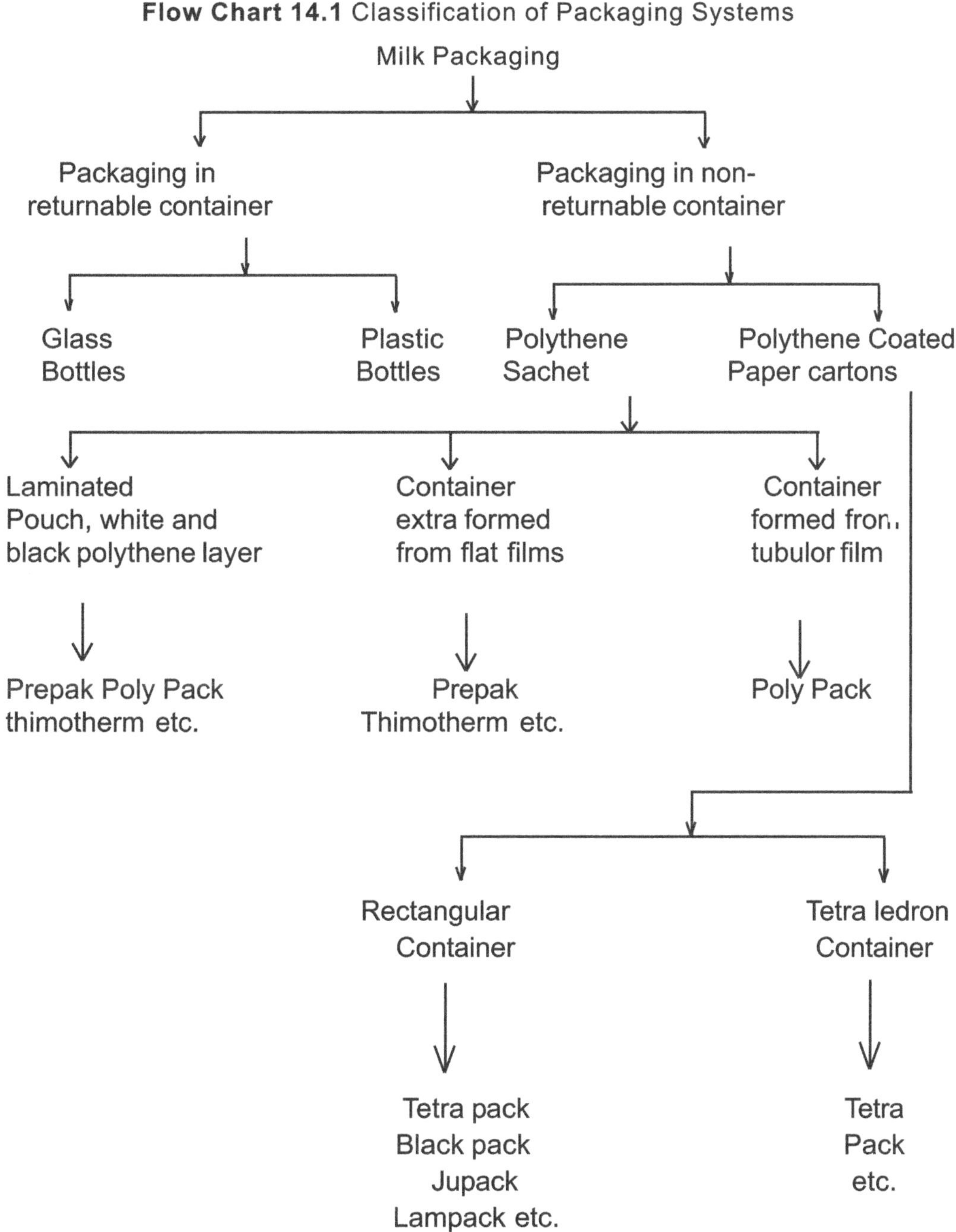

Source : Milk and milk products technology, Subbasish Biswas & Subhash Kumar Bhattacharya.

Sterilization of the packaging material should not impair that material. In majority cases sterilization with hot and concentrated H_2O_2 solution is applied and hot air readily removes the residues of H_2O_2 and also provides an additional sterilizing effect. Further, the H_2O_2 has an added advantage over the other liquid disinfectants due to the fact that it poses no serious problems with respect to the residues left in the milk or milk products. In addition, due to the rinsing effect the H_2O_2 produces better effects in removing the dust particles if any that have been attached on the packaging material.

14.4 New Concepts of Packaging

Depending upon the need and the economic considerations, the following packaging concepts, are also available for adoption by the dairy industry.

14.4.1 MAP Technology

MAP may be defined as the enclosure of food products in gas barrier materials, in which gaseous environment has been changed. The two major spoilage agents, namely aerobic bacteria and oxidative reactions require oxygen. Therefore, the non-availability of oxygen will inhibit spoilage and thus maximize the quality and/or storage life of the product. However, some deterioration might occur due to anaerobic/ micro aerophilic organisms and non-oxidative reactions. Product compression is unavoidable and makes vacuum packaging unsuitable for many products.

14.4.2 Advantages and Disadvantages of MAP

The advantages and disadvantages of MAP system are:

Advantages

1. Fresh in appearance.
2. Potential shelf life increase.
3. Reduced economic loss due to longer shelf life.
4. Product can be distributed over long distances and increase in market area.
5. Provides high quality product.
6. Easier separation of slices (Vacuum Vs. MAP)
7. Easy to open i.e. user friendly.

Disadvantages

1. Visible added cost.
2. Temperature control necessary.
3. Different types of gas formulations are required for each product.
4. Special equipment and training of the personnel is required.
5. Product safety is to be established.
6. There is a scope of the loss of benefits, when the pack is opened or leaked.
7. CO_2 if dissolved into the product/food could lead to the collapse of the pack.

14.4.3 Gases for MAP

Generally three major gases namely O_2, N_2 and CO_2 or their mixture is being used in the MAP for foods. Several other gases such as CO, Ozone, Ethylene Oxide, Nitrous oxide, Helium, Neon, Argon, Propylene Oxide, Ethanol Vapour, Hydrogen Sulphur dioxide and Chlorine have been used experimentally or commercially to extend the shelf life of a number of food products.

The properties of the three major gases has been described below:

O_2 : Maintains 'bloom' (colour) of fresh meat. Sustains basic metabolism, prevents anaerobic spoilage.

N_2 : Chemically inert, prevents oxidation and rancidity.

CO_2 : Inhibits bacterial and mold growth, shelf life prevents insects attack.

14.4.4 Methods of MAP

Methods of atmosphere modification within a packaged food product can be sub-devided into two main categories.

1. Passive modification.
2. Active modification

In commodity generated or passive modification the product is modified as a result of the consumption of CO_2 and generation of CO_2 through respiration of the product. Passive modification is commonly used to modify the gaseous atmosphere of packed fruits and vegetables. Several methods can be used to actively modify the gas atmosphere within the packaged product. This includes vacuum packaging, the use of Oxygen or gas also bents or scavanges CO_2 generators, ethanol vapour generators, and injection of gas mixtures. MAP is simply as an extention of vacuum packaging technology which involves the evacuation of air followed by the injection of the appropriate gas mixture.

14.4.5 Packaging Material for MAP

The examples of laminated films used for MAP foods have been presented in the table14.3.

Table 14.3 Laminated films used in MAP Foods

Laminates	Guage * Microns *	Permeability (ml / m² / 24H / ltr		
		O_2	CO_2	N_2
PET/PVDC/PE	12** / 3** / 50*	8-10	30	8
PVC/LDPE	400*/75*	15	30	4
Nylon/PVdC/PE	60* / 5** / 100*	9	--	-
PVdC-Coated				
PET/PE	15** / 60*	2-4	--	-
PVC/PE	400* / 75*	15	--	-
Nylon/EVAL	25** / 10** / 25**			
Nylon/PE	100*	20	20	1

Note : Guage - Single Star (*) Micron - Double Star (**)

Source : Dairy Science & Technology-Pieter Walstra et.al

14.4.6 Typical MAP Machines

The gas packaging techniques involves removing air from the pack and replacing it with a mixture of gases, the pressure of gas inside the package usually reaching about 1 (one) atmosphere i.e. equal to external pressure. This is usually achieved by one of the following types of packaging equipments.

1. Horizontal form - fill seal (HFFS)
2. Vertical form - fill seal (VFFS)
3. Thermoform - fill seal (TFFS)
4. Preformed Tray and lidding film (PTLF)
5. Three Web - Thermoform - fill seal (TWTFFS)
6. Vacuum machine with gas injection.

MAP has the potential to increase the shelf life of a number of dairy products including milk powder, cheeses, and fat spreads, fermented products, yoghurt milk etc. In general, these products are spoiled due to the development of oxidative rancidity in the case of powders and/or the growth of microorganisms particularly yeasts and moulds in the case of cheese.

14.5 Packaging Machines

Selection of packaging machines depends upon the products to be packaged. Packaging material, size of package, and desired speed of filling and the extent of automation required. Other important factors to be considered are :

1. Cost of machine
2. Efficiency of the machine with regard to power consumption
3. Requirements of sanitation i.e. CIP or manual cleaning and
4. Facility to replace the worn out or damaged parts

The imperative of manufacturing, Indian milk products under hygienic conditions require taking recourse to automatic packaging and sealing machines. However, in view of the highly specialized precision engineering skills coupled with high designing costs of new packaging machines, it is more practicable to adopt the available machinery for packaging the milk products. Performance of the existing packaging machines is being studied to incorporate modifications for improving their efficiency. Some of the packaging machines that are being in use have been listed below.

1. Tin can filling machine
2. Screwing Machine
3. Empty Tin Sterilizer
4. Form Fill and Seal (FFS) machine
5. Ghee Pouch (FFS) machine
6. Pasteurized Dahi/Lassi (FFS) machine
7. Cup Thermofill and sealing machine
8. Pre-formed cup filling and sealing machine
9. Multi fill packaging machine
10. Vacuum and gas packaging machine
    (a) Nozzle Vacuumizing
    (b) Chamber Vacuumizing
11. Cartoning machine and
12. Shrink Wrapping machine

Fig.14.1 Vacuum Packaging Machine

Several types of packaging machinery are available, which includes, those for the production of basic packaging materials and converting them into finished packages for filling, sealing, handling and storage, printing, testing etc. In order to improve the productivity and trade both at National and International levels, adoption of the standard packaging materials and techniques pertaining to its testing and usage will be much more beneficial and go a long way. The BIS has already been engaged in the development of the standards that are suitable for use in our country both for packaging materials and filled packages. The empty packages, have necessarily to be discarded after the usage of the product, which constitute a fair proportion of the solid waste especially in the cities. The disposal of these empty packages poses problems and is a great concern of the public health authorities and care should be taken in selecting the disposal methods in order to avoid subsequent atmospheric or land pollution and should meet the standards that have been laid down by the pollution control board.

14.6 Disposal of Packages

The growing trend of using the single service plastic container has resulted in an alarming increase in the volume of waste because plastic materials are not recyclable or biodegradable. The current practices of waste disposal are:

1. Dump filling
2. Incineration and
3. Re-use

Recycling of packaging materials:

Different types of packaging materials are recycled as under:

1. Tin plate
2. Aluminium
3. Plastic and
4. Paper

An attempt has been made to explain and present the storage temperature and shelf life of different traditional with products in different packages has been presented in table 14.4

Table 14.4 Storage Temperature and shelf life of traditional milk products in different packagings

Product	Type of package	Pack size (gm)	Type of package	Storage temp. (^oC)	Shelf life* (days)
Fermented Milk Products Shrikhand	Cups	100 200 500	Polystyrene (PS) and Polypropylene (PP)	-10	180
Dahi	Cups	100 200 400	Polystyrene (PS) and Polypropylene (PP)	+5	7
Mishti Doi	Cups	100 200	Polystyrene (PS) and Polypropylene (PP)	+5	7
Lassi	Pouch	200ml	Low density Polyethylene (LDPE)	+5	7

Table 14.4 ***Contd....***

Product	Type of package	Pack size (grams)	Type of package	Storage temp. (^oC)	Shelf life* (days)
Condensed Milk Products					
Kheer	Pouch	--	Aluminium foil/LDPE	+5	30
Peda	Carton	250 500 1000	Cardboard-Inner pearl film, outer, laminated	+5	30
Burfi	Carton	250 500 1000	Cardboard-Inner pearl film Outer laminated	+5	30
Kalakand	Carton	250 500 1000	Cardboard-Inner pearly film Outer laminated	+5	30
Gulabjamun	Cups Tins	500 500/1000	Polystyrene (PS) Laquered Tin	+5 Room Temp.	30 270
Coagulated Milk Products					
Rasogolla	Tins	500 1000	Laquered Tin	Room Temp.	180
Sandesh	Cartons	250 500 1000	Cardboard - Inner pearl film	+5	7
Paneer	Pouch Wrap	200/1000 200	PS and LDPE	-18 +5	180 30
Fat-rich Milk Products					
Ghee	Tins	1 kg/2kg	Laquered Tin	Room Temp. +5	180 270
	Pouch	500	LDPE Coexctruded nylon + Polystyrene film	Room Temp. 27	15 120
	Bag-in-box Outer carton, Inner pouch	500 1000	Pouches HD/LD/HD laminates and LDPE/Bonding	Room Temp. Room Temp.	180 120
	Cartons	500 1000	Duplex board Inner lined metalized PET or LDPE	Room Temp. Room Temp.	180 120
Milk-based Desserts					
Kheer, Pal Payasam, Rabri	Cups	100 200 400	Polystyrene (PS)	+5	NA

NA : Not Available. **Source :** Technology of Indian Milk Products - Dairy India.

*The shelf life depends on the product characteristics, quality of raw material, manufacturing process, hygienic standards maintained in the plant, type of packaging material, storage temperature. Here, shelf life is given based on information received from manufacturers.

CHAPTER 15
Cleaning and Disinfection

The safety and quality of the products depends to a large extent on proper cleaning and disinfection of machines, equipments, tools, installations, process lines and rooms. A considerable effort is to be made to dispose of the used cleaning and disinfection agents in the form of ecologically acceptable waste. In order to use these agents economically and obtain an optimal design of the cleaning and disinfection process the following aspects must be considered.

(a) Type and composition of the dirt.

(b) Shape, material and surface characteristics of the area to be cleaned.

(c) Requirements type and effectiveness of the cleaning and disinfection agents.

(d) Process of cleaning and disinfection and,

(e) The quality of the water used.

Cleaning is defined as the complete removal of all residues of milk, milk components and other dirt, whereby complete coverage of the cleaned areas with cold water must be achieved.

Disinfection is defined as the complete inactivation of all pathogenic microorganisms and the reduction of all non-pathogenic microorganisms to such a low level, they do not detract from the quality of the products. The dirt encountered in the dairies consists mainly of adhering products and product particles made of milk components such as fat, protein and minerals. Lactose plays only a secondary role. The different concentrations of these components in the product residues and their different adhesion characteristics and build up on surface must be considered. Any thing from a soft deposit to burnt on stone like deposits may be encountered. The extent of and problems in cleaning depend on factors such as whether there has been a heat or acid coagulation of the protein residues and the fat content of the dirt. Almost all kinds of dirt have very high level of harmful gems such as coliform bacteria, spoilage agents etc. Other types of dirt are organic and inorganic dirt particles such as sand and dust, which have found their way into the dairy in different ways.

Requirements

The general requirements for the cleaning and disinfection agents are:

(a) Assurance of the required cleaning and disinfection effect.

(b) Compliance with physiological and toxicological requirements and

(c) Approval and certification.

No agent can meet all cleaning and disinfection requirements, different reagents have been developed and used for different areas of application. Cleaning and disinfection agents can be classified into two broad groups according to their mode of action.

(a) Mechanical means

(b) Chemical means

Fats are emulsified by alkaline cleaning agents and turn into small amounts of soap which makes them easily transportable, and on the other hand soluble components such as soap and glycerine are also formed, due to the low concentration of lye the soap reaction is not pronounced. The emulsifying action is increased by the surfactants. Stone like deposits of milk minerals and salts of the water are best dissolved by acidic reagents. Peptization is the reconversion of a gel into a colloidal solution or a solution. The peptization of proteins, emulsification and soaping of fat and removal of stone like deposits are best supported by mechanical means. The cleaning can be separated into six phases.

(a) Contact phase - cleaning solution comes into contact with the surface.

(b) Coating phase - the surface active components coat the dirt. In this phase the temperature of the cleaning agent are decisive.

(c) Penetration phase - Particles penetrate into the dirt layer and larger agglomerates are formed.

(d) Dispersion phase - Particles are finely distributed, some are already solubilized in ion form.

(e) Emulsion phase - Insoluble compounds are emulsified and kept floating the extent of the removal depends on hydrodynamic influences (rinsing action).

(f) Post cleaning phase - Displaced compounds are carried away and should not sediment.

Cleaning Speed

It indicates the amount of dirt removed from the surface per unit time and is important for the cleaning process and is influenced by:

(a) Temperature of the cleaning solution.
(b) Type and concentration of the cleaning solution.
(c) Type of dirt.
(d) Mechanical energy applied which includes flow velocity, turbulence and impact pressure.

The disinfection agents that are generally used for the inactivation of the microorganisms can be classified into physical and chemical.

Mode of Action

Heat as a basic agent for the inactivation of microorganisms is applied to the disinfection areas in the form of hot water, steam or hot air.

15.1 Cleaning and Disinfection Processes

These processes are based on the principles described earlier and in many cases must be designed, tested and modified *in situ* for the specific situation and requirements. However, the following principles remain the same for the basic operations and can be modified on a case to case basis. The basic operations have been presented in the flow diagram 15.1.

Flow Diagram 15.1 Basic operations of cleaning process

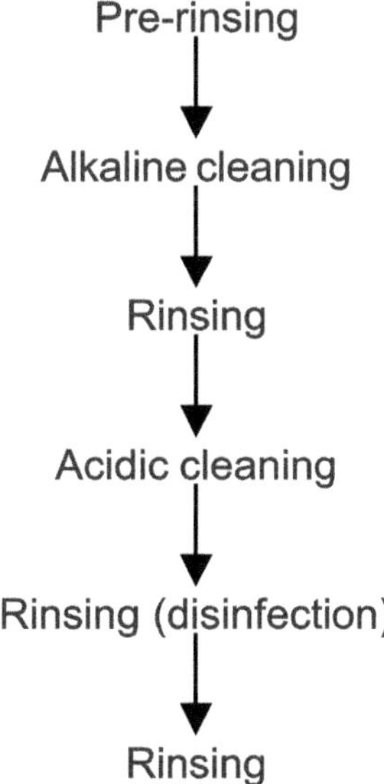

15.1.1 Check on Cleaning Efficiency

The evaluation criteria for cleaning and disinfection efficiency has been presented in the table 15.1.

Table 15.1 Evaluation criteria for cleaning and disinfection efficiency

Cleaning Object	Sequence	Means or type	Period Min	Temperature in °C	Remarks
Heat ex-changer and stainless steel pipelines in the circuit	Rinsing	Water	Until water is clear	15-30	The velocity of the cleaning solution must be double that of milk. After rinsing, the plate heat exchanger can be opened to relieve internal pressure
	Cleaning	CP or CDPI	30-40	70-75	
	Rinsing	Water	Until absence of alkali	70-75	
	Cleaning	CAP		70-80	
	Rinsing	water	20-30 until absence of acid	10-20	
	Disinfection (before start-up)	Water	10	74	
Separator disks of non-self-cleaning centrifuges with washing machines	Cleaning	CP	20-30	60-100	The velocity of the cleaning solution must be double that of milk. After rinsing, the plate heat exchanger can be opened to relieve internal pressure
	Rinsing	Water	Until absence of alkali	60-80	
	Drying	Air			
Continuous buttering machine circuit	Rinsing	Steam		>100	Dissolution of butter
	Rinsing	Water	Until clear discharge	50-70	Residues Rinsing of the water butter oil mix
	Cleaning	CP or CAP	20-30	40-50	
	Rinsing	Water	Until absence of alkali	10-20	
	Disinfection	Water	10	75	Before start-up
Storage tanks	Rinsing	Water		10-20	
	Cleaning	CDP	10-20	50-60	
	Rinsing	Water	Until absence of alkali	10-15	
	Disinfection	Steam	10	>100	
Floors, tiles	Cleaning	CP brush or scraper	Until dirt is removed	40-50	
	Rinsing	Water	Until absence of alkali	10-20	

Source : Dairy Science & Technology-Pieter Walstra et.al

Note : CP: Alkaline reagents for general and manual cleaning in the dairy, cleaning of transport and storage installations as well as process lines.

CAP: Cleaning agents for acid cleaning of milk processing lines and transport and storage processing lines and transport and storage facilities.

CDP: Cleaning and disinfection agents for milk processing lines and transport and storage facilities.

The evaluation criteria for the cleaning and disinfection efficiency based on the microbial population has been given in table15.2.

Table 15.2 Evaluation Criteria for the cleaning and disinfection efficiency

Qualitative evaluation	Maximum tolerance in count/cm	Coliform bacteria in/ml of test solution
Good	< 5	Negative
Acceptable	6-20	Negative
Not acceptable	720	Positive

Source : Dairy Science & Technology-Pieter Walstra et.al

The CIP process may also be called circulation cleaning and in many cases it is an automatic operation. The following requirements must be complied with:

(a) Utilization of a control system, separate from the production control system.

(b) Utilization of valves that prevent blending of product with cleaning solutions.

(c) Design and conception of plants in such a way that all parts in contact with product are reached and covered by the cleaning solution.

(d) Corrosion-resistant material for all parts in contact with the product.

This process is applied, where there are circuits eg : storage tanks, process lines, milk production, butter section, cheese section, milk drying and others.

The automatic cleaning process and its phases are controlled from a control panel. According to the required cleaning protocol a programme can be started, which executes the cleaning steps/as indicated in the table 15.3.

Table 15.3 Cleaning Programme

Cleaning step	Temperature in °C	Length of operation in min.	
		Possible	Optimal
1. Pre-rinsing	30-40	1-15	3 - 5
2. Discharge	1-10	2-3	
3. Caustic cleaning	70 - 90	1 -60	15-20
4. Discharge		1 -10	2-3
5. Rinsing	30-40	1-15	3-5
6. Discharge		1-10	2-3
7. Acid cleaning	60-80	1-45	15-20
8. Discharge		1-10	2-3
9. Rinsing	30 - 40	1-15	3-4
10. Dishcarge		1-10	2-3

Source : Dairy Science & Technology-Pieter Walstra et.al

15.2 Water for Dairy Industry

Water is used in very huge volume in various diverse purposes/functions in the dairy industry such as chilling (refrigeration-brine tanks) heating (steam generation), reconstitution of dried milks, cleaning of equipment, premises etc for sanitizing and rinsing and finally the effluent treatment and disposal. In fact, the easy availability of sufficiently good quality water is one of the prime considerations while choosing a dairy site. The water that is being used in a dairy should be of a very good quality both chemically and bacteriologically and should be free from organic and suspended matter and should not contain any toxic or corrosive substances including harmful microorganisms. To ensure that these requirements are met, water supply to a dairy plant is subjected to various treatments as detailed below.

15.2.1 Coagulation

In this method, the finely suspended material in the form of colloidal suspension does not settle on standing and has to be precipitated with a flocculating agent like alum. Aluminium sulphate alone or in combination with lime sodium aluminate, alumino, ferric or iron salts are commonly used. Alum or aluminium sulphate reacts with calcium biocarbonate.

$$Al_2(S0_4)_3 + 3\ Ca\ (HCO_3)_2 \longrightarrow 2\ Al\ (OH)_3 \text{ -- } 3\ CaSO_4 + 6\ CO_2$$

Along with Soda ash or lime the reactions are:

$$Al_2\ (SO_4)_3 + 3\ Ca\ (OH)_2 \longrightarrow 2\ Al\ (OH)_3 + 3\ Ca\ SO_4$$

$$Al_2\ (SO_4)_3 + 3\ Ma_2\ CO_3 + 3\ H_2O \longrightarrow 2\ Al\ (OH)_3 + 3\ Na_2SO_4 + 3CO_2$$

The flocculent precipitate of aluminium hydroxide entrains or adsorbs the suspended matter in the water. Silica solutions when used in combination with other coagulants increases the rate of coagulation. The coagulated impurities in the water can be removed in a setting tank as sludge which is removed from time to time. The water is then allowed to pass through the sand filter during which most of the suspended impurities are removed.

15.2.2 Chlorination

Normally the water supply is disinfected by the addition of regulated amounts of chlorine and the efficiency of disinfection depends on a number of factors such as:

(a) Concentration of chlorine

(b) Exposure time

(c) Temperature

(d) pH of the water

(e) Amount of organic matter and

(f) The type of the microorganisms present.

After treatment a residual concentration of 0.2 to 0.4 ppm is normal and sufficient.

The chlorine demand of water is the difference between the chlorine applied and the amount of chlorine either free or combined as chloramines or total available remaining at the end of the contact period. It is measured easily by treating a series of samples of the water with varying known dosages of chlorine or hypochlorite. After the desired contact period, determination of residual chlorine in the sample will demonstrate which of the dosage satisfies the requirement of chlorine demand. The residual chlorine is determined by the ortho toludine (0. Toludine) method. This aromatic compound is oxidised by chlorine, chloramines in an acidic medium to a yellow coloured holoquinone and the intensity of the yellow colour is measured in a "Lovibond Comparator".

15.3 Hardness of Water

The capability of soap to form a stable, coprious foam or lather with water, varies considerably from place to place depending on the type of water. This characteristic feature of water is called hardness, which is caused by divalent metallic cations, which react with soap to form precipitate and with certain anions forming the scales. The cations of calcium and magnesium which are responsible for the hardness of the water must be precipitated by sodium oleate in the soap before a lather can be produced. This requires more soap and also leaves a ring of deposit or precipitate in the container. Most of the hardness that is being encountered normally is due to the presence of clacium and magnesium and other cations like strontium, ferrous and manganese ions.

Hard water reduces the efficiency of cleaning and the economy of the detergents. It also forms scales on boilers, thus reducing their efficiency considerably and in some extreme cases it acts as a corrosive also. Therefore, the hardness of water must be determined and the remedial measures/steps are to be taken to soften it to an acceptable degree before its usage. The hardness of water does not pose any health hazard and hard water has not been implicated in any problem related to health. Water is often classified according to its degree of hardness; for uniformity of expression hardness of water is always represented as milligram per litre (mg/l) or ppm of clacium carbonate, whatever, may be the cation or anion, responsible for the hardness. Generally accepted classification of water according to its hardness is as follows:

Soft	-	Hardness less than 50 ppm
Moderately hard	-	50-150 ppm
Hard	-	150-300 ppm
Very hard	-	Hardness more than 300 ppm

The salts causing the hardness determine the steps to be adopted for softening or the removal of hardness. Based on this consideration water hardness is categorised as either temporary or permanent hardness.

15.3.1 Temporary Hardness

Temporary hardness in water arises mainly due to the presence of bicarbonate of calcium and magnesium. The alkalinity of natural water is due to hydroxides, carbonate or bicarbonates, when the pH of the water is less than 8.3 the alkalinity generally consists and due mainly to the bicarbonates. If the total alkalinity is more than or equal to the amount of calcium and magnesium present then the hardness is temporary because the bicarbonates decompose to CO_2 on boiling.

$$Ca\,(HCO)_3 \longrightarrow Ca\,CO_3 + H_2O + CO_2$$

$$Mg\,(HCO_3)_2 \longrightarrow Mg(OH)_2 + 2CO_2$$

The solubility of $CaCO_3$ is 14 mg/litre in cold and 19 mg/litre in hot water, whereas the $Mg(OH)_2$ is soluble up to 9 mg/litre in cold and 4.0 mg/litre the hot water as compared to the calcium. Therefore, when the water is boiled and cooled $CaCO_3$ and $Mg\,(OH)_2$ are precipitated and the temporary hardness is removed. The carbon dioxide released is capable of forming carbonic acid with water.

$$CO_2 + H_2O \longrightarrow H_2\,CO_3$$

Although carbonic acid is a very weak acid and is present in a very dilute form, it is capable of causing corrosion due to continuous slow action.

15.3.2 Permanent Hardness

Hardness which can not be removed by boiling is called permanent hardness. It is caused by the presence of chlorides or sulphates and other radicals of calcium and magnesium. Here the total alkalinity is less than the calcium and magnesium in terms of equivalent. When hard water is used in the boiler the salts causing hardness mainly the sulphates get concentrated and eventually crystallize out to form a thick, hard scale on the surface of the boiler coils which is called scale formation and the boilers have to be descaled periodically.

15.3.3 Estimation of Hardness

It is essential to know the amount of calcium and magnesium present in water in order to take measures for its removal. The hardness due to divalent ions like Ca++, Mg++ etc is determined by making use of a strong chelating agent like EDTA which forms stable complexes. In an alkaline buffer medium these cations form weak complexes with the indicator Erochrome Black T giving a winered colour. On titration with EDTA, which forms much stronger complexes with the cations and at the end point reverts the indicator to its original blue colour. Hardness is always expressed in terms of parts per million of $CaCO_3$, irrespective of the salts causing hardness to the water.

15.3.4 Treatment of Hard Water

The process of removing the hardness from water is called softening. This step or measure is essential to increase the efficiency and economy of dairy processor. Scale formation in the boiler is a persistent serious problem leading to loss of efficiency in heat exchange and requires costly descaling operations from time to time. Similarly, the foaming capability of the detergents is very much reduced, due to the formation of flocculation. More soap or detergent is required to give the same cleaning efficiency and these problems can be removed/solved by softening the water before use.

15.4 Removing Temporary Hardness

15.4.1 Lime Soda Process

The calcium salts are precipitated as calcium carbonate and the magnesium salts as magnesium hydroxide, when the hydrated lime and sodium carbonate are added to the water in sufficient amounts. This process is more efficient in the hot than in cold. At high temperature of about 70 °C the hardness can be reduced to very low levels. The amount of reagents to be added is determined by the hardness and the volume of water to be treated. The reagents are suspended in water as milk for a period of 2-3 hours, for the precipitate to settle before filtration. Due to this treatment, the calcium and magnesium in water are replaced by equivalent quantities of sodium. Hence, the amount of sodium present in the water after treatment will depend on its initial hardness. The process of precipitation can be facilitated or speeded up by the addition of preformed slurry seed.

15.4.2 Treatment with Ion Exchange Resins

Zeolites are Ion-exchange resins, which are capable of interchanging the cations or anions with the salts in the water. Commercially used zeolites comprise of natural glauconite, treated clays, or synthetic resins, prepared from sodium aluminate or sodium silicate. Base exchange zeolites substitute the calcium and magnesium from water with sodium. The zeolite is packed in a column or bed through which the water to be treated is allowed to pass at a controlled rate. The hardness of the water is almost completely removed and the treated water can be used for almost all the purposes. In the course of time all the sodium in the resin is completely depleted and substituted by calcium and magnesium from the water. At this stage the zeolite column can be regenerated for further use by treatment with sodium chloride or brine solution. The calcium and magnesium are again replaced with sodium and the column is ready for its usage again.

The ion exchange process offers several advantages. The plant occupies comparatively less space and the softened water with not more than 10 ppm hardness can be obtained, since this procedure involves anion-cation exchange, there is no sedimentation and hence no sludge removal is involved. An optimum size column can be installed to deliver sufficient softened water at a steady flow rate. The zeolites can be regenerated and used repeatedly thus making the process more economical. The most important factor to be borne in mind in this process is that the incoming water should be free from colloidal and suspended matter as these adversely affect the efficiency of the Zeolite.

The primary requisite for the production of good quality milk is the cleanliness and sterility of the equipment and implies the removal of "soil" from the surface of each machine used. Cleaning and sterilisation are complimentary processes very often used at random to clean the residue of milk and to reduce the bacterial numbers as much as possible. Cleaning process in the dairy industry normally comprise the following stages.

1. Pre-rinsing with cold tap water to remove most of the loose soil and to wet the surface.
2. Removal of soil from the surface by solution emulsification, saponification or mechanical action or by a combination of all these methods.
3. Dispersion of the undissolved soiling matter.
4. Removal of the used detergent solution together with the suspended and dissolved soil.
5. Final rinsing to remove the last traces of detergent.

Different types of detergents are available but the type of detergent to be used and its concentration depends on the specific requirement and method of washing. Broadly all detergents are either organic or inorganic in nature. The type and concentration of the detergents used largely depends on the nature of the soil to be removed and the method of cleaning. The soil deposited on the surface of the equipment varies with the processing operation involved. In the dairy industry and in general the soil could be in the form of an air dried milk film or heat precipitated film or hardened film or in the form of milk stones and foreign matter. The soils essentially consists of the milk constituents and could be rich in fat protein or minerals in varying amounts depending upon the processing operations involved and the type of equipment used.

15.5 Milk Stone

It is an accumulation of dried milk solids and salts from hard water and washing solutions. It consists largely of calcium phosphate, milk protein, precipitated, coagulated and baked on by heat and insoluble calcium salts from water and washing solutions. The typical chemical composition of some common type of the milk soils have been presented in the table 15.4.

Table 15.4 Chemical composition of typical milk soils

Types of Soil				
Component	**Cold milk**	**Hot Milk swept surface**	**Tubular heater**	**Milk stone**
Lactose	38.11	--	--	--
Fat	29.9	43.0	23.0	3.6 - 17.7
Protein	26.6	41.1	30.3	4.4 - 43.8
Ash	5.3	11.9	46.6	42.0 - 67.3

Source : Text Book of Dairy Chemistry - M.P. Mathur, D.D. Roy & P. Dinakar

The strength and type of detergent used is determined by the solubility and nature of the soils. Sugars are water soluble and easily cleaned unless they undergo caramelisation on heating. Fats are insoluble but can be cleaned with alkaline detergents. Proteins are insoluble in water and require acid or alkali cleaning but are difficult to remove when they are denatured due to heating. The proportion of these constituents in the soil varies, therefore the type of detergent used should also be changed according to the specific need.

15.6 Methods of Cleaning

Three methods of cleaning are generally employed in the dairy industry namely.

1. Manual or hand cleaning
2. Semi-mechanical and
3. Mechanical cleaning

Removal of soil is achieved by rubbing the surface with suitable detergent solution in the manual cleaning. Normally a detergent concentration of 0.25 to 1.0 percent at 40 - 50 °C is used.

In the semi-mechanical cleaning process, bottle, cans, crates, etc are placed in the water containing a system of detergents revolving brushes, water rinsing etc.

Cleaning in place (CIP) or In place cleaning is a completely mechanical method, where all parts of the equipment can be thoroughly cleaned without dismantling. In this method there is no human contact is involved hence more stronger detergent and higher temperatures can be employed. Further high pressure jets or sprays can be used for the effective removal of the soils. The concentration/strengths of the detergent and the temperature of cleaning etc are determined by the method adopted for cleaning and the specific need of the equipment or situation.

15.7 Classification of Detergents

Detergents are generally classified as alkalies, acids, sequesting agents and surface active agents. These are the main class of detergents generally used and comprise caustic soda, soda ash, trisodium phosphate, and sodium meta silicate. Most of these detergents have specific characteristics and the requirements. Caustic soda is a strong alkali with excellent dissolving, saponifying and peptizing property. It is a poor deflocculating, emulsifying and wetting agent and is highly corrosive to metals and is avoided in cases of hand cleaning.

Sodium carbonate or washing soda is the most common and inexpensive ingredient used in the dairy detergents. It has a fair deflocculating and emulsifying action and is a good emulsifier. It maintains its pH through good buffering action and is employed extensively in solution used over long period like bottle washer etc. It is not harsh on the skin or hand and can be handled safely. When sodium Carbonate is used with hard water, it forms precipitate of calcium carbonate which promotes the formation of stone and soils. This draw back can be overcome by using soda ash in combination with sequesting agents which bind up the hardness in water.

Trisodium phosphate (Na_3PO_4) is readily soluble and has deflocculating and emulsifying power. It has a fair water softener and forms insoluble phosphates with calcium and magnesium. It is corrosive towards tin but this can be countered by the presence of sodium metasilicate. It is relatively inexpensive and is widely used in the formulations of detergents. Sodium metasilicate is strongly alkaline but not as corrosive as caustic soda. It protects the metals from the corrosive action of other alkalies and has an excellent deflocculating and emulsifying properties and also fairly good water softener. The silicate formed with calcium and magnesium in hard water are insoluble and flocculant and can be removed easily. The deflocculating property of metasilicate keeps the soil in suspension during washing and thus facilitates proper cleaning.

Acids

Several organic acid (inorganic) mineral acids are used as detergents which include tartaric, citric, gluconic, acetic, nitric and phosphoric acids, approximately at 0.1 to 0.5 percent concentration. Acids are mainly used in high temperature heat exchange equipment. The acid dissolves the deposits of mineral salts and denatured or heat coagulated proteins which are difficult to remove with regular alkaline detergents. They also prevents the precipitation of water hardening salts like carbonates by maintaining the pH on the acidic side. For optimum efficiency a pH of 2.5 or lower for a 0.1 percent solution is desirable and should not be corrosive on the metals of the equipment.

15.7.1 Sequestering/Chelating Agents

The main function of the sequestering agent is to soften water by combining with calcium and magnesium to form soluble complexes instead of insoluble precipitate. Sequestering agents are usually inorganic linear polyphosphate like sodium tetraphosphate, sodium hexametaphosphate, sodium tripolyphosphate etc., whose efficiency in binding the calcium or magnesium could be selective and varying and their relative merits have been presented in the table 15.5.

Table 15.5 Relative merits of some common sequestering agents

Chemical	Formula	Sequestering activity towards Calcium	Magnesium	Remarks
Tetra sodium pyrophosphate	$Na_4P_2O_7$	Poor	Very Good	Cheap, stable at high temperature and alkalinity, slow to dissolve
Sodium tri polyphosphate	$Na_5P_3O_{10}$	Good	Good	Soluble in warm water, unstable at high temperature and inert to lower forms in hot alkaline solutions
Sodium tetra Phosphate	$Na_6P_4O_{13}$	Good	Good	Capacity to sequester calcium is decreased in presence of Magnesium
Sodium hexa meta phosphate	$(NaPO_2)$	Excellent	Poor	Expensive, unstable under high temperature and alkaline conditions

Source: Text Book of Dairy Chemistry - M.P. Mathur, D.D. Roy & P. Dinakar

The organic sequestering agents which form ring like structures are termed as chelating agents. The sequestering efficiency of organic chelating agent in hard water is much more than that of polyphosphate. Most of the commonly used chelating agents are sodium salts of ethylene diamine tetra acetate (EDTA). Unlike polyphosphates, these are stable in solution and their efficiency increases with the pH.

15.7.2 Surface Active/Wetting Agents

These are capable of influencing the interfacial conditions and possess varying degrees of emulsifying, dispersing, or wetting property. They improve the wetting and penetration of soil particles and surfaces. They are chemically stable and are effective under acid as well as alkaline conditions and in hard water also. The water soluble portion of the surfactant attaches itself to water molecules and oil soluble portion attaches to the fat molecules thus providing a link between two immiseble compounds. Surfactants are mainly three types - anionic, cationic and non-ionic.

The hydrophilic property of anionic surfactants is derived from the acid, sulphate or sulphonate groups while those of the cationic is provided by the amine groups and is regulated by the length of the carbon chain. Non-ionic surfactants depend upon the terminal hydroxyl group and hydrogen bonding to water molecules for their hydrophilic property.

Eg : Teepol, Acinol, N det-10, common soaps etc.

Another group of surfactants are the amphoteric agents which have multifunctional groups which are cationic in acidic solution and anionic in alkaline solution. Normally a balance between hydrophilic and lypophylic values called HLB is maintained. The HLB represents the varying proportions of water soluble and alcohol groups on numerical basis. Surfactants are considered more effective in removing fatty soils than protein deposits.

15.8 Properties of Detergent

A detergent is expected to possess some or all of the following properties.

1. Wetting or establishment of intimate contact between the surface and cleaning agent.
2. Deflocculation or holding in suspension, removed soils down to micro particle size and preventing them from clumping or coagulating.
3. Emulsification holding un saponified fat and oil in solution.
4. Hydrolysis-The break up of fats, fatty acids, glycerides and soaps in manageable globules and simpler compounds.

Apart from these properties good rinsability and water softening property is also essential.

The Bereau of Indian Standards (BIS) has proposed that a good cleaning or washing agent should have the following properties.

1. Saponifying power is, making soap out of the fatty acid of oils and fats.
2. Quick and complete solubility.
3. Non-corrosive to metal surfaces.
4. Stable during storage.
5. Mould on hands if used for hand washing.
6. Germicidal action.

No single detergent has all the desirable properties, hence a blend or combination of several compounds is often required in their usage for different operations for efficient cleaning with safety.

In addition to these factors, the easy availability and the cost of the cleaning compounds also has to be considered. Depending upon the specific requirements, different detergent combinations are blended or formulated for the various cleaning operations in the dairy industry. The properties of different detergents have been presented in the table15.6.

Table 15.6 Properties of detergents

Property	Strong alkali	Mild alkali	Mild acid	Strong acid	Surfactants
Corrosion	++++	++++	++	++++	0
Sequestening	0	+	0	0	0
Wetting	+	++	+	0	++++
Emulsifying	+	++	+	0	++++
Dissolving	++++	++++	+++	++++	+
Saponifying	++++	+++	0	0	+
Reptizing	+++	+++	++	+++	0
Dispension	++	+++	+	0	+++
Rinsing	++	+++	+	0	++++

Key : Degree of acidity

Extreme	++++
High	+++
Medium	++
Low	+
None	0

Source : Text Book of Dairy Chemistry - M.P. Mathur, D.D. Roy & P. Dinakar

15.8.1 Sanitizers

Destruction of all the pathogens and most of the non-pathogenic micro-organisms is commonly referred to as sanitisation. This destruction of microorganisms can be brought about either by physical means or by the use of chemicals. Some of the commonly used methods of sanitization are discussed below:

15.8.2 Heat

Heat is the most effective and reliable method of sanitisation when both temperature and time are controlled. Heat can be applied either through steam or hot water or hot air. Sanitisation by steam or scalding water is effectively done at 15 PSI for 5 minutes or Zero PSI for 15 minutes. Steam is readily available in the dairy plants and widely used for sanitising the equipments. However, it suffers from certain disadvantages that heat stresses may cause cracks in soldered seams and welds, and cause rapid and extensive damage to rubber parts and hoses. It consumes a lot of energy since steam is dissipated into the atmosphere. It is also quite noisy and increases humidity. Unlike with chemical sanitizers and detergents steam does not require extensive rinsing before use of the equipment.

15.8.3 Hot Water

In small scale dairy establishment where steam generation is not possible, hot water is often used for sanitisation. The equipment either dipped in scalding water or the water is poured over the equipment. The temperature of the water should be as high as possible and should not be less than 85 °C for sanitisation to be effective.

15.8.4 Hot Air

Hot air is an effective sanitising agent but its use is limited due to high cost and difficulty in its application. Hot air cabinets are sometimes used. As a sanitizer, hot air is very effective since it has a high penetrating power and can destroy bacterial spores also. The effect of heat on microorganism is principally on the heat liable proteins within the cells.

15.8.5 Irradiation

Ultra Violet light is very effective in destroying the microorganisms and can be used for sanitisation. But it is not easy to use and hence finds a limited application. However, packaging materials etc., can be exposed to U.V. light before use.

15.8.6 Chemical Sanitisation

A large number of chemicals exhibit bactericidal action and are used as sanitizers. Prominent among these are the halogen bearing compounds like chlorine, iodine, or bromine compounds, quaternary ammonium compounds and others.

15.8.7 Chlorine Compounds

Free or available chlorine is the active agent responsible for the bactericidal action of chlorine compounds. The free chlorine forms hypochlorous acid (HOCL) in solution which destroys microorganism by protein denaturation and enzyme inactivation. For smooth metal surfaces a one minute contact with a solution containing 200 ppm of available chlorine or a 7 minute contact with a 25 ppm chlorine solution is sufficient for sanitisation. Sufficient contact time should be allowed for proper concentration of solution should be used for effective sanitization. Increasing the contact time or the concentration beyond reasonable limits serves no practical purpose. The properties of commonly available chlorine sanitizers have been summarised in the table 15.7.

Table 15.7 properties of chlorine sanitizers

Compounds	Solubility	Percentage available chlorine	Remarks
Inorganic			
Gaseous Chlorine	0.716 g/100 g	100	Difficult to handle, regulate or use. Limited application
Calcium hypochlorite	6.9 g	100	Forms calcium precipitate, easily available as bleaching powder, widely used.
Sodium hypochlorite	15 g	100	Liquid, easy to use directly, leaves no sediment, widely used, likely to loose strength on extended storage.
Organic			
Chloramine T	1.2	25	Organic sanitizers, lower solubility with varying percentage of available chlorine
Dichloro Cyanuric acid	2.6	70	Slower rate of chlorine release through hydrolysis
Trichloro cyanuric acid	1.2	90	Slower rate of chlorine release through hydrolysis
Dichlorodimethyl hydration		66	Not very freely available nor as widely utilised

Source : Text Book of Dairy Chemistry - M.P. Mathur, D.D. Roy & P. Dinakar

15.8.8 Iodine Compounds

Iodine is an excellent bactericide but its corrosiveness, toxicity and low solubility makes it unsuitable for use in the food industry. It is less effective against bacterial spores and phages than chlorine and may also cause 'off' flavours apart from causing discolouration of the equipment. These difficulties have been overcome in the recent products called "Iodophores" in which the iodine is loosely combined with a suitable non-ionic wetting agent which acts as a carrier. Commonly the phosphoric acid is also added to increase the germicidal activity probably by the decrease in pH to increase stability and to prevent the formation of milk stone. The carrier or the wetting agent acts as a solubilising medicines for the iodine. In this combination the germicidal activity of iodine is enhanced and the decreased vapour pressure increases the stability. Iodophores reduce the characteristic iodine odour, prevents permanent staining and corrosive action. It also retains its germicidal activity in hard water and similar to chlorine, the sanitising action of iodine compounds is related to the free iodine which form HI_3 which further produces IO_3^- and H^+ ions.

15.9 Quaternary Ammonium Compounds (QAC)

These are cationic compounds which are finding an increasing application in the dairy industry. Out of the several hundreds of QAC available, only a few have the distinct properties. The most common QAC used is Cetyl Trimethyl Ammonium Bromide (CTAB) whose structural formula is as under :

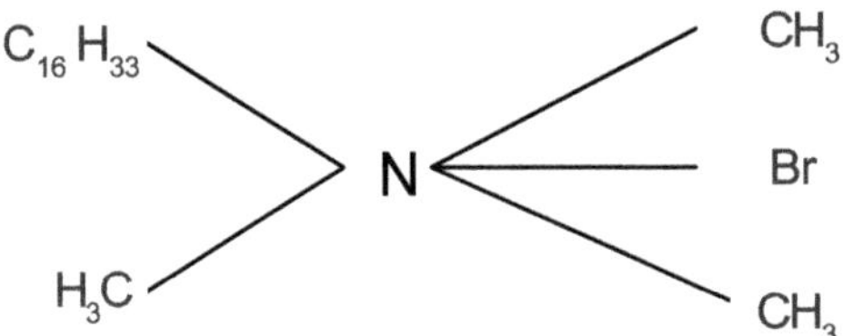

These compounds are colourless, odourless, tasteless and non-corrosive, making them ideal for use as disinfectants. They are highly surface active and reduce the surface and interfacial tension considerably. The action of QAC is different from that of the halogen compounds. The cationic part of the molecule is hydrophobic and tends to form a layer on the surface of the bacterial cell wall and ultimately leads to its destruction. The cationic portion concentrates the disinfectant at the required points. The action of QAC is strong and lasts for a longer period of time after treatment. It is natural that the QAC are inhibited by certain anionic detergents because they are cationic in nature. Soda ash is usually compatible while the common wetting agent inhibit QAC action and they are more effective against the gram positive than gram negative organisms especially in low concentration.

15.10 Procedure of Cleaning and Sanitizing

Principle

In the course of selection of any particular detergent due consideration must be given to type of soil, quality of water supply, material of surface and the type of equipment which is to be cleaned and the method of cleaning, namely soaking, brushing, spraying and/or recirculation. Detergents are invariably used as an aqueous solution. In the selection of dairy sanitizers, the following considerations are to be borne in mind.

(a) ***High Temperature Sanitizing***: The main advantages are penetrating ability and quick drying of the equipment. Heat is the most reliable sanitizer, especially when both temperature and time are controlled. Thus effective sanitization can be done by steam at 15 PSI for 5 minutes or Zero PSI for 15 minutes or scalding water at 90-95 °C for 10 minutes.

(b) ***Low temperature Sanitizing***:

The main advantages of this method are:

1. Permits sanitizing immediately before the equipment is used, because the hot equipment is injurious to the quality of milk and milk products.
2. Avoids excessive strain on equipment such as ice cream freezers and
3. Permits flushing out of equipment immediately before use, thereby removing any possibility of the dust that might have entered. Generally Chlorine solution at 15-20 °C containing 150-200 ppm of available chlorine is used for a contact time of 1-2 minutes.

The usual procedure for cleaning and sanitization of major items of dairy equipments should consists of the following:

1. **Draining:** It is carried out to remove any residual loose milk and any other extraneous matter.
2. **Pre-rinsing:** This is done with cold or tepid water in order to remove as much milk residue and other matter as possible.
3. **Warm to hot detergent washing:** This is done with detergent solution of 0.15 to 0.60 percent alkalinity, in order to remove the remaining milk solids.
4. **Hot water rinsing:** This is carried out to remove the traces of detergents that have been used previously.
5. **Sanitizing:** It is carried out to destroy all the pathogens and all most all non pathogens. Generally it is carried out just before the usage of the equipment to ensure more safety.
6. **Draining and Drying:** This will help to prevent the bacterial growth and corrossion of the equipment. Drying is readily accomplished by heat and ventilation. Never use any cloth or towel of any kind. Drying is not necessary, when the equipment is refilled with the dairy product immediately.

At pH 5.0 it is entirely present as $HOCl$ but at pH 10.0 it is almost entirely OCl. Organic compounds hydrolyse to produce one or more molecules of $HOCl$. The rate of hydrolysis is dependant on pH, being slower at high pH values. The release of Cl_2 from organic compounds is generally slower than from in organic sanitizers.

The selection and precautions in the usage of detergents and sanitizers for different surface materials of dairy equipment and the choice of detergents for general purpose or organised dairies have been given in the tables 15.8 and 15.9.

Table 15.8 Selection of detergents and Sanitizers

Material	Cleaning	Sanitization
Stainless steel	All alkalis may be used. Care should be taken with acids	All sanitizers may be used.
Mild steel	All alkalis may be used. Acids should be used together with inhibitors	- do -
Tinned steel/ copper	Weak alkalis, together with sodium sulphite as inhibitor, should be used.	All sanitizers may be used.
Bronze	- do -	- do -
Galvanized	- do -	- do -
Aluminium alloy	Weak alkalis, together with sodium silicate as inhibitor, should be used.	- do -
Glass	All alkalis and acids may be used	- do -
Vitreous enamel	Weak alkalis, together with sodium silicate as inhibitor, should be used.	- do -
Plastics	Cleaning temperatures should not be above the softening point of plastic	Only chemical sanitizes should be used.
Rubber	Strong alkalis should be used to remove any fatty material struck to the surface.	- do -

Source : Text Book of Dairy Chemistry - M.P. Mathur, D.D. Roy & P. Dinakar

Table 15.9 Choice of detergents generally used in organized dairies

S.No.	Ingredients	Quantity	Remarks
1.	Tri-sodium Phosphate Wetting agent	850 g 150 g.	For general use
2.	Tri-sodium phosphate Sodium meta-silicate Wetting agent	650 g. 200 g. 100 g.	For aluminium utensils
3.	Tri-sodium phosphate Sodium sulphite Wetting agent	750 g. 100 g. 150 g.	For tinned utensils

Source : Outlines of Dairy Technology - S.K. De

15.11 Methods

15.11.1 Hand Washing

(A) The normal cleaning and sanitization of hand washed dairy equipment in organised dairies should be carried out as detailed below:

1. Prepare 0.8 to 1.0 percent of the detergent mixture in tap water so as to give a minimum alkalinity of 0.5 percent (pH over 11.0) in a wash up tank and maintain the temperature at about 50 °C.
2. Thoroughly rinse the utensils with clean cold water or tepid water in winter.
3. Introduce the detergent solution into the equipment, quantity of solution to be determined by requirement and experience. Thoroughly brush the equipment surface inside and outside with a clean brush.
4. Wash the utensils with enough fresh cold water (tepid water in winter) using a clean brush again if needed to remove all the traces of the detergent.
5. Allow the equipment to drain thoroughly and let it dry for at least one to two hours.
6. Sanitize the equipment surface by steam or hot water after cleaning and/or by rinsing with chlorine solution (200 ppm available chlorine) just before using.

Precautions

(i) Use rubber gloves to avoid skin injury from detergent action.

(ii) Two compartment wash-up tank, one for warm detergent solution and the other for hot water with a drain outlet for each tank together with a steam jet/ chest is helpful.

(iii) Use fresh detergent and sanitizing solution every time and discard the previously used solutions.

(B) Bottles may be hand washed as follows:

The operator uses either a hand brush or a motor driven brush and a mild alkaline solution, which will not be injurious to the hand. A three compartment tank with a drainage outlet for each is helpful and table space at both ends should also be provided. Two thirds of the first compartment is filled with water at 50-55 °C containing alkali detergent. The second compartment is filled similarly with water only at 50-55 °C. The third compartment contains enough clean and cold water with 150-200 ppm available chlorine. The drained bottles are first put into the first compartment and allowed to soak there for a few minutes, then brushed with a clean bottle brush both inside and outside. After brushing, the bottles are placed in the second compartment, where they remain until all the bottles in the batch are washed. Then a new batch of bottles are placed in the first compartment to be soaked, and the bottles in the second compartment after carefully emptying are placed in the third compartment and the bottles remain here while the second batch bottles are brushed and transferred from first to the second compartment. The crates are also carefully cleaned inside and outside and the bottles from the third compartment kept up side down in the crates to be drained and dried. Use rubber gloves to avoid skin injury from the action of the detergent and also as a protection against the broken bottles. Use fresh detergent and sanitizing solutions when needed.

15.11.2 Mechanical Washing

This consists of mainly can and bottle washing:

(A) Can Washing: Cans may be cleaned and sanitized either manually or mechanically. The mechanical can washer may be of either the Rotary or straight through/Tunnel type. The rotary can washer is used in small plants. In this the loading and unloading are done at the same point, and the cans move in a circular manner. The capacity usually ranges from 2-6 cans and lids per minute.

Advantages

1. It occupies little space.
2. Machine can be operated by a single worker.

The straight through type is used in bigger plants; in this loading is done at one end and unloading at the other and the cans move in a straight line. The capacity usually ranges from 4-12 cans and lids per minute. It has greater capacity.

The cleaning and sanitization procedure for mechanical can washing (can and lid) consists of the following stages:

1. Drainage stage for liquid milk residues.
2. Pump-fed pre-rinsing with cold or luke warm water.
3. Drainage stage.
4. Pump-fed jetting with detergent at not less than 70 °C
5. Drainage stage.
6. Rinsing stage - pump fed or by steam and water ejector at not less than 88 °C.
7. Final fresh water rinsing with steam and water ejector at 88-93 °C
8. Live steam injection.
9. Hot air drying at 95-115 °C

Note :

(a) Detergent mixture used should be suitable for the can metal. Not more than 0.5 percent alkalinity desirable.

(b) Sanitization with chlorine is not recommended for tinned milk cans since it attacks tin surfaces especially if left in contact with the tin for longer periods of time.

(B) Bottle Washing: The mechanical bottle washing may either be a soaker (soaking) or Hydro (Jetting) or Soaker-Hydro (Partly soaking and partly jetting). It may also be of the come back or straight through type. In the former the loading and unloading take place at the same end, whereas in the later they are done at the opposite ends. Generally soaker hydro-come back types are popular in small capacities. In all the machines, bottles are loaded manually or semi-automatically with manual assistance and are discharged automatically on to the one or more bottle conveyors. The various stages of treatment in a mechanical bottle washer are given below:

1. Pre-rinse using water at 32-38 °C
2. Detergent wash usually 1-3 percent caustic soda, together with chelating and wetting agents, given preferably in two stages at different temperatures within 60 to 75 °C sanitizes the bottles as well.
3. **Warm Water rinse:** It removes all the traces of detergent. Reduces the temperature of the bottles for next stage. Water temperature varies from 25 to 45 °C and is usually re-circulated.
4. **Cold Water Rinse:** Normally, re-circulated chlorinated water containing 35 to 50 ppm available chlorine is used to prevent the recontamination of the bottles.
5. **Draining:** It is carried out when the bottles come out of the machine.

Important

Visual inspection of washed bottles is important. Under Indian conditions, this is done by workers examining each bottle through a large magnification lens. In developed dairying countries, an electronic device called Roto tector is used for detection of foreign matter in glass bottles, although it is not satisfactory entirely.

15.11.3 Cleaning in Place (CIP)

Definition

It is also called, In Place Cleaning (IPC) system which refers to that system of cleaning and sanitization solution circulates in a circuit and does not require the dismantling of the dairy equipment everyday for this activity.

Merits

The advantages or merits of the CIP system are as follows:

(a) This system ensures that all equipment receives uniform treatment day after day by eliminating the human factor.

(b) Less damage to the equipment due to daily dismantling and assembling.

(c) There is a saving of about 25 percent or more in the total clean up costs including the man hours.

(d) It reduces the contamination possibility through human error.

(e) The plant can be utilised with the maximum efficiency and also presents a better and clean appearance.

Success Factors

The success of the CIP system depends upon:

(a) Proper selection of pipes and fittings, installation and development of circuits.

(b) Proper temperature of cleaning solution.

(c) Adequate velocity of cleaning solution.

(d) Use of detergents designed specifically for recirculation, cleaning.

(e) Proper concentration of detergent solution and

(f) Sufficient cleaning time.

Types

There are two types of CIP system in use namely.

1. **Manual Control:** In this method the completion and setting up of the product and CIP circuits is carried out manually where the valves are hand operated and the entire process is controlled by the plant operator and the use of key pieces are recommended for safety.
2. **Automation :** This method consists of three levels and broadly speaking they are.
   - (a) **Low level:** Setting up of CIP and product circuits is carried out automatically.
   - (b) **Medium level:** Setting up of CIP and product circuits as well as the different types of treatment are all controlled automatically.
   - (c) **High level:** Use of computer for complete control of the entire product manufacture and CIP system in large plants.

Procedure:

(A) HTST Pasteurizer

The CIP method of cleaning and sanitization of HTST pasteurizer is as follows:

(a) **Pre-rinse:** It is carried out with cold or tepid water till the water discharge becomes clear.

(b) **Acid rinse:** Rinsing with acid mainly phosphoric acid solution of 0.15 to 0.60 percent acidity re-circulated at 65 to 11 °C for 20 to 30 minutes. Wetting agent may be added to increase the efficiency of cleaning.

(c) Drain out the acid solution.

(d) Hot water rinse with water at 65-71°C for 5 to 7 minutes and drain the rinse water.

(e) **Alkali Rinse:** Rinsing with alkali detergent solution of 0.15 to 0.60 percent alkalinity re-circulated at 65-71°C for 20-30 minutes wetting agent may be added to increase the efficiency of cleaning.

(f) Drain out alkali solution.

(g) Final hot water rinse. Rinsing with hot water at 71-82 °C till the whole system has been heated. Rinse water must be drained out, then slightly loosen the plates for drainage and drying.

Important

1. Nitric acid may be used with stainless steel plates.
2. At regular intervals, the equipment may be dismantled for thorough cleaning including the gaskets and inspection of all milk contact surfaces.

B. Milk storage tank/Milk tankers

The method of cleaning and sanitizing the milk storage tanks and milk tankers by the CIP method is as follows. Especially designed spray arms and nozzles either turbine or ball spray varieties are generally used to ensure uniform spraying of detergent and sanitizing solutions over the entire surface area.

1. Pre-rinse with tap water.
2. Drain completely for 3 to 5 minutes.
3. Hot detergent wash with sodium hydroxide solution consisting of sodium hydroxide 90 parts, sodium thiosulphate 9 parts and wetting agent 1 part of 0.35 to 0.50 percent strength at 71 °C for 15 to 20 minutes. Once or twice a week an-acid-alkali programme may be used. The acid may be phosphoric or nitric and it should be followed by alkali as mentioned above.
4. Drain for 3-5 minutes
5. Post rinse with hot water at 65-70 °C
6. Drain completely for 3-5 minutes.
7. Sanitize usually with hot water at 90 °C for 2-3 minutes otherwise chlorine solution at 15-20 °C containing 150-200 ppm of available chlorine for a contact period of 1 to 2 minutes.
8. Drain the contents completely for 3-5 minutes.
9. Hot air blow for 1-2 minutes.

Systems

There are two types of systems prevalent and those are:

1. **Total loss system:** In this method, the detergent solutions are drained out completely after its usage. The advantages of this method are
   (a) Less steam is required for heating the water
   (b) A compact CIP unit and
   (c) Fewer pipelines
2. **Saving Solution System:** In this method the detergent solutions are returned to their respective tanks and the strengths of the solutions however is maintained automatically. The main advantage of this system is the saving of the detergent solutions, thereby reducing the cost of the detergents and thus making it economically more viable.

Note: The system to be adopted/used will depend largely on the hardness of water and the type of soil.

Practical Manual

EXERCISE - 1

The Care and use of Common Volumetric Glassware

Relevant Information

The common volumetric glassware used in a chemical laboratory requires some special techniques to handle them as well as to maintain them for their accuracy and efficient use.

A. Measuring Cylinder or Graduated Cylinder

Measuring cylinder is a volumetric glassware usually used for delivering a certain amount of liquid when great accuracy is not necessary. It is a cylindrical glass jar and it is graduated starting from its bottom up to a height very close to the top end. Usually there is a beak at the top for pouring out the liquid. Some types of measuring cylinders are closed at the top with a ground glass stopper and the cylinder is used for containing a known volume of a liquid. A measuring cylinder can have a capacity ranging from 10 ml to 2 liters.

Procedure to use a Measuring Cylinder

1. Clean thoroughly the measuring cylinder and dry by draining.
2. Rinse the cylinder with the same liquid which is to be poured out.
3. Place the cylinder on a horizontal table-top and fill the cylinder with the liquid till the lower meniscus (upper meniscus for a coloured liquid) is tangential to the desired graduation.
4. Allow sufficient time for the liquid meniscus to become stationary and make final adjustment by adding or pouring out the liquid drop by drop taking utmost care.

5. Pour out the liquid to the container by gently tilting the cylinder and finally inverting it, allowing sufficient time to drain out maximum amount of the liquid.
6. Wash the cylinder after use and allow it to get dry by draining.

Precautions

1. Clean the cylinder thoroughly.
2. Do not heat a measuring cylinder.
3. Do not add either very hot or very cold liquid to the cylinder.
4. Ascertain whether the cylinder is to be used to contain or to deliver and use it accordingly.
5. Do not use a measuring cylinder for making neither a standard solution up to a certain accurate volume nor for delivering a very accurate volume of a liquid. Use it for approximation work.

B. Measuring flasks (also called volumetric flasks)

These volumetric glassware are used either to contain or to deliver a stated volume of a liquid at a certain temperature. The volume may range from 10 ml to 2 liters or more. The level, at which the upper surface of the liquid stands is in the narrow neck of the vessel to ensure accurate readings. Sometimes there are two marks etched on the neck, the lower one indicating the volume to contain and the higher one indicating the volume to deliver. The flask may be slightly closed with ground glass stoppers or rubber stoppers or corks.

Procedure for use of a measuring Flask

1. Pour the liquid into the clean, dry volumetric flask (refer to the procedure for cleaning) with the help of a clean dry glass funnel, the stem of which reaches to the mark on the neck of the flask.
2. Before final adjustment, remove the funnel and add the liquid through a long glass dropper or a pipette till the lower meniscus of the liquid is tangential to the circular mark on the neck of the flask (in case of a deep coloured liquid read the upper meniscus).
3. Close the flask carefully with the stopper and mix the contents thoroughly, if it is a solution or a mixture of liquids, by inverting the flask and shaking the contents.

Precautions

1. Never place any liquid which is either too hot or too cold in a measuring flask.
2. Never heat or cool a volumetric flask (for example, keeping in the refrigerator).
3. In order to prepare a solution which requires heating, prepare the solution in a

beaker and cool it to room temperature, transfer quantitatively into the volumetric flask and finally make the volume up to the mark.

4. Volumetric flasks should be scrupulously clean, one of the criteria of such cleanliness is that water runs smoothly down the wall of the flask and does not adhere to it as droplets.
5. When the use of the volumetric flask is over, clean the flask thoroughly (see general cleaning procedure) and keep it inverted for draining out water, so that it can be used next time readily.

C. Pipette

The pipette is a glass tube used to deliver fixed quantities of liquid. A pipette for delivering a fixed quantity of liquid is known as bulb or simple mark pipette, whereas the one for delivering varying quantities is known as graduated pipette. Pipettes may be graduated for volumes varying from 1 ml to 100 ml. A bulb pipette has only one graduation mark somewhere in the uniform bore portion of the tube above the bulb, whereas the end portion tube on the other side of the bulb is narrowed down to a well ground nozzle. A graduated pipette has graduation marks almost throughout its uniform wide bore cylindrical portion, which at its lower end narrows down to an aperture and at its upper end is drawn to a cylinder of narrow bore.

Procedure to use a pipette

1. Clean thoroughly the pipette (see general procedure for cleaning) and rinse it with the experimental liquid.
2. Dip the lower end of the pipette below the surface of the experimental liquid and apply suction from the upper end of the pipette by placing it in between the lips.
3. When the liquid rises over the top most graduation mark stop suction, close the tube with the middle of the tip of your index finger and take out the pipette from the liquid.
4. Wipe out any adhering liquid from the outer portion of the lower end of the pipette with a piece of clean cloth, absorbant cotton or filter paper.
5. Touch the lower end of the pipette to the inner wall of the vessel containing the liquid from the pipette to run down by adequately releasing the finger pressure till the meaniscus of the liquid is tangential to the graduation mark.
6. Take out the pipette, place the lower end inside the vessel into which the liquid is to be delivered and release the finger pressure so that the liquid slowly runs down.
7. In case of bulb pipette after the liquid has run down, wait for a few seconds more and place the tip of the pipette in contact with the junction of the liquid

surface and the wall of the vessel; some more liquid will run down leaving a small drop inside the tip, gently draw out the pipette, allow the liquid to run down till the liquid meniscus reaches any desired graduation mark. In case the last graduation mark extends to the tip of the lower end blowout the last drop of liquid to the container.

Precautions

1. Do not heat or cool a pipette.
2. Do not draw into a pipette any hot or cold liquid, or any concentrated acid or poisonous solution e.g. potassium cyanide solution.
3. Take care not to draw any liquid much above the top graduation mark and not to allow it to enter your mouth.
4. For a bulb pipette, do not hold around the bulb.
5. For a graduated pipette first ascertain whether the graduation starts from the top or from the bottom.
6. Always close the upper end of the pipette with your index finger and never with the thumb.
7. Take care to keep the tip of the pipette sufficiently below the surface of the liquid while sucking the liquid, this is also ensured by taking sufficiently more liquid in the drawing vessel than the capacity of the pipette.

D. Burette

The burette is an instrument by which a small accurately known amount of liquid is delivered in the process of titration. It is a glass tube of uniform diameter throughout a certain length which is graduated also. The lower end of the tube is terminated after a construction by a glass tap, or a piece of rubber tubing provided with a Mohr's clip and a glass jet through which the liquid is delivered. The upper end of the burette is usually drawn into a rim or into a small funnel to facilitate pouring in a liquid. A burette is generally made to deliver 50 or 25 ml (sometimes even up to 2 ml) of a liquid and it is graduated from top to bottom in divisions of 1 ml along with subdivisions of one-tenths of a milliliter.

Procedure to use a burette

1. Clean the burette thoroughly (see general cleaning procedure) and dry it by draining.
2. Add a very little amount of grease to the wider part of the barrel of the glass tap and to the narrow part of the barrel of the glass tap and to the narrow part of the cavity to which the tap fits in. Rotate the tap so that the grease spreads into a uniformly thin layer.

3. Clamp the burette in a burette stand and see that it stands vertical.
4. Close the glass tap and pour into the burette, with the help of a small clean funnel, the liquid to be used. Take out the burette clamp, wash the burette by inversion and reinversion and run out the liquid by operating the tap. This is known as rinsing. Again clamp the burette, close the tap and add liquid till the burette has been filled about 5 cm above the zero graduation mark.
5. Remove the funnel and open the tap to run out some liquid till the lower meniscus of the liquid coincides tangentially with the zero mark; for coloured liquid read the upper menicus, take care that when the liquid falls it removes all air bubbles and the whole burette length, right up to the tip of the delivery jet, is filled with the liquid.
6. Read the burrette by placing eyes on a level with the liquid meniscus thus avoiding parallax.
7. To deliver any amount of liquid from the burette take the initial reading of the position of the liquid menicus. Open the glass tap manipulating it with the thumb and the index and the middle fingers and protecting the glass tap and the cavity into which it is filled with the palm and the other two fingers. At any desired length of time close the tap and read the position of the liquid meniscus. Difference between the two readings gives the volume of the liquid delivered.

Precautions

1. As far as possible always start delivering a liquid from the zero position of the burette which is always at the top end.
2. Read the burette avoiding parallax preferably against a white background.
3. Use a scrupulously clean burette.
4. Do not use an excess of grease that does not enter into the hole in the barrel of the tap.
5. Allow sufficient time to remove any air bubble from the liquid in the burette.
6. See that the burette is completely filled, from zero graduation to tip of the delivery jet with the liquid and no air gap is present.
7. Do not use any hot or cold liquid in the burrette.
8. Do not dry burette by heating or drawing hot air.
9. Rinse the burette with the liquid to be used at least once before filling in for final use.
10. Do not run out the liquid from a burette at a very fast rate; a speed of 1 ml in 2 seconds is desirable.
11. Always remove the funnel before you adjust the zero level.

General procedure for cleaning the volumetric glassware

1. Fill the vessel completely with a good commercial detergent solution (e.g. teepol) or with a strong alcoholic caustic soda solution. In case of a pipette it is preferable to dip the pipette completely into the detergent solution taken in a sufficiently tall measuring cylinder.
2. Remove the detergent solution from the apparatus after about an hour and wash out the same with running tap water.
3. Fill the apparatus with chromic acid (in case of pipette, place the same in chromic acid) and allow it to stand for some time, preferably overnight.
4. Remove chromic acid and wash thoroughly with running tap water followed by distilled water.
5. Place the glassware in such place and position as to allow them to get dried by draining of the water. In case a dry apparatus is required urgently, dry it by either blowing or drawing in dry cold air or by washing successively with ethanol and diethyl ether.
6. Whenever permitted rinse the volumetric glass ware with the liquid which is to be used for measuring before it is used for the actual measurement.
7. As far as possible clean a volumetric glassware immediately after its use is over.

Exercise - 2

Sampling of Milk/Milk Products for Analysis

Principle

The sampling procedure differs according to the nature of material and the purpose for which it is needed. The sample should be such that it should be truly representative of the bulk.

Relevant Information

Correct sampling of milk and milk products is very important. Errors in sampling will lead to erroneous result. The sample may be required for chemical, bacteriological and physical analysis.

Materials Required

1. Milk
2. Preservatives
   (a) **Formalin -** 36% @ 0.4 ml for every 100 ml of milk
   (b) **Plunger:** It should have sufficient area to produce adequate disturbance of the product, and sufficiently light in weight for the operation for the operator to be able to move them rapidly through the liquid. For buckets and cans, it consists of disc 150 mm in diameter, perforated with six holes each 12.5 mm in diameter or a pitch circle of 100 mm diameter, the disc being fixed centrally to a metal rod, the other end of which forms a loop handle. The length of the rod, including the handle, should be approximately one metre (BIS Specification). For road and rail tankers, the specifications are different.
   (c) **Sampling Dippers :** Sampling dippers shall be fitted with a solid handle, at least 150 mm long. The capacity shall not be less than 80 ml (BIS specification).

(d) **Sampling Tube:** Straight seamless metal tube about 600 mm long, 6 mm inside diameter and about 1.6 mm thick may be used.

(e) **Sample Bottle:** Good quality glass suitable for sterilization, wide mouthed. It shall be provided with well fitting caps or bark corks (for chemical analysis) and glass stoppered (for bacteriological analysis). Capacity varies from 100 ml, 150 ml or 250 ml.

(f) Hot water bath.

(g) Oven, Autoclave etc.

Procedure

Milk churns easily at 26.5°C to 29.5°C and agitation near this temperature shall be avoided.

The different procedure to be followed in collecting milk sample are given below:

1. Milk Sample

(i) ***Sampling of milk from single container:*** The operation involved are mixing and then immediate taking of the sample. Mixing can be done either by using plunger which should be moved up and down in the milk vigorously for about ten times. If it is in a small container, it should be poured from one vessel to the other and shaking it. Care should be taken to avoid fat separation.

(ii) ***Collecting a composite sample from a number of containers*:** Pour all the milk from the different containers into a vat if feasible and then take a single sample. If the cans are of the same diameter the following procedure can be applied:

Mix the milk in each of the cans with the help of a plunger, and then equalize the level of milk in all these cans and withdraw equal quantities from them and put together.

OR

The tube method of sampling can also applied in the above instance. Mix the milk with the help of a plunger. Introduce a sampling tube of uniform bore, which is opened at one end, and having provision for closing the other end. Close the tube can will be remaining in the tube. Transfer this to a container. Repeat the process with the other remaining cans. Collect together in the sample tube.

(iii) ***Sampling bulk units*:** When milk of inform quality is supplied in bulk units (for example, cans filled from storage tanks), the number of random units to be sampled shall be as follows:

Total No.of units	Number of units to be selected
1	1
2 - 5	2
6 - 20	3
21 - 60	4
61 - 100	5
Over 100	5 Plus one for each additional units or fraction thereof

(iv) ***Sampling from storage tanks, rail and road tankers***: Mix the milk thoroughly using a large plunger or a mechanical agitator or by compressed air. When a plunger is used for mixing, then insert the plunger through the main hole and move it forward, downward & backward and each time bring it to the surface. This should be done for 15 minutes. After proper mixing of milk, the sample may be taken from tank.

(v) ***Old Canned milk***: In old canned milk, there is formation of cream line. Break the cream line before sampling. Pour the milk from the can into a vat. Record the temperature of the milk. Adjust the temperature of the milk at 90-104 °F < (40 °C). Mix the milk with the help of an agitator. Take samples at three or more different places from the vat.

(vi) ***Partially Churned milk***: Heat the milk 90°F to 104°F. Mix the milk thoroughly and take representative sample.

2. Milk Product

(i) ***Cream***: Warm the sample to 40-50 °C if the sample is very viscous. Then pour from one vessel to another three to four times to mix well and take a representative sample.

(ii) ***Butter***: Remove three or four cores with a sampler from various parts of the sample. Each core should be extending from the exterior to the center. Keep the combined core in a wide-necked bottle and stopper it. Mix these and analyze.

(iii) ***Cheese***: Insert the sampler from a point on the circumference midway between the top and bottom of the center, by keeping the cheese, on its flattened surface. Remove about 3/4 inch at the rind end of each sample to plug the hole. Transfer the rest to a sample bottle.

(iv) ***Dahi*** : Entire dahi should be rendered homogenous by mixing/stirring. Take representative sample.

Labeling and transport of samples

Each sample container should be sealed airtight after filling and marked with particulars regarding the purpose of sampling, the name of supplier, etc., the date and time of sampling, the nature of bacteriological examination shall be marked distinctively.

Milk samples, which are to be examined for flavour should be protected from light and should not be exposed to odour, which may be absorbed during transport:

It is desirable that samples should be transported on the same day, they are taken. Samples should be stored in refrigerator at a temperature of 0 to 5°C.

Precautions

1. All precautions should be taken to prevent contamination and adulteration.
2. For chemical examination the sampling equipment should be clean and dry and plastic bottles may also used.
3. For bacteriological purposes all equipment including plungers, sample bottles and rubber stoppers should be sterilized and sample should be collected under aseptic conditions. Equipment should be sterilized by one of the following methods.
   (i) Heating in a hot air oven, not less than 2 hours at 160°C
   (ii) Autoclaving for not less than 15 minutes at 120°C.
   (iii) Under field conditions, equipment may be sterilized by immersion in boiling water for at least 5 minutes or 10 minutes for rubber stoppers and used immediately.
4. When samples are collected for bacteriological examination, it is desirable to avoid air space by filling the bottles to the top leaving however, sufficient space for expansion of rubber stopper.
5. Bark corks should not be used for closing milk sample bottles for bacteriological examination.
6. Sample bottles containing milk, which are to be examined for flavour subsequently, shall be closed with greaseproof non-absorbent stoppers.

Questions

1. What is a sample ?
2. What is the purpose of sample ?
3. What is a composite sample ?
4. Write the precautions you will take, while sampling.
5. Draw the figure/diagram of equipments used in sampling and name them.

EXERCISE - 3

Specific Gravity of Milk

The Specific Gravity of Milk may be determined by any one of the following methods:

1. by lactometer
2. by Specific Gravity bottle or Pycno meter and
3. by Westphal balance
4. by Philips method

Principle

The method based upon the law of floatation, which states that when a solid is immersed in a liquid, it is subjected to an upward thrust equal to the weight of the liquid displaced by it and acting vertically upward.;

Relevant Information

Lactometers are used for rapid determination of specific gravity. The lactometer has a scale graduated from 0 to 40 corresponding to specific gravities of 1.000 to 1.040. It is used to detect the adulteration of milk. Since milk has a specific gravity of about 1.030 and that of water is 1.000. Any addition of water will lower the lactometer reading. Skim milk has a specific gravity of 1.036 and hence, its addition will raise the specific gravity. Addition of skim milk and water in the right quantity will give normal lactometer reading. So, a better technique for determining the addition of water is by freezing point method.

The types of lactometer generally used are:

1. Quevenne (60°F)
2. Zeal (60°F or 84°F)
3. ISI / BIS lactometer (27°C)

The temperature of milk should be close to the temperature of which lactometer is calibrated or at the most + 1.0. If it is less or more than the calibration temperature, then correction factor is used.

Correction Factor

Add 0.1 to lactometer reading for each degree above 60°F/84°F and subtract 0.1 0.1 for each degree below 60°F/84°F.

Materials Required: Milk Sample

Apparatus

1. Lactometer (84°F/60°F)
2. Lactometer Jar
3. Thermometer
4. Beaker
5. Petridish

Procedure

1. Adjust the milk temperature between 75°F to 95°F for zeal lactometer (84°F).
2. Pour enough milk (properly sampled) into the plastic jar taking care to avoid the formation of air bubbles, so that some milk overflows when the lactometer is inserted.
3. The lactometer should float freely and should not touch the side of the jar.
4. Allow the lactometer to remain steady in the milk.
5. Take reading, within 30 seconds. Note the reading of the lactometer corresponding to the top of the meniscus of the stem.
6. Note the temperature of the milk simultaneously.
7. Apply correction factor, if needed.

Observation

S. No.	Temp. of Milk	Observed Lactometer Reading (OLR)	Correction factor	Corrected Lactometer Reading (C.L.R.)
1				
2				

Calculation

$$\text{Specific Gravity} = 1 + \frac{CLR}{1000}$$

Interpretation

Result

Precautions

1. The milk to be tested should be 2 to 3 hours old after milking.
2. This will allow air, gases to escape from the sample, thereby, helping to get accurate specific gravity.
3. The temperature of the milk should be as close to the calibrated lactometer temperature as possible.
4. Sample should be thoroughly mixed.
5. Lactometer should not touch the side of the jar.

2. Specific Gravity by a Pycnometer: This method is based on the gravimetric estimation of the specific gravity of milk by means of a specific gravity bottle or pycnometer, many patterns of which are available. Fundamentally it is a bottle fitted with a glass stopper with a side hole in it for the discharge of the expanding fluids. The method is specially suitable when small quantities of milk are available for examination.

The bottle is first washed with distilled water, alcohol and ether, and then placed in the dessicator to dry and weighed empty. It is then filled with distilled water and placed in a hot water bath maintained at 60°F till the water inside is no longer alters its volume. It is then viped dry from outside and weighed. The difference between the weights of the filled and empty bottle gives the weight of water in grammes at 60°F. The bottle is emptied again and rinsed with milk several times and then filled with the milk which is to be tested. It is then placed in the water both at 60°F till the volume of the milk remains constant and dried from outside and weighed as indicated above. The difference between the full milk and the empty bottle gives the weight of the milk at the standard temperature. The specific gravity is then calculated as follws:

$$\text{Specific Gravity of milk} = \frac{\text{Weight of the volume of the milk at } 60^{\circ}\text{ F}}{\text{Weight of the same volume of water at } 60^{\circ}\text{F}}$$

Note : The volume of bottle and the weight of the empty pycnometer once determined need not be determined every time.

3. Specific Gravity of the milk by the Westphal balance: In this method the weight of a known volume of a liquid is determined by noting the loss of weight which occurs through displacement. The westphal balance consists of a pivoted steel yard or beam with a glass plummet hanging from one end. It is adjusted in such a way that when hanging in air the pointer indicates zero. A weight which is provided when hooked on to the end, will cause the pointer to return to zero, when the plummet is immersed in water at 60°F. The beam is devided into 10 parts indicated by the notches and riders weighing 1/10th, 1/100th and 1/1000th of the said weights are also provided. The milk to be tested is placed in the glass cylinder and the plummet is allowed to let down into the milk at 60°F. The weights are then removed until an equilibrium is established. The specific gravity is indicated by the position of the weights. The weight on the end is counted as 1, 1/10th rider indicates when in the notches 1/100th of the weight, 1/100th rider indicates when in notches 1/1000 and so on. Milk of an average specific gravity of 1.032 4 would have the weights and riders in the following position. 1 end weight 1/10th rider on the third notch 1/100th rider on the second notch and 1/100th rider on the fourth notch.

Use of Formulae in relation to Specific Gravity of milk

Simple determination of the specific gravity of milk is not enough to detect the adulteration especially of the nature of watering and skimming or both. It must be conjoined with the determination of the percentage of total solids, solids not fat and fat in milk in order to establish definite proof of any adulteration. In this connection several formulae are in use by the help of which it is possible to calculate with a close degree of approximation, the total solids of milk, if the percentage of fat and the lactometer reading are known. These formulae are:

1. **Babcock's Formulae:**

$$\text{Percentage of solids not fat} = \frac{L+0.7F}{38}$$

2. **Troy's Formulae:**

Percentage of Solids not fat =

where L = Lactometer reading

F = Percentage of Fat

The percentage of Total solids can be made out by adding the fat percentage of solids not fat.

Note : In working out these formulae, the points to be borne in mind are that.

1. Specific gravity reading cannot be used and
2. The percentage of fat expressed in hundredth cannot be used.

3. Total solids and SNF formulae

   Richmonds formula

$$\text{Percentage of total solids} = \frac{L}{4} + \frac{6F}{4} + 0.14$$

4. Babcocks Formula $\frac{100S - SF}{100 - 1.0753SF - 1} \times (100 - F)2.5$

$$\text{Percentage of SNF} = \frac{1.2F + 2.665 \times 100S - 100}{S}$$

5. Fleischman's Formula $\frac{L}{4} + 0.2F$

   Percentage of T.S. =

   Percentage of SNF =

where L = Lacto meter reading

F = percentage of fat

S = Specific Gravity

6. Richmonds Scale or Slide rule

4. **Phillips method of Estimation of Specific Gravity of Milk:** This method consists of adding drops of milk to a series of $CuSO_4$ solutions graded according to the specific gravity. On entering the $CuSO_4$ solution the milk droplet becomes encased in copper proteinate and its rise and fall in the solution is observed. When the milk droplet remains stationary for not less than 10 seconds, the specific gravity of the milk sample corresponds to that of the $CuSO_4$ standard.

Questions

1. Explain the importance of specific gravity of milk.
2. What are the factor, that affects the specific gravity of milk ?
3. What happens to the specific gravity when milk is adulterated with water?
4. What is the principle of this experiment ?
5. Enlist the precautions, which are to be taken in this experiment.

EXERCISE - 4

Sediment in Milk

AIM

To test for sediment in the given sample of milk. Visible foreign matter in milk is objectionable. Dirt or sediment usually indicates carelessness in cleaning the udders and other sanitary practices. Cows with mastitic udders constitute leucocyte and epithelial cells in the form of flaky material. Dirt that appears as visible and insoluble sediment may or may not have any indication to the number of bacteria present in milk. In addition to the objection to the presence of dirt, the price is also fixed depending on the results got in the test.

Principle

It is determined by filtering a definite quantity of milk through a white cotton disc.

Procedure

There are two methods (1) the off-the bottom method and (ii) the mixed sample method. Only the second method is done.

The milk is thoroughly mixed. One pint of milk is taken and filtered through a lintine cotton disc. The diameter of the disc measures 1.25 inches. It is held over a flat wire screen in the tester so as to expose a filteration area of diameter 1/8".

The disc is allowed to dry and the amount of visible dirt is estimated by comparing it with a set of standard disc showing all degrees of clenliness from "clean" to "dirty". Definite scores may be established for each standard and the results are recorded.

But actually one liter of milk is strained and the results recorded.

Result

Amount of sediment present.

EXERCISE - 5
Organoleptic Test

Relevant Information

Judging the quality of milk organoleptically for its flavor and odour requires considerable skill, which can be acquired only by practice. An experienced person can pick out bad samples of milk with a fair degree of accuracy. Many times milk supplies are accepted or rejected on the receiving platform merely by smelling but this method is not always reliable.

Materials Required

1. Samples of milk in cans or other containers.
2. Milk plungers.

Procedure

Remove the lid from the container and immediately smell the milk, stir the milk thoroughly with a plunger and again note the smell. Observe the color of the milk. (Abnormal colors are reddish, bloody, bluish etc).

Observation

Sr. No.	Characteristics	Sample A	Sample B
1	Odour / Smell		
2	Color		

Interpretation

Milk sample giving any foul or abnormal smell (acid or sour smell, rancid, putrid, metallic flavor, oxidized flavor, weed and stable flavor) and any abnormal color should be rejected or held over for further examination.

Result

Questions

1. Write the characteristics of normal milk.
2. Narrate the reasons for spoilage of milk.

Exercise - 6
Clot on Boiling (COB) Test

Relevant Information

COB test is based on the observation that milk sample having high-developed acidity (above 0.17%) due to bacterial growth are coagulated when heated to boiling temperature. Sample of milk having poor heat stability due to disturbed salt balance also show signs of coagulation when heated to boiling.

Materials required : Milk Samples :

Apparatus

1. Test tubes
2. Water Bath

Procedure

Transfer 5 ml of milk into the test tube. Note the smell, place the test tubes in a water bath at boiling temperature for 5-6 minutes. Remove the tubes from the bath and rotate them in an almost horizontal position. Examine the film of milk or sides of the test tube for any precipitated particles constitutes a positive COB test.

Observation

Sample No.	C.O.B.	
	(+) ive	(-) ive
1		
2		

Interpretation

Sample showing +ve test are considered to have developed acidity due to bacterial growth. Milk sample in which slight sweet curdling has taken place due to growth of certain rennet producing bacteria, may also give +ve COB test without much developed acidity. All the samples showing +ve COB test are unsuitable for processing or for distribution as fluid milk.

Result

Precautions

1. Avoid direct heating of milk.
2. Use hot water bath.

Exercise - 7
Alcohol Test

Relevant Information

The alcohol test is used for rapid assessment of stability of milk to processing, particularly for condensing and sterilization. The test is useful as an indication of the mineral balance of milk and not as much as the index of developed acidity. The test aids in detecting abnormal milk; such as colostrums, milk from animals suffering from mastitis and milk in which the mineral balance has been disturbed.

Materials Required

1. Test tubes preferably with graduation at 5 and 10 ml.
2. Measure for alcohol 5 ml.
3. Milk sample.

Reagents

Ethyl alcohol 68 percent by weight or 75 percent by volume (density 0.8675 gm/ml at 27°C).

Procedure

Place 5 ml of milk in a test tube and add an equal quantity of alcohol. Mix the contents of the test tube by inverting several times. Note flakes or clot. The presence of a flake or clot denotes a positive test.

Observation

Sample No.	Alchol Test		Conclusion	
	(+) ve	(-) ve	Accepted	Rejected

Interpretation

A negative test indicates low acidity and good heat stability of the milk sample. Milk showing positive test is not considered suitable for the manufacturer of evaporated milk, which has to be sterilized to ensure its keeping quality.

Precautions : Avoid direct heating of milk.

Questions

1. What is the significance of alcohol test ?
2. How this test differs from C.O.B. test ?

EXERCISE - 8
Alizarin-alcohol Test

Relevant Information

The test is similar to the alcohol test and the incorporation of alizarin helps to indicate the approximate percentage of acidity.

Materials Required

1. Test tubes.
2. Measure for alcohol.
3. Milk sample.

Reagents

Alizarin solution - 0.2 percent in alcohol (68 percent by weight or 75 percent by volume, density 0.8675 gr/ml at 27°C).

Procedure

Place 5 ml of milk in a test tube and equal quantity of the alizarin solution. Mix the contents of the test tube by inverting several times. Note the color of the mixture and presence of flakes or clots.

Observation

Sample No.	Alizarin Alcohol Test		Conclusion	
	Colour	Size of Flake	Accepted	Rejected

Interpretation

The general interpretation of the results is an indication for the alcohol test. For acidity of 0.14 percent upward the gradation in size of the flakes and color is approximately as follows:

Colour	Size of Flake	Approximate Acidity (% lactic acid)
Lilac	--	Up to 0.14
Pale Red	--	0.14 to 0.17
Reddish brown to brown	Small flakes	0.17 to 0.20
Brownish yellow to yellow	Large Flakes	Over 0.20

If acidity has not developed and yet coagulation occurs, it indicates the presence of rennet producing bacteria (sweet curdling). Milk from animals suffering from mastitis is alkaline in reaction and when mixed with alizarin-alcohol solution, a violet or purple color is produced. From the practical point of view, it is of little material difference, whether milk clots through the production of acid or the production of rennin by bacteria as in either case it is unsuitable to process.

Precaution

Avoid direct heating of milk

Questions

what is the significance of this test ?

Exercise - 9

Half-hour Methylene Blue Reduction (M.B.R) Test

Principle/Relevant Information

This test is based on the principle that Methylene blue (an oxidation reduction dye indicator), which is blue in its oxidized state, is reduced to a colorless liquid (leucoform) as a result of the metabolic activities of bacteria in milk. When a solution of methylene blue and similar indicators are reduced or decolourised, the number and types of bacteria growing in milk influence the time taken for the reduction of the dye. The greater the number of organisms present in milk and greater their activity, the more rapidly the dye is reduced. The MBR time, thus, gives an indication of bacterial numbers and activity in milk. This test is, therefore, used for:

1. Judging the hygienic quality of milk and grading raw milk supplies.
2. For accessing the probable quality of milk.
3. For detecting post pasteurization contamination in milk.

Materials Required

1. Thermostatically controlled water bath maintained at 37 °C + 0.5 °C.
2. Sterile test tube without rim preferably with marking at 10 ml.
3. Sterilized rubber stoppers: For closing the test tube. The stoppers are sterilized by immersing them in a boiling water bath for not less than 10 minutes.
4. 10 ml and 1 ml pipette.
5. Clock watch.
6. Forceps, beakers and flasks.
7. Sample of milk (One fresh milk and other of poor quality sample).

8. **Methylene Blue Solution :** A standard solution of methylene blue is prepared by dissolving one tablet of approved methylene blue thiocyanate of chloride in 200 ml of cold sterile glass distilled water in a sterile flask by gentle heating in water bath or by allowing the mixture to stand for several hours to facilitate complete solution and then adding 600 ml of sterile glass distilled water. 1 ml of this solution is mixed with 10 ml milk results in obtaining a final concentration of 11300000 for the dye, which has been found to be satisfactory for the test. The stock solution must be stored in a sterile glass stoppered amber colored bottle in a dark place. Fresh solution must be prepared once in two months.

Procedure

Thoroughly mix the sample (shake at least 25 times) of milk. Transfer 10 ml of each sample of milk into a test tube (sterile). Add 1 ml methylene blue solution. Close the test tube with sterile rubber stopper held and inserted by sterile forceps or by the tips of the fingers on the extreme end. While transferring methylene blue solution care should be taken not to contaminate the pipette by touching the milk. Mix the dye and the milk by inverting the tubes twice. Place the tubes in water bath.

Put the following control tubes with each batch/sample.

(a) 10 ml mixed milk + 1 ml of tap water.

(b) 10 ml mixed milk + 1 ml of methylene blue solution.

Fit the control tubes (a) and (b) with stoppers and immerse for 3 minutes in boiling water in order to destroy the natural reducing system present in the milk. Comparison of the experimental tubes with control tube (b) will show when decolourization begins and comparison with control tube (a) will show when it is complete.

Observe the test tube after 30 minutes

Record the milk as decolourized when the whole column of milk is completely decolourized or is completely decolourized up to within 5 mm of the surface. If a trace of color persists at the bottom of the tube and does not extend for more than 5 mm, it may be ignored.

Observation

Milk Sample	MBR Time	Inference
1	After 30 minutes	
2	After 30 minutes	
3	After 30 minutes	
4	After 30 minutes	

Result

Precautions

1. Methylene blue solution, when not in use, should be kept in the dark.
2. The interior of the water bath during the progress of the test should be completely dark.
3. The precautions against the contamination of the milk sample described in the method shall be observed or followed carefully.

Questions

1. Why is methylene blue reduction test not used for testing pasteurized milk?
2. How do you prepare and store methylene blue solution ?
3. How is the test useful in the quality control programme ?
4. What is O-R potential ?
5. What is the O-R potential of milk ?

EXERCISE - 10
Ten Minutes Resazurin Test

Relevant Information

This test provides rapid measure of the sanitary condition and keeping equality of milk. Resazurin is an oxidation-reduction indicator like Methylene blue. Resazurin reductions occur in two stages - the first an irreversible change from the blue resazurin to the pink resorufin, and second a reversible change from the pink resorufin to the colorless dithydroresorufin as a result of bacterial activity. In the course of reduction of resazurin in milk there will be series of color changes from blue to lilac, mauve, purple, pink and finally colorless which can be compared with standard color discs in lovibond comparator disc, numbers (6 to 0).

Materials Required

1. Milk Sample
2. Resazurin - either powder or tablet.

In the case of powder, 0.05% (w/v) is first prepared by dissolving 0.05 g powder in 100 ml of sterilized glass - distilled water. The mixture is steamed or boiled for 30 minutes in semi-dark place and transferred to semi amber/amber colored bottle and finally in the refrigerator. Whenever the test is required to be carried, the bench reagent solution (0.005%) is prepared by adding 1 ml of stock solution to 9 ml of sterile water. This bench solution should be used within 8 to 12 hours of its preparation.

In the case of tablets, each tablet is dissolved in 50 ml sterile distilled water and heated to boil which will give you a bench solution for direct use. This solution is used within 6 hours of its preparation.

Apparatus

1. Comparator with resazurin disc.

2. Sterile test tubes without rims (150 × 16 mm)
3. Sterilized rubber stopper (immersing in boiling water for not less than 10 minutes.
4. Sterile 1 ml pipette
5. Water bath - maintain 37.5 °C + 0.5 °C.
6. Stop Clock or Watch

Procedure

Mix the sample thoroughly. Pour 10 ml of sample into two test tubes by sterile pipette. Add 1 ml resazurin (bench) solution into one of the test tubes by 1 ml sterile pipette with a pair of forceps insert sterile rubber stopper into the tubes. Invert the tubes twice so as to mix the contents well and immediately transfer them to water bath maintained at 37.5 °C + 0.5 °C. The level of water in water bath should be higher than of milk. At the end of 10 minutes, remove the tubes and match the color with the resazurin disc in the comparator. The milk tube in which resazurin is not added, is used as control & placed in the left slot of the comparator to compensate for the natural color of the milk. If color falls between two disc numbers, record half value.

The quality of milk is judged as follows

Resazurin DiscNo.	Color	Quality of Milk
6	Blue	Excellent
5	Lilac	Very Good
4	Mauve	Good
3	Pink Mauve (Purple)	Fair
2	Pink	Poor
1	Deep Pink	Bad
0	Colorless	Very Bad
	OR	
4 or higher	-	Satisfactory
3.5 to 1	-	Doubtful
0.5 to 0	-	Unsatisfactory

Precautions

1. The test should be carried out aseptically.
2. The water bath should be covered during the test.

Questions

1. What is the practical utility of resazurin test ?
2. What are the advantages of the test ?

Exercise - 11
Phosphatase Test

The Phosphatase test of pasteurised milk may be determined by any of the following methods.

1. Modified key graham method
2. Lovi Bond comparator with standard phosphotase disc

Lovi Bond comparator Method

Relevant Information

The test is used to judge the efficiency of pasteurization of milk. To test whether the heat treatment (LTLT or HTST) was properly carried out or not; the treated milk is subjected to the phosphatase test, which helps to indicate the presence or absence of phosphatase enzyme. In properly pasteurized milk, the enzyme is absent. If phosphatase is present, it indicates that the milk either has been contaminated after the heating process or the heat treatment has not been properly carried out.

Principle

There are two enzymes in milk, which are capable of hydrolyzing phosphoric ester. They are alkaline & acid phosphatases. Alkaline phosphatases require alkaline condition while acid phosphatases requires acid condition. Moreover alkaline phosphatase is inactivated at just below the temperature of pasteurization. Hence, its presence in pasteurized milk shows the inefficiency of pasteurization. When P-nitrophenyl disodium orthophosphate is added to milk, and if alkaline phosphatase is present in milk due to under pasteurization/post pasturization contamination with raw milk, the enzyme acts on substrate and release phenol (P-nitrophenol) turning it to yellow under alkaline condition. The intensity of yellow color gives an idea regarding efficiency of pasteurization.

Apparatus

1. The all purpose lovibond comparator complete with stand for work in reflected light.
2. Standard phosphatase discs : Giving 0, 6, 10, 18, 42, or 0, 6, 10, 14, 18, 25, 42 reading.
3. Two 25 mm fused glass cell
4. Water bath - 37.5°C + 0.5°C
5. Pipette - 5 ml, 1 ml
6. Graduation flask - 1000 ml
7. Measuring cylinder - 100 ml
8. Test tubes with rubber stoppers.

Reagent

1. **Buffer solution:** Dissolve 3.5 g anhydrous Sodium Carbonate (A.R.) and 1.5 gm of Sodium Bicarbonate (A.R.) in 1 litre of distilled water.
2. **Substrate:** P-nitro-phenyl disodium orthophosphate.
3. **Buffers Substrate solution:** Transfer 0.15 gm of the substrate into a 100 ml measuring cylinder or stoppered graduated flask and make up to the mark with the buffer solution. The solution should not be stored for long period but may normally be kept protected from light in a refrigerator for up to one week. The solution is practically colorless when viewed through a 25 mm cell in the all-purpose comparator, it should give a reading of less than 10 on the disc.

Procedure

Pipette 10 ml (5 ml may be used) of buffer substrate solution into test tube (marked at 10 ml). Add 2 ml (1 ml in 5 ml of buffer substrate) of milk, close the tubes with rubber stoppers and invert to mix and place in the water bath. Prepare in the same way a blank from boiled milk of the same type as that under test. Incubate all the tubes at 37°C to 38°C. Read the yellow color after 30 minutes, return to the bath, and take a second reading after incubation for a further 90 minutes. The yellow colour is read in a lovibond all purpose comparator on a resazurin stand, fitted with the disc calibrated in microgram of paranitro-phenol. The blank is placed on the left of the stand and the sample on the right. Reading is taken by looking down on to the two apertures with the comparator facing a good source of northern day light; the disc is revolved until the sample is matched, readings falling between two standards are recorded by adding a plus (+) sign or a minus (–) sign to the figure of the nearest standard.

Disc reading after 30 minutes incubation

Reading	Interpretation
0 or trace	Properly pasteurized
6	Doubtful
10 or over	Under pasteurized

Disc reading after 2 hours incubation

Reading	Interpretation
0 - 10	Properly pasteurized
Over 10 to 18	Slightly under Pasteurized
Over 18 to 42	Under pasteurized
Over 42	Grossly under pasteurized

The 30 minutes test will reveal any serious fault in pasteurization, but to enable minor errors to be detected, reading shall be taken after further incubation for 90 minutes.

Precautions

1. After use, test tubes should be emptied, rinsed in water, washed in hot water containing an alkaline detergent followed by thorough rinsing in clean hot water. Finally, they should be rinsed in distilled water & dried before use.
2. Pipettes should be thoroughly rinsed in clean cold water immediately after use followed by rinsing in distilled & dried before use.
3. The test tube stoppers should be thoroughly rinsed in hot water immediately after use followed by boiling for 2 minutes in distilled water.

Key Graham Method

AIM

To detect the presence of phosphatase in milk - modified Key-Graham method.

Principle

Normal milk contains two kinds of phosphatase enzymes - the alkaline phosphatase and acid phosphatase. These hydrolyze the phosphate ester, but they differ in their optimum activity with the difference of pH. Alkaline phosphatase acts in the alkaline medium (pH 4). Again they also differ in their resistance to heat. Acid phosphatase is more resistant while alkaline phosphatase is inactivated at the time-temperature combination of pasteurization of milk. So detection of alkaline phosphatase has been adopted to detect whether the time-temperature combination of pasteurization given has been perfect or properly pasteurized milk has been mixed with raw milk afterwards. Secondly Microbacterium tuberculosis are destroyed at pasteurization temperature where the alkaline phosphatase also is inactivated.

The milk to be tested is incubated with a buffer substrate which contains disodium phenyl phosphate. If phosphatase is present in the milk, it will liberate phenol from the substrate. Later, when the Folin and ciocalteu's phenol reagent is added along with Na_2CO_3 solution, phenol reacts with the reagent and form indophenol which gives blue colour. From the intensity of blue colour, the degrees of phosphatase present can be determined.

REAGENTS

1. **Buffer substrate**

   Contains disodium phenylphosphate (0.109%) and sodium diethyl barbiturate i.e. sodium vernol (1.154%) in 1% by volume of chloroform or BDH tablets can be used.

2. **Folin and Ciocalteu's Phenol reagent**

   100 g of sodium tungstate, 25 g of sodium molybdate in 700 ml of water mixed with 50 ml of 85% phosphoric acid and 100 ml hydrochloric acid and refluxed for 10 hours. Cooled and 150 g lithium sulphate, 50 ml of water and 4 to 6 drops of liquid bromine were added. Excess bromine was removed by boiling the mixture for 15 minutes. Taking a litre flask, volume was made up to the mark and then filtered. This is stock solution, to be used as reagent by adding 2 volumes of 5% solution of sodium hexametaphosphate.

3. 5% sodium hexameta phosphate solution.
4. 14% $Na_2\ CO_3$ solution.

Procedure

1. **Short Test:** This test can be carried out in $1^1/_2$ hours and is useful for determining whether a milk has been heated or not, and for observing gross errors of pasteurization technique

   (a) **Determination:** To 10 ml of buffer substrate solution contained in a test tube 0.5 ml of milk was added and was mixed thoroughly. The mixture was incubated in a water bath at 47°C + 2 for 10 minutes. The tube was then removed, cooled to 15 to 20°C by immersing in water. 4.5 ml of the diluted Folin and ciocalteu's phenol reagent was added and allowed to stand for 3 minutes and then filtered. To 10 ml of the filtrate, 2 ml of Na_2CO_3 solution was added, mixed and heated in boiling water for 2 minutes. The test was carried out in duplicate along with duplicate control tubes.

   (b) **Control test:** 10 ml of buffered substrate was mixed thoroughly with 0.5 ml of milk sample and 4.5 ml of dilute folin and ciocalteu's reagent, allowed to stand for 3 minutes and filtered. To 10 ml of filtrate 2 ml of Na_2CO_3 solution was added and boiled in water bath for 2 minutes. All the 4 tubes were examined with standard phosphatase disc for the development of colour.

(c) **Interpretation:** If the control show more than a trace of blue colour, the reagents contained no free phenol, it is probable that a phenol producing organism is present in the milk. This does not occur in pasteurised milk which has been kept at a satisfactory low temperature or in freshly pasteurized milk. If with such controls, the indicator tubes show a definite blue colour over 2.3 units, it can be said with certainity that milk was not adequately pasteurized and if the color is over 6.0 blue units, probably the milk was not heated at all.

2. **Control laboratory test:** This test detects small as well as gross errors in processing and is the one most generally used in laboratory.

(a) **Determination:** Procedure is same as short test except that incubation period is 24 hours at 37°C with addition 3 drops of $CH\ Cl_3$.

(b) **Control test:** It is done when necessary. The milk sample should be kept in refrigerator for 24 hours and then should be proceeded as previously described.

(c) **Interpretation:** Milk which with controls below 1.5, given colors with 2.3 blue units may be classified as properly pasteurized. Those giving colour between 2.3 and 6.0 blue units should be classified as improperly pasteurized while those giving above 6.0 units show evidence of serious error in pasteurization. Raw milk actually gives more than 30 units of colour.

The limits of accuracy claimed for test (2) *are as follows*:

(a) Holding at 145°F for 20 minutes instead of 30 minutes

(b) Holding at 1.5°F below the recommended temperature i.e. 143.5°F

(c) admixture of 0.2% raw milk with correctly pasteurized milk.

Precautions

(i) When possible, test should be carried out on samples within 18 hr after pasteurization, although milk samples can be kept at low storage temperature for two weeks.

(ii) Any materials having phenol should be kept away from the vicinity where test is carried on.

(iii) Glassware should be cleaned and properly washed.

RESULTS

For example

1. **Short Test :**

(a) LTLT (145°F or 63 °C.30 min) treated sample 1.5 units

(b) HTST (161°F or 72°C. 16 sec) treated sample 1.5 units

(c) Raw milk > 6.0 units

2. 24 Hours Test

(a) LTLT (145°F/30 min) treated sample 1.5 units

(b) HTST (161°F/16 sec) treated sample 1.5 units

(c) Raw milk > 6.0 units

Hence properly pasteurized.

Questions

1. What are the time temperature combinations for destruction of phosphatase?
2. Does the destruction of enzyme phosphatase indicate the destruction of pathogenic organism ? If so, why ?
3. At what pH is the enzyme phosphatase maximum activate ?

EXERCISE - 12
Determination of Fat in Milk

AIM

To determine the fat content in milk by

1. Analytical method and
2. commercial method.

Principle

The methods for determining fat are conveniently divided into two:

1. Analytical and
2. commercial method.

   For commercial purposes a quick with reasonable accuracy is necessary which can be conducted inside the laboratory. The accuracy of result in commercial method is again dependant on the constructed glass apparatus (a) The Gerber test and (b) the Babcock test. On the other hand in analytical method the result that is got is accurate since it is completely under the control of the skilful operator. The method includes (a) Rose-Gottlieb method and (b) Adams method. A modified method of Rose-Gottlieb method called Mojannier method is also included under analytical methods.

In Rose-Gottlieb method the principle is the fat from the milk is extracted in fat solvent, after dispersing the protective film surrounding the fat globule and then determined.

In Adams method the fat in milk is determined by the milk spreading out to a greater extent on an absorbing paper and then extracting fat from it by fat solvent through a process of distillation.

In Gerber test, the fat alone is separated from milk by subjecting the latter to digestion by sulphuric acid and by a process of centrifugation.

In Babcock test, the principle is same as above.

Procedure

I. Rose-Gottlieb Method: This method is primarily used for milk products. This can also be used for milk.

The extraction cylinder used in this test is weighed with the siphon tubes, fitted through a tight bark cork. The milk is thoroughly mixed and 10 to 11 gm of it is transferred to the cylinder. Again it is reweighed. 1 ml of concentrated Ammonia of density 0.888 + 0.003 at 20 °C or of sp. gravity 0.885 + 0.003 at 60 °F is added. Ammonia serves to disperse the protective layer surrounding the fat globules. Otherwise it is not possible to effect the quantitative separation of fat from milk by shaking it with fat solvent. It is then mixed by gentle rotation. 10 ml of rectified spirit (95% ethyl alcohol) are added and again mixed. After well mixing, 25 ml of peroxide free ether is placed in the cylinder. After closing with cork it is shaken for one minute. 25 ml of Petroleum spirit whose boiling point ranges 40 to 60 °C are added and again the mixture is shaken for 30 seconds. These mixed solvent petroleum and ether effect a clean separation of fat from the non-fatty solids due to their property of being completely non-miscible with water. As these two solvents are very volatile and inflammable, it is necessary to extinguish any burners in the viscinity.

The cylinder is set aside for not less than 30 min. Now the contents separate into two distinct layers, the upper one being a solution of fat in mixed solvents and the lower one being the aqueous layer containing the non-fat milk constituents.

In the meantime a 150 ml flask is weighed and kept ready. When the upper layer is quite clear the cork is removed, washed with a little of the mixed solvents to transfer any adhering fat to the main bulb of the solution. The siphon tubes are then adjusted, so that the mouth of the covered inlet is 2 to 3 mm above the aqueous layer. Complete transfer is neither possible nor necessary. The ethereal layer is transferred to the weighed flask. 5 ml of ether is added to the extraction cylinder to wash the inside of the cylinder and the interior of the siphon tubes without mixing or shaking the cylinder, this washed ether is also transferred to the flask. This is repeated again.

The extraction is also repeated with 15 ml of ether shaking for 15 seconds followed by 15 ml of Petroleum spirit, shaken for 15 seconds and then the extract is transferred to the flask. More number of extractions with small amount of solvents for each time will be more effective than the use of greater amount of solvents for a few number of extractions.

1. Wt. of empty clean Fat-flask.
2. Wt. of empty beaker with milk before absorption
3. Wt. of empty beaker with milk after absorption
4. Wt. of milk sample taken for test (2-3)
5. Wt. of Fat-flask with ether extract

6. Wt. of Ether Extract (5-1)
7. % of Fat in the given milk sample.

The fat flask containing the combined extracts is connected to a distillation apparatus and the solvents are removed by slow distillation. The residue is dried to constant weight in an oven at 98 to 100 °C.

It is important also to carry out a blank determination on the reagents used and to make any corrections.

The PRECAUTIONS which are considered here are:

(1)The cork used should be previously solvent extracted to free it from any fat.

(2) Ether used should be free from peroxide since its presence is responsible for a number of explosions which occur during distillation and again peroxides increase the weight in fatty residues.

II. The Adams Method: This is devised by Dr. M.A. Adams of Kent for determination of fat in milk. 5 to 10 ml of milk are taken in a small beaker and weighed. Fat free Adam's strips are coiled up loosely and it is dropped into beaker of milk so that the milk is absorbed as completely as possible and uniformly.

Filter paper or Blotting paper should not be used for this purpose as they have ether extractable substances like resines and calcium salts in them.

When absorption is complete the coil is transferred to oven at 100 °C for drying and left therein till it ceases to lose weight. Meanwhile, to find out the weight of milk that has been absorbed by the coil, the beaker is weighed again and the difference in weights is found.

Next the Soxhelt flask is weighed before it fitted on to the soxhlet extractor. The dried coil is then placed in the thimble and the fat in that is extracted with ether. The ether extract that is collected in the flask finally is dried at 100 °C and the weight of it is found.

If filter paper is used, first it is treated with alcohol and 10% acetic acid and then used for absorption.

III. Gerber method :

Principle

When concentrated H_2SO_4 is mixed in correct proportion, will digest the protein of the milk, breaking them down into simpler constituents in a form in which they are no longer capable of maintaining the fat droplets in the emulsified state of dispersion. The fat globules thus freed may rise to the surface of the liquid and join together as mass.

The heat generated by the reaction between the acid and milk causes the melting of fat globules and permits them to come together. The heating of the liquid also lowers the viscosity of the acid mixture, which favours the movement of fat droplets to the surface.

Relevant Information

Fat is the most important constituent of milk as it is used as a basis for fixing the purchase and sale price of milk. Fat percentage can be determined either by gravimetric or volumetric method. Though gravimetric method (like Mojonnier & Rose-Gottlieb) are accurate but time consuming. So, for routine analysis methods are used. Volumetric methods are of two types, Babcock and Gerber. Gerber's method is commonly used in our country.

Apparatus

1. Gerber centrifuge
2. Gerber Butyrometer for milk (0-10% scale with 0.1 percent mark)
3. Hot water bath maintained at 65 °C + 2 °C
4. 10 ml automatic tilt measure for acid
5. 1 ml automatic tilt measure for alcohol
6. 10.75 ml pipette
7. Butyrometer stopper (lock and key)
8. Butyrometer stand

Reagents

1. Gerber sulphuric acid density 1.807 to 1.812 gm/ml at 27 °C corresponding with a concentration of sulphuric acid from 90-91 RO by mass.
2. Amly alcohol 95% of clear colorless liquid shall distill between 130 °C to 132 °C, density 0.803 to 0.805 gm/ml at 29 °C.

Procedure

Take 10 ml of Gerber sulphuric acid from automatic tilt measure into butyrometer. Pipette out 10.75 ml of the well-mixed sample of milk and transfer it to the butyrometer carefully without allowing it to mix with the acid. This is done by allowing the jet of milk from the pipette to hit the inside wall of the butyrometer by holding the pipette in a slanting manner and resting the tip end on the mouth of the butyrometer. With the help of the automatic tilt measure add 1 ml alcohol to the above butyrometer. Tighten the stopper and mix the contents by shaking the butyrometer at a 45° angle until all the curd has been dissolved. Keep the butyrometer in the water bath at 65° + 2°C for 5 minutes. Place the butyrometer in the centrifuge and balance the machine, centrifuge for 5 minutes (1000-1200 rpm). After centrifuging, temper the butyrometer in hot watter bath at 65 °C + 2 °C for 5 minutes. Adjust the fat column within the scale on butyrometer and take the reading.

Observation

Sr. No.	Initial Reading of the Butyrometer	Final Reading of the Butyrometer	Fat %	Average
1.				
2.				
3.				
4.				

Conclusion

The given sample of milk contains ______ percent of fat.

Precautions

1. Care must be taken not to wet the neck of the butyrometer while adding sulphuric acid, milk and amyl alcohol.
2. The test must be repeated if particles of curd are observed.
3. Thorough mixing of milk is necessary.
4. The three fluids should be added gently, so that they form three distinct layers.
5. Centrifuge machine must be properly balanced.
6. Always carry out the test in duplicate.
7. Rubber stopper should be dry, clean and without cracks.
8. Amyl alcohol must be pure.
9. Before centrifuging see that there is no curdy white material left undissolved.

Advantages of H_2SO_4 (Density - 1.807 to 1.812 gm/ ml at 27 °C i.e. 90 to 91% by mass) :

1. It quickly dissolves the SNF and destroys the viscosity, thus liberating the fat.
2. It generates much heat and so keeps the fat in liquid condition which is necessary for complete separation.
3. It increases the specific gravity of the moisture surrounding the fat thereby aiding the separation of fat.
4. It is non-volatile while hot and so produces no injurious gas.
5. It is comparatively cheap and easy to obtain.

Advantages of Amyl Alcohol (Density - 0.803 to 0.805 gm / ml at 27 °C) :

By adding amyl alcohol, the surface tension is reduced around the fat globules thereby fat liberation is easy, in villages we add Kerosene oil. Addition of amyl alcohol will help to prevent charring and to get a clear fat column.

IV. **Babcock Test:** This test was published by S.M. Babcock. This test is conducted in special milk bottles of capacity 50 ml. The stem in graduated to read usually 0 to 8% or 10% of fat. By adopting this method there are manifold advantages. They being:

(a) cheap
(b) quick
(c) more accurate
(d) easy to operate and
(e) this test can be used not only for milk but also for other milk-products.

The volume internally in the stem of the space of 1% is 0.20 ml. In other words the volume of 1% of fat is 0.20 ml. Assuming the density of fat as 0.9, the weight of 1% fat will be 0.18 g. If 1% represent 0.18 g, 100% will be 18 g which is the weight of milk to be taken for analysis. Assuming the specific gravity of milk at 60 °F as 1.032, the volume of milk to be taken to get the weight of 18 g of milk is 17.44 ml. A pipette having a capacity of 17.6 ml will deliver the required amount of 17.44 ml of milk at 60°F.

At 60°F the specific gravity of milk ranges from 1.029 to 1.035. The volume fixed to as per the above calculations is based on specific gravity of 1.032 at 60 °F which is 17.44 ml. But with change in specific gravity, the pipette delivers less or more of milk, not 17.44 of milk, which in turn alters the weight of milk used in the test. Variations in weights of milk ranges from 18.054 g to17.946 g.

The weight of milk again depends upon the temperature of milk. A sample pipetted at 100 °F weight less approximately by 0.09 g and gives a value of 3.98% for 4.0%. Sulphuric acid (specific gravity 1.820 to 1.830) at 60°F is used. 17.5 ml of this acid is found to be just sufficient, to produce the best required results as stated for Gerber's test. It is possible to produce the necessary heat by using an acid of low specific gravity by increasing its volume. But when volume of acid is increased, it does not give a satisfactory result.

The rate and time of centrifuging is also fixed. The rate of centrifuging varies with the diameter of the centrifuge. Lesser the speed of the centrifuge, the lesser the separation of fat. The first centrifuging period is not critical since the very smallest fat globules that escape recovery during the first period can not be recovered when the specific gravity of the mixture has been lowered by the addition of water. Temperature of centrifuge affects the fat %. Cold machine reduces the fat %.

Tempering of fat is necessary before reading since it is known that fat expands rapidly than glass with the temperature. As such higher the temperature, the volume will be more, resulting in lesser density. It is assumed for calculation the density of fat as 0.9%. It has been found that at 140°F the density of fat avaerages 0.889. So a temperature of 140°F or 60°C is used for tempering before taking the reading.

In Babcock test the fat column is read from lower meniscus at the bottom to the upper meniscus at the top. This is practised to compensate the fat loss which has settled down at the botom of the bottle (about 0.2%).

Sources of Error

By adding too much acid to strong acid or warm milk or defective mixing or improper addition of water produces black charred particles. Weak acid, less or acid and cold milk gives rise to cloudy light coloured fat. The normal colour of fat is golden yellow which is translucent. When hard water is used due to liberation of Carbon di-oxide air bubbles will be formed.

17.6 ml (18 grams) of well mixed milk is taken in a test bottle after cooling the milk to 60 °F. The milk in the pippette is blown out. 17.5 ml of sulphuric acid which is also cooled to 60 °F is taken in a special cylinder and added in small quantities at a time down the neck of the bottle which is simultaneously and gently rotated. It is shaken until all traces of curd have disappeared. Then it is heated to 60°C by keeping in water bath and then placed in a centrifuge. It is whirled or centrifused at 910 rpm for 5 minutes after balancing the same. After centrifusing distilled water at 60 °C is added to bring the liquid to the base of the stem. Again it is whirled for 2 minutes. Then water at 60 °C is added to bring the fat column near the top of graduation, while adding water care should be taken to make the water to flow along with the side and not over the fat. Otherwise the fat will be pushed down to the base resulting in loss of fat. Then it is whirled for 1 minute and then placed for tempering in the water bath at 60 °C for 3 minutes and the fat column is measured with dividers.

V. **Soxhlet's Method:** Take about 5 cc of milk in a beaker and weight it. A fat free thimble is then weighed accurately and brought in contact with the milk until as much of milk as possible is absorbed by the thimble. The beaker is weighed again and the exact amount of milk absorbed is calculated. The thimble is then dried first in the open air and then in an oven at a temperature not exceeding 100 °C. Afterwards the thimble is transferred to the soxhlet ether extraction apparatus in which the flask is previously weighted and extraction is continued for at least 20 times. Then the ether is evaporated carefully and the residue left in the flask is weighed. The difference between the weights of empty flask and the same flask with the residue gives the weight of the fat present in the amount of milk taken from which the percentage is easily calculated. This method of determination of fats in milk is very accurate, but is very expensive and takes a long time to be carried out. Therefore, it is seldom employed in routine examination of milk.

VI. **Werner Schmidt's Method:**

The specific gravity of the sample of milk to be tested must be determined first, I0 cc of the milk is pippetted into the 'Stokes ' tube and then I0 cc **Soxhlet Assembly & Kjeldal Apparatus**

conc. hydrochloric acid is added. The contents are boiled with constant shaking until the liquid forms or turns into a dark brown solution and the fat collects at the top. The tube with its contents is then cooled rapidly by placing it in a cold water. Then 30 cc of the ether is added and the contents are shaken vigorously for half a minute and the tube is allowed to stand for 5 minutes. The ether separates out as a clear layer at the top and is removed as much as pssible 20 cc of ether is again added and the process is repeated. This process is repeated once again and the ether thus collected is distilled and evaporated to dryness. The residue is weighed and the percentage is calculated. However, this method is not accurate.

VII. Chemical method :

This method has been devised by Late Balwant Singh, Agricultural Research Chemist. Military farms Laboratory, Lahore, in which the use of centrifuge has been avoided altogether.

Method: Measure 11 cc of milk and deliver the same into the gerber's butyrometer, add 4.5 cc of Fehling's B Solut ion (50 - 55 gm NaOH depending upon the quality and 17.3 gm of Rochelle salt dissolved in half litre of water). Then add 1 cc of the alcohol mixture consisting of 55 parts of Isobutyl alcohol and 45 parts of methyl alcohol and cork it. If Isobutyl alcohol is not available, then a mixture of 3 parts of methyl alcohol and 1 part of any alcohol will serve the purpose which is equally good. Invert the tube and then holding horizontally shake it in a gentle manner, rotating the tube between the two palms. Place the tube, keeping the cork upwards in a water bath at a temperature of about 80 °C. See that the tube is immersed completely and totally. Take out the tube when the colour of the liquid inside is turned to deep orange, which will take about 3 minutes time, shake it as before and replace in the hot water bath for another 3 minutes. The fat will now begin to appear as a distinct layer at the top; open the cork and add about 5 cc of hot water very gently down the sides of the tube without disturbing the fat layer and also taking care not to wet the neck of the tube. If by chance the cork and neck are get wet, then these should be dried thoroughly before inserting the cork again which is most important because, the cork may slip if it is wet. Invert the tube gently and stand it again in the hot water bath this time with the graduations at the top or upper side. The fat column will become distinct and easily readable within a short time.

Questions

1. What is the correct method of sampling milk ? What is its importance in the fat test ?
2. Why 10.75 ml of milk is taken for fat test ?
3. Why is sulphuric acid preferred to other acids in the fat test by Gerber method?
4. What precaution would you take while conducting the experiment?

Exercise - 13
Determination of Total Solids of Milk

Relevant Information

Total solid content is the entire residue left after complete evaporation of water from milk. It includes fat, protein lactose, and mineral matter. Total solids can be determined by:

1. Gravimetric Method
2. By using formulae
3. By using Richmond scale
4. By T.S. apparatus

Materials Required: Milk Sample.

Apparatus

1. Fat testing apparatus
2. Specific gravity testing apparatus
3. Richmond Scale

Procedure

(a) **Gravimetric Method:** Though very accurate, but not routinely used. It is determined by weighing a small sample (5 ml) in an aluminium dish of known weight. The dish then is heated on boiling water bath for at least 30 minutes or until most of the moisture in the sample has been evaporated. Then the dish is placed in a hot air oven (98-100 °C) for about 3 hours. After cooling, it is again weighed and from the loss of weight the amount of water evaporated is calculated.

(b) **By use of forumlae:**

(i) Determine the fat percentage of milk sample by Gerber method.

Tray Dryer

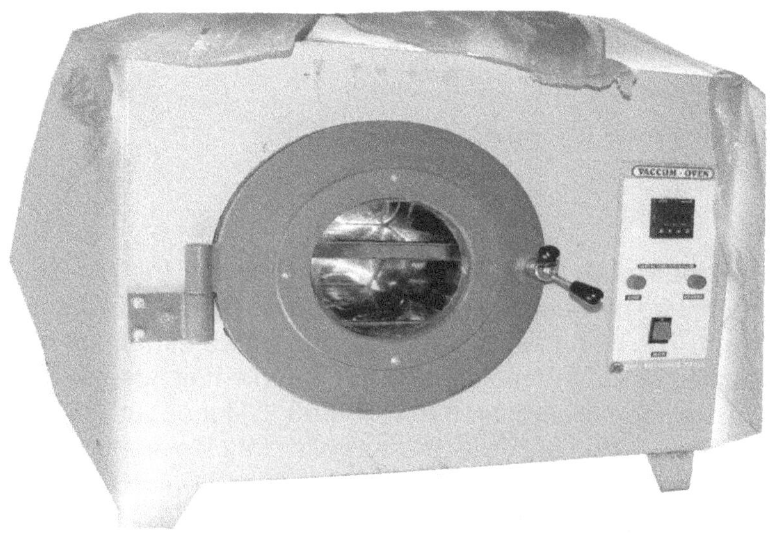

Vacuum Oven

(ii) Determine the lactometer reading and temperature of milk and calculate CLR.

(iii) Place the figure of fat and CLR in the following formula :

Richmond's formulae for different types of lactometer :

Zeal lactometer (84 °F or 29 °C)

$$\% \text{ SNF} = \frac{\text{CLR}}{4} + 0.25\text{F} + 0.50$$

$$\% \text{ TS} = \frac{\text{CLR}}{4} + 1.25\,\text{F} + 0.50$$

ISI / BIS lactometer (27 °C)

$$\% \text{ SNF} = \frac{\text{CLR}}{4} + 0.25\,\text{F} + 0.60$$

$$\% \text{ TS} = \frac{\text{CLR}}{4} + 0.25\,\text{F} + 0.60$$

Quevenne lactometer (15.5 °C or 60 °F)

$$\% \text{ SNF} = \frac{\text{CLR}}{4} + 0.20\,\text{F} + 0.14$$

$$\% \text{ TS} = \frac{\text{CLR}}{4} + 0.20\,\text{F} + 0.14$$

(c) By Richmond Scale: To get immediate result, one can use this scale. Mere, in this scale, one movable and another fixed scale is present. TS can be determined, if lactometer reading, temperature of milk and fat percentage is known.

Operation of the Scale: Place the OLR reading against the observed temperature. Read CLR against standard temperature (20 °C). Put the black arrow behind the observed fat percentage. Read TS in the bottom part of the fixed scale against CLR.

(d) By T.S. apparatus: It is designed to determine T.S. in raw milk, cheese, ghee, yoghurt, flavoured milk, condensed milk. etc and accuracy to the tune 0.05%. It is housed in robust metallic cabinet. The panel has knobs for intensity control, timer etc. The I.R. lamp is adjustable in well air-cooled shade. The instrument operates by drying the sample with infrared radiation. Drying times is within 10 minutes depending upon the nature and quantity of sample. Rest of the procedure is that of gravimetric method.

Observation and Calculation

(a) Gravimetric Method:

(i) Weight of the crucible = W_1

(ii) Weight of milk + Crucible = W_2

(iii) Weight of milk + Crucible (oven dry) = W_3

Therefore, Wt. of milk $W_2 - W_1 = X$

Wt. of milk (oven dry) $W_3 - W_1 = Y$

$$\% \text{ T.S.} = \frac{Y}{X} \times 100$$

(b) Using Richmond formulae:

Fat = OLR = CLR =

$$\% \text{ SNF} = \frac{CLR}{4} + 0.25\,F + 0.50$$

(c) Using Richmond's Scale:

Fat = ______________

Temperature = ______________

OLR = ______________

TS = ______________

(d) T.S. Apparatus:

(i) Weight of the crucible = W_1

(ii) Weight of Milk + crucible = W_2

(iii) Weight of Milk crucible (dried) = W_3

Therefore,

Wt. of milk $W_2 - W_1 = x$

Wt. of Milk (dried) $W_3 - W_1 = Y$

$$\% \text{ T.S.} = \frac{Y}{X} \times 100$$

Precautions

1. Care must be taken not wet the neck of the butyrometer while adding sulphuric acid, milk and amyl alcohol.
2. The test must be repeated if particles of curd are observed.

3. Thorough mixing of milk is necessary.
4. The three fluids should be added gently so that they form three distinct layers.
5. Centrifuge machine must be properly balanced.
6. Always carry out the test in duplicate.
7. Rubber stopper should be dry, clean and without crack.
8. Amyl alcohol must be pure.
9. Before centrifuging see that there is no curdy white material left undissolved.
10. Lactometer should not touch the side of the jar.
11. The milk to be tested should be 2 to 3 hours old after milking. This will allow air and gases to escape from the sample, thereby helping to get accurate specific gravity.
12. The temperature of the milk should be as close to the lactometer temperature as possible.
13. Sample should be thoroughly mixed.

EXERCISE - 14
Determination of Acidity of Milk

Principle/Relevant Information

Acidity of milk is of two types: Natural (apparent) acidity and developed (real) acidity. Natural acidity in milk is due to its constituents such as casein, albumin citrates, phosphates and carbon dioxide. Developed acidity is due to conversion of lactose to lactic acid by the action of microorganisms. Titratable acidity is the measure of both, natural and developed acidity. The acidity can be measured by titrating milk against a standard alkali using an indicator like phenolphthalein and is expressed in terms of lactic acid.

Materials Required

1. Milk Sample
2. N/10 NaOH solution
3. Phenolphthalein indicator

Apparatus

1. Burette with stand
2. Porcelain dish - white Flat bottomed of apparently 100 ml capacity
3. Pipette
4. Tile
5. Stirring glass rod

Procedure

Thoroughly mix the sample and transfer 10 ml milk with the pipette to the porcelain dish. Add few drops of indicator. Now titrate against N/10 NaOH, drop by drop, till a faint pink color develops which remain constant for 10-15 seconds. Note down the burette reading. Repeat the titration for confirmation.

Observation

Sr.No.	Burette reading	Titration No.1	Titration No.2	Titration No.3	Average
1.	Final reading				
2.	Initial reading				
3.	Vol. delivered in ml				

Calculation

Lactic acid reacts with sodium hydroxide as shown below :

$$CH_3CH(OH)COOH + NaOH \longrightarrow CH_3CH(OH)COONa + H_2O$$

According to the above equation, one molecular weight of NaOH neutralizes one molecular weight of lactic acid.

That is

40 gm of NaOH	=	90 gm of lactic acid
1lit of 1N NaOH	=	90 gm of lactic acid
Or 1 c.c. of N/10 NaOH	=	0.009 gm of lactic acid

$$\%\ \text{Acidity} = \frac{\text{N/10 NaOH used} \times 0.009 \times 100}{\text{Wt. of milk (Volume} \times \text{Sp. Gravity of milk)}}$$

Result

Precaution

1. Thorough mixing should be done during titration.
2. Titration should be carried out within 20 seconds.

Questions

1. What substances are responsible for the titratable acidity of
   (a) fresh milk,
   (b) milk which has been stored at room temperature for a few hours ?
2. Which will have higher titratable acidity, buffalo or cow milk ? Why ?
3. How does the stage of lactation affect the titratable acidity of milk?
4. Do you expect any change in acidity of mastitis milk ? If so, why ?

Exercise - 15

Proteins in Milk

AIM

To determine the Total nitrogen, Casein and albumin in the given sample of milk.

Principle

1. **Kjeldahl's Method:** Nitrogen content of total Protein as well as various fractions of nitrogenous matter are estimated by the method described below. Assuming that the milk proteins contain 15.67% of nitrogen, a factor of 6.38 is used for multiplication of the percentage of the Nitrogen estimated actually from various fractions to obtain the percentage of various fractions of nitrogenous matter and total proteins present in the milk.
2. Only the total proteins can be estimated. Proteins of milk are amphoteric in reactions. When neutral formaldehyde is added to milk, CHO (Aldehyde) group of the same contains with the free amino (NH_2) group of the amino acids of proteins, thus allowing the COOH to exert more fully its acidic nature. As a result of which the titratable acidity is increased which is proportional to the quantity of amino acids present and therefore to the percentage of protein. Potassium oxalate is added to eliminate the disturbing effects of soluble Ca salts. A factor of 1.7 is used to get the percentage of protein for multiplication of the increased titratable acidity by formaldehyde (pyne's factor).

$$R.C.H.\ NH_2\ COOH + H.CHO \rightarrow R.C_4\ (NH)\ CH.COOH + H_2O$$

KJELDAHL'S METHOD

Wt. of milk taken for digestion	10 g
Strength of H_2SO_4 used	0.1 N
Strength of NaOH used	0.1 N

		Vol. made	Aliquot taken	Acid used	NaOH required ml	Acid consumed	Acid for white Digest ml
1.	Total Nitrogen	100	25	10	0.6	9.4	97.6
2.	Casein Nitrogen	100	25	10	2.9	7.1	28.4
3.	Albumin Nitrogen	Whole volume taken	Whole volume taken	10	6.1	3.9	3.9

1. Total Protein =

2. Casein $= \dfrac{0.0014 \times 28.4 \times 100 \times 6.8}{10} = 2.54\%$

3. Albumin $= \dfrac{0.0014 \times 3.9 \times 100 \times 6.8}{10} = 0.35\%$

A. Aldehyde Method

1. Burette reading before the additions of formaldehyde.
2. After the addition of formaldehyde
3. Vol. of 0.1N NaOH required by protein (2.1)
4. Percentage of proteins present.

A. Procedure

1. Total Nitrogen (A) Kjeldahl's Method:

10 ml of milk of the given sample was weighed and taken in a kjeldahl digestion flask. A pinch of catalyst consisting of potassium sulphate, copper sulphate and selenium oxide in the proportion of (15 : 2 : 1) was added, 25 ml of concentrated sulphuric acid was added so as to immerse the contents of the flask completely. The content was digested by boiling over a flame, care being taken to avoid any spurting to happen, until a clean liquid was obtained.

The contents cooled and transferred to a 100 ml volumetric flask. The volume was made up to the mark with distilled water.

Micro-Kjeldahl Estimation or Distillation :

25 ml of the diluted digest from the volumetric flask was pipetted into the micro-kjeldahl. It was washed down with water 2 drops of phenolphthalein followed by 15 ml of 40% NaOH were added. They were also washed down

with distilled water. At the tail end of the micro kjeldahl a 100 ml beaker containing 10 ml of known standard H_2SO_4 with a drop of methyl red was kept to absorb the NH_3 released from the digest when the steam was made to bubble through the mixture. After 10 minutes by which time all NH_3 would have been driven off into the acid, kept at the tail end. The beaker was lowered, the tip of the tail end was washed with distilled water into the beaker. The contents of the beaker were then boiled and titrated hot against standard NaOH. The nitrogen content was then calculated from which the total protein and other fractions of proteins were calculated from respective nitrogen content.

B. **Aldehyde Method:** 10 ml of milk was taken in 250 ml conical flask to which 10 drops of phenolphthalein indicator were added, followed by 0.4 ml neutral saturated potassium oxalate solution. After 2 minutes, it was titrated against 0.1N NaOH to a faint pink color. Then 2 ml of neutral formalin was added and well mixed. Titration was done again to a faint pink color with 0.1N

2. **Casein Nitrogen (Moir's Method):** 10 ml of milk are weighed in 100 ml beaker, 50 ml distilled water at 40-42 °C were added to it. Immediately followed by 1.5 ml of 10% acetic acid (1:67N) and these stirred gently by means of a glass rod. The beaker was allowed to stand for 20 minutes after which 4.5 ml of 0.25 N Sodium acetate solution were added. It was allowed to stand for one hour. The precipitated casein solution was then filtered through 9 cm No.42 whatman filter paper. After washing the precipitate in the beaker three times, it was transferred on to the filter paper. Finally the rim of the filter paper was rinsed with a fire at ream of water. After complete filtration the precipitate with filter paper was transferred into the Kjeldahl digestion flask. Catalyst was then added, followed by 5 to 7 ml of concentrated sulphuric acid. Then as described before, it was digested and the volume made up. Nitrogen was estimated by Micro-Kjeldahl apparatus.

Albumin Nitrogen (A.O.A.C Method)

The filtrate obtained under casein in neutralised with 10% NaOH solution and 0.3 ml of acetic acid (1+9) was added and heated on a steam bath till the albumin is completely precipitated. The precipitate is then collected on acid washed filter, and washed with cold water. The precipitate was transferred to the kjeldahl flask and proceeded as already described.

Results

***Percentage of Proteins by Kjeldahl's method*:**

1. Total protein
2. Casein
3. Albumin
4. By Aldehyde method Total protein

Exercise - 16

Lactose in Milk

AIM

To determine the lactose content in the fluid milk by

1. Munson and Walker gravimetric method
2. Polarimetric method and
3. Bock's method

Principle

Lactose solution has the property of reducing the copper sulphate to cuprous oxide. This reaction is proportional to the amount of lactose present in the solution. Hence by finding the amount of cuprous oxide obtained by treating copper sulphate solution into the lactose solution, prepared from the sample of milk, the lactose content is determined from a standard table.

Reagents

1. Copper sulphate solution - 34.639 g of $CuSO_4 . 5H_2O$ dissolved in 500 ml of distilled water.
2. Alkaline tartrate solution - 173 g of Rochelle salt and 50 g of NaOH in 500 ml of distilled water.
3. 0.5 N NaOH solution.

Procedure

25 g of the sample was taken in a 500 ml of volumetric flask to which were added 400 ml of water, 100 ml of copper sulphate solution and 8.8 ml of 0.5 N NaOH solution. The volume was then made up to the mark with distilled water. The contents were mixed well. Casein got precipitated and it was then filtered and the filtrate collected.

50 ml of the filtrate was taken in a beaker and to which 25 ml of copper sulphate solution and 25 ml of alkaline tartrate solution were added. The beaker was heated in controlled flame so that boiling began in four minutes and then boiling was continued exactly for 2 minutes. The beaker was kept covered with watch glass. During heating, the hot solution was then filtered through asbestos mat in a porcelin Gooch crucible using a suction pump. The precipitate of cupric oxide was washed thoroughly with hot water (at 60 °C). The crucible was then dried in hot air oven until 2 consecutive weights agree. The amount of cupricoxide is determined by weighing. The percentage of lactose was determined by Munson and Walker table for lactose.

Munson and Walker gravimetric method

Wt. of copper oxide = 188 mg

From the Table 188 mg of copperoxide = 12.72 mg of Lactose

50 ml of the filtrate contain = 12.72 mg of lactose

500 ml (or 25 g of milk) contain = 12.72 mg of Lactose

$$\therefore \text{ 100 gm of milk will contain} = \frac{127.2 \times 100}{25 \times 1000} = 0.5088$$

II. Polarimetric Method

Reagents

1. Acid mercurie nitrate solution - Dissolve mercury twice its weight of HNO_3 and dilute with 5 - fold volume of water.
2. Mercurie-iodide solution - Dissolve 33.2 g of KI, 13.5 of Hg Cl_2 in 200 ml of acetic acid and 640 ml of water.

Determination

Weight 65.8 g (twice the normal weight) of milk into each of two volumetric flasks, 100 ml and 200 ml respectively. Add 20 ml of acid Hg $(NO_3)_2$ solution or 30 ml of the Hg I_2 solution to each flask, followed by 15 ml of 5% phosphating solution to the 200 ml flask, and sufficient phosphotungstic acid solution to the 100 ml flask to make to mark with H_2O_2 shake both flasks frequently during 15 minutes, filter through dry filter, and polarize (It is preferable to read solution from 200 ml flask in 400 mm tube to reduce error of reading, solution from 100 ml flask may consequently be read in 200 mm tube). Compute percent lactose in sample as follows:

(a) Substract reading of solution from 200 ml flask solution from 100 ml flask (using 200 mm tube)

(b) Multiply difference by two

(c) Subtract result from reading of solution from 100 ml flask

(d) Divide result by 2

III. Bock's Method

Principle

Benedicts solution for the estimation of reducing sugars contains potassium thiocyanate, as well as copper sulphate. In the presence of the KCNS a white precipitate of cuprous thiocyanate is formed in reduction instead of the usual red precipitate of cuprous oxide. The small amount of potassium ferrocyanide also acids in keeping the traces of cuprous oxide in solution. As the precipitate formed is white, the loss of all blue first in the solution, indicating complete reduction of the copper, is readily observed. The alkali used is sodium carbonate which has the advantage over the hydroxides of being less likely to cause destruction of small amounts of sugar. The solution also has the great advantage of being stable for indefinite length of time.

Appraratus required

Burette, 25, 20 and 12 ml, 2 pipettes, 100 ml volumetric flask, conical flask, procelin bits, filter paper, beaker and heating equipment.

Procedure

20 ml of milk is pipetted into 100 ml volumetric flask. From a pipette or a burette 12 ml of 10% sodium tungstate and 12 ml of 2/3 N suphuric acid are added. Mixed by rotating the flask, diluted to mark and filtered. Lactose is determined in the filtrate by titration with Benedict's solution. The titration should be done with the boiling benedicts solution. Calculation in the above titration method is done bearing in mind that 20 ml of Benedict's solution are completely reduced by 0.067 g of lactose.

$$\text{100 g of milk contains} = \frac{0.067 \times 100 \times 100}{X \times \text{sp. gravity of milk} \times 20}$$

where X = the volume of lactose solution required to reduce 25 ml Benedict's solution.

Results

Percentage of lactose by

1. Munson and Walker gravimetric method
2. Bock's method

BOCK'S METHOD

Volume of milk taken 20 ml x

sp. gravity of milk 1.026 y

25 ml of Benedicts solution as reduced by 8.05 ml of solution z

$$100 \text{ ml of solution (or 20 ml of milk)} = \frac{0.067}{8.05} \times 100 \text{ g lactose}$$

$$100 \text{ gm of milk will contain } \frac{Y}{X}$$

$$100 \text{ gm of milk will contain } \frac{0.067}{8.05} \times \frac{100}{20} \times \frac{100}{1.026}$$

$$= 4.24 \text{ g lactose}$$

Percentage of lactose = 4.24

Preparation of Benedict's Solution

$CuSO_4$ (Crystallised)	18.0 g
Na_2CO_3 (Crystallised)	200 g
or	
Anhydrous Na_2CO_3	100 g
Na or K citrate	125 g
KCNS	200 g
Pot. ferrocyanide	
5.0% solution	5 ml
Dist. water to make up	1.0 Lit.

With the aid of heat dissolve the carbonate, citrate and thiocyanate in enough water to make about 80 ml of the mix and filter, if necessary. Dissolve the $CuSO_4$ in 100 ml water separately and pour the solution slowly into the other liquid with constant stirring. Add the ferro cyanide solution and make up to volume to 1.0 L.

Sod. tungstate Solution

10% solution and sodium tungstate in water is prepared. If the salt is not dissolving prepare a hot 10% solution, cool and titrate 25 ml with 10% NaOH and phenolphthalein to a pink colour which lasts for 3 minutes. Add proportional amount of NaOH to the main solution.

Comment

Lactose content in normal milk is 4.5%. In mastitis the lactose content is decreased because lactose is synthesised in the udder tissue. In order to maintain the Isotonicity of the affected milk with blood the chloride content increases. The lactose chloride number or kosteler number chloride% / lactose % should be in the ratio of 1:35. If the ratio of chloride increases above 1 mastitis can be suspected.

EXERCISE - 17

Citric Acid in Milk

AIM

To determine the citric acid content of the milk.

Reagents Used :

1. Potassium bromide solution - 15 g dissolved in 40 ml of water
2. Ferrous sulphate solution 40 g of $FeSO_4$, 7 H_2 is dissolved in 100 ml of water containing 1 ml of H_2SO_4.
3. Potassium permanganate solution : 5 g dissolved in 100 ml of water (5%).
4. H_2SO_4 solution (1 + 1)
5. Lead acetate solution 75 g normal lead acetate is dissolved in 250 ml of distilled water, contained 1 ml of glacial acetic acid.

Procedure

1. **Preparation of Sample:** 50 g of milk is taken in a 150 ml beaker to which100 mg of tartaric acid and 6 ml of H_2SO_4 were added. Place the mixture on a steam bath for 15 minutes. The 3 ml of 20% phosphotungstic acid solution was added, mixed well and returned to steam bath for 5 minutes. The contents are transferred to a 250 ml volumetric flask with 95 percent alcohol, cooled and diluted to the mark with alcohol. The contents of the volumetric flask are well mixed and filtered 200 ml of the clear filtrate is taken into a centrifuge bottle.
2. **Determination:** To the solution in the centrifuge bottle 10 ml of lead acetate solution is added and shaken vigorously for 2 minutes and centrifuged at about 1000 r.p.m. for 15 minutes. The supernatant liquid is carefully decanted from the precipitated lead salts and tested with a small quantity of lead solution. If

a precipitate forms, again it is centrifuged. If sediment is left centrifuging is repeated increasing the time and speed. The bottle is drained thoroughly by inverting it for few minutes. To the lead salts in the centrifuge bottle, 150 ml of water is added, mixed well and H_2S is passed to saturation. The content is transferred to a 250 ml volumetric flask and the volume is made up and filtered. 200 ml of the filtrate is pipetted into 500 ml Erlenmeyer flask and evaporate to 75 ml. Then cooled to which 10 ml of H_2SO_4 (1+1) and 5 ml of KBr solution are added. The mixture is heated to 48-50 °C, allowed to stand for 5 minutes and 50 ml of $KMnO_4$ solution was added, mixed and allowed to stand for one minute. Then the flask is stoppered, mixed for 1 minute and allowed to stand for 3 minutes. During this time there should be a heavy deposit of MnO_2. If necessary, add to assure an excess of oxidising agent. If at any time during the oxidation the precipitated MnO_2 disappears, discard the determination and repeat, using more K MnO_4. MnO_2 is removed with $FeSO_4$ solution (about 20 ml), cooled 15°C, stoppered, mixed vigorously for several minutes and placed in the refrigerator overnight. The supernatent fluid is decanted into a thin tight stamped pad of asbestos in a Gooch crucible. It is important that filtration is completed as quickly as possible. The volume of filtrate is noted (S in the formula) and the filtrate is used to transfer the precipitate to the crucible. The contents of the crucible is washed with 50 ml of ice cold water. Then it is dried by aspirating with dry air in a vacuum desiccator and weighed. The contents are treated to remove the penta Bromoacetone with 3 portions 20 ml each of alcohol and 3 portions 20 ml each of ether. It is dried and weighed. The difference in the two weights represents the weight of Penta Bromoacetone.

Then the mg of citric acid in the portion is calculated by the formula

$$x = 0.695 P + 0.028 S,$$

where x = mg of citric acid in the portion taken for analysis P = mg of Penta Bromoacetone, S = ml of the filtrate.

Reaction

Citric acid in milk serum was converted into Pentabromoacetone which is precipitated out. This is filtered through Gooch crucible in which the pentabromoacetone precipitate remained. When alcohol and ether were added, pentabromoacetone got dissolved; difference in weights gave the weight of pentabromoacetone formed.

K MnO_4 first oxidised the citric acid to acetone dicarboxylic acid with liberation of CO_2 as follows:

acetone dicarboxylic acid reacts with bromine above forming Pentabromoacetone.

For Example:

Wt. of milk taken	=	50 g
Volume made up to	=	250 ml
Volume taken for analysis	=	50 ml
Difference in weights or Wt. of Pentabromoacetone	=	0.0352 g
Citric acid in 50 ml of serum	=	$0.424\left(0.0352 + \frac{0.005 \times 50}{100}\right)$
	=	0.0160 g
Percentage of citric acid	=	0.0160 × 5 × 2 = 0.160 g%

EXERCISE - 18

Ash in Milk

AIM

To determine the ash-content in the given sample of milk.

Principle

The milk is evaporated to dryness and the dry residues is ignited until all the carbon is burnt away and a white ash is obtained. The white ash represents the mineral matter of milk.

Procedure

About 25 ml of well mixed milk is taken in a weighted silica dish and the weight of milk is determined by reweighing. The dish is heated over a flame, adjusting the flame to avoid spurting and skin formation. The material was ignited till it is smoke free and transferred to a muffle furnace at 500 °C, till a white ash is obtained. From the weight of the ash, the percentage can be calculated.

During the ignition some loss of alkali chloride occurs by volatalisation. To prevent this the initial ash is dissolved in about 50 ml of distilled water and warmed gently. It is filtered through a number 40, 9 cm whatman filter paper, collecting the filtrate in a small beaker. Wash the filter paper with few ml of distilled water, return the paper and contents to the basin and carefully dry over the burner. The dish and contents are placed in the furnace and ignited at bright red heat, until a white ash is obtained. The dish is removed and cooled and the filtrate is added and evaporated to dryness, cooled and weighed. This procedure prevents volatilisation of alkali chlorides. Extraction of partially ignited solids removes the alkalies and allows the residue to be fully ignited without loss.

Wt. of milk taken

Wt. of ash obtained

Percentage of ash

Loss of alkali chlorides can also be prevented by adding 6 ml of Nitric acid to 20 ml of milk sample and evaporate to dryness and finally ignited at a temperature below redness until the ash is free from carbon. The nitric acid fixes the alkali chlorides.

Limitations

Ashing or incineration of milk destroys the organic compounds leaving the mineral constituents behind. Most tabulations of the gross compositions of milk include a figure for milk ash of approximately 0.7%. Actually the ash does not truly represent the salt composition of milk for the following reasons.

1. Organic radicals such as citrates are destroyed by incineration.
2. Phosphorus and sulphur of the milk proteins and lipids appear in the ash. The organic sulphur is present in a non-salt form and the phosphorus occurs in a sort of "Semi salt" form having part of its potentiality ionizable groups involving ester linkage with hydroxyl groups of the protein or lipids.
3. Carbonates appear in milk ash arise in part from carbondioxide produced by decomposition of the organic materials.
4. Oxidation during incineration results in the formation of oxides of the metals and
5. Some salt materials such as Sodium Chlorides and Carbonates may be lost by volatalization. But this complication can be avoided by properly regulating the temperature to not over 600 °C.

The ash content gives a rough idea of the total mineral content and of course, ashing is an useful technique for getting rid of organic matter as a preclude to analysis for many of the mineral elements.

Result

The percentage of ash in the given sample is

Basic TCP is held by casein to stabilise the colloidal solution. As basic TCP is likely to interfere with the reaction it is precipitated by neutral Pot, oxalate.

$$Ca_3P_2O_8 + 3.K_2 \text{ oxalate} \rightarrow 3\text{Ca oxalate} + 2K_3PO_4$$

Colloidal TCP   neutral   ppt   (alkaline in reaction)

Exercise - 19

Calcium in Milk

AIM

To determine the calcium content in the given sample of milk.

Principle

The calcium in the soluble ash fraction is precipitated as calcium oxalate using saturated ammonium oxalic acid liberated from the calcium-oxalate by the action of sulphuric acid is titrated against standard potassium permanganate solution and from the titre value the calcium content can be calculated.

Procedure

The ash obtained in the previous experiment was dissolved in HCl (1+ 4) and transferred into 100 ml volumetric flask and the volume was made up to the mark. 20 ml of this solution was taken in 250 ml beaker and diluted with water to 100 ml. Few drops of methyl orange indicator (0.05%) was added. The solution was made slightly alkaline by adding ammonia (1+ 4). Then with HCl (1+ 4) solution was adjusted to faintly acidic and then 10 ml of 0.5N HCl was added. Next 10 ml of 2.5% oxalic acid solution was added and the content was boiled for 15 minutes when 15 ml of saturated of ammonium oxalate solution was slowly added with constant stirring. When the precipitate became granular the heating was stopped and cooled. 8 ml of 20% sodium acetate solution was then added with constant stirring and then the beaker was allowed to stand for 12 hours. The content was then filtered. The precipitate was washed with hot water until it was free from oxalates. Then the filter paper was pierced with stirring rod and the precipitate was washed down into the beaker with hot water. Then 10 ml of H_2SO_4 (1+1) was added, the content was heated to boiling. Titration was done with standard K MnO_4 solution, while the solution was still hot. Finally the filter paper was added to the solution and titration was completed.

Reactions

HCl is added to provide an acid medium in which the precipitation of Ca as $Ca_3\ (PO_4)_2$ is prevented and $CaCl_2$ is formed oxalic acid is added to produce ca-oxalase ppt.

$$CaO + 2\ HCl \rightarrow CaCl_2 + H_2O$$

$$CaCl_2 + (COOH)_2 \qquad CaC_2O_4 + 2\ HCl$$

$(NH_4)_2C_2O_4$ is added to provide medium which unfavourable for co-precipitation of Mg salts by forming NH_4Cl

$$(NH_4)_2\ C_2O_4 + 2\ HCl \qquad 2\ NH_4Cl + (COOH)_2$$

Sodium acetate is added to prevent the formation of $(NH_4)_2C_2O_4$ crystals which is in saturated state at that temperature. Boiling before titration with $K\ MnO_4$ is done to make the oxidation easy.

$$5\ CaC_2O_4 + 5H_2SO_4 \qquad 5\ CaSO_4 + 5\ (COOH)_2$$

$$2KMnO_4 + 3H_2SO_4 \qquad K_2\ SO_4 + 2\ MnSO_4 + 3\ H_2O + 5[O]$$

$$5\ (COOH)_2 + 5\ [O] \qquad 10\ CO_2 + 5\ H_2O$$

Wt. of milk taken	= 20.54 g
Vol. of ash solution made	= 100 ml
Vol. of solution taken for analysis	= 20 ml
Strength of $KMnO_4$ soln. used	= 0.05 N
Vol. of $K\ Mn\ O_4$ soln. used	= 5.5 ml
1 ml of 0.05 N $KMnO_4$ solution	= 0.001 g calcium
Calcium content in the given sample of milk	$= \dfrac{5.5 \times 0.001 \times 5 \times 100}{20.54}$
	= 0.134%

Result

Calcium content in the given sample of milk is = 0.134%

EXERCISE - 20

Phosphorus in Milk

AIM

To determine the phosphorus content of milk.

Principle

The phosphorus in the soluble ash fraction is precipitated as phospho ammonium molybdate using ammonium molybdate solution. The precipitate is then dissolved in excess of known amount of N/10 NaOH solution and this excess is back titrated against N/10 HNO_3 solution. From the titre value phosphorus content is estimated.

Procedure

20 ml of the ash solution prepared previously was taken in a beaker and 10 ml of HNO_3 was added to it. Ammonia was then added with simultaneous stirring. The content was diluted to 75 to 100 ml and the temperature was adjusted to 25 to 30 °C. Freshly prepared molybdate solution 20-25 ml was added and vigorously stirred for 15 to 20 minutes and then allowed to stand for few hours. Filtration was done and the precipitate was washed twice with water 25 to 30 ml. From the beaker the precipitate was transferred to the filter paper and washed with cold water, until the filtrate was free from molybdic acid, which was found by adding a drop of phenolphthalein to 1/2 test tube of filtrate and then to it by adding 2 drops of standard alkali (pink colour indicates free from acid). The precipitate was then transferred with the filter paper to the original beaker and dissolved with a known volume of excess of standard alkali with phenolphthalein as indicator. The excess of alkali was titrated back by standard HNO_3 solution.

Molybdate reagent is prepared as follows. 100 g of Mo O_4 is dissolved in a mixture of 144 ml of NH_4 OH and 271 ml of water. Then it is cooled and poured slowly with constant stirring into a cool mixture of 489 ml of HNO_3 and 1148 ml of water. It is kept in a warm place for several days, or the portion heated to 40 °C deposits no yellow precipitate of NH_4 phosphomolybdate. Then it is decanted from any sediment and preserved in glass stoppered vessel. To 100 ml of molybdate solution 5 ml of HNO_3 are added and filtered before use.

Reactions

$$H_3PO_4 + 12\,(NH_4)_2MoO_4 + 21\,HNO_3 \rightarrow (NH_4)_3PO_4 + 12\,MoO_3 + 21\,NH_4NO_3 + 12\,H_2O$$

$$(NH_4)_3PO_4 12\,MoO_3 + 23\,NaOH \quad 11\,Na_2MoO_4 + (NH_4)_2MoO_4 + Na(NH_4)HPO_4 + 11\,H_2O$$

The temperature of precipitation is important as only at that temperature the ammonium phosphomolybdate form, 1 molecule of phosphorus requires 23 molecules of NaOH.

Equivalent weight of phosphorus is 23.

1000 ml of N solution should contain 30.98/23 = 1.347 g of P,
or 1 ml of 0.1 N NaOH will contain 0.0001347 g of P.

Wt. of milk taken		20.54 g
Vol. of ash solution made up		100 ml
Vol. of ash solution taken		20 ml
Strength of NaOH solution used		0.1 N
Strength of HNO_3 solution used		0.1 N
Volume of standard Na OH added		10 ml
Volume of Std HNO_3 used back titration		74 ml
Volume of NaOH used up (10-7)		3 ml
1 ml of 0.1N NaOH solution	=	0.0001347 g of P
or		0.000307 g of P_2O_5

$$\text{Percentage of } P_2O_5 \text{ in milk} = \frac{3 \times 0.000307 \times 5 \times 100}{20.54} = 0.022$$

Result

The percentage of P_2O_5 in the given sample of milk.

EXERCISE - 21

Magnesium in Milk

AIM

To detect the Magnesium content of milk.

Procedure

To the filtrate obtained from calcium determination 40 ml of HNO_3 was added and evaporated to dryness to decompose the ammonium salts. The dry mass was dissolved in HCl (1+4) and the volume was made to 100 ml with water. 5 ml of 10% sodium citrate solution and 10 ml of 10% Na_2HPO_4 solution was then added. The solution was made faintly alkaline with NH_3(1+4) and then 25 ml of ammonia was added and the content was allowed to stand in a cool place for about 12 hours. The content was filterated and the precipitate was washed with ammonia (1+9). Precipitate was dissolved in HCl (1+ 4) and again precipitated as before. After some hours the content was filtered and the precipitate was washed until free from chloride with ammonia (1+ 9). The precipitate was then ignited and weighed as $Mg_2 P_2 O_7$. The result was reported as grams MgO/100 g or Mg/100 g of milk.

Reactions

Calcium should be removed to prevent its co-precipitation with magnesium.

HNO_3 is used for decomposing the ammonium salts. Sodium citrate is added for its buffer action.

$$MgO + 2\,HCl \rightarrow MgCl_2 + H_2O$$

$$Na_2HPO_4 + MgCl_2 + NH_4OH \rightarrow Mg(NH_4)PO_4 + 2\,NaCl + H_2O$$

(i) To get $Mg(NH_4)PO_4$ only, which give on ignition $Mg_2 P_2 O_7$

$$2\,Mg(NH_4)PO_4 \rightarrow Mg_2P_2O_7 + 2\,NH_3 + H_2O$$

(ii) To avoid precipitation of Mg ++ as $Mg_3\ (PO_4)_2$ and also co-precipitation of $(NH_4)_3\ PO_4$

(iii) In the presence of excess of NH_4, precipitation of $Mg_3\ (PO_4)$ and $Mg(NH_4)4\ (PO_4)_2$ may also take place. The former is unchanged on ignition and only the latter changes to $Mg_2P_2O_7$ (desired).

(iv) If original solution contains too much of Na ions, precipitation of $MgNaPO_4$ may also happen. Heating carefully at low temperature avoids this.

Result

Percentage of Magnesium is - 0.02 g%

Wt. of milk taken - 20.54 g

Vol. of ash solution made up - 100 ml

Vol. of ash solution taken - 20 ml

Wt. of ignited mass (in 20.54 g of milk) - 19.2 mg or 0.0192 g of $Mg_2P_2O_7$

100 g of milk contains $\frac{0.0192 \times 100}{20.54} = 0.0934$ g of $Mg_2P_2O_7$

Percentage of Magnesium = 0.0934 × 0.21622 = 0.02

To convert $Mg_2\ P_2O_7$ to Magnesium, use the factor 0.21622

to Magnesium oxide the factor 0.36036

Exercise - 22

Chlorides in Milk

AIM

To determine the chlorine content in the given sample of milk.

Principle

A known volume of excess of $AgNO_3$ solution of known strength is added to milk, when chlorine of milk combines with silver of $AgNO_3$ solution to form AgCl; amount of $AgNO_3$ thus utilised is obtained by titration with standard KCNS solution, using iron alum as indicator. KCNS solution, after exhausting the surplus $AgNO_3$ left reacts with iron alum to give brown color which is taken as end point.

Reagents

1. Standard silver nitrate solution (0.05N)
2. Standard potassium thiocyanate solution (0.05N)
3. Standard potassium chloride (0.05N) to be used in standardization of silver nitrate and KCNS solutions
4. Concentrated Nitric acid
5. Saturated Iron-alum in 10% nitric acid, prepared by boiling excess of iron alum in 10% nitric acid cooling and filtering.

Procedure

10 g of the given sample of milk is taken in a 250 ml Erlenmeyer's flask and 10 ml of standard silver nitrate solution is added to it. Then 10 ml of concentrated $H\ NO_3$ is added to the mixture. The contents in the flask is then digested until all the casein got dissolved and the evolution of reddish brown fumes were stopped.

The excess of $Ag\ NO_3$ was titrated with standard potassium thiocyanate, using Iron alum solution as indicator to permanent brown colour. The amount of chlorine was calculated from the volume of $Ag\ NO_3$ solution used up by chlorine in milk.

Reactions

$Cl^- + AgNO_3 \rightarrow AgCl + NO_3$

$AgNO_3 + KCNS \rightarrow AgCNS + KNO_3$

Ferric alum + KCNS → Ferric sulphocyanide + Pot. alum

(Brown - end pt)

Result

The percentage of chlorine in the given sample of milk is 0.104%

1 ml of 0.05N Ag NO_3 solution = 0.00177 g of chlorine

Volume of Ag NO_3 (0.05N) solution added 10 ml

Volume of KCNS (0.05N) solution utilised 4.1 ml

∴ vol. of Ag NO_3 (0.05N) combined with

10 g of milk 5.9 ml

5.9 ml of Ag NO_3 (0.05N) represents 0.00177 × 5.9 = 0.0104 g chlorine

Hence 10 g of milk contain 0.0104 g chlorine

∴ 100 g will contain 0.104 g of chlorine

$$\text{Percentage of chloride present in the milk} = \frac{0.104 \times 58.48}{35.46} = 0.172 \text{ g chlorine}$$

Exercise - 23
Copper in Milk

AIM

To determine the copper content in the given sample of milk.

Procedure

100 g of milk sample was weighed accurately to which was added 5 ml of H_2SO_4 (1+1) in a proceline dish and evaporated to dryness and heated to until copious fumes of SO_3 was given off. Pb SO_4 (if any) was filtered and washed in dilute alcohol (50% by volume) to free it from Fe. The filtrate was collected into 250 ml beaker, boiled to remove alcohol, adjusted to 200 ml, 1 gm of NH_4 Cl added. Heated to boil, saturated with H_2S and to remove precipitated sulphur and allowed to stand for 2 hours (until the supernatant fluid became clear). CuS was filtered and washed with water. The precipitate was dissolved in hot HNO_3 (1+3), cooled and the content was made slightly alkaline by adding NH_4 OH with phenolphthalien as indicator. 10 ml of NH_4 NO_3 was added and the volume was adjusted to 100 ml and boiled gently until the solution showed neutral reaction (to litmus paper). The solution was then filtered to remove any Fe that might be present and the volume of the filtrate was made up to 100 ml. To an aliquot, added 3 drops of K_4 $Fe(CN)_6$ solution. The colour obtained was compared with standards containing 0.1, 0.2, 0.3, 0.4 and 0.5 mg of copper. The standards were prepared by assuring the corresponding quantity of the standard solution adding phenolphthalien indicator, and slight excess of NH_4OH solution and 10 ml of NH_4NO_3 solution, boiling the solution until neutral to red litmus, cooling and adding 3 drops of K_4 $Fe(CN)_6$ solution. The colorimetric comparison was made in 100 ml nessler jar.

Reagents

1. **NH_4 NO_3 solution:** Dissolve 10 g in 100 ml of water.
2. **Pot. ferro cyanide solution:** Dissolve 3.5 g of K_4 Fe $(CN)_3$ H_2O in 100 ml of distilled water. This should be prepared fresh.

3. **Standard copper solution:** Dissolve 20 g of $CuSO_4$ $5H_2O$ in 500 ml of water containing 1 ml H_2 SO_4. Determine Cu in 50 ml of this solution as CuO by precipitation with KOH. Dilute stock solution so that 1 ml contain 0.1 mg of copper.

Reactions

During heating with H_2SO_4 all metals in milk get converted to their sulphate salts. $PbSO_4$ being insoluble filtered out. Next alcohol is added to precipitate $FeSO_4$. Both $PbSO_4$ and $FeSO_4$ should be removed to avoid remaining Fe will be in solution. Cu by the action of H_2S gets converted to CuS and gets precipitated. Also remaining FeS gets precipitated. Precipitate taken and dissolved in HNO_3 so that CuS is transformed to $Cu(NO_3)_2$ on adding NH_4 OH_2. Fe changed to Fe $(OH)_3$ in precipitate and on adding NH_4 NO_3 to neutralize the medium Fe $(OH)_3$ is thrown away. The filtrate contain Ammonium Copper complex which is formed by decomposition of Cu $(NO_3)_2$ by NH_4 OH. On adding the neutral filtrate K_4 $Fe(CN)_6$, Cu salt change to Cu_2 Fe $(CN)_4$ to give a brown colour

K_4 $Fe(CN)_6$ acts only in neutral medium

$$CuSO_4 + H_2S \rightarrow CuS + H_2 SO_4$$

$$CuS + 2\ HNO_3 \rightarrow Cu\ (NO_3)_2 + H_2S$$

Exercise - 24
Freezing Point of Milk

AIM

To determine the freezing point of milk using Hortvet Cryoscope.

Principle

The freezing point of any solution is proportional to its osmotic pressure or the concentration of the dissolved particles. The osmotic pressure and so the freezing point of milk is controlled chiefly by, the salts and lactose in milk and is remarkably constant. Addition of water will alter the osmatic pressure and so alter the freezing point. As the alteration of the freezing point is proportional to the amount of water added, the freezing point of milk gives a fairly accurate measure of the amount of water added. So this test is the only reliable test for detecting the adulteration of milk with water.

The technique is to super cool the milk and then seed with ice crystals when the temperature rapidly rises to the freezing point as ice separates out. In case of pure water the temperature remains constant until all the water is frozen. But when milk is super cooled and seeded the temperature rises to a point of momentary stability, after which it gradually falls. This different behaviour of milk is due to the separation of ice crystals which causes a concentration of solutes, resulting in further depression in freezing point.

In milk lactose is the principal contributor to freezing point depression. Souring involves a net increase in the number of molecules in solution due to degradation of lactose, results in lowering of freezing point.

Correction of thermometer

Standardizing liquid	Actual reading	Intervals	Known correct reading	Intervals
Distilled water	+ 0.121	0.439	0.000	0.422
7 g sucrose soln.	– 0.318	–0.422		
10 g sucrose soln.	– 0.523	0.205	– 0.621	0.199

Example

Reading - 0.298°C

Apparent freezing point =– (0.298 + 0.12°C) = – 0.419°C

Correction factor for any reading would be 0.422/0.439 = 0.9613

Actual freezing point of milk = – 0.419°C × 0.9613 = – 0.413°C

If reading is – 0.451°C

The reading is lower than 7 g sucrose solution – 0.318°C – (– 0.451°C)

= 0.133°C

Correction factor = 0.199/0.205 = 0.9709

So the correction is 0.133 × 0.9707 = 0.129

∴ Corrected freezing point = –0.422 - 0.129 = –0.551°C

It may be done as follows

Apparent freezing point = – (0.451°C + 0.121°C) = – .572°C

Correction factor = 621 / 644 = 0.964

Actual Freezing Point = –0.572°C × 0.964 = – 0.551°C

Procedure

The freezing point determination should be made only on milks that have an acidity not more than 0.18% as lactic acid.

400 ml of previous cooled Ether (to less than 10 °C) was poured through the vertical portion of the T-tube using a funnel. The apparatus was made airtight and connection was made to the suction pump so as to facilitate continuous vaporization of ether which lowered down the temperature in the flask. The pump was worked until the control thermometer registered a temperature –3°C. At this point the volume of ether in the flask was measured and requirement was again added to bring the volume to 400 ml. Sufficient water (<10 °C, 40 ml) was poured into the freezing tube so that the bulb of the thermometer was completely submerged. The

thermometer with the stirrer at their respective positions in the cork were inserted into the freezing tube and the freezing tube was lowered into the metallic case. A little quantity of alcohol is sufficient to fill the space between the tube and the case was poured for better conduction. The stirrer was worked continuously at an uniform rate (1 or 2 / sec). The temperature of the cooling bath (ether) was kept at –2.5 °C by manipulating the suction pump. The stirrer was worked until a super cooling of sample of –1 to – 1.2 °C was obtained. At that time a piece of ice crystal was inserted into the freezing tube. The liquid began to freeze at this time as noted by the rapid rise of the mercury column. The stirrer was manipulated slowly and carefully 3 or 4 times as the mercury column approached its highest point. By means of a mallet the top of the thermometer was tapped continuously a number of times until the mercury column remains stationery for at least 1 minute. The exact reading of the thermometer was noted carefully.

Next the tube was rinsed with the cold sample of milk (<10 °C) to be tested and the same was poured to sufficient level in the tube (40 ml). The freezing point was determined as described above after the equipment was arranged properly. Supper cooling was done at –1.5 to –1.7 °C (i.e. 1-1–2 °C below the probable freezing point). The reading was taken carefully.

To determine the correction factor due to variation in the internal bore of the thermometer which usually happens, calibration was made by using 7% and 10% sucrose solution whose freezing point depression being – 0.422 °C and 0.621°C respectively. This range covers the freezing point of normal milk, so any fluctuation in the thermometer can easily be detected. The method followed was the same as described for water.

Result

Freezing point of the milk sample is - 0.545 °C

	Observed	**Actual**
Freezing point of water	0.000 °C	0.000 °C
Freezing point of 7% sucrose soln.	– 0.421 °C	– 0.422 °C
Freezing point of 10% sucrose soln.	– 0.624 °C	– 0.621 °C
Interval between the readings	– 0.203 °C	– 0.199 °C
Freezing point of milk sample	– 0.546 °C	

Correction factor = 0.199 / 0.203 = 0.980

Actual F.P. of the sample of milk = 0.422 + (0.546 - 0.421) × 0.980

= –0.545 °C

Exercise - 25
Detection of Adulterants in Milk

Materials Required

1. Milk sample
2. Test tubes
3. Different reagents

Procedure

1. **Urea Detection**
   (a) Take 2 ml milk
   (b) Add 2 ml Urea reagent and mix well
   Distinct yellow Color - Urea has been added to milk
2. **Ammonia Fertilizers Detection**
   (a) Take 1 ml milk
   (b) Add 2 ml Ammonia reagent and mix well
   Brownish Color - Ammonia Fertilizers have been added to milk
3. **Nitrate Fertilizers/Pond Water Detection**
   (a) Take 1 ml milk
   (b) Add 2 ml Nitrate reagent along the sides of the tube
   Blue ring at the junction - Nitrate Fertilizers/pond water have been added to milk.
4. **Starch and Cereal Flours Detection**
   (a) Boil 5 ml of well-mixed milk
   (b) Cool and add few drops of Starch reagent
   Blue Color - Starch and Cereal flours have been added to milk.

5. **Sugar/Sucrose Detection**
   (a) Take 1 ml milk
   b) Add 1 ml sugar reagent
   c) Place in boiling water for 3-5 minutes
   Red Colour - Sugar has been added to milk
6. **Glucose Detection**
   (a) Take 1 ml milk
   (b) Add 1 ml Glucose reagent - I
   (c) Place in boiling water for 3 minutes and cool
   (d) Add 1 ml Glucose reagent - II
   Deep Blue Color - Glucose has been added to milk
7. **Salt Detection**
   (a) Take 5 ml salt reagent - I
   (b) Add a few drops of Salt reagent - 11 (red color develops)
   (c) Add 1 ml milk and mix well
   Yellow color - Salt has been added to milk
8. **Neutralizers Detection**
   (a) Take 5 ml milk
   (b) Add a few drops of Neutralizer reagent - II and mix well
   Rose Red / pink color - Neutralizers have been added to milk
9. **Hydrogen Peroxide Detection**
   (a) Take 1 ml milk
   (b) Add a few drops of Hydrogen peroxide reagent and mix well
   Pink/Red color - Hydrogen peroxide has been added to milk
10. **Formalin Detection**
    (a) Take 10 ml milk
    (b) Add 5 ml Formalin reagent along the sides of the tube
    Deep purple/Violet ring at the junction - Formalin has been added to milk

Procedure

1. **Cane Sugar (Resorcin Hydrochloride Reaction)**

   To about 2 ml of milk, 1 ml of HCl (sp.gr. 1.19) and about 0.1 g of resorcinol were added and the mixture was boiled. In the presence of cane sugar red colour would develop.

2. **Starch**

   About 10 ml of milk was taken in a test tube and brought to boiling then cooled and 1 ml of 5% solution of iodine was added. In the presence of starch a blue colour developed.

3. **Milk-Powder**

   Take about 10 ml of sample in a test tube and add 1 drop of formalin and place it in the incubator at 60 °C for about 10 minutes. In the presence of milk powder a peculiar odour is felt. In the case of pure milk there is no odour.

4. **Calcium Carbonate**

   To 10 ml of milk in a test tube 1 ml of strong HCl was added. The presence of carbonate was indicated by effervescence.

Observation and Result

Serial No.	Tests	Results
1.	Urea detection test	
2.	Ammonia fertilizer detection test	
3.	Nitrate fertilizer test	
4.	Starch & Cereal Detection Test	
5.	Sugar/Sucrose detection	
6.	Glucose Test	
7.	Salt Test	
8.	Neutralizer detection test	
9.	Hydrogen Peroxide detection test	
10.	Formalin detection test	

EXERCISE - 26
Preservatives in Milk

AIM

To detect the preservatives used in milk.

Procedure

1. (a) **Formalin**

   To 15 ml of milk in a test tube 1 ml of 1% phenyl hydrazin hydrochloride solution, few drops of 1% $FeCl_3$ solution and HCl was added. The presence of HCHO was indicated by the formation of a red colour which changed to orange yellow after sometime.

   (b) **Hehner's Sulphuric Acid Test**

   Take 10 ml of milk in a test tube, add 1 drop of $FeCl_3$ solution and dilute the milk with 30 ml of water. Pour pure concentrated sulphuric acid slowly down the sides of the tube. A violet colouration appears at the junction of the liquids in the presence of formalin. This test is fairly delicate and will detect 1 part of HCHO in 200,000 parts of milk.

   (c) **Amidol Test (Magnat and Marion)**

   A little of amidol or amido phenol is sprinkled on the surface of a little warm milk in a porcelain basin. If formalin is present milk becomes yellow and in its absence milk is pinkish in colour.

   (d) **Leach's Test**

   Take 5 ml of milk in porcelain dish and add an equal quantity of water. Add 15 ml of concentrated HCl acid and a trace of ferric chloride and heat gently. The presence of HCHO is indicated by production of violet colouration.

2. **Salicylic Acid**

To 10 ml of milk, 5 ml of HCl (1 + 3) was added and mixed. Then filtered and the curd was extracted with 50 to 100 ml of ether. The ether layer was washed twice with 5 ml of water. Greater portion of ether was evaporated in a porcelain dish on a steam bath and the remainder was allowed to evaporate spontaneously. A drop of 0.5% neutral $FeCl_3$ solution was added to the dish. A violet colour indicates salicylic acid.

3. **Benzoic Acid:**

100 ml of milk was acidified, filtered and extracted with ether as in the case of salicylic acid, crystals of benzoic acid formed in ether. The crystalline deposit was dissolved in water and to which few drops of NH_4 OH was added to make the solution alkaline. The excess of NH_3 was evaporated out. The residue was dissolved in hot water and then a few drops of 0.5% $FeCl_3$ solution was added. Brown precipitate of Ferric benzoate was formed.

4. (a) **Boric Acid**

A strip of turmeric paper was immersed in a sample of milk acidified with HCl in the proportion of 7 ml of con.acid in 100 ml of sample. The paper was allowed to dry spontaneously. In case of boric acid the paper became red coloured which changed to dark blue green colour by NH_4OH_3 but restored by acid.

(b) **Glycerine Test**

To a test tube 3/4 full of milk and phenolphthalein and the N/10 NaOH added until a faint pink colouration is developed. Divide the milk into 2 equal portions. Add to one an equal volume of distlled water. Add to the other an equal vol. of neutral 50% glycerol (50% neutral aqueous solution of glycerol). Mixed thoroughly and compare the two colours. In the presence of boric acid there will be distinct decolouration of milk in the glycerol tube, while the water tube becomes somewhat deeper in shade. Should the test be inconclusive, it should be repeated using in addition for comparative purposes a sample of pure fresh milk known to be free from Boric acid.

(c) **Flame test**

A little of milk is evaporated to dryness in a porcelain crucible and the residue burnt to ash. Then a little dilute H_2SO_4 acid and some methyl-alcohol are added and light applied. The presence of boric acid and borates causes the production of green flame.

5. **Hydrogen Peroxide**

Its presence may be detected by shaking 5 ml of milk with a few crystals of paraphenylenediamine hydrochloride. A blue colour will be developed almost immediately owing to the liberation of atomic oxygen released by the enzyme peroxidase. This test fails if the milk has been heated to temperature sufficiently high to destroy the enzyme. The test is uncertain if the milk has been heated to 165°F or more.

6. **Tests for carbonates and bi-carbonates**

(a) About 10 ml of milk are mixed with equal volume of 95% alcohol in a test tube and a few drops of 1% solution of resolic acid added and mixed. If an alkali is present a rose-red color is produced. Otherwise a bluish yellow colour is produced.

(b) The ash from milk to be tested if treated with drop of HCl acid will effervesce if $NaCO_3$ is present.

Exercise - 27
Colouring Matter in Milk

Annatto and coal-tar dyes (Azo dyes) are added to milk to make it look rich, in the case of skim milk or watered milk to cover the sign of such adulterants.

Test for Annatto

To 10 ml of milk in a test tube add 10 ml of ether, shake vigorously and let stand, until ether separates on the top. If annatto is present the layer of ether will be yellow.

Test for Coal Tar Dyes

The azo colors are commonly used in coloring milk may be detected by adding 10 ml of milk to 10 ml of strong HCl. On mixing pink colouration appears.

Test for Blood in Milk

Portion of sample at 50°C is centrifuged. If blood is present, bright red deposit is seen at the bottom. As a confirmatory test portion of the red deposit is treated with a drop of glacial acetic acid, evaporated very gently, while nearly dry the slide is examined under microscope, the presence of characteristic brown rhomboid crystals of haemin hydrochloride indicates blood.

EXERCISE - 28
Microscopic Examination of Milk

AIM

To judge the quality of a sample of milk by estimating the bacterial population by direct microscopic examination.

Principle

An exactly known volume of milk (0.01 ml) is spread over an exact area (1 sq.cm) so that each microscopic field examined represents a quantitative aliquot of the sample. The bacteria is stained with methylene blue. The area of the microscopic field is found out by stage micrometer. The area of the milk film in 100 sq.mm (1 sq.cm) divided by the area of the microscopic field yields the number of microscopic fields in the film. Since the film contains 0.01 ml, the number of fields per film multiplied by 100 gives the number of fields per ml of milk. This is termed as microscopic factor. The bacteria in an microscopic field is taken by counting a few fields and average is taken and multiplied by the microscopic factor (MF) gives the total number of bacteria per ml.

Procedure

Calculation of Microscopic Factor

With the help of the stage micrometer, using oil immersion objective, the diameter of the field is found out. The micrometer is calibrated to read each mm in 100 divisions, giving each division as 0.01 mm. From this the M.F. is calculated by the formula $XY/\pi r^2$ where X is the 100 sq.mm are covered by 0.01 ml of milk. Y is 100, which is the number of 0.01 ml portion of milk in 1 ml and r is the radius in mm of the microscopic field. Field diameter measuring approximately 0.206 mm, 0.178 mm, 0.160 mm and 0.146 mm; the MF of 300,000, 400,000, 500,000 and 600,000 respectively.

A clean and grease free dry slide was placed over a guide plate and marked properly for identification. Then from the well mixed sample of milk 0.01 ml was drawn with the Breads pipette. Stainless steel syringe type pipette or loops can be used to deliver 0.01 ml of sample instead of Breeds pipette. The exterior of the pipette was wiped with a clean blotting paper. The milk was delivered carefully by gentle blowing over the slide and it was spread uniformly over 1 sq.cm area by the help of a guide plate and sterile needle. The slide was quickly dried up approximately within 5 minutes, taking care that the film is not cracked. The slide was then dipped in xylene or xylon to remove the fat for 1 minute and then drained and dried off. Next it was immersed in 90% ethyl alcohol for 2 minutes for fixing and then transferred to attaining bath containing methylene blue for at least 1 minute (prepare the stain by adding 0.2 g of methylene blue powder to 30 ml of 95% ethyl or soluble denatured alcohol). Add this solution to 100 ml of distilled water. It was removed and rinced in water, decolorized in alcohol if over stained, dried and examined under oil. The number of fields to be examined depends upon the number of bacteria per field and the diameter of the field. When clump count was made clumps were counted as one.

Fields to be Examined are

No.of Organisms/Field		No.of fields to field diameter 0.206 mm	Examined 0.146 mm
Less than	1	60	120
	1	30	60
Upto	10	15	30
More than	10	8	15

Precautions

1. All glassware and equipments should be clean and sterile.
2. Sample should be spread uniformly
3. The milk should be kept cold before examining.

Merits

(1)Results are obtained quickly (2) After counting, the slides can be stored as permanent records (3) The equipments required are simple and less expensive (4) with excessive bacteria, their morphology gives clues for the source of contamination. Clumps usually indicate dirty vessels, rods indicate dust and soil contamination, chains indicate mastitis milk and pairs indicate improperly cooled milk (5) High test milk can only be tested.

Limitations

(1) Not useful for low counts (2) The error will be too greatly magnified as the M.F. is too high (3) Dead bacteria will also be counted.

RESULT

No. of organisms in the given sample is /ML

Exercise - 29
Microbiological Methods - Milk

Standard Plate Count

AIM

To judge the quality of milk by estimating the bacterial count (colony count) by the standard plate count method.

Principle

A known volume of milk is added to nutrient agar and incubated. The number of visible colonies developed are counted. This number is multiplied by the dilution to get the bacterial population and not the bacterial numbers. The factors like temperature of incubation, pH of the media, relative humidity will influence the count greatly. All the errors involved have been standardized to minimise the error.

Reagents

1. **Buffered phospate solution**

   This is for making dilutions of milk. Stock solution is prepared by dissolving 34 g of KH_2PO_4 in 500 ml of distilled water, adding about 175 ml of normal NaOH solution. The pH is adjusted to 7.2 and diluted to 1 litre. 1 ml of this stock solution is diluted to 800 ml with boiled and cooled distilled water. Tap water can be used provided no toxic or bacteriostatic elements are present.

2. **Medium (Standard Nutrient agar tryptone - glucose extract agar)**

   Composed of tryptone 5 g, Glucose 1 g, Beef extract 3 g, Agar 15 g and distilled water 1000 ml.

Method of Preparing Dilutions

For Standard Plate Count

Final pH is adjusted to 7 + 0.1 and autoclave at 15 lbs pressure for 20 minutes.

3. **Dilution Blanks**

   Buffered solution was taken up to the mark (slightly in excess to compensate the evaporation) in 99 ml dilution bottles or 9 ml in test tubes and autoclave at 15 lbs pressure for 20 minutes.

Procedure

The sample was mixed thoroughly with a sterilized pipette, 11 ml was added to 99 ml blank taking all aseptic precautions, giving a dilution of 1 in 10. The diluted sample was shaken 25 times over 1 foot during a period of 7 seconds in order to break the clumps and to distribute the bacterial cells uniformly. From this 1 ml is transferred to the next blank to give a dilution of 1 in 1000 and 0.1 ml from this give 1 in 10,000. Thus serial dilutions are made according to requirements. Meanwhile the nutrient agar is melted in a water bath and cooled and maintained at 42 °C. Before starting the operation the work table should be cleaned with the lotion. The stoppers of the blanks were loosened before pipetting. The petri dishes were marked with respective dilutions and arranged in order. The pipettes and petri dishes when removed were operated near the flame.

1 ml of 0.1 ml of the diluted sample depending on the requirements were transferred into the stride petri dishes. The tempered agar was then poured (10-12 ml) after flaming the neck of medium. The medium was mixed thoroughly with the sample by rotating and moving to and fro gently so as a complete film cover the bottom of the dish. The medium was allowed to solidity. Then the plates were inverted and incubated at 35 °C for 48 hours + 3 hours.

No.of colonies in petri dish - I	114
No.of colonies in petri dish - II	97
Average number of colonies per dish	105.6
Dilution used	1 in 100,000
No.of organism/ml of milk	10,550,000 or
	11,000,000/ml of milk

Counting of plates should be done within 3 hours after incubation. Plates with 30 to 300 colonies were selected and counted. When the colonies exceed 300, 4 representative squares were counted and the average is multiplied by appropriate factor. For 90 mm diameter petri dish the factor is 63.5. The results were reported as S.P.C./ml of milk at 35 °C. In the case of spreaders in single chains or colony count each chain as one.

Precautions

All glasswares are sterilized at 170 °C for 1 hour. The dilution blanks should be filled 2 or 3 ml above the mark and the cork slightly loosened before autoclaving. In making dilutions and also in transferring the dilluted sample to the petri dish the last drop should be blown off. Inversion of plates were done to prevent condensation of water vapour on the agar and spreading. The plates are so arranged that they are 1 inch apart from each other and from the side walls. To reduce the tendency for spreader formation excessive humidity in the incubator is avoided and excessive drying is also controlled by proper ventilation.

Merits

Live organism only could be detected. From all nature of the colonies the types of organisms predominant is found on. It is specially useful in low count milk.

Limitations

The incubation temperature, the medium used, period of incubation etc. are not condusive to the all types of organisms. The method is cumbersome, involving much of glassware. The result is available only after 48 hours of incubation.

Result

The given sample contained /ml SPC at 35 °C

Exercise - 30
Preparation of Khoa/Mawa

Title : To prepare Khoa/Mawa from buffalo milk/cow milk by standard method.

Relevant Information

In India, the quality of milk available in flush season far exceeds to actual requirement. Under the tropical climate of our country and with the limited facilities of storage, the excess milk cannot be stored or transported for marketing. Hence, it is imperative to convert milk into suitable concentrated or desiccated products of comparatively longer shelf life. Among such products, Khoa/Mawa is one.

Mawa is an ancient indigenous milk product, which constitutes an important base for the preparation of various Indian sweet meats. The surplus milk produced during flush season is converted into mawa by the producers/halwai. There are 3 varieties of mawa viz pindi, dhap and danedar. Pindi is used for burfi and peda preparation, Dhap for gulab jaamun and Danedar for Kalakand.

Materials Required

Cow/Buffalo milk

Apparatus

1. Iron Khadai / Saucepan
2. Khunti (Scrapper)
3. Thermometer
4. Fat, T.S. and Acidity determination apparatus.
5. Balance

Procedure

Clean and dry the kettle/Saucepan/Iron Khadai. Take a representative sample of milk for fat & SNF test. Weigh the quantity of milk taken and record its weight. Pour the raw milk in the container after straining & bring it to boiling. Stirring continued with circular motion by the flattened end of the stirrer. During operation, scrap all parts of pan with which the milk comes into contact. Note the time when heating started. Continue the vigorous stirring and desiccation to later stage. The final product is considered to be ready when it leaves the sides of the pan sticking together to make pat formation possible. Close the steam valve/put off the gas stove and leave it to cool down to room temperature. Take out the Khoa in Vegetable parchment paper/s plate. Determine the weight of the Khoa obtained and take a sample for moisture and fat test.

Production Chart

I. Particulars of milk:

Type :______________ Quantity :_______________

Acidity :_____________ % lactic Lactometer reading : ___ at___ °C

Total Solids : _______________%

Solids not Fat : _____________%

II. Processing :

Room Temperature : __________ Time started :_____ hrs ____°C

Temperature at boiling :______°C Temperature at finish :______°C

Time finished : ____ Hr ____ min Type of stove & fuel used:______

III. Making pats :

Time started :_______________ Time finished : _______________

Total time taken for heating & making pats : _______ hr ________min

IV. Physical characteristics of Khoa :

Flavour : ____________________

Body & Texture :_______________

Color :______________________

V. Chemical composition of Khoa :

Moisture : __________ % Fat : __________ %

Total solids : __________ %

VI. Yield of Khoa :

Quantity obtained : __________ %

Out turn : _____________ %

Precautions

Take utmost care while stirring the milk during desiccation to avoid slashing and excessive charring.

Remarks

Observation

Quantity of Milk : = ____________________

Total solids in milk : = ____________________%

Amount of T.S. present : = ____________________

Quantity of Khoa obtained : = __________ = __________%

T.S. in Khoa : = ____________________%

Amount of T.S. recovered = __________ = __________

Loss of T.S. in handing = ____________________

Questions

1. Name the varieties of khoa and their uses.
2. Indicate the stage of khoa when precautions are needed.
3. Describe the desirable/undesirable physical attributes of khoa.

EXERCISE - 31
Preparation of Paneer/Chhana

Title : Preparation of Paneer/Chhana.

Relevant Information

Paneer and Chhana are acid and heat coagulated indigenous milk product. Paneer is used for the preparation of large number of culinary dishes, whereas chhana is used for the manufacture of different sweets e.g. Rasogolla, Sandesh, etc. Paneer and Chhana should contain not less than 50 percent fat on dry matter basis and moisture should not be more than 70 percent.

Materials Required

1. Milk
2. Coagulant (1% Citric Acid)
3. Saucepan
4. Muslin cloth
5. Stirrer
6. Weight
7. Fat & SNF Testing apparatus etc.

Procedure

Test the milk for its fat & SNF content. Filter it through muslin cloth into a saucepan. Put the saucepan on the gas stove. Heat the content up to 90 °C. Cool buffalo milk to 70 °C (for paneer) and cow milk up to 80 °C (for chhana). Add 1 per cent citric acid solution (having the same temperature i.e. 70 °C or 80 °C) and continue slow stirring till the desired coagulation is achieved. Spread a clean muslin cloth over empty vessel to receive the whey. Pour the content from the saucepan over the cloth immediately. Then follow different methods for paneer and chhana.

For Paneer

After complete coagulation, stop the stirring and allow the curd to settle down. Drain the whey using muslin cloth. The temperature of the coagulam should not go below 63 °C until this stage. Collect the curd into the muslin cloth and apply pressure on the top by 40-50 kg weight per 100 kg milk used for 5-10 minutes. Then, take out the paneer block from the muslin cloth and put in chilled water for 2 hours. After 2 hours, take it out and put on a plate so that the free water drains away. Paneer is ready.

For Chhana

Collect the four ends of the cloth piece and tie it into a knot & hang it for 30 minutes on a hook together with the whey container kept underneath to receive further drops of whey. Give one to two mild jerks to chhana whilst it is hung. Then take out the chhana from the muslin cloth & put it on tray, Chhana is ready.

Production Chart

A. Paneer

I. Particulars of milk

Type : ______________ Quantity :________________

Acidity : _____________ L.Reading : ________ at _______ °C

Fat : ________________ T.S. ______________ %

SNF ________________

II. Coagulation :

Type of Coagulant used : ________________ Amount of coagulant added : ___________ °C

Acidity of Coagulant ___________ %

Temperature of coagulant : ____________ °C

III. Straining :

Type of strainer used : ___________ Time taken for straining :_________

Temperature of coagulant : ____________ °C

IV. Physical Quality :

Flavour : _____________

Body & Texture : _______________

Color : _____________

V. Chemical Quality :

Moisture : _____________ % Fat : ___________ % TS : _________ %

VI. Yield : _____________ g Out turn : ______________ %

VII. **Whey :**

Amount : ______________ ml Acidity : ______________ % LA

Fat : ______________ % Total fat loss : ____________ g

Remarks

B. Chhana :

I. **Particulars of milk :**

Type : ____________ Quantity : ________________

Acidity : ___________ L. Reading : ________ at ________ °C

Fat : _____________ T.S. : ____________ %

SNF : ____________

II. **Coagulation :**

Type of Coagulant used : ________ Acidity of coagulant added : ______

Amount of coagulant added : __________ %

Temperature of coagulant :_________ °C

III. **Straining :**

Type of strainer used ___________ Time taken for straining _________

Temperature when straining started : ___________ °C

IV. **Physical Quality :**

Flavour : _____________

Body & texture : ________________

Color : ______________

V. **Chemical Quality :**

Moisture : ________ % Fat : ________% TS :_______%

VI. **Yield :** ____________ g Outturn : ____________ %

VII. **Whey :**

Amount : ____________ ml Acidity : ____________ % LA

Fat : _____________ % Total fat loss : ___________ g

Remarks :

Observation :

A. Paneer

Quantity of milk taken : _______________

T.S. content of milk :________________

Amount of T.S. in milk : ______________

Quantity of product obtained : ______________

T.S. of product : ______________

Amount of T.S. : ______________

Loss of T.S. during handling : ______________

B. Chhana

Quantity of milk taken : _______________

T.S. content of milk :________________

Amount of T.S. in milk : ______________

Quantity of product obtained : ______________

T.S. of product : ______________

Amount of T.S. : ______________

Loss of T.S. during handling : ______________

Precautions

1. The temperature of coagulation should be properly checked.
2. Burning of milk should be avoided during heating.

Questions

1. What is the basis for selecting the temperature of coagulation/heating ?
2. During filtration the temp, should not go below 63°C. Why ?
3. Name three coagulants used for paneer/channa manufacture.

EXERCISE - 32
Preparation of Dahi

Title : Preparation of dahi.

Relevant Information

Dahi is a fermented milk product consumed throughout India. There are mainly two types of dahi prevalent in the country viz. 1) Sweet/mildly, sour whole milk/skim milk dahi. 2) Sweetened dahi (mist dahi. eai dahi. payodhi). There is another type known as yoghurt (short set dahi).

Materials Required

1. Whole milk/Skim milk.
2. Mixed Starter culture (portion of previous day's dahi).
3. Long handled spoon to transfer the culture.
4. Plastic cup/Beaker.
5. Saucepan.
6. Gas Stove.
7. Thermometer, B.O.D. Incubator.

Procedure

Take 1-2 litres of milk. Boil the milk for 5-10 minutes. Cool the milk to 22-25 °C. Fill the milk in the plastic cup/beaker. Inoculate the milk with mixed starter culture (portion of previous day's dahi) @ 2%. Mix the milk thoroughly. Incubate the cup/beaker at temperature 22-25 °C for 16-17 hours (overnight) in B.O.D. incubator. After incubation, cup is removed and stored in refrigerator.

Observation

Quantity of milk taken = ____________________

Heating time of milk = ____________________ minutes.

Temperature of milk at the time of inoculation = _____________ °C.

Quantity of culture =______________

Incubation temperature = ___________________ °C

Physical Characteristics

Color - Yellowish, heavy white, free from browning

Appearance - Smooth & glassy surface, top creamy layer

Flavour - Pleasant, clean of acidic, free from off flavour

Body - Soft, firm, free from gas holes and whey pocket

Precautions

Starter culture of dahi must be free from undesirable and harmful bacteria.

Questions

1. Why milk is converted into dahi ?
2. Describe the merits of dahi.
3. State different types of dahi.
4. Write down the care to be taken in the preparation of dahi.
5. How dahi differs from yoghurt ?
6. What are the uses of dahi ?

Exercise - 33
Preparation of Lassi

Title : Preparation of lassi.

Relevant Information

Lassi is a fermented milk product from dahi/curd obtained from whole milk/skim milk. It is consumed as a part of daily diet or as a refreshing beverage almost in every part of the country. Another product that resembles to that of lassi is chhas/malha which is obtained by churning dahi during butter preparation.

Materials required

1. Thermometer
2. Curd beater/wooden handle / big spoon

Procedure

Take well set curd in a container. Add little quantity of water. Mix the curd with wooden handle/big spoon. Dilute the contents with sugar solution (sugar @ 15% of dahi (chilled) equal amount of water). Mix the contents. Add essence (optional) (@ 2-3 ml/10 litres of dahi).

Observation

Quantity of curd =

Quantity of sugar solution =

Weight of lassi =

Physical Characteristics

Appearance - Wholesome, free from extraneous matter

Consistency - Thin, Thick

Flavour - Off flavour, pleasant flavour, clean, acidic smell

Taste - Acidic, Intense acidity

Precautions

1. Use pasteurized chilled water for preparing lassi.
2. Don't use off flavoured curd for lassi preparation.
3. Use clean utensil for preparation.

Questions

1. Comment on preparation of lassi.
2. Can lassi be poor man's beverage/drink ? How ?
3. Write down the factors responsible for quality of lassi.
4. What is the difference between lassi & chhas ?

EXERCISE - 34
Separation of Cream

Title : Separation of cream from milk using cream separator.

Relevant Information

When milk fat is concentrated into a fraction of the original milk, that portion is known as cream. Cream is rich in energy giving fat and soluble vitamins A.D.E. and K. It is used as:

1. for direct consumption as table/coffee/whipping cream,
2. in the preparation of special dishes,
3. in the production of plastic, frozen and cultured cream,
4. in the manufacture of butter, ghee etc.

Principle

The basic principle of cream separation is based on the fact that milk fat is lighter than skim milk portion. At 16 °C(60 °F), the average specific gravity of milk fat is 1.036. Hence, when milk, which may be considered to be a mixture of fat (as cream) and skim milk, is subjected to centrifugal force, the two components viz., cream and skim milk enters the rapidly revolving bowl of a cream separator, the heavier portion i.e. skim milk is forced to the periphery while the lighter portion i.e. fat (cream) moves towards the center. The skim milk and cream both form vertical walls within the bowl and are separated being led through separate outlets.

Assembling of Cream Separator

Place the bowl base on the bowl support and place the rubber ring in the groove in the bowl bottom. Slip the discs on the distributor one by one. Make sure that the bottom disc, which is caulked on both sides, comes next to the bowl button. Turn the discs so that these must right over the guiding post of the distributor. Put on the top disc and bowl hood. The cream screw plug and the guide pin of the later should fit into the notch in the bowl bottom. The bowl is designed in such a way that it cannot be wrongly assembled unless it is forced. Screw and tighten the bowl nut. The last tightening should be done by aid of the bowl nut wrench. Having mounted the bowl put on the skim milk cover over the gearbox and insert the bowl in the skim milk cover over the spindle carefully. Then, cream cover is to be placed over the bowl connecting the skim milk cover. Put on the regulator cup, insert the float and set up the milk tank. All the parts of the separator should be cleaned in hot water with little soda before use. Do not assemble the bowl until the separator is to be used. All parts must be dried.

Cream Separation

Pour the milk (38-40 °C) into the milk tank. Turn the handle slowly and smoothly and increase the speed up to 50-60 rpm. Let it take 2 to 3 minutes to attain this speed. The speed should be maintained constant. Then open the faucet fully, The cream and skim milk will come separately through their respective covers. Do not interrupt the feed of milk during separation. Go on pouring more milk in the tank before it is empty. The skimming should be done regularly. When all milk has been separated and whilst the cream is still flowing, pour about 2 litres of skim milk or warm water in the tank taking care that the correct speed is still maintained. The objective of this is to assure that last remains of cream go out of the bowl. Allow the bowl to come to rest on its own accord. Till then do not remove the spout covers. They should be removed only after separator has come to stop.

Regulating the Cream

The cream adjusting screw is created in the top disc. It is already set correctly but if we want more thick / thin cream, then for thick cream turn the screw into the bowl a little to the right and for thinner cream, turn the screw out of a little left. The slightest adjustment of the screw is sufficient to alter the flow of cream. The adjustment can be done with the help of 'L' type tong.

Observation

Amount of milk taken = ____________ litre

Amount of cream obtained = ____________ gm

Amount of skim milk obtained = ____________ litre

Result

gm of cream obtained from __________ litre of milk.

Precautions

1. Whenever possible separation should be carried out directly after milking otherwise the milk must be heated to a temperature of approximately 35 °C before separating.

2. Make sure the faucet is shut before milk is poured into the milk tank.
3. It is advisable to filter the milk thoroughly before separation.
4. Bowl part must be warmed in severe winter before separation is carried out.

Questions

1. Enlist different types of cream.
2. Differentiate between cream and malai.
3. What is plastic cream?

Exercise - 35
Preparation of Ghee

Title : Preparation of ghee.

Relevant Information

Ghee is clarified butter fat. It is prepared on both small and large scale. Desi method of ghee making comprises more than 90% of Indian ghee production. Ghee is the richest source of animal fat in vegetarian diet. It has long keeping quality under tropical storage conditions. It is prepared through different methods. In India, ghee is, generally, prepared from dahi. This ghee has long keeping quality. A new method of ghee making has been started which is known as prestarification method.

Materials required

1. Cream
2. Muslin Cloth
3. Sauce pan
4. Stirrer with flat end
5. Gas stove
6. Thermometer
7. Balance
8. Beakers etc.

Procedure

Clean and dry the saucepan and stirrer. Mix thoroughly the cream and take representative sample for fat content. Transfer the cream after recording the weight into the saucepan. Put on the gas stove to heat the contents of saucepan. Use stirrer continuously varying the speed of stirring as required. Note the different stages and observe temperature at each stage. Stop heating as soon as the temperature reaches 110-115 °C. Record temperature. Slow down the heat (a characteristic ghee smell develops and curd particles become light brown, foaming is also decreases.)

Filter the ghee through muslin cloth in a beaker/container, squeeze the cloth carefully to extract as much ghee as possible. Weight the beaker/container and record the out come of ghee and residue.

Observation

Quantity of cream taken = ____________________

Fat content in cream = ____________________

Amount of fat in cream = ____________________

Quantity of ghee obtained = ____________________

Fat content of ghee obtained = ____________________

Amount of fat loss in ghee = ____________________

Percentage of fat loss = ____________________

Production Chart

I. Particulars of Cream :

Type - __________ Quantity - ____________

Fat (%) - __________ Amount of fat - ____________

II. Heating of Cream :

Room temperature -______ Type of heating - ___________

Time heating started - ______________ hr ____________ min.

Progressive time temperature - ________________

Temperature and time foaming was observed - ____________

Temperature and time foaming subsided - ________________

Time heating finished - _________________

Highest temperature reached - __________________

III. Straining :

Type of strainer - _____ Temp. of ghee (When straining started) - ______

Quantity of residue obtained : ______ Color of residue - _______

IV. Ghee :

Quantity obtained - _____________ Yield - ________________

Color - _____________ Total fat recovered _________

Mode and temperature of storage -

V. Sensory evaluation

(a)	Sensory Characteristics	Max. Score	Code number of sample		
		1	2		
	Flavour				
	Body & Texture				
	Color				
	Ghee residue				
(b)	Characteristics	Defects	Degree of defects		
	Flavour		Suspicious	Slight	Pronounced
Texture					
Color					
Freedom from suspended					
Impurities					

VI. Grading :

Quality	Score	Grade
Excellent	90 or more	A
Good	80 - 89	B
Fair	70 - 79	C
Poor	Below 70	D

Judging

Precautions: Prevent overflow due to foaming by regulation of heating and with careful stirring.

Questions

1. What is the principle of ghee making in indigenous method ?
2. Comment on ghee prepared from cow and buffaloe milk.

EXERCISE - 36
Preparation of Ice Cream

Title : Preparation of Ice Cream

Relevant Information:

Ice cream is defined as a frozen dairy product made by suitable blending and processing of cream and other milk products together with sugar and flavour, with or without stabilizer or color, and with the incorporation of air during the freezing process. It is an excellent source of food energy. Having two to three times the fat content of milk, and more than half of its total solids being sugar (lactose and sucrose), the energy value of ice cream is very high. It is a desirable item for all.

Materials Required

1.	Cream	2.	Whole Milk	3.	Skim Milk
4.	S.M.P.	5.	Sugar	6.	Stabilizer operated
7.	Color (optional)	8.	Flavour	9.	Ice cream freezer
10.	Deep fridge	11.	Ice cream scoop (electrical/hand operated)		
12.	Cups	13.	Gas Stove	14.	Sauce pan
15.	Scrapper	16.	Thermometer etc.		

Procedure

I. **Calculation of mix**

II. **Preparation of Mix:** Place the calculated amounts of liquid ingredients in a saucepan. Put on the gas stove. Dried ingredients like milk powder, sugar, stabilizer, etc. are added while the liquid material is agitated and before the temperature reaches 49 °C. Proper suspension to avoid lumpiness of the dried ingredients may be obtained by either mixing the dry materials thoroughly with part of the sugar before slowly adding it to the liquid. If gelatin is the stabilizer used, it is best added after it is thoroughly mixed with an equal volume of sugar, and before the liquid material reaches 49 °C, alternatively, it can be sprinkled on the surface of the cold liquid and allowed to soak before the mixture is heated, or soaked in water and then the mixture heated to completely dissolve the gelatin, which is usually added to a warm (38 - 49 °C) mix. If Sodium Alginate is used as a stabilizer, it should not be added until the temperature of the liquid material has reached at least 66 °C. The dry Sodium alginate is not allowed to soak but is stirred up with water and immediately dumped into the hot mix. With a few exceptions, coloring and flavouring materials are added when the mix is frozen.

III. **Pasteurization of Mix:** Heat the mix to 68 - 69 °C for 30 minutes or 80 °C for not less than 25 seconds and then rapidly cooling it to below 5 °C. Age the mix at 4 °C for at least 3 - 4 hours, except for Sodium alginate, which requires no ageing.

IV. **Preparation of ice cream freezer [hand operated / electrically operated (Softel) & freezing the mix:** Ensure that the freezer is clean and if not, open the freezer, take out the dasher. Wash and scrub thoroughly including dasher and also inside the freezer. Assemble the parts into their respective positions.

Proper amount of coloring and flavouring material is generally added to the mix in the freezer. The total volume of the mix, flavour and color should be about half the total capacity of the freezing chamber. Close the freezer. Along the sides of the freezer, add equal amounts of salt crystals and ice cubes, layer by layer. Put on the switch for electrically operated Softel. It takes roughly 20-25 minutes to freeze. For hand operated freezer, it takes roughly 7-8 minutes to freeze. End point is indicated when the handle cannot move easily in hand operated freezer. Transfer the ice cream mix either into different cups of desired volume with ice cream scoop or transfer the freezing chamber into deep fridge maintained at 20 °C for 6-8 hours (minimum) for hardening. After the ice cream is hardened, it can be immediately marketed or it may be stored for a week or two at the most. The overrun for softy type ice cream is 30-50%, for bulk ice cream 90-100% and for packaged ice cream 70-80%.

Judging and Grading of Ice Cream

A. Score Card :

Flavour	45
Body and Texture	30
Bacteria	15
Color and Package	5
Melting	5

B. Procedure for Examination : Since the condition of ice cream changes rapidly following sequence of observation must be carried out.

1. Observe the type and condition of container. Note defects if any.
2. Observe the color of product for uniformity.
3. Sample of ice cream usually; Feel how it cuts. Observe the body and texture. Take out chipper full of ice cream in a flat-bottomed dish and observe its melting quality. High quality ice cream shows little resistance towards melting at room temperature.
4. Go for flavour of melted sample.
5. Confirm your observation by placing, spoonful of frozen product to the mouth. Quickly move it between teeth and tongue. Note the body and texture for any defects.

Observation

A. Sensory Evaluation :

Serial No.	Sample No.	Percent

B. Overrun :

Serial No.	Sample No.	Percent

Precautions

1. If Sodium alginate is the stabilizer, add it to the mix at temperature 66-67 °C.
2. Take care that while freezing, salt does not go inside the ice cream chamber of the freezer.
3. While preparing the mix, add powder and emulsifier slowly, not at once.

Questions

1. Write the advantage of ageing.
2. Write the names of the two most popular flavours used in ice cream making.
3. Write down the functions of stabilizers and emulsifiers used in ice cream manufacturing.

Exercise - 37
Preparation of Butter

Title : Preparation of butter.

Relevant Information

Butter may be defined as a fat concentrate. Which is obtained by churning cream, gathering the fat into a compact mass and then working on it. PFA rules say, it must contain not less than 80 per cent by weight of milk fat, not more than 1.5 per cent by weight of curd and not more than 3.0 per cent by weight of common salt.

Materials Required

1. Pasteurized & ripened cream
2. Chilled and luke warm water
3. Butter color, etc.

Apparatus

1. Butter churn
2. Cream ladle
3. Butter scoop
4. Dairy Floating thermometer
5. Butter worker
6. Weighing balance etc.

Procedure

Weigh the required quantity of ripened cream. Note its temperature. The temperature should be below 9-11 ^{0}C. Transfer the cream to the churn (Old churn should be properly sanitized and cooled). New churn should be filled with cold water for 24 hours, so that it becomes leak proof. Water is then drained out and scrubbed with hot water or chlorine solution filled up to half of its volume. Calculated amount of butter color is added (0 to 250 ml more per 100 kg of butter fat). Rotate the churn for 5-10 minutes. The liberated gas is removed once, or twice by opening the vent. Then, churning process starts by rotating the churn. During the churning process, there is invariably rise in temperature by about 1-3 °C. Churning is accompanied by foaming. Then comes out the braking stage, when the cream breaks away at this stage the fat in cream emulsion breaks and very small butter granules are formed (pin head size). Then add break water at this stage to reduce the temperature of the cream contents, thereby control the body and texture (In the tropics, addition of break water can be avoided by providing an A.C. butter making room and/or chilled water spray over the butter churn). Continue churning till fat globules attain the size of pea, which is observed through glass window. As soon as the churning is completed, the butter milk is drained off. Then, equal quantity of water (temperature 1-2 °C lower than the churning temperature of cream) added. After few revolutions, wash water is removed. One washing is sufficient for good quality butter. Then butter is spread over the butter worker. Calculated amount of salt is sprinkled (2 - 2.5% of butter fat). The working should be continued until the butter has compact body, a closely-knit grain, a tough wax texture & an even distribution of salt and moisture is achieved. The butter is moulded into various shapes. Food grade weight of the butter is taken. Overrun is calculated as follows :

$$\% \text{ OR} = \frac{B - F}{F} \times 100$$

F - Fat in churn (kg)

OR - Overrun

B - Butter made (kg)

Score Card (Table butter)		**Grade of Butter**
Flavour	- 45	AA - Butter scoring 93 points
Body & Texture	- 30	A - Butter scoring 92 points
Colour	- 10	B - Butter scoring 90 points
Salt	- 10	C - Butter scoring 89 points
Packet	- 5	

Precautions

1. Maintain proper temperature during churning.
2. Over ripening or cream should be avoided
3. It should be prepared by adopting hygienic condition

Questions

1. Write basic difference between cream and butter.
2. Write importance of washing.
3. Write the aim of adding salt.

Index

D

E

F

G

J

K

L

M

N

O

P

Q

R

S

T

U

www.ingramcontent.com/pod-product-compliance
Ingram Content Group UK Ltd.
Pitfield, Milton Keynes, MK11 3LW, UK
UKHW060019200726
13845UKWH00046B/110

9 789389 354133